Mishchenko / Shatalov / Sternin
Lagrangian Manifolds and the Maslov Operator

A. S. Mishchenko V. E. Shatalov B. Yu. Sternin

Lagrangian Manifolds and the Maslov Operator

Translated from the Russian
by Dana Mackenzie

With 13 Figures

Springer-Verlag
Berlin Heidelberg New York
London Paris Tokyo Hong Kong

A. S. Mishchenko
Moscow State University, Leninskie Gory, Moscow, USSR

V. E. Shatalov
MIIGA, Pulkovskaya 6a, Moscow, USSR

B. Yu. Sternin
MIEM, B. Vusovskij per. 3/12, Moscow, USSR

Title of the Russian original edition:
Lagranzhevy mnogoobraziya i metod kanonicheskogo operatora
Publisher Nauka, Moscow 1978

This volume is part of the *Springer Series in Soviet Mathematics*
Advisers: L. D. Faddeev (Leningrad), R. V. Gamkrelidze (Moscow)

Mathematics Subject Classification (1980):
58G15, 58G17, 58F05, 58F06, 34E, 35P20, 35L

ISBN 3-540-13613-4 Springer-Verlag Berlin Heidelberg New York
ISBN 0-387-13613-4 Springer-Verlag New York Berlin Heidelberg

Library of Congress Cataloging-in-Publication Data
Mishchenko, Aleksandr Sergeevich.
[Lagranzhevy mnogoobraziya i metod kanonicheskogo operatora. English]
Lagrangian Manifolds and the Maslov Operator / A. S. Mishchenko, V. E. Shatalov. B. Yu. Sternin; translated from the Russian by Dana Mackenzie. p. cm. – (Springer series in Soviet mathematics) Translation of: Lagranzhevy mnogoobraziya i metod kanonicheskogo operatora. Bibliography: p. Includes index.
ISBN 0-387-13613-4 (U.S.: alk. paper)
1. Manifolds (Mathematics) 2. Operator theory 3. Differential equations – Asymptotic theory.
I. Shatalov, V. E. (Viktor Evgen'evich) II. Sternin, B. IU. III. Title IV. Series
QA613.M5713 1990 514'.223–dc20 89-6363

Printed in the United States of America

2141/3140-543210 – Printed on acid-free paper

Foreword

This book presents Maslov's canonical operator method for finding asymptotic solutions of pseudodifferential equations. The classical WKB method, so named in honor of its authors: Wentzel, Kramers and Brillouin, was created for finding quasiclassical approximations in quantum mechanics. The simplicity, obviousness and "physicalness" of this method quickly made it popular: specialists in mathematical physics accepted it unequivocally as one of the weapons in their arsenal. The number of publications which are connected with the WKB method in one way or another can probably no longer be counted.

The alternative name of the WKB method in diffraction problems – the ray method or the method of geometric optics – indicates that the approximations in the WKB method are constructed by means of rays. More precisely, the first approximation of the WKB method is constructed by means of rays (isolating the singular part), after which the usual methods of the (regular) theory of perturbations are applied.

However, the ray method is not applicable at the points of space where the rays focus or form a caustic. Mathematically this fact expresses itself in the fact that the amplitude of the waves at such points become infinite. Physicists knew certain "canonical" equations (for example, the Airy equation) whose solution were defined everywhere, including at focal points. It is natural, therefore, that the first and most successful efforts to deal with the difficulties arising at focal (caustic) points were the so-called method of canonical equations, whose essence consists of replacing an arbitrary equation in the neighborhood of a focal point by a certain standard ("canonical") one. The method of canonical equations is appealing in its simplicity and the explicitness of its results: indeed, even in a neighborhood of focal points the solution is expressed in terms of functions whose values have, by and large, been tabulated. An excellent presentation of the method of canonical equations is given by the book of V.M. Babich and V.S. Buldyriev [1]. But the method of canonical equations, which works nicely in the classical problems of mathematical physics, turns out to be not general enough to obtain asymptotic solutions of many interesting problems in the theory of differential and pseudodifferential equations with partial derivatives.

A powerful instrument for solving problems of this kind is the essentially new method for finding asymptotic solutions introduced by V.P. Maslov in 1965 – Maslov's canonical operator method [1]. Without getting into the details of the method (see the Introduction for this), we can point out here its universality. Maslov's canonical operator method works in such apparently distant areas as the quasiclassical asymptotics of quantum mechanics and the problem of the stability of difference schemes, the problem of the propagation "in the large" of a discontinuity in solutions of hyperbolic equations and the short-wave asymptotics in diffraction problems, the propagation of waves in the ionosphere and the existence and uniqueness problem in the general theory of pseudodifferential equations.

Maslov's canonical operator method also revealed the topological nature of the well-known effect of the phase shift of the Jacobian under passage through a focal point. In particular, V.P. Maslov [1] calculated a characteristic class which enters into the quantization condition and realizes the so-called Maslov index on a Lagrangian manifold. The Maslov index, as well as the quantization condition, plays a fundamental role in finding the asymptotic behavior of spectra.

At the present time Maslov's canonical operator method is used widely both by Soviet and foreign mathematicians in various problems associated with asymptotic decompositions.

Today it doesn't seem possible to give a more or less full list of works in which Maslov's theory is being developed or used. We mention here only those nearest in their subjects to the present book. These are: V.P. Maslov, V.I. Arnol'd, V.S. Buslaev, B.R. Vainberg, A.M. Vinogradov, V.G. Voropaeva, V.G. Danilov, S.Yu. Dobrokhotov, V.L. Dubnov, M.V. Karasev, A.D. Krakhnov, Yu.A. Kravtsov, V.V. Kucherenko, V.V. Lychagin, A.S. Mishchenko, V.E. Nazaikinskij, V.A. Oshchmian, V.E. Shatalov, B.Yu. Sternin, V.A. Tsupin, F. Cardoso, P. Dazord, J.P. Eckmann, J.J. Duistermaat, V. Guillemin, J. Leray, A. Melin, R. Seneor, D. Schaeffer, J. Sjöstrand, M.J. Souriau, F. Treves, G.A. Uhlmann, V. Verghe, A. Voros, A. Weinstein, B. Yoshikawa et al. (cf. Bibliography).

Several subtle and elegant investigations in the mathematical problems of diffraction which have been carried out by the Leningrad mathematicians V.M. Babich and V.S. Buldyriev [1] and others (see Bibliography) are also of great interest from the viewpoint of the Maslov's canonical operator method.

Finally, we mention one more aspect of the development of Maslov's canonical operator – the formula for the asymptotic decomposition of the integral of a rapidly oscillating function. This formula plays a fundamental role in the proof of the invariance of the canonical operator; it is used in deriving the formula for the commutation of a pseudodifferential operator and a rapidly oscillating exponential, and it is an important tool of the theory. The fundamental results pertaining to the decomposition of the integral of a rapidly oscillating function by the stationary phase method belong to M.V. Fedoriuk [1]–[3]. We will present here a complete derivation of the formula for the asymptotic expansion of the integral of a rapidly oscillating function for a complex phase, taking into account the universal character of the coefficients.

In his recent works [1]–[3], V.I. Arnol'd studied the asymptotic behavior of the integral in the "non-Morse" case and found relationships between the rate of decrease and the Coxeter numbers.

Part I of this book (Chapter 1, 2 and 3) was written by A.S. Mishchenko (except for sections 3.3 and 3.4, which were written by B.Yu. Sternin); Chapters 4, 6, and sections 7.2 and 7.3 were written by B.Yu. Sternin; Chapter 5 and section 7.1 were written by V.E. Shatalov.

The authors

Table of Contents

Introduction

In this book we present the construction of asymptotic solutions of differential (and pseudodifferential) equations with respect to a small parameter h. These asymptotic solutions are constructed by Maslov's canonical operator method, which allows one to obtain not only local but also global solutions of the equations under consideration. The goal of this introduction is to illustrate some of the important phenomena which arise in constructing asymptotic solutions of $1/h$-pseudodifferential equations, both with real and complex symbols (the Hamiltonian). This partially accounts for the style of exposition in the material in the Introduction. We present the material "heuristically", not in strict logical sequence, and sometimes leave out certain details in our definitions and proofs in those cases when these details are of a technical nature and lead to more involved formulations. To illustrate the method we consider a series of examples, beginning with the simplest.

1. As our first example we consider an ordinary differential equation with constant coefficients. To be precise, let

$$P_n(p) = \sum_{j=0}^{n} a_j p^j$$

be a certain polynomial in one variable of degree n. Formally replacing the variable p in the polynomial $P_n(p)$ by the operator

$$\widehat{p} = -ih\frac{d}{dx}$$

we arrive at the differential operator

$$P_n(\widehat{p}) = P_n\left(-ih\frac{d}{dx}\right) = \sum a_j (-ih)^j \frac{d^j}{(dx)^j}$$

with constant coefficients (depending on h).

We consider the problem of constructing the solutions of the homogeneous equation

$$P_n(\widehat{p})\,\psi(x) = \sum a_j\,(-ih)^j\,\frac{d^j\psi}{(dx)^j} = 0 \tag{1}$$

In agreement with Euler's method we will seek a solution of this problem of the form

$$\psi(x) = \exp\left\{i\frac{\lambda}{h}x\right\}c\,, \qquad c = \text{const.} \tag{2}$$

Since

$$\widehat{p}\,\exp\left\{i\frac{\lambda}{h}x\right\}c = -ih\frac{d}{dx}\left[\exp\left\{\frac{i}{h}\lambda x\right\}c\right] = \lambda\,\exp\left\{\frac{i}{h}\lambda x\right\}c\,,$$

substituting function (2) into Eq. (1), we obtain the relation

$$P_n(\widehat{p})\,(x) = P_n(\lambda)\,\exp\left\{\frac{i}{h}\lambda x\right\}c = 0\,. \tag{3}$$

In order to satisfy Eq. (3), we must require the number λ to be a solution of the (algebraic) equation

$$P_n(\lambda) = 0\,, \tag{4}$$

which is usually called the characteristic equation for the differential Eq. (1). We will call Eq. (4) and analogous equations, which we will arrive at in the future, *Hamilton-Jacobi equations.*

We note that in the customary terminology of mechanics, one can rewrite the solution of Eq. (1) in the form

$$\psi(x) = \exp\left\{\frac{i}{h}S(x)\right\}c\,,$$

where the function $S(x) = \lambda x$ is called the *action.* We will also adopt this terminology in the sequel. Observing that, under this convention

$$\lambda = \partial S/dx\,,$$

we can write Eq. (4) in a form coinciding with the classical form of the Hamilton-Jacobi equation:

$$P_n\left(\partial S(x)/\partial x\right) = 0\,. \tag{5}$$

The fact that the solution of the Hamilton-Jacobi equation is a linear function is associated with the particular form of the equation under consideration (an equation with constant coefficients). Moreover, we note that the solution of the Hamilton-Jacobi equation is defined for all x ("in the large"); moreover, in the given case we obtained not an asymptotic but an exact solution of Eq. (1).

2. As our next example we consider the linear differential operator

$$\widehat{E} + P_n(\widehat{p}) = -ih\frac{\partial}{\partial t} + P_n\left(-ih\frac{\partial}{\partial x}\right)$$

with partial derivatives, where $P_n(p)$ is polynomial (1). We have introduced here the notation

$$\widehat{E} = -ih\frac{\partial}{\partial t} , \quad \widehat{p} = -ih\frac{\partial}{\partial x} .$$

We consider the Cauchy problem for the differential operator

$$\widehat{E} + P_n(\widehat{p})$$

with initial conditions of a special type:

$$\begin{gathered}\left[\widehat{E} + P_n(\widehat{p})\right]\psi(x,t) = \left[-ih\frac{\partial}{\partial t} + P_n\left(-ih\frac{\partial}{\partial x}\right)\right]\psi(x,t) = 0 , \\ \psi(x,0) = \exp\left\{\frac{i}{h}\alpha x\right\}c_0 , \quad \alpha, c_0 = \text{const.}\end{gathered} \tag{6}$$

By analogy with Eq. (1) we seek a solution of Problem (6) in the form

$$\psi(x,t) = \exp\left\{\frac{i}{h}(\lambda_1 x + \lambda_2 t)\right\}c . \tag{7}$$

Substituting function (7) into the system (6) and taking into account the relations

$$\widehat{E}\psi(x,t) = \left(-ih\frac{\partial}{\partial t}\right)\exp\left\{\frac{i}{h}(\lambda_1 x + \lambda_2 t)\right\}c = \exp\left\{\frac{i}{h}(\lambda_1 x + \lambda_2 t)\right\}\lambda_2 c ,$$

$$\widehat{p}\psi(x,t) = \left(-ih\frac{\partial}{\partial x}\right)\exp\left\{\frac{i}{h}(\lambda_1 x + \lambda_2 t)\right\}c = \exp\left\{\frac{i}{h}(\lambda_1 x + \lambda_2 t)\right\}\lambda_1 c ,$$

we obtain

$$\left[\widehat{E} + P_n(\widehat{p})\right]\psi(x,t) = \exp\left\{\frac{i}{h}(\lambda_1 x + \lambda_2 t)\right\}[\lambda_2 + P_n(\lambda_1)]c = 0 ,$$

$$\psi(x,0) = \exp\left\{\frac{i}{h}\lambda_1 x\right\}c = \exp\left\{\frac{i}{h}\alpha x\right\}c_0 .$$

These equations show that, in order for function (7) to be a solution of Eq. (6), it is necessary to require the conditions

$$\lambda_2 + P_n(\lambda_1) = 0 , \quad \lambda_1 = \alpha , \quad c = c_0 . \tag{8}$$

to be fulfilled. The first of these is the Hamilton-Jacobi equation for the system (6). If we denote the action appearing in Formula (7) by $S(x,t)$:

$$S(x,t) = \lambda_1 x + \lambda_2 t ,$$

then we can write the Hamilton-Jacobi equation in the form

$$\frac{\partial S}{\partial t} + P_n\left(\frac{\partial S}{\partial x}\right) = 0 \; . \tag{9}$$

The condition $\lambda_1 = \alpha$ gives the inital conditions for Eq. (9), which can be written in the form

$$S\,(x,0) = S_0(x) = \alpha x \; . \tag{10}$$

Equation (9), together with the initial conditions (10), is the Cauchy problem for the nonlinear Hamilton-Jacobi equation. In the given case the solution of this problem can easily be obtained directly. Indeed, conditions (8) give an expression for the function $S(x,t)$:

$$S(x,t) = \alpha x - P_n(\alpha)t$$

and a formula for the solution of problem (6):

$$\psi(x,t) = \exp\left\{\frac{i}{h}\left(\alpha x - P_n(\alpha)t\right)\right\} c_0 \; .$$

We note that in this case a solution of the Hamilton-Jacobi equation has been obtained in the large (in the form of a linear function), and that the solution $\psi(x,t)$ determined by the last formula is an exact (not an asymptotic) solution of problem (6).

3. To illustrate some further particulars of the theory, we replace the initial conditions in Problem (6) by more general ones. Indeed, we consider the Cauchy problem

$$\begin{aligned} &\left[\widehat{E} + P_n\left(\widehat{p}\right)\right] \psi\,(x,t) = 0 \; , \\ &\qquad \psi\,(x,0) = \exp\left\{\frac{i}{h}\alpha x\right\} \varphi_0(x) \; . \end{aligned} \tag{11}$$

We will seek a solution in a form analogous to (7) but with obvious changes associated with the form of the initial conditions:

$$\psi(x,t) = \exp\left\{\frac{i}{h}\left(\lambda_1 x + \lambda_2 t\right)\right\} \varphi(x,t) \; . \tag{12}$$

On investigation of Eq. (11), some substantial differences from the two cases considered earlier appear. We will carry out a substitution of function (12) into Eq. (11). For this purpose we use the conditions

$$\widehat{E}\psi(x,t) = \left((-ih)\frac{\partial}{\partial t}\right) \exp\left\{\frac{i}{h}(\lambda_1 x + \lambda_2 t)\right\}\varphi(x,t)$$
$$= \exp\left\{\frac{i}{h}(\lambda_1 x + \lambda_2 t)\right\}\left[\lambda_2\varphi(x,t) - ih\frac{\partial\varphi(x,t)}{\partial t}\right]$$
$$= \exp\left\{\frac{i}{h}(\lambda_1 x + \lambda_2 t)\right\}\left[\lambda_2 - ih\frac{\partial}{\partial t}\right]\varphi(x,t) ,$$

$$\widehat{p}\psi(x,t) = \left(-ih\frac{\partial}{\partial x}\right) \exp\left\{\frac{i}{h}(\lambda_1 x + \lambda_2 t)\right\}\varphi(x,t)$$
$$= \exp\left\{\frac{i}{h}(\lambda_1 x + \lambda_2 t)\right\}\left[\lambda_1\varphi(x,t) - ih\frac{\partial\varphi(x,t)}{\partial x}\right]$$
$$= \exp\left\{\frac{i}{h}(\lambda_1 x + \lambda_2 t)\right\}\left[\lambda_1 - ih\frac{\partial}{\partial x}\right]\varphi(x,t) .$$

We obtain the result

$$\left[\widehat{E} + P_n(\widehat{p})\right]\psi(x,t) = \exp\left\{\frac{i}{h}(\lambda_1 x + \lambda_2 t)\right\} \cdot \left[\left(\lambda_2 - ih\frac{\partial}{\partial t}\right) + P_n\left(\lambda_1 - ih\frac{\partial}{\partial x}\right)\right]\varphi(x,t) = 0 , \tag{13}$$

where

$$P_n\left(\lambda_1 - ih\frac{\partial}{\partial x}\right) = \sum a_j\left(\lambda_1 - ih\frac{\partial}{\partial x}\right)^j .$$

The initial conditions of system (11) give

$$\psi(x,0) = \exp\left\{\frac{i}{h}\lambda_1 x\right\}\varphi(x,0) = \exp\left\{\frac{i}{h}\alpha x\right\}\varphi_0(x) ,$$

and it is sufficient to require the conditions

$$\lambda_1 = \alpha , \quad \varphi(x,0) = \varphi_0(x) .$$

to hold.

We note now that the expression in square brackets on the right-hand side of condition (13) is a polynomial in the variable h, whose coefficients are differential operators. To calculate the coefficients we note that the expression under consideration,

$$\left(\lambda_2 - ih\frac{\partial}{\partial t}\right) + P_n\left(\lambda_1 - ih\frac{\partial}{\partial x}\right)$$

can be obtained from the function

$$(\lambda_2 - ihE') + P_n(\lambda - ihp') = F(h) \tag{14}$$

by formally replacing E' by $\partial/\partial t$ and p' by $\partial/\partial x$. We expand function (14) as a Taylor series in the variable h. Since P_n is a polynomial of degree n, the Taylor series obtained breaks off at the n-th term:

$$F(h) = \sum_{h=0}^{n} \frac{h^k}{k!} \frac{d^k}{dh^k} F(h)\Big|_{h=0} = (\lambda_2 + P_n(\lambda_1))$$
$$+ h\left(-iE' - i\frac{\partial P_n}{\partial p}(\lambda_1)\, p'\right) + \sum_{k=2}^{n} \frac{h^k}{k!} (-i)^k \frac{\partial^k P_n}{\partial p^k}(\lambda_1)\,(p')^k \ .$$

If in expression (14) we carry out a substitution of $\partial/\partial t$ for E' and $\partial/\partial x$ for p', we obtain the equation

$$\left(\lambda_2 - ih\frac{\partial}{\partial t}\right) + P_n\left(\lambda_1 - ih\frac{\partial}{\partial x}\right) = (\lambda_2 + P_n(\lambda_1)) - ih\left(\frac{\partial}{\partial t} + \frac{\partial P_n}{\partial p}(\lambda_1)\frac{\partial}{\partial x}\right)$$
$$+ \sum_{k=2}^{n} \frac{(-ih)^k}{k!} \frac{\partial^k P_n}{\partial p^k}(\lambda_1) \frac{\partial^k}{\partial x^k} \ .$$

We note that in this argument the fact that the coefficients of the operator $\widehat{E} + P_n(\widehat{p})$ are constant is used in an essential way, since under this assumption all the operators considered commute. Taking into account the last formula, we can write condition (13) in the form

$$\exp\left\{\frac{i}{h}(\lambda_1 x + \lambda_2 t)\right\}\left\{(\lambda_2 + P_n(\lambda_1))\,\varphi(x,t) \right.$$
$$- ih\left(\frac{\partial\varphi(x,t)}{\partial t} + \frac{\partial P_n}{\partial p}(\lambda_1)\frac{\partial\varphi(x,t)}{\partial x}\right) \tag{15}$$
$$\left. + \sum_{k=2}^{n} \frac{(-ih)^k}{k!} \frac{\partial^k P_n}{\partial p^k}(\lambda_1) \frac{\partial^k \varphi(x,t)}{\partial x^k}\right\} = 0 \ .$$

From Formula (15) it is evident that it is impossible to obtain an exact solution of Problem (11) by choosing the constants λ_1, λ_2 and the function $\varphi(x,t)$. However, setting as many terms as possible in the expansion (15) equal to zero, we obtain a solution of Problem 11 asymptotic in h.

Thus we obtain the equation

$$\lambda_2 + P_n(\lambda_1) = 0 \tag{16}$$

for determining the constants λ_1 and λ_2 and the equation

$$\frac{\partial\varphi(x,t)}{\partial t} + \frac{\partial P_n(\lambda_1)}{\partial p}\frac{\partial\varphi(x,t)}{\partial x} = 0 \tag{17}$$

for determining the function $\varphi(x,t)$. Equation (16) is called the *Hamilton-Jacobi equation*, and Eq. (17) is called the *equation of transport.*

The Hamilton-Jacobi equation is solved in precisely the same way as in the previous case. The solution of this equation is written in the form

$$\lambda_1 = \alpha\ , \quad \lambda_2 = -P_n(\alpha)\ , \quad S(x,t) = \alpha x - P_n(\alpha)t\ .$$

The equation of transport can be written in a more convenient way equivalent to (17). We consider a vector field (x,t) on the plane, whose coordinates are independent of x and t and are equal to $\mathbf{v} = (\partial P_n/\partial p\,(\lambda_1)\,,1)$ (Fig. 1). Then the left-hand side of the equation is the derivative along the trajectory of the vector field $\mathbf{v}$. The equation of transport can thus be written in the form

$$d\varphi/dt = 0\ , \tag{18}$$

where d/dt is the derivative along the field $\mathbf{v}$. From Eq. (18) it is evident that the function φ must be constant along the trajectory of the vector field $\mathbf{v}$ (this circumstance explains the terminology "equation of transport": the amplitude $\varphi(x,t)$ of function (12) is transported along the trajectory of the vector field $\mathbf{v}$).

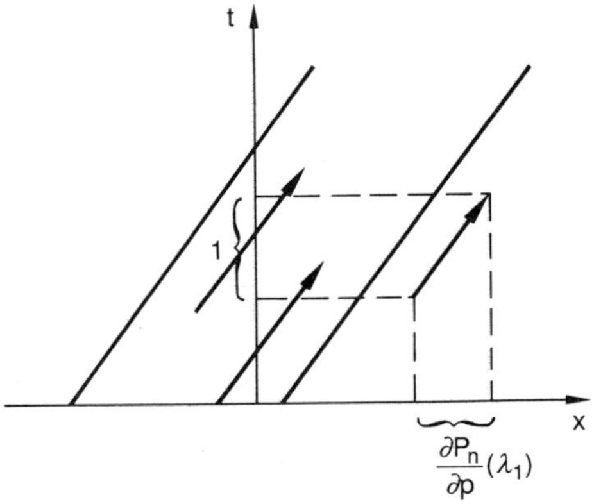

Fig. 1

We note that, in contrast to the two cases considered before, we did not arrive at an exact solution of the problem. Moreover, the right-hand side of Eq. (11) approaches zero with an accuracy up to terms of order $O\left(h^2\right)$. In order to obtain the following terms of the asymptotic expansion, one must seek a function $\varphi(x,t)$ in the form of a formal power series in the variable h:

$$\varphi(x,t) \equiv \sum_{l=0}^{\infty} h^l \varphi_l(x,t)\ . \tag{19}$$

As is obvious from Formula (15), for the function $\varphi_0(x,t)$ we again obtain the equation of transport (18). It is not hard to show that the functions $\varphi_l(x,t)$ can subsequently be determined in such a way that the right-hand side of Eq. (11) has order $O(h^s)$ for any integer $s > 0$, if in the expansion (19) one takes a large enough partial sum of the series.

We consider, for example, the equation for the function $\varphi_1(x,t)$. Picking out the terms of order h^2 in Eq. (15), we obtain the equation

$$\frac{\partial \varphi_1}{\partial t} + \frac{\partial P_n}{\partial p}(\lambda_1)\frac{\partial \varphi_1}{\partial x} = -\frac{i}{2}\frac{\partial^2 P_n}{\partial p^2}(\lambda_1)\frac{\partial^2 \varphi_0}{\partial x^2},$$

which, in view of the definition of the derivative d/dt, can be written in the form

$$\frac{d\varphi_1}{dt} = -\frac{i}{2}\frac{\partial^2 P_n}{\partial p^2}(\lambda_1)\frac{\partial^2 \varphi_0}{\partial x^2}. \tag{20}$$

If one determines the function $\varphi_0(x,t)$ from Eq. (18), Eq. (20) allows one to calculate the function $\varphi_1(x,t)$ by integrating along the trajectories of the vector field $\mathbf{v}$.

One can obtain a recursive system for the definition of the functions $\varphi_l(x,t)$ by considering the terms of Eq. (15) of order $h^3, h^4 \ldots$, in which each successive function is obtained from the preceding ones by integration along the trajectories of the vector field $\mathbf{v}$.

4. As the simplest example of an equation with variable coefficients we consider the non-stationary equation for a quantum-mechanical oscillator:

$$\left[\widehat{E} + \frac{1}{2}\left(x^2 + \widehat{p}^2\right)\right]\psi(x,t) = 0\,, \quad \psi(x,0) = \exp\left\{\frac{i}{h}\alpha x\right\}\varphi_0(x)\,. \tag{21}$$

Since the coefficients in Eq. (21) are non-constant, there is no basis for supposing that the function $S(x,t)$ will depend linearly on x and t. Therefore we will seek the asymptotic solution of problem (21) in the form

$$\psi(x,t) = \exp\left\{\frac{i}{h}S(x,t)\right\}\varphi(x,t)\,, \tag{22}$$

where for simplicity we will limit ourselves to finding only the first term asymptotic in h. When we substitute function (22) into problem (21), as before, we have

$$
\begin{aligned}
&\left[\widehat{E} + \frac{1}{2}\left(x^2 + \widehat{p}^2\right)\right] \psi(x,t) \\
&\quad = \exp\left\{\frac{i}{h} S(x,t)\right\} \cdot \left\{\left[\frac{\partial S}{\partial t} + \frac{1}{2}\left(x^2 + \left(\frac{\partial S}{\partial x_1}\right)^2\right)\right]\right. \\
&\qquad \left. - ih\left[\frac{\partial}{\partial t} + \frac{\partial S}{\partial x}\frac{\partial}{\partial x} + \frac{1}{2}\frac{\partial^2 S}{\partial x^2}\right]\right\} \cdot \varphi(x,t) + O\left(h^2\right) = 0 \;, \\
&\psi(x,0) = \exp\left\{\frac{i}{h} S(x,0)\right\} \varphi(x,0) = \exp\left\{\frac{i}{h}\alpha x\right\} \varphi_0(x) \;,
\end{aligned} \tag{23}
$$

from which we obtain the Hamilton-Jacobi equation

$$
\frac{\partial S}{\partial t} + \frac{1}{2}\left(x^2 + \left(\frac{\partial S}{\partial x}\right)^2\right) = 0 \tag{24}
$$

and an equation relating to the function φ:

$$
\frac{\partial \varphi(x,t)}{\partial t} + \frac{\partial S}{\partial x}\frac{\partial \varphi(x,t)}{\partial x} + \frac{1}{2}\frac{\partial^2 S}{\partial x^2}\varphi = 0 \;. \tag{25}
$$

The initial conditions for Eqs. (24) and (25) are obtained from the second of the conditions (23):

$$
S(x,0) = \alpha x \;, \quad \varphi(x,0) = \varphi_0(x) \;.
$$

The Hamilton-Jacobi equation (24), as before, is a nonlinear partial differential equation (not containing the sought-after function S as a term). However, in distinction from the cases we have already considered, this equation does not admit a trivial solution in form of a linear function in x. To solve this equation we use the method of characteristics. We write the corresponding Hamiltonian system of ordinary differential equations, corresponding to the Hamilton function $H(x,p) = \left(x^2 + p^2\right)/2$:

$$
\dot{x} = H_p = p \;, \quad \dot{p} = -H_x = -x \tag{26}
$$

with the initial conditions

$$
x(0) = x_0 \;, \quad p(0) = \frac{\partial S_0}{\partial x}(x_0) = \alpha = p_0{}^{1} \;. \tag{27}
$$

One can solve the system (26), (27) explicitly:

[1] We use here mechanical analogues suggested by the form of Eq. (24).

$$x\,(x_0,t) = x_0 \cos t + \alpha \sin t \;, \quad p(x_0,t) = \alpha \cos t - x_0 \sin t \;. \tag{28}$$

From mechanics it is known that the solution $S(x,t)$ of Eq. (24) can be written in the form

$$S(x,t) = S_0\,(x_0) + \int_0^t (p\,dx - H\,dt)\big|_{x_0 = x_0(x,t)} \;, \tag{29}$$

where the integration is taken along the trajectories of the system (28), and $x_0(x,t)$ is a function determined by the first of conditions (28).

We will give now a somewhat different interpretation of conditions (29). We consider the set of points (x,p) given by conditions (29) when x_0 travels along the line R. This set of points is a submanifold L_0 in the plane (x,p) (a line parallel to the axis Ox). We will consider Eq. (28) to be a relation governing the evolution of the manifold L_0 in time. It is not hard to see that the line L_0 rotates around the origin of the coordinates with a constant angular speed, so that the vector $(x\,(x_0,t)\,, p\,(x_0,t))$ is obtained from the vector (x_0, p_0) by the orthogonal transformation whose matrix is

$$\begin{pmatrix} \cos t & \sin t \\ -\sin t & \cos t \end{pmatrix} .$$

If we portray the trajectory of motion of the line L_0 in coordinates (x,p,t), we arrive at the two-dimensional surface L in $\mathbb{R}^3$ (a ruled surface; see Fig. 2).

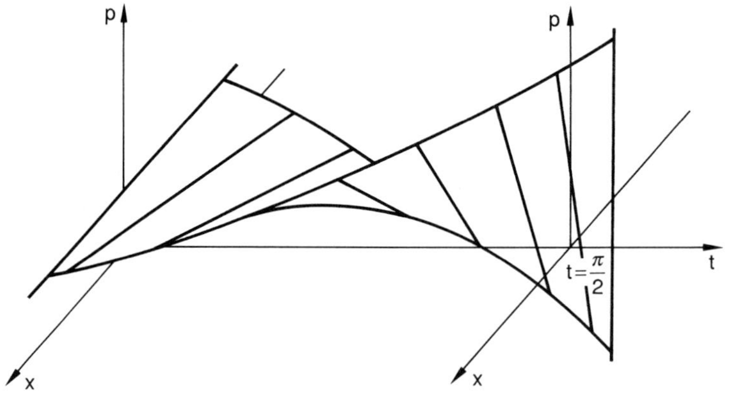

Fig. 2

We will now interpret Formula (29) in terms of the surface L. Since the functions (x_0,t) are global coordinates on the surface L, the existence of a solution $x_0(x,t)$ of the first of the relations (28) means that we are considering that part of the surface L which projects diffeomorphically onto the plane (x, t). As is clear from Fig. 2, we must thus restrict ourselves to values of the variable t less than $\pi/2$. Thus, in contrast to the cases considered earlier, we do not obtain a global solution of Eq. (24).

We can, however, "lift" the solution $S(x, t)$ of Eq. (24) onto the surface L [2]. Here the function S on the surface L will satisfy the condition

$$dS = p dx - H dt \ . \tag{30}$$

We note in particular that a solution of (30) on the surface L already exists globally, that is for all $t \geq 0$.

Next we consider Eq. (25). To simplify this equation we perform the substitution

$$\varphi(x,t) = \frac{\varphi_1(x,t)}{\sqrt{\partial x/\partial x_0}} \ .$$

We note that Eq. (25) can be rewritten in the form

$$\begin{aligned} \frac{\partial \varphi}{\partial dt} + \frac{\partial S}{\partial x}\frac{\partial \varphi}{\partial x} + \frac{1}{2}\frac{\partial^2 S}{\partial x^2}\varphi &= \frac{\partial \varphi}{\partial t} + \dot{x}\frac{\partial \varphi}{\partial x} + \frac{1}{2}\frac{\partial^2 S}{\partial x^2}\varphi \\ &= \frac{d\varphi}{dt} + \frac{1}{2}\frac{\partial^2 S}{\partial x^2}\varphi = 0 \ , \end{aligned}$$

where d/dt is the derivative along the trajectories of the Hamiltonian system (26), projected onto the plane (x, t). Further we have

$$\frac{d}{dt}\left(\frac{\partial x}{\partial x_0}\right) = \frac{\partial}{\partial x_0}(\dot{x}) = \frac{\partial}{\partial x_0}(p) = \frac{\partial}{\partial x_0}\left(\frac{\partial S}{\partial x}\right) = \frac{\partial^2 S}{\partial x^2}\frac{\partial x}{\partial x_0} \ .$$

Using the latter condition, we perform the necessary substitution

$$\begin{aligned} 0 &= \frac{d}{dt}\left(\frac{\varphi_1}{\sqrt{\partial x/\partial x_0}}\right) + \frac{1}{2}\frac{\partial^2 S}{\partial x^2}\frac{\varphi_1}{\sqrt{\partial x/\partial x_0}} \\ &= \frac{1}{\sqrt{\partial x/\partial x_0}}\frac{d\varphi_1}{dt} - \frac{1}{2}\frac{1}{(\partial x/\partial x_0)^{3/2}}\frac{\partial^2 S}{\partial x^2}\frac{\partial x}{\partial x_0} + \frac{1}{2}\frac{\partial^2 S}{\partial x^2}\frac{\varphi_1}{\sqrt{\partial x/\partial x_0}} \\ &= \left(\frac{\partial x}{\partial x_0}\right)^{-1/2}\frac{d\varphi_1}{dt} \ . \end{aligned}$$

In this way we arrive at the equation of transport relative to the function φ_1:

[2] We will denote the function S "lifted" onto the surface by the same letter S.

$$\frac{d\varphi_1}{dt} = 0 \,, \tag{31}$$

which we will consider as an equation pertaining to the function φ_1 lifted onto the surface L. The derivative d/dt can also be considered to be the derivative along the vector field

$$V(H) = p\frac{\partial}{\partial x} - x\frac{\partial}{\partial p} + \frac{\partial}{\partial t} \,,$$

tangent to the surface L. However, in contrast to Eq. (20), the vector field now lies on the surface L in phase space (x, p, t) (and not in the space (x, t)). We note that the equation of transport in the form (31) likewise has a global solution on the surface L.

Despite the existence on the surface of global functions S, φ_1, we cannot yet determine a global solution ψ of Eq. (21), since the surface L no longer projects diffeomorphically onto the (x, t) space when $t = \pi/2$. However, these functions can be used to construct a global solution if in a neighborhood of the point $t = \pi/2$ we go to the Fourier transform in Eq. (21):

$$\psi(x,t) = \left(\overline{F}^{1/h}_{p\to x}\widetilde{\psi}\right)(x,t) = \left(\frac{-1}{2\pi i h}\right)^{1/2} \int \exp\left\{\frac{i}{h}px\right\} \widetilde{\psi}(p,t)dp \,.$$

By this Eq. (21) becomes the equation

$$\left[\widehat{E} + \frac{1}{2}\left(\widehat{x}^2 + p^2\right)\right] \widetilde{\psi}(p,t) = 0 \,,$$

where $\widehat{x} = ih\partial/\partial p$. We will again seek a solution of this equation in a form analogous to (22):

$$\widetilde{\psi}(p,t) = \exp\left\{\frac{i}{h}\widetilde{S}(p,t)\right\} \widetilde{\varphi}(p,t)$$

in a neighborhood of the point $t = \pi/2$. In order to obtain a solution which continues the solution $\psi(x,t)$ obtained earlier, it is necessary to require that for $0 < \delta < t < \pi/2 - \delta$,

$$\left(\frac{1}{2\pi i h}\right)^{1/2} \int_{-\infty}^{+\infty} \exp\left\{\frac{i}{h}\left(S(x,t) - px\right)\right\} \varphi(x,t)dx = \exp\left\{\frac{i}{h}\widetilde{S}(p,t)\right\} \widetilde{\varphi}(p,t) \,. \tag{32}$$

Condition (32) cannot in general be satisfied exactly. However, we can choose the functions $\widetilde{S}$ and $\widetilde{\varphi}$ in such a way that (32) holds up to terms of order $O(h)$; this is sufficient for obtaining the first term of the asymptotic series. According to a well-known formula for the asymptotic expansion, the zero-order term of the asymptotic expansion of the integral on the left-hand side of Eq. (32) equals

$$(-1)^{1/2} \exp\left\{\frac{i}{h}(S(x,t)-px)\right\} \varphi(x,t)\big|_{x=x(p,t)} \frac{1}{\sqrt{-\frac{\partial^2 S(x,t)}{\partial x^2}}},$$

where the argument of the expression under the square-root sign is either equal to 0 or $-\pi$, depending on the sign of that expression, and the function $x(p,t)$ is a solution of the equation

$$p - \frac{\partial S(x,t)}{\partial x} = 0 \tag{33}$$

relative to x.

In order for the right- and left-hand sides of condition (32) to coincide with an accuracy up to terms of order $O(h)$, we must require that

$$\widetilde{S}(p,t) = S(x,t) - px\big|_{x=x(p,t)},$$

$$\widetilde{\varphi}(p,t) = \left.\frac{\varphi(x,t)}{\sqrt{\frac{\partial^2 S(x,t)}{\partial x^2}}}\right|_{x=x(p,t)}.$$

Since the equation $\partial S/\partial x = p$ holds on the surface L, a solution of Eq. (33) is the function of transition from (p,t) coordinates to (x,t) coordinates on the surface L for $\delta < t < \pi/2 - \delta$. Here it is necessary to set $\widehat{S} = S - px$ on the surface L. Furthermore,

$$\widetilde{\varphi} = \frac{\varphi}{\sqrt{\frac{\partial^2 S(x,t)}{\partial x^2}}} = \frac{\varphi_1}{\sqrt{\frac{\partial^2 S(x,t)}{\partial x^2}\frac{\partial x}{\partial x_0}}} = \frac{\varphi_1}{\sqrt{\frac{\partial p}{\partial x}\frac{\partial x}{\partial x_0}}} = \frac{\varphi_1}{\sqrt{\frac{\partial p}{\partial x_0}}}.$$

Here it is necessary to set the argument $\arg \partial p/\partial x_0$ equal to

$$\arg \frac{\partial p}{\partial x_0} = \arg \frac{\partial x}{\partial x_0} + \arg\left(-\frac{\partial^2 S}{\partial x^2}\right) + \pi i \tag{34}$$

for each choice of the argument of the function $-\partial^2 S/\partial x^2$.

Thus, in a neighborhood of the point $\pi/2$ the solution of Eq. (21) is represented in the form of the Fourier transform of the function

$$\exp\left\{\frac{i}{h}(S - xp)\right\} \frac{\varphi_1}{\sqrt{\partial p/\partial x_0}}.$$

5. We now go on to the general case. We have chosen the Cauchy problem to illustrate the methods we adopt because its solution demonstrates

the ideas and methods used in this work in an explicit and geometric fashion. Starting from the example of the Cauchy problem we would like to begin our exposition from "ground zero" and build up Maslov's canonical operator directly in the process of solving a differential equation.

Thus we consider the following Cauchy problem:

$$\begin{aligned} -ih\frac{\partial\psi}{dt} + H(x,\widehat{p},t)\psi = 0 \\ \psi\big|_{t=0} = \exp\left\{\frac{i}{h}S_0(x)\right\}\varphi_0(x)\ . \end{aligned} \tag{35}$$

Here the function $H(x,p,t)$ satisfies the estimate

$$\left|D_x^\alpha D_p^\beta H(x,p,t)\right| \le C_{\alpha,\beta}\left(1+|x|+|p|\right)^m$$

for every multi-index (α,β), and the expression $H(x,\widehat{p},t)\psi(x)$ means the following by definition.[3]

$$H(x,\widehat{p},t)\psi(x) = \left\{\overline{F}^{1/h}_{p\to y}\left[H(x,p,t)F^{1/h}_{x\to p}\varphi(x)\right]\right\}\big|_{x=y}\ ,$$

where

$$F^{1/h}_{x\to p}\varphi = \left(\frac{-i}{2\pi h}\right)^{n/2}\int\limits_{\mathbb{R}^n}\exp\left\{-i\frac{\langle p,x\rangle}{h}\right\}\varphi(x)dx\ ,$$

and $\overline{F}^{1/h}_{p\to x}$ is the dual transform. Thus, for example, $\widehat{p} = -ih\partial/\partial x$.

We will seek asymptotic solutions as $h\to 0$ of this problem in the following sense. We call a function $\psi(x,t,h)$ an asymptotic solution of order $r\ge 0$ if, when we substitute this function in the left-hand side of Eq. (35) we

[3] For simplicity one may consider the function $H(x,p,t)$ here to be a polynomial in the variable p, in other words $H(x,p.t) = \sum_{|\alpha|\le m} a_\alpha(x,t)p^\alpha$; here $\alpha = (\alpha_1,\dots,\alpha_n)$ is a multiindex, $|\alpha| = \alpha_1+\ldots+\alpha_n$, $p^\alpha = p_1^{\alpha_1}\ldots p_n^{\alpha_n}$. Then denoting

$$\begin{aligned} \left(\frac{\partial}{\partial x}\right)^\alpha &= \left[\frac{\partial}{\partial x^1}\right]^{\alpha_1}\cdots\left[\frac{\partial}{\partial x^n}\right]^{\alpha_n} \\ \left(\frac{\partial S}{\partial x} - ih\frac{\partial}{\partial x}\right)^\alpha &= \left[\frac{\partial S}{\partial x^1} - ih\frac{\partial}{\partial x^1}\right]^{\alpha_1}\cdots\left[\frac{\partial S}{\partial x^n} - ih\frac{\partial}{\partial x^n}\right]^{\alpha_n}\ , \end{aligned}$$

we have

$$\begin{aligned} H\left(x,\widehat{p},t\right) &= \sum_{|\alpha|\le m} a_\alpha(x,t)(-ih)^{|\alpha|}\left(\frac{\partial}{\partial x}\right)^\alpha\ ; \\ H\left(x,\frac{\partial S}{\partial x}+\widehat{p},t\right) &= \sum_{|\alpha|\le m} a_\alpha(x,t)\left(\frac{\partial S}{\partial x} - ih\frac{\partial}{\partial x}\right)^\alpha\ . \end{aligned}$$

In this case the following formulas can be obtained by direct calculation. In the general case see Proposition 2.4 in Section 4.2.

obtain a quantity of order $O(h^r)$. Here the norm of a function is evaluated in the norm of a certain function space (precise formulations will be given in the main text). For simplicity we will seek asymptotics of the second order. The scheme of the solution is as follows. We seek the solution of problem (35) in the form

$$\psi(x,t,h) = \exp\left\{\frac{i}{h}S(x,t)\right\}\varphi(x,t) . \tag{36}$$

Substituting expression (36) into the system (35), we have

$$\exp\left\{\frac{i}{h}S(x,t)\right\}\left[\frac{\partial S}{\partial t} - ih\frac{\partial}{\partial t} + H\left(x, \frac{\partial S}{\partial x} - ih\frac{\partial}{\partial x}, t\right)\right]\varphi(x,t) = O\left(h^2\right) .$$

Taking into consideration that $\exp\left\{\frac{i}{h}S(x,t)\right\} = O(1)$, and expanding the obtained expression by degree of h, we obtain from (35)

$$\frac{\partial S}{\partial t} + H\left(x, \frac{\partial S}{\partial x}, t\right) = 0 , \quad S\big|_{t=0} = S_0(x) , \tag{37}$$

$$\left[\frac{\partial}{\partial t} + \frac{\partial H}{\partial p_i}\frac{\partial}{\partial x^i} + \frac{1}{2}\frac{\partial^2 S}{\partial x^i \partial x^j}\frac{\partial^2 H}{\partial p_i \partial p_j}\right]\varphi(x,t) = 0 , \quad \varphi\big|_{t=0} = \varphi_0(x) . \tag{38}$$

Here and below we use the convention that repeated indices should be summed.

To solve Eq. (37) we use the method of characteristics. We form the Hamiltonian system in the space $\mathbb{R}^n_x \oplus \mathbb{R}_{np} \oplus \mathbb{R}^1_t$:

$$\dot{x} = H_p , \quad \dot{p} = -H_x , \quad x(0) = x_0 , \quad p(0) = \frac{\partial S_0(x)}{\partial x}\Big|_{x=x_0} . \tag{39}$$

Let $x(x_0,t)$ and $p(x_0,t)$ be solutions of the Hamiltonian system. Then the equations

$$x = x(x_0,t) , \quad p = p(x_0,t) ,$$

$$x = \left(x^1, \ldots, x^n\right) , \quad p = (p_1, \ldots, p_n) , \quad x_0 \in \mathbb{R}^n , \quad t \in \mathbb{R}^1_+ ,$$

determine a manifold L of dimension $n+1$ in the space $\mathbb{R}^n_x \oplus \mathbb{R}_{np} \oplus \mathbb{R}^1_t$ with boundary $\{t = 0\}$. It can be shown that the form

$$p_i dx^i - H(x,p,t)dt = \omega$$

is closed on this manifold. Since the manifold L is simply-connected, the form ω is exact; moreover, for $t = 0$ the equality

$$dS_0 = \frac{\partial S_o}{\partial x^i}\bigg|_{x=x_0} dx^i = p_i dx^i$$

holds. Therefore there exists a solution of the problem

$$dS' = \omega\big|_{L'} , \quad S'\big|_{t=0} = S_0(x) .$$

Now let U be an open neighborhood of the set $\{t = 0\}$ in the space $\mathbb{R}^n_x \oplus \mathbb{R}^1_{t+}$ onto which the manifold L projects in one-to-one fashion, i.e. such that the equations $x = x(x_0, t)$ are solvable with respect to $x_0 : x_0 = x_0(x, t)$. Then the function

$$S(x,t) = S'(x_0,t)\big|_{x_0=x_0(x,t)} = \left(\pi_x^{-1}\right)^* S'(x_0,t) ,$$

where $\pi_x : L \to \mathbb{R}^n_x \oplus \mathbb{R}^1_t$ is the projection, is a solution of problem (37) in U.

We seek a solution of Eq. (38) in the domain U with the help of the substitution

$$\varphi(x,t) = \varphi'(x_0,t) \left[\frac{Dx}{Dx_0}\right]^{-1/2}\Bigg|_{x_0=x_0(x,t)} .$$

Since the function $J = Dx/Dx_0$ satisfies the Liouville equation

$$\frac{d}{dt}(\ln J) = \mathrm{div} V(H) = \frac{\partial}{\partial x^i}\left(\frac{\partial H}{\partial p_i}\right) ,$$

the function φ, given on the manifold L, must satisfy the conditions

$$\left[\frac{d}{dt} + \frac{1}{2}\frac{\partial^2 H}{\partial x^i \partial p_i}\right]\Bigg|_L \varphi = 0 , \quad \varphi\big|_{t=t_0} = \varphi_0(x) , \tag{40}$$

where d/dt is the restriction of the vector field

$$\frac{\partial}{\partial t} + V(H) = \frac{\partial}{\partial t} + \frac{\partial H}{\partial p_i}\frac{\partial}{\partial x^i} - \frac{\partial H}{\partial x^i}\frac{\partial}{\partial p_i} \tag{41}$$

onto the manifold L.

From (39) it follows that d/dt is the derivative along the trajectories of the Hamiltonian system (39). We also note that from the method of construction of the manifold L it follows it is invariant with respect to the vector field (41).

It can be shown that in a neighborhood of those points of the manifold L for which the projection π_x has a singularity, there exists a set of indices $I \subset \{1, 2, \ldots, n\}$ such that the equations

$$\begin{aligned} x^i &= x^i(x_0,t) , \quad i \in I , \\ p_j &= p_j(x_0,t) , \quad j \in \bar{I} = \{1,2,\ldots,n\} \setminus I , \end{aligned} \tag{42}$$

can be solved in terms of x_0:

$$x_0 = x_0\left(x^I, p_{\overline{I}}, t\right) .$$

In a neighborhood of such points we apply a Fourier transform in the variables $x^{\overline{I}}$ to the equation:

$$\psi(x,t,h) = \overline{F}^{1/h}_{p_{\overline{I}} \to x^{\overline{I}}} \psi_I\left(x^I, p_{\overline{I}}, t, h\right) ; \tag{43}$$

obtaining the equation

$$-ih\frac{\partial \psi_I}{\partial t} + H\left(x^I, \widehat{x}^{\overline{I}}, \widehat{p}_I, p_{\overline{I}}, t\right)\psi_I = O\left(h^2\right) , \quad \widehat{x}^i = ih\frac{\partial}{\partial p_i} .$$

for the function ψ_I.

Analogously to our previous considerations, we seek a function ψ_I in the form

$$\psi_I\left(x^I, p_{\overline{I}}, t, h\right) = \exp\left\{\frac{i}{h}S_I\left(x^I, p_{\overline{I}}, t\right)\right\}\left[\frac{D\left(x^I, p_I\right)}{Dx_0}\right]^{-1/2} \varphi .$$

Here one shows that as the function S_I one can take the function

$$S_I\left(x^I, p_{\overline{I}}, t\right) = \left(\pi_I^{-1}\right)^*\left\{S(x_0,t) - x^{\overline{I}}(x_0,t)p_{\overline{I}}(x_0,t)\right\} ,$$

and the function φ on the manifold L must satisfy Eq. (40). Here π_I: $L \to \mathbb{R}^I_x \oplus \mathbb{R}_{\overline{I}p} \oplus \mathbb{R}^1_t$ is the projection map. π_I^{-1} exists in the neighborhood under consideration by virtue of the existence of a solution of the system (42).

We will denote by U_I the region on the manifold L on which the mapping π_I is nonsingular. In order for Formulas (36) and (43) to determine the same solution (accurate up to h) of Eq. (35), it is necessary that the condition

$$\begin{aligned} &\exp\left\{\frac{i}{h}S(x,t)\right\}\varphi(x,t)\left[\frac{Dx}{Dx_0}\right]^{-1/2} \\ &= \overline{F}^{1/h}_{p_{\overline{I}} \to x^{\overline{I}}}\left\{\exp\left\{\frac{i}{h}S_I\left(x^I, p_{\overline{I}}, t\right)\right\} \cdot \left[\frac{D\left(x^I, p_{\overline{I}}\right)}{Dx_0}\right]^{-1/2} \varphi\right\} + O(h) \end{aligned} \tag{44}$$

hold at those points of U_I where the projection π_x is nonsingular. It turns out that (in the simply-connected case) one can choose the arguments of the functions $D\left(x^I, p_{\overline{I}}\right)/Dx_0$ in each of the regions U_I in such a way that conditions (44) are satisfied. To wit (comp. (34)), $\arg D\left(x^I, p_{\overline{I}}\right)/Dx_0$ and $\arg D\left(x^J, p_{\overline{J}}/Dx_0\right)$ in the intersection $U_I \cap U_J$ must be associated by the relation

$$\arg \frac{D\left(x^I, p_{\overline{I}}\right)}{D x_0} - \arg \frac{D\left(x^J, p_{\overline{J}}\right)}{D x_0} = \sum_k \arg \lambda_k + |J \setminus I| \pi ,$$

where $\{\lambda_k\}$ are the eigenvalues of the matrix

$$\frac{\partial\left(-p_{I_2}, x^{I_3}\right)}{\partial\left(x^{I_3}, p_{I_2}\right)} = \mathrm{Hess}_{x^{I_3}, p_{I_2}}\left(-S_I\right) , \quad I_2 = I \setminus J , \quad I_3 = J \setminus I .$$

This concept is due to Maslov [1].

We now consider the general problem

$$H\left(x, \widehat{p}\right) u(x, h) = 0 , \quad \widehat{p} = -ih \frac{\partial}{\partial x} .$$

Let L be an arbitrary Lagrangian manifold in phase space $\mathbb{R}_x^n \oplus \mathbb{R}_n^p$ lying on the null level surface of the Hamiltonian $H(x,p)$. (The fact that the manifold L is Lagrangian means that the form $p\,dx\big|_L$ is closed on this manifold. In other words, the equation $dS = pdx$ is locally solvable.) To begin with, we suppose that the manifold L projects bijectively onto the subspace $\mathbb{R}_x^n$, i.e. the x-coordinates are a system of coordinates on the manifold L and the p-coordinates are functions of the coordinates $x, p = p(x)$, on the manifold L. In this case an asymptotic solution (accurate up to $O(h^2)$) is constructed by the formula

$$\psi(x) = \exp\left\{\frac{i}{h} S(x)\right\} \varphi(x) ,$$

where the action $S(x)$ is found from the equation $dS(x) = pdx = p_1(x)dx^1 + \ldots + p_n(x)dx^n$, and the function $\varphi(x)$ is found from the equation of transport

$$\left[\frac{d}{dt} + \frac{1}{2} \frac{\partial^2 H}{\partial p_j \partial p_i} \frac{\partial^2 S}{\partial x^i \partial x^j}\right] \varphi = 0 ,$$

where d/dt is the derivative along the Hamiltonian vector field $\mathbf{v}(H)$. Of course, there are many such asymptotic solutions as indicated, and the choice of an appropriate one depends on the initial or boundary conditions.

In the general case we cannot assert that the x-coordinates are a system of coordinates on the Lagrangian manifold L. In general, if a manifold L of dimension n is imbedded in phase space $\mathbb{R}_x^n \oplus \mathbb{R}_n^p$, then at best we can indicate a local system of coordinates in a neighborhood of each point of the manifold L. Moreover, as the local system of coordinates it is always possible to take some group of coordinates of the ambient phase space $\mathbb{R}_x^n \oplus \mathbb{R}_n^p$, say $\left(x^{i_1}, \ldots, x^{i_k}, p_{i_{k+1}}, \ldots, p_{i_n}\right)$. If we know, furthermore, that L is Lagrangian, then it is possible to select such a local system of coordinates $\left(x^{i_1}, \ldots, x^{i_k}, p_{i_{k+1}}, \ldots, p_{i_n}\right)$ in such a way that the numbers $(i_1, \ldots, i_k)$

and $(i_{k+1}, \ldots, i_n)$ are complementary to each other, i.e. so that one never encounters coordinates x^j and p_j with the same index on the Lagrangian manifold.

We drop for the time being the requirement that the function φ satisfy the equation of transport. Then, using the local system of coordinates $(x^1, \ldots, x^n)$ on the Lagrangian manifold L, it would be possible to construct an asymptotic solution in the same form

$$\psi(x) = e^{\frac{i}{h}S} \cdot \varphi ,$$

considering S to be a function on the Lagrangian manifold satisfying the equation $dS = (p\,dx)\,|L$, and φ to be a smooth function whose support lies entirely in the region of the manifold L where the coordinates $(x^1, \ldots, x^n)$ are defined. If in a certain neighborhood of the Lagrangian manifold L it is necessary to take coordinates of mixed type $(x^{i_1}, \ldots, x^{i_k}, p_{i_{k+1}}, \ldots, p_{i_n}) = (x^I, p_{\overline{I}})$, $I = (i_1, \ldots, i_k)$, $\overline{I} = (i_{k+1}, \ldots, i_n)$, $I \cap \overline{I} = \emptyset$, $I \cup \overline{I} = \{1, 2, \ldots, n\}$, then as an asymptotic solution one can take the function

$$\psi(x) = \overline{F}^{1/h}_{p_{\overline{I}} \to x^{\overline{I}}} \left[\exp\left\{ \frac{i}{h} S \right\} \varphi \right] , \tag{45}$$

where S is a function on the Lagrangian manifold satisfying the equation

$$dS = \left(p_I dx^I - x^{\overline{I}} dp_{\overline{I}} \right) |L ,$$

and φ is a smooth function with support lying in the given neighborhood. Here we will consider the expression to which we apply the Fourier transform to be a function in the variables $(x^I, p_{\overline{I}})$ – the coordinates on the Lagrangian manifold L.

Now we can organize all the observations mentioned in the following way. Given a Lagrangian manifold L. We cover the manifold L by an atlas of charts U_I, for which in each chart U_I we give as a local system of coordinates the coordinates of the phase space $(x^I, p_{\overline{I}})$. We choose in each chart U_I two functions: S_I, satisfying the equation

$$dS_I = \left(p_I dx^I - x^{\overline{I}} dp_{\overline{I}} \right) |L ,$$

and φ_I, whose support lies in the chart. We construct according to Formula (45) the functions

$$\psi_I(x) = \overline{F}^{1/h}_{p_{\overline{I}} \to x^{\overline{I}}} \left[\exp\left\{ \frac{i}{h} S_I \right\} D_I \cdot \varphi_I \right]$$

as the values of an operator K_I on the function φ_I:

$$\psi_I(x) = K_I(\varphi_I) \ .$$

Here D_I are certain, for now arbitrary, smooth functions, given on each chart. We require the operators K_I to agree on the intersections of charts, i.e. if the support of a function φ lies in the intersection of charts $U_I \cap U_J$, then

$$K_I(\varphi) - K_J(\varphi) \equiv 0 \left(\operatorname{mod} h^2\right) \ . \tag{46}$$

The consistency condition (46), with the help of the stationary phase method, can be transformed into certain conditions of agreement between the action functions S_I and the functions D_I.

It turns out that, in order for the relations linking the operators K_I on different charts to hold, it is necessary for the functions S_I and D_I to satisfy the following equations:

$$S_I - S_J = x^{\overline{J}} p_{\overline{J}} - x^{\overline{I}} p_{\overline{I}} \ , \tag{47}$$

$$\frac{D_I}{D_J} = \left[\frac{\partial\left(-p_{I_2}, x^{I_3}\right)}{\partial\left(x^{I_2}, p_{I_3}\right)}\right]^{1/2} \exp\left\{\frac{i\pi}{2}|I_2|\right\} \ , \tag{48}$$

where $I_2 = I \setminus J$, $I_3 = J \setminus I$.

We arrive at the natural problem of eliminating the obstructions given by relations (47), (48). These obstructions are one-dimensional cocycles in the cohomology of the Lagrangian manifold L. These two obstructions are called "the Maslov conditions of quantization" of the Lagrangian manifold L and depend only on the topological structure of L and the way it is imbedded in phase space $\mathbb{R}^n_x \oplus \mathbb{R}^p_n$.

If the Maslov conditions of quantization hold on the Lagrangian manifold L, then the operators K_I "agree" on the intersections of charts. This means that with help of the operators K_I one can construct a single operator K, which acts on functions φ on the Lagrangian manifold L with arbitrary supports, and one can use it to seek asymptotic solutions in the form

$$\psi(x) = K(\varphi) \ . \tag{49}$$

The proof of all the facts we have stated can be found in V.P. Maslov's book [1]. (We note, however, that the terminology used in that book is somewhat different from ours.)

We now turn our attention to the complex theory of the canonical operator. In 1970, at the international mathematics conference in Nice, V.P. Maslov announced a certain variant of the complex theory of the canonical operator, which he called the theory of the canonical operator on a Lagrangian manifold with a complex germ. The fundamental object in this work is a real Lagrangian manifold with a complex-valued function defined on it – a complex germ. V.P. Maslov's idea in the complex theory of the canonical operator was developed further in the work of V.G. Danilov, V.P. Kucherenko, Le Vu An, A.S. Mishchenko, V.E. Nazaikinski, A.G. Prudkovski, B. Yu. Sternin, V.E. Shatalov, A. Melin, J. Sjöstrand, etc.

6. We now consider the problem

$$\begin{aligned}&\left[-ih\frac{\partial}{\partial t}+H\left(x,-ih\frac{\partial}{\partial x},t\right)\right]\psi(x,t)=0\\&\qquad\qquad\qquad\qquad\qquad\qquad \psi\big|_{t=0}=e^{iS_0(x)/h}\varphi_0(x)\ ,\end{aligned}\tag{50}$$

where $H(x,p,t) = H_1(x,p,t)+iH_2(x,p,t)$ is a complex-valued function such that $H_2(x,p,t) \le 0$. We will also assume that the functions $S_0(x)$, $\varphi_0(x)$ are also complex-valued:

$$\begin{aligned}S_0(x) &= S_{01}(x) + iS_{02}(x)\ ,\\ \varphi_0(x) &= \varphi_{01}(x) + i\varphi_{02}(x)\ .\end{aligned}\tag{51}$$

We observe that when $S_{02}(x) < 0$ the function $e^{iS_0(x)/h}$ grows exponentially as $h \to 0$. Under these circumstances it becomes unnatural to pose the problem of finding asymptotic solutions accurate to order h. For this reason we will assume that $S_{02}(x) \ge 0$.

We seek a solution of problem (51) in the form

$$\psi(x,t,h) = e^{iS(x,t)/h}\varphi(x,t)\ , \quad S(x,t) = S_1(x,t) + iS_2(x,t)\ . \tag{52}$$

For reasons analogous to those above, we require $S_2(x,t) \ge 0$.

Formally repeating the arguments made in the real case leads to equations analogous to (37) and (38):

$$\frac{\partial S}{\partial t} + H\left(x,\frac{\partial S}{\partial x},t\right) = 0\ ; \quad S\big|_{t=0} = S_0(x)\ ; \tag{53}$$

$$\left[\frac{\partial}{\partial t} + \frac{\partial H}{\partial p_i}\frac{\partial}{\partial x^i} + \frac{1}{2}\frac{\partial^2 S}{\partial x^i \partial x^j}\frac{\partial^2 H}{\partial p_i \partial p_j}\right]\varphi = 0\ ; \quad \varphi\big|_{t=0} = \varphi_0(x)\ . \tag{54}$$

However, in the complex case certain difficulties already arise when we try to interpret the formal Eqs. (33)–(34). Indeed, if the function $H(x,p,t)$ is not analytic in p, it is not clear how to understand the substitution of a complex number for p. But if we demand analyticity in the variable p, then we must also require analyticity in the variable x, since even in the real case it is necessary to take a p-Fourier transform (see p. 17); when this is done the variables p and x change roles.

Secondly, even if the function $H(x,p,t)$ is analytic, it is impossible to solve the Hamiltonian system by the method of characteristics (at least in the same form as in the real case), since the Hamiltonian system corresponding to the Hamiltonian function $H = H_1 + iH_2$ does not, in general, have real trajectories.

We will consider first the case when the function $H(x,p,t)$ is analytic in the variables x and p [4], in other words, when there exists a function $H(z,\zeta,t)$ which is analytic in a neighborhood of the real subspace and which coincides with the function $H(x,p,t)$ when $z = x$, $\zeta = p$. All of the peculiarities of the complex theory are evident in this case, while difficulties of a technical nature do not arise.

In order to solve Eq. (53) using the method of characteristics, we will look for a function analytic in $z = x + iy$, satisfying the equation

$$\frac{\partial S}{\partial t} + H\left(z, \frac{\partial S}{\partial z}, t\right) = 0 , \quad S\big|_{t=0} = S_0(z) . \tag{55}$$

If Eq. (55) is solved, then the functions

$$(x,t) = S(z,t)\big|_{z=x}$$

obviously satisfy Eq. (53). Equation (54) is correspondingly replaced by the equation

$$\left[\frac{\partial}{\partial t} + \frac{\partial H}{\partial \zeta_i}\frac{\partial}{\partial z^i} + \frac{1}{2}\frac{\partial^2 S}{\partial z^i \partial z^j}\frac{\partial^2 H}{\partial \zeta_i \partial \zeta_j}\right]\varphi(z,t) = 0 ; \quad \varphi\big|_{t=0} = \varphi_0(z) . \tag{56}$$

Equations (55)–(56) are solved by the method of characteristics in the same way as Eqs. (37)–(38), with x, p replaced by $z = x + iy$, $\zeta = p + i\eta$. After solving them one must verify the condition

$$\operatorname{Im} S(x,t) = S_2(x,t) = S_2(z,t)\big|_{y=0} \geq 0$$

(if we are solving the corresponding equations in a chart U_I then this condition is obviously replaced by the condition $\operatorname{Im} S_I\left(x^I, p_{\bar{I}}, t\right) = S_{2I}\left(x^I, p_{\bar{I}}, t\right) \geq 0$). The following objects arise in solving problem (50):

1) The manifold $M \subset \mathbb{C}^n_z \oplus \mathbb{C}_{n\zeta} \oplus \mathbb{R}^1_t$, which is analytic in the variables z, ζ for each fixed t and is invariant under translation along the trajectories of the Hamiltonian system

$$\dot{z} = \frac{\partial H}{\partial \zeta} , \quad \dot{\zeta} = -\frac{\partial H}{\partial z} . \tag{57}$$

2) A function $S(z,t)$ on the manifold M such that

$$dS(z,t) = \zeta_i dz_i - H(z,\zeta,t)dt ; \quad S(z,0) = S_0(z) ; \tag{58}$$

[4] The functions $S_0(x), \varphi_0(x)$ are also assumed to be analytic.

$$\operatorname{Im}\left[S\left(z^I,\zeta_{\bar I},t\right)-z^{\bar I}\zeta_{\bar I}\right]_{y^I=0,\,\eta_{\bar I}=0}\geq 0 \quad \text{on each } U_I{}^5 \ . \tag{59}$$

3) On each set U_I, a branch of the argument of the complex-valued function $D\left(z^I,\zeta_{\bar I}\right)/Dz_0$.

4) A function $\varphi(z,t)$ which satisfies the transport equation

$$\frac{d\varphi}{dt}+\frac{1}{2}\frac{\partial^2 H}{\partial z^i\partial\zeta_i}\varphi=0\ ;\quad \varphi\big|_{t=0}=\varphi_0(z)\ . \tag{60}$$

Along with this, as in the real case, the manifold M is Lagrangian:

$$d\zeta\wedge dz-dH\wedge dt\big|_M=0\ ,$$

and the functions $\left[D\left(z^I,\zeta_{\bar I}\right)/Dz_0\right]^{-1}$ define a global measure $dz_0\wedge dt$, which is invariant with respect to translation along the trajectories of the Hamiltonian system (57).

In order to proceed to the case of a non-analytic Hamiltonian $H(x,p,t)$, we pose, first of all, the following question: *is it necessary for the construction of the asymptotics of* $\psi(x,t,h)$ *for the Cauchy problem* (50) *to have an exact solution of Eqs.* (53)–(54)?

(We note that in the non-analytic case a solution of problem (53) does not always exist, even if the function $H(x,\zeta,t)$ is defined for complex values of ζ.)

Upon substituting the function (52) in the left-hand side of Eq. (50) we obtain

$$\begin{aligned}-ih\frac{\partial\psi}{\partial t}+H\left(x,\widehat{p}\right)\psi&=\mathrm{e}^{iS(x,t)/h}\left[\frac{\partial S}{\partial t}+H\left(x,\frac{\partial S}{\partial x},t\right)\right]\varphi(x,t)\\&\quad-ih\left[\frac{\partial}{\partial t}+\frac{\partial H}{\partial\zeta_i}\frac{\partial}{\partial x^i}+\frac{1}{2}\frac{\partial^2 S}{\partial x^i\partial x^j}\frac{\partial^2 H}{\partial\zeta_i\partial\zeta_j}\right]\\&\quad\cdot\varphi(x,t)+O\left(h^2\right)\ .\end{aligned} \tag{61}$$

The following assertions follow from Eq. (61):

1. If $\Omega=\cup\Omega_I$; $\Omega_I=\left\{\alpha\in U_I\middle|y^I(\alpha)=0,\ \eta_{\bar I}(\alpha)=0,\ \operatorname{Im}S_I(\alpha)=0\right\}$, then outside some open neighborhood of the set Ω

$$\psi(x,t,h)=O\left(h^N\right)$$

for any $N>0$. (As a consequence we see that it is possible to define the set Ω invariantly.)

---

[5] Here U_I is a region in the *complex* manifold M.

2. If Eqs. (53), (54) are solved up to terms of order $O\left(\sqrt{\operatorname{Im} S(x,t)}\right)^s$, then the first two terms on the right-hand side of Eq. (61) are of order $O\left(h^{s/2}\right)$.

To make assertion 2 global, we introduce a nonnegative real function $\rho(\alpha)$ on the manifold M, such that:
a) $\rho^2(\alpha) \in C^\infty(M)$.
b) On every neighborhood U_I this inequality holds:

$$c \operatorname{Im} S_I\left(x^I, p_{\bar{I}}, t\right) \le \rho^2\left(x^I, p_{\bar{I}}, t\right) \le C \operatorname{Im} S_I\left(x^I, p_{\bar{I}}, t\right) . \tag{62}$$

A manifold M and a function ρ satisfying (62) are called positive. Moreover, by assertion 1 we may consider all functions only in some neighborhood of the set

$$\Omega = \left\{\alpha \in M \middle| \rho(\alpha) = 0\right\} , \tag{63}$$

and by assertion 2 it is possible to solve Eqs. (55)–(56) with accuracy up to terms of order $O\left(\rho^s\right)$ for $s \ge 3$.

Assertion 2 allows us to obtain the following important corollary: *In Eqs.* (53)–(54) *(as well as* (55)–(56)*) it is possible to replace the function* $H(z,\zeta,t)$ *by a finite partial sum of the Taylor series expansion of* $H(z,\zeta,t)$ *in the variables* $y = \operatorname{Im} z, \eta = \operatorname{Im} \zeta$.

Indeed, from the condition $\operatorname{Im} S(x,t) \ge 0$ it follows that

$$|\operatorname{grad}_x(\operatorname{Im} S(x,t))| \le C\sqrt{\operatorname{Im} S(x,t)} ,$$

and therefore

$$\begin{aligned} H\left(x, \frac{\partial S}{\partial x}, t\right) &= \sum_{|\alpha| \le s} \frac{1}{\alpha!}\left[i \operatorname{Im} \frac{\partial S}{\partial x}\right]^\alpha \left[\frac{\partial}{\partial p}\right]^\alpha H\left(x, \operatorname{Re}\frac{\partial S}{\partial x}, t\right) + O\left(\operatorname{Im} S\right)^{s/2} \\ &= {}^sH\left(x, \frac{\partial S}{\partial x}, t\right) + O\left(\operatorname{Im} S\right)^{s/2} . \end{aligned} \tag{64}$$

The function ${}^sH(x,\zeta,t)$ is called the s-analytic continuation of the function $H(x,p,t)$ in the variable p. Since we will have to consider the Hamiltonian function in an arbitrary complex chart U_I with coordinates $\left(z^I, \zeta_{\bar{I}}\right)$, it is natural to consider s-analytic continuations in the variable x as well:

$${}^sH(z,\zeta,t) = \sum_{k=0}^{s} \frac{1}{k!}\left[iy\frac{\partial}{\partial x} + i\eta\frac{\partial}{\partial p}\right]^k H(x,p,t) .$$

Furthermore, from the same considerations it is clear that the requirement of analyticity in the variables (z,ζ) of the manifold M for a fixed t may[6] be weakened to s-analyticity, understood in the following sense. If

[6] And even must, since, for example, analytic partition of unity on an analytic manifold does not exist, while an s-analytic partition of unity does indeed exist.

$$\alpha : U \to \mathbb{C}^n \ , \quad \beta : V \to \mathbb{C}^n$$

are coordinate mappings on M for a fixed t, then the mapping $\beta \circ \alpha^{-1}$ is given by a function satisfying the condition

$$\frac{\partial}{\partial \bar{z}} \left(\beta \circ \alpha^{-1}\right) = \frac{1}{2} \left[\frac{\partial}{\partial x} + i \frac{\partial}{\partial y}\right] \left(\beta \circ \alpha^{-1}\right) = O\left(\rho^s\right) \ .$$

Thus the construction of an asymptotic solution to the Cauchy problem (50) may be reduced to the construction of the following objects:

1. *An s-analytic manifold M which is invariant with respect to the trajectories of the Hamiltonian system* (37) *and is Lagrangian:* $dz \Lambda d\zeta + dH(z,\zeta,t)\, dt\big|_M = O(\rho^s)$.

2. *An s-analytic function $S(\alpha)$ on M such that*

$$dS = \zeta dz - H dt + O\left(\rho^s\right) \tag{65}$$

$$c \,\mathrm{Im}\, S_I\left(x^I, p_{\bar{I}}, t\right) \le \rho^2\left(x^I, p_{\bar{I}}, t\right) \le C \,\mathrm{Im}\, S_I\left(x^I, p_{\bar{I}}, t\right) \ \text{on } U_I \ . \tag{66}$$

3. *An s-analytic measure μ which is invariant with respect to translation along the trajectories of the Hamiltonian system* (57).

4. *A choice of a branch of the argument of the density of the measure* $\mu_I = D\mu / D\left(z_I, \zeta_{\bar{I}}\right) = D\left(dz_0 \wedge dt\right) / D\left(z^I, \zeta_{\bar{I}}\right)$ *in each chart U_I of the manifold M.*

Under these assumptions an asymptotic solution of the Cauchy problem (50) can be written in the form

$$\begin{aligned} \psi(x,t,h) = \sum_{\{U_I\}} \overline{F}^{1/h}_{p_{\bar{I}} \to x^{\bar{I}}} & \left[e^{iS(x^I, p_{\bar{I}}, t)/h} \sqrt{\mu_I\left(x^I, p_{\bar{I}}, t\right)} \cdot \varphi\left(x^I, p_{\bar{I}}, t\right) \right. \\ & \left. \times\ l_I\left(x^I, p_{\bar{I}}, t\right) \right] = K\varphi \end{aligned} \tag{67}$$

where $\{l_I\}$ is an s-analytic partition of unity subordinate to the covering $\{U_I\}$ and the function $\varphi(\alpha)$ satisfies the transport equation (60) on the manifold M up to terms of order $O\left(\rho^s\right)$.

In the book it is shown that the construction of objects (1)–(4) is possible whenever the Hamiltonian function $H(x,p,t)$ belongs to the class $C^\infty\left(\mathbb{R}^n_x \oplus \mathbb{R}_{np} \oplus \mathbb{R}^1_t\right)$. For this purpose we have constructed the analysis of general s-analytic manifolds; we study Lagrangian s-analytic manifolds in § 3.3 and § 3.4.

Here we will touch on one more question which has to do with formula (67). Upon careful consideration of this formula one notices that the s-analytic functions $S_I\left(x^I, \zeta_{\bar{I}}, t\right)$, $\mu_I\left(x^I, \zeta_{\bar{I}}, t\right)$, $\varphi\left(x^I, \zeta_{\bar{I}}, t\right)$ in each chart U_I carry *information about rate of decay* which is not necessary for the construction of the canonical operator K *in the chart U_I*. For such a construction it is only necessary to know the values of these functions on the submanifold U_I^0 of the manifold U_I defined by the equations $z^I = x^I$, $\zeta_{\bar{I}} = p_{\bar{I}}$ or, equivalently,

$$y^I = 0 \,, \quad \eta_{\bar{I}} = 0 \,. \tag{68}$$

Here the manifold U_I^0 has *real* dimension $n+1$. However, this information is already insufficient for the agreement of the operator on the intersection of the charts U_I and U_J, since the corresponding manifolds U_I^0 and U_J^0 do not coincide on the intersection of the charts U_I and U_J.

On the other hand, using the manifold U_I^0 in constructing the canonical operator meets with other obstacles as well. Indeed, if we consider the phase flow of the manifold

$$(x, \zeta) = \left(x, \frac{\partial S_0(x)}{\partial x}\right) ,$$

which is a manifold of type U_I^0 on the initial Lagrangian manifold

$$(z, \zeta) = \left(z, \frac{\partial S_0(z)}{\partial z}\right)$$

for $I = \{1, \ldots, n\}$, then we discover that for $t > 0$ the manifold we obtain is no longer a manifold of this type in any coordinate system of the form $(U_I; z^I, \zeta_{\bar{I}}, t)$.

In connection with this, the question arises whether it is possible to define a real submanifold of dimension $n+1$ in each domain U_I which would be free from these drawbacks.

With this in mind we describe in § 3.3 (subheading 5) a class of $(n+1)$-dimensional submanifolds of the chart U_I, which in a certain (not necessarily canonical!) coordinate system α on U_I satisfy the equations

$$\operatorname{Im} \alpha = 0 \,. \tag{69}$$

We denote by $U[\alpha]$ the submanifold defined by formulas (69) for a given s-analytic system of coordinates. An important property of s-analytic functions is the existence and uniqueness (up to modulus $O\left(\rho^{s+1}\right)$) of an s-analytic continuation of functions from the submanifold $U[\alpha]$ to the entire chart U_I. Let us formulate here the concept of s-analytic continuation, ignoring, for simplicity, functions of order $O\left(\rho^{s+1}\right)$ (precise definitions and proofs will be given in § 3.3). Let ${}^s\mathcal{O}(U_I, \rho)$ be the ring of s-analytic functions, and

$$R[\alpha] : \ {}^s\mathcal{O}(U_I, \rho) \to C^\infty(U[\alpha]) \tag{70}$$

be the restriction homomorphism. The following assertion is true:

The homomorphism $R[\alpha]$ is a ring isomorphism. We denote the inverse homomorphism to $R[\alpha]$ by

$$A[\alpha] : C^{\infty}(U[\alpha]) \to {}^{s}\mathcal{O}(U_I, \rho) \ . \tag{71}$$

We give an intrinsic description of all manifolds of the type $U[\alpha]$ in terms of the s-analytic coordinates $(z^I, \zeta_{\overline{I}})$. Indeed, for each s-analytic coordinate system α in the chart U_I there exists a mapping

$$g : U_I^0 \to \mathbb{C}^I \oplus \mathbb{C}_{\overline{I}}$$

which has the following properties:

$$1) \ \left| g\left(x^I, p_{\overline{I}}\right) \right| \leq C\rho\left(x^I, p_{\overline{I}}\right) , \tag{72}$$

$$2) \ \det \left. \frac{\partial\left(x^I + ig^I\left(x^I, p_{\overline{I}}\right),\ p_{\overline{I}} + ig_{\overline{I}}\left(x^I, p_{\overline{I}}\right)\right)}{\partial\left(x^I, p_{\overline{I}}\right)} \right|_{\Omega} \neq 0 , \tag{73}$$

3) the manifold $U[\alpha]$ is defined by the formulas

$$z^I = x^I + ig^I\left(x^I, p_{\overline{I}}\right) , \quad \zeta_{\overline{I}} = p_{\overline{I}} + ig_{\overline{I}}\left(x^I, p_{\overline{I}}\right) \ . \tag{74}$$

Conversely, if the map g satisfies the relations (72) and (73), the formulas (74) define a submanifold in the chart U_I for which there exist s-analytic coordinates α in the chart U_I such that this submanifold is defined by Eq. (69). We will call a mapping which satisfies relations (72) and (73) a *nonsingular germ*, and will denote the manifold (74) by U^g. The operators (70) and (71) will be denoted R^g and A^g from now on.

From what has been said it is clear that the class of submanifolds of type U^g is invariant with respect to the type of the chart; this means that for any germ g in the chart U_I there exists a germ g' in the chart U_J such that in the intersection $U_I \cap U_J$ the submanifolds U^g and $U^{g'}$ coincide.

To define the canonical operator (67) using the values of the functions S, μ_I and φ on the set U^g only, we must know how to "translate" the values of these functions to the manifold U_I^0, in other words be able to calculate the operator T^g appearing in the commutative diagram

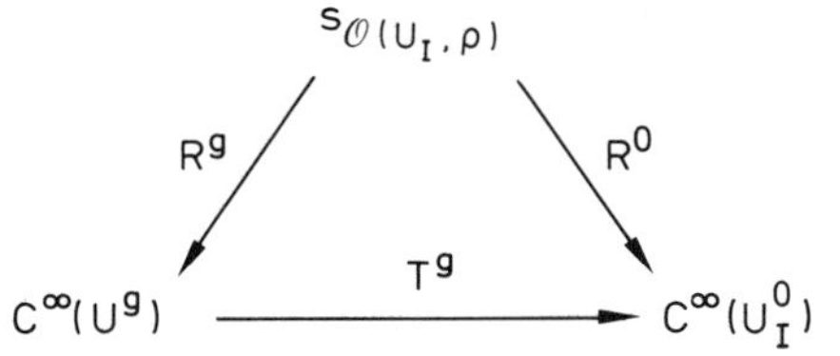

(75)

In § 3.3 we give explicit formulas for computing the operator T^g. Using this operator the canonical operator (67) can be rewritten in the form

$$K\varphi = \sum_{\{U_I\}} \overline{F}^{1/h}_{p_{\overline{I}} \to x^{\overline{I}}} \left[e^{\frac{i}{h} T^g S_I} T^g \left(\mu_I^{1/2}\right) T^g \left(\varphi e_I\right) \right] \ . \tag{76}$$

The importance of manifolds of the type described is clear from the following proposition.

If the submanifold U' of the initial Lagrangian manifold is nonsingular, then its phase flow along the trajectories of the Hamiltonian system (57) *is also nonsingular; here it is assumed that the imaginary part of the Hamiltonian function is nonpositive.*

We will not present the proof of this proposition, which can be found in the book of A.S. Mishchenko, B.Yu. Sternin, and V.E. Shatalov [1]. The methods of the proof are close to the work of F. Trèves [1].

The examples of finding solutions of the Cauchy problem for a differential (or pseudodifferential) equation, asymptotic with respect to a small parameter h, that we have given above, illustrate some general rules of the method of finding a solution. First of all, only differential (or pseudodifferential) equations which are homogeneous of degree zero in terms of the variables $(h^{-1}, \partial/\partial x)$ are considered. The class of such operators can be given with the help of the Hamilton functions $H(x,p)$, depending on the two groups of variables $x = (x^1, \ldots, x^n)$, $p = (p_1, \ldots, p_n)$. In order to obtain the operator $H(x, \widehat{p})$ itself, it is necessary in the case H is a polynomial function in the variables $p = (p_1, \ldots, p_n)$ to simply substitute the operators $-ih\partial/\partial x^k$ in place of the variables p_k, considering the operation of differentiation to be the first operation and the operation of multiplication by a function – the coefficients of the polynomial – to be the second. In the case of pseudodifferential operators one must use the Fourier transform according to the formula

$$H(x, \widehat{p})\psi(x) = \left\{ \overline{F}^{1/h}_{p\to y} \left[H(x,p) F^{1/h}_{x\to p} \psi(x) \right] \right\}_{x=y} .$$

The latter formula gives an equivalent definition of the operator in case the Hamiltonian $H(x,p)$ is polynomial in the variables p.

Then any asymptotic solution of the equation $H(x,\widehat{p})\psi(x) = 0$ is associated with a certain submanifold L lying in phase space $\mathbb{R}^n_x \oplus \mathbb{R}^p_n$ (for simplicity we consider here the real case). The phase space $\mathbb{R}^n_x \oplus \mathbb{R}^p_n$ has coordinates $(x,p) = (x^1, \ldots, x^n, p_1 \ldots, p_n)$, and the dimension of the submanifold L equals n, i.e. half the dimension of the phase space. Moreover, the submanifold L is not arbitrary, but such that the 2-form

$$\omega = dp \wedge dx = dp_1 \wedge dx^1 + dp_2 \wedge dx^2 + \ldots + dp_n \wedge dx^n$$

is identically zero when restricted to the submanifold L. Such submanifolds L are called *Lagrangian manifolds*. The operator K is called *Maslov's canonical operator*.

In this way, the problem of construction of an asymptotic solution for a given Lagrangian manifold L decomposes into two problems: a) the topological problem of calculating the conditions of quantization of the Lagrangian manifold and b) finding a function φ on the Lagrangian manifold which satisfies the equation of transport and constructing a solution by evaluating Maslov's canonical operator $\psi(x) = K(\varphi)$.

In agreement with the two problems indicated we will present in Part I the topological properties of Lagrangian manifolds and the methods of calculating the Maslov conditions of quantization. In Part II we present all the analytical questions: the construction of Maslov's canonical operator, the formulas governing the interaction of Maslov's canonical operator with pseudodifferential operators and applications to various problems of the theory of partial differential equations.

Part I
The Topology of Lagrangian Manifolds

Chapter 1
Some Topological Considerations

In this chapter we present a collection of concepts and theorems from topology and geometry which will be encountered in this book. The material of this chapter can be divided into two groups. The first group consists of the concepts and theorems which occur in most of our arguments and constructions. The second group consists of concepts and theorems of a purely subsidiary, computational nature. Although we expect these concepts and theorems to be familiar to the reader, we nevertheless treat them in detail for two reasons. First, we attempt to make the exposition depend as little as possible on supplementary literature; and second, the presentation of the material is based on a maximal exploitation of topological methods of reasoning, enabling us to make the problem of constructing global asymptotic solutions of equations with small parameter as transparent as possible. A detailed exposition can be found in the books of L.S. Pontryagin [2], N. Steenrod [1], and D.B. Fuks, A.T. Fomenko, V.L. Gutenmacher [1].

1.1 Manifolds and Bundles

The fundamental geometrical objects with which we shall work throughout the book are smooth manifolds. Let M be a separable topological space with a countable basis of open sets. The space M is called a *smooth manifold of class C^∞ and dimension n* if there exists a covering of M by open sets $U_1, U_2, \dots$ satisfying the following conditions.

(1) For each set U_1 there exists a homeomorphism $\varphi_i : U_i \to V_i$, where V_i is an open neighborhood in n-dimensional Euclidean space $\mathbb{R}^n$;

(2) if $U_i \cap U_j \neq \emptyset$, then the homeomorphism

$$\varphi_{ji} = \varphi_j \varphi_i^{-1} : \varphi_i(U_i \cap U_j) \to \varphi_j(U_i \cap U_j)$$

is a smooth (C^∞) transformation of open sets of $\mathbb{R}^n$.

If y is a point on the manifold M, $y \in U_k$, then the coordinates of the point $\varphi_k(y) \in \mathbb{R}^n$ are called *local coordinates* of the point $y \in M$. A pair (U_k, φ_k) is called a *local chart* of the manifold M, and the collection of charts $\{(U_k, \varphi_k)\}$ is called an *atlas of local charts* of M. Thus in each local chart (U_k, φ_k), the coordinates of $\varphi_k(y)$ determine certain (continuous) functions $\alpha_k^1(y), \ldots, \alpha_k^n(y)$ in the open set U_k, which are called a *local system of coordinates of* M in the chart (U_k, φ_k). In the following, we will for the sake of simplicity denote the local chart (U_k, φ_k) by U_k, considering the corresponding map φ_k to be fixed. On the intersection U_{kj} of two charts U_k, U_j we are given two local systems of coordinates $\{\alpha_k^i(y)\}$ and $\{\alpha_j^i(y)\}$, for which by definition there exists a C^∞ map of open sets of Euclidean space φ_{kj}, i.e. n smooth functions in n variables, such that

$$\{a_j^i(y)\} = \varphi_{kj}\left(a_k^1(y), \ldots, a_k^n(y)\right) .$$

Two atlases of local charts $\{U_k, \varphi_k\}$ and $\{U'_s, \varphi'_s\}$ on the manifold M are called *equivalent* if in the intersection $U_k \cap U'_s$ the local systems of coordinates $\{\alpha_k^i(y)\}$ of the chart U_k and $\left\{\alpha_s'^i\right\}$ of the chart U'_s are related by the identity

$$a_k^i(y) \equiv \psi_{ks}^i\left(\alpha_s'^1(y), \ldots, \alpha_s'^n(y)\right)$$

for some smooth functions ψ_{ks}^i. In this case it is said that the transformation from one local system of coordinates to the other is accomplished by a smooth change of coordinates. The condition of equivalence of two atlases of local charts is an equivalence relation. A family of atlases of local charts, all pairwise equivalent, on the manifold M will be called a smooth structure on the manifold M.

A function f on a manifold M is called a *smooth function* of class C^∞ if the functions

$$f \circ \varphi_k^{-1} : V_k \to \mathbb{R}^1 , \quad V_k = \varphi_k(U_k) \subset \mathbb{R}^n , \tag{1.1}$$

are smooth functions on the domain $V_k \subset \mathbb{R}^n$. The definition of a smooth function f is independent of the choice of the atlas of local charts for a given smooth structure on M.

Indeed, if $\{U'_s, \varphi'_s\}$ is another atlas of local charts, then the function $f \circ \varphi_s'^{-1} : V'_s \to \mathbb{R}^1$ in a neighborhood of each point of the domain V'_s can be factored into a composition of the form

$$f \circ \varphi_s'^{-1} = (f \circ \varphi_k^{-1}) \circ \left(\varphi_k \circ \varphi_s'^{-1}\right) ,$$

where the maps $f \circ \varphi_k^{-1}$ and $\varphi_k \circ \varphi_s'^{-1}$ are smooth maps. This means the function $f \circ \varphi_s'^{-1}$ is also a smooth function.

Using the composition (1.1) we obtain that any smooth function f on the manifold M can be represented in the form

$$f(y) = g_k\left(\alpha_k'^1(y), \dots, \alpha_k'^n(y)\right) ,$$

where g_k is a smooth function in n variables.

By covering the manifold M with finer local charts, one may always choose an atlas of charts (U_k, φ_k) such that $V_k \cong \mathbb{R}^n$ for any k. Moreover, one can obtain (less trivially) that any intersection $U_{i_1 \dots i_s} = U_{i_1} \cap \dots \cap U_{i_s}$ is also homeomorphic to Euclidean space.

We will also consider a more general class of objects: smooth manifolds with boundary. To distinguish manifolds without boundary from manifolds with boundary, we will call the former *closed* manifolds. A *smooth manifold with boundary* is a separable topological space M with a subspace N and a countable open covering $U_1, U_2, \dots$ such that there exist homeomorphisms

$$\varphi_i : U_i \to V_i \subset \mathbb{R}^n ,$$

satisfying the following conditions:

(1) If the set U_i of the given covering is contained in $M \setminus N$, then the corresponding homeomorphism $\varphi_i : U_i \to V_i$ maps the set U_i onto an open ball V_i in Euclidean space $\mathbb{R}^n$; otherwise one is given a homeomorphism $\varphi_i : U_i \to V_i$, where V_i is a hemisphere of the form

$$\sum_{i=1}^{n} x_i^2 < 1 , \quad x_n \geq 0 ,$$

under which the set $U_i \cap N$ is mapped onto the subset of V_i of points for which $x_n = 0$;

(2) if U_i and U_j are two sets of the given covering and $U_i \cap U_j \neq \emptyset$, then the mapping $\varphi_i \circ \varphi_j^{-1} = \varphi_{ij}$ is a smooth map of the set $\varphi_j(U_i \cap U_j)$ onto the set $\varphi_i(U_i \cap U_j)$.

Since the homeomorphisms φ_{ij} must map interior points to interior points and boundary points to boundary points, it is obvious that the sets $U_i \cap N$ form a covering of N, providing it with the structure of a smooth manifold; N is called the *boundary of the manifold* M and is denoted by ∂M. Since the concept of a smooth function also has meaning in case the function is defined only on a half-space $\mathbb{R}^n_+$, smooth changes of local coordinates and a smooth structure of a manifold with boundary can be defined for manifolds with boundary.

For smooth manifolds the following statement holds.

Theorem 1.1. *Let M be a smooth manifold, $\{W_k\}$ be some covering by open sets, i.e. such that $\cup W_k = M$. Then there exist real-valued smooth functions φ_k on the manifold M such that*

a) $0 \leq \varphi_k(y) \leq 1$, $y \in M$;
b) $\operatorname{supp} \varphi_k \subset W_k$;
c) $\sum_k \varphi_k(y) \equiv 1$.

The collection of functions $\{\varphi_k\}$ is called a *partition of unity*, subordinate to the covering $\{W_k\}$.

An example of a partition of unity is illustrated in Fig. 3.

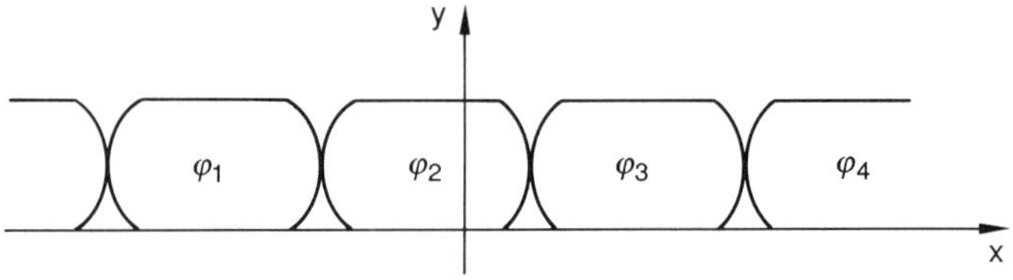

Fig. 3

Partitions of unity are useful for separating any function f into a sum of functions, $f = \sum_i f_i$, such that each term f_i has a sufficiently small support. For this it is sufficient to set $f_i = \varphi_i f$. In precisely the same way one can break up other objects into sums of terms with small support – vector functions, vector fields, differential forms, etc.

The simplest example of a smooth manifold is Euclidean space $\mathbb{R}^n$. In this case the atlas of charts consists of one chart U, coinciding with the space $\mathbb{R}^n$ itself, and the local system of coordinates is a linear system of coordinates in Euclidean space. Consequently, it is a *global* system of coordinates on the manifold $\mathbb{R}^n$.

There are only two one-dimensional closed manifolds: the real line $\mathbb{R}^1$ and the circle S^1. On the circle there does not exist a global system of coordinates. Local coordinates in a neighborhood of any point $x \in S^1$ can be determined by one angular parameter φ, a mutiple-valued function on the entire circle whose value at each point is determined up to a multiple of 2π.

In two dimensions there are already infinitely many closed manifolds (surfaces). The simplest two-dimensional manifold (except for Euclidean space $\mathbb{R}^2$) is the two-dimensional sphere S^2, which one can express in three-space by the equation $x^2 + y^2 + z^2 = 1$. It is impossible to give a global system of coordinates on the two-sphere. The atlas must consist of at least two charts. But the simplest is the atlas of six charts given by the inequalities $x > 0$, $x < 0$, $y > 0$, $y < 0$ $z > 0$, $z < 0$. In each of these six charts the two Euclidean coordinates not appearing in the inequality serve as a local system of coordinates. For example, in the chart $x > 0$ we obtain $x = \left(1 - y^2 - z^2\right)^{1/2}$, while in the chart $x < 0$ we obtain $x = -\left(1 - y^2 - z^2\right)^{1/2}$. This demonstrates a homeomorphism of these charts with the domain in $\mathbb{R}^2$ given by the inequality $y^2 + z^2 < 1$. It is easy to check that the transition functions from one local system of coordinates to another are smooth. Thus, for example, if one chooses coordinates (y, z) in the chart $x > 0$ and coordinates (x, z) in the chart $y > 0$, then the transition functions take the form

$$y = \sqrt{1 - x^2 - z^2} \; ,$$
$$z = z \; ,$$

and hence are smooth functions for $x^2 + z^2 < 1$.

Another important example of a surface is the two-dimensional torus T^2, which one can define as the Cartesian product of two copies of the circle, $T^2 = S^1 \times S^1$, or as the 2-plane $\mathbb{R}^2$, in which points are identified by means of the integral lattice $\mathbb{Z} \times \mathbb{Z} \subset \mathbb{R}^2$. On the torus there exist two several-valued parameters φ_1 and φ_2, which become single-valued local coordinates in a small enough neighborhood of each point $x \in T^2$.

In general one can obtain many surfaces in an analogous way, identifying points in $\mathbb{R}^2$ by the action of some discrete group on the plane $\mathbb{R}^2$.

There are several ways to represent a manifold. The simplest of the gives a smooth manifold as the *graph* of some *smooth function* or *vector function.* Let $f : \mathbb{R}^n \to \mathbb{R}^m$ be a smooth vector function, $M \subset \mathbb{R}^{n+m}$ its graph, i.e. the set of points of the form $(x, f(x))$, $x \in \mathbb{R}^n$, $f(x) \in \mathbb{R}^m$, $\mathbb{R}^{n+m} = \mathbb{R}^n \oplus \mathbb{R}^m$. It is completely obvious that M is a smooth manifold. Indeed, we choose as the unique chart the manifold M itself, and as the homeomorphism $\varphi : M \to \mathbb{R}^n$ we take the restriction of the projection of $\mathbb{R}^{n+m}$ onto its first factor $\mathbb{R}^n$.

Another common way of representing a manifold as a subset of Euclidean space $\mathbb{R}^n$ is the *parametric method.* Let $G \subset \mathbb{R}^k$ be an open set; $f : G \to \mathbb{R}^n$ be a smooth map which is a homeomorphism on its image $M = f(G)$. Then M is a smooth manifold. Generally curves and two-dimensional surfaces in three-space are given in this way. The case of the graph of a vector function is a special case of the parametric representation of a manifold.

The parametric representation of a manifold allows us to examine only the simplest manifolds, the homeomorphic images of Euclidean space. Such simple examples as the circle S^1 and the sphere S^2 cannot be represented parametrically. In general one expresses manifolds by an equation or system of equations in Euclidean space:

$$f_1\left(x^1, \ldots, x^n\right) = 0 \; , \; \ldots$$
$$f_k\left(x^1, \ldots, x^n\right) = 0 \; .$$

The sphere S^{n-1} in $\mathbb{R}^n$ is given by the single equation $(x^1)^2 + \ldots + (x^n)^2 - 1 = 0$. The two-dimensional torus $T = S^1 \times S^1$ can be expressed by the system of two equations in $\mathbb{R}^4$:

$$\begin{aligned}(x^1)^2 + (x^2)^2 - 1 = 0 \; , \\ (x^3)^2 + (x^4)^2 - 1 = 0 \; .\end{aligned}$$

However, not every system of equations determines a manifold. For example, the equation $xy = a$ gives a 1-manifold in two-space for $a \neq 0$. For $a = 0$ the solution set of the equation $xy = 0$ forms a cross of coordinate axes, which is not a manifold. For this reason one usually requires of a system of equations that at every solution point $(x_0^1, \ldots, x_0^n)$ the hypotheses of the implicit function theorem be satisfied: the determinant of the matrix of partial derivatives of the functions $f_1, \ldots, f_k$ with respect to certain variables $x^{i_1}, \ldots x^{i_k}$ must be nonzero:

$$\det \begin{pmatrix} \frac{\partial f_1}{\partial x^{i_1}} & \cdots & \frac{\partial f_k}{\partial x^{i_1}} \\ \vdots & & \vdots \\ \frac{\partial f_1}{\partial x^{i_k}} & \cdots & \frac{\partial f_k}{\partial x^{i_k}} \end{pmatrix} \neq 0 \; .$$

According to the implicit function theorem, such a system of equations can be solved uniquely in a neighborhood of the point $(x_0^1, \ldots, x_0^n)$ with respect to the variables $(x^{i_1}, \ldots, x^{i_k})$. This means that the solution set of the system in a neighborhood of the point $(x_0^1, \ldots, x_0^n)$ can be represented as the graph of a vector function

$$\begin{aligned} x^{i_1} &= x^{i_1}\left(x_{i_{k+1}}, \ldots, x_{i_n}\right) \; , \\ &\vdots \\ x^{i_k} &= x^{i_k}\left(x_{i_{k+1}}, \ldots, x_{i_n}\right) \; , \end{aligned}$$

i.e. the solution set of the system of equations is a smooth manifold. Since the choice of coordinates $(x^{i_1}, \ldots, x^{i_k})$, with respect to which the system of equations is solved, is irrelevant, it is simpler to formulate the hypothesis of the implicit function theorem in terms of the rank of the Jacobian matrix of all partial derivatives of the functions $f_1, \ldots, f_k$:

$$\operatorname{rank} \begin{pmatrix} \frac{\partial f_1}{\partial x^1} & \cdots & \frac{\partial f_k}{\partial x^1} \\ \vdots & & \vdots \\ \frac{\partial f_1}{\partial x^n} & \cdots & \frac{\partial f_k}{\partial x^n} \end{pmatrix} = k \; .$$

In the next section this condition, as well as the implicit function theorem itself, will be formulated in invariant terms and in a more general situation.

As an example we consider the equation of the sphere $(x^1)^2 + \ldots + (x^n)^2 - 1 = 0$. The matrix of pertial derivatives in this case has the form

$$\left(2x^1, \ldots, 2x^n\right) .$$

The rank of this matrix equals either zero or one. If the rank is zero, then $2x^1 = \ldots = 2x^n = 0$. Since the point $x^1 = \ldots = x^n = 0$ is not a solution of the equation of the sphere, the rank of the Jacobian matrix of the equation is one at all points. Hence by the implicit function theorem the sphere is a smooth manifold.

Many classical spaces are smooth manifolds. The most important of them are matrix groups and their homogenous spaces. We consider first the group $GL(n, \mathbb{R})$ of all nondegenerate square matrices of order n. Each element $A \in GL(n, \mathbb{R})$ is a matrix

$$A = \begin{pmatrix} a_{11} & \ldots & a_{1n} \\ \vdots & & \vdots \\ a_{n1} & \ldots & a_{nn} \end{pmatrix} , \qquad \det A \neq 0 .$$

Thus we can consider the group $GL(n, \mathbb{R})$ to be a submanifold in the n^2-dimensional vector space $\mathbb{R}^{n^2}$ (according to the number of elements a_{ij} appearing in the matrix). Since a matrix A belongs to the group $GL(n, \mathbb{R})$ if and only if $\det A \neq 0$, the set $GL(n, \mathbb{R}) \subset \mathbb{R}^{n^2}$ is open. Consequently the group $GL(n, \mathbb{R})$ is a smooth manifold, whose atlas of charts consists of one chart.

The subgroup of $GL(n, \mathbb{R})$ consisting of matrices A whose determinant is one is denoted $SL(n, \mathbb{R})$. Thus the group $SL(n, \mathbb{R})$ is determined as the solution set of one equation $\det A - 1 = 0$. The function $f(A) = \det A - 1$ is a polynomial in the variables $\{a_{ij}\}$. To show that $SL(n, \mathbb{R})$ is a manifold, it suffices to show that on the group $SL(n, \mathbb{R})$ the gradient of $f(A)$ is nonzero. We compute first the gradient of $f(A)$ at the point $A = E$, the identity matrix. Denote by A_{ij} the matrix obtained by eliminating the i-th row and the j-th column from A. Then

$$f(A) = \det A - 1 = \sum_{k=1}^{n} (-1)^{1+k} a_{1k} \det A_{1k} - 1 ,$$

Therefore, if $A = E$,

$$\frac{\partial f}{\partial a_{11}} = \det A_{11} = 1 .$$

Thus the gradient of the function $f(A)$ is nonzero at the point $A = E$. Now let $A_0 \in SL(n, \mathbb{R})$ be an arbitrary matrix. We consider the mapping $\varphi : GL(n, \mathbb{R}) \to GL(n, \mathbb{R})$ given by the formula

$$\varphi(A) = A_0^{-1} A \ .$$

The mapping φ is a smooth homeomorphism. Then the composition $h(A) = f(\varphi(A))$ is a smooth function, with $h(A) = \det\left(A_0^{-1}A\right) - 1 = f(A)$. Consequently, $\frac{\partial h}{\partial a_{ij}} = \frac{\partial f}{\partial a_{ij}}$. On the other hand,

$$\frac{\partial h}{\partial a_{ij}}(A) = \sum_{\alpha,\beta} \frac{\partial f}{\partial a_{\alpha\beta}}(\varphi(A)) \cdot \frac{\partial \varphi_{\alpha\beta}(A)}{\partial a_{ij}} \ .$$

In particular,

$$\frac{\partial h}{\partial a_{ij}}(A_0) = \sum_{\alpha\beta} \frac{\partial f}{\partial a_{\alpha\beta}}(E) \cdot \frac{\partial \varphi_{\alpha\beta}(A_0)}{\partial a_{ij}} \ .$$

Inasmuch as φ is a smooth homeomorphism, the Jacobian matrix of order n^2 composed of the partial derivatives $\left(\frac{\partial \varphi_{\alpha\beta}(A_0)}{\partial a_{ij}}\right)$ is nondegenerate. Thus the nontriviality of the gradient of the function f at point A_0 follows from the nontriviality of the function f at the point E. We have shown that the gradient of f is nonzero at every point of the group $SL(n, \mathbb{R})$. By the implicit function theorem, this means that the group $SL(n, \mathbb{R})$ is a smooth manifold.

We now turn to a less trivial example – the group $O(n)$ of *orthogonal matrices*, i.e. matrices A which satisfy the matrix relation $A^t A = E$. Here A^t denotes the transposed matrix. This relation on orthogonal matrices can be understood as a system of equations on the space $\mathbb{R}^{n^2}$, consisting of n^2 equations. We are unable to apply the implicit function theorem directly to the given system, because if it were applicable we would obtain that the set of solutions is a zero-dimensional manifold, which is untrue. This actually means that there are dependent equations in the system of equations $A^t A = E$. Thus one should first reduce to a maximal subsystem of independent equations, and check the hypotheses of the implicit function theorem for the latter.

To us, such an approach seems awkward and does not clarify the essence of the matter. Thus we will consider another method, which constructs in explicit form a special atlas of charts in the group $O(n)$. We begin, first of all, with the construction of a special atlas of charts on the group $GL(n, \mathbb{R})$. We consider the map

$$\exp : \mathbb{R}^{n^2} \to GL(n, \mathbb{R}) \ ,$$

given by the formula

$$\exp(X) = \sum_{k=0}^{\infty} \frac{1}{k!} X^k \ .$$

The map exp is a smooth mapping, under which

$$\exp(X+Y) = \exp(X)\exp(Y) ,$$

if the matrices X and Y commute, i.e. $XY = YX$. Here the null matrix 0 is mapped into the unit matrix $E : \exp(0) = E$. We calculate the determinant of the Jacobian matrix of the map exp at the point 0. If $Y = \exp(X)$, $Y = (y_{ij})$, $X = (x_{ij})$, then

$$\frac{\partial y_{ij}}{\partial x_{\alpha\beta}} = \begin{cases} 1 , & \text{if } (i,j) = (\alpha,\beta) , \\ 0 , & \text{if } (i,j) \neq (\alpha,\beta) , \end{cases}$$

at the point $X = 0$. Consequently the determinant of the Jacobian matrix of f equals 1. Applying the implicit function theorem to the system of equations $Y = \exp(X)$, we obtain that the map is a smooth homeomorphism in a certain small neighborhood U of the null matrix. We denote $V_E = \exp(U)$. The inverse transformation of exp on the open set V_E into the open set U is denoted by ln:

$$\ln(A) = \sum_{k=1}^{\infty} \frac{(-1)^{k+1}}{k} (A-E)^k , \quad A \in V_E .$$

If the matrices A and B commute, then $\ln(AB) = \ln(A) + \ln(B)$.

We define an atlas of charts $\{V_A\}$ on the group $GL(n, \mathbb{R})$ by setting $V_A = A \cdot V_E$, $\varphi_A : V_A \to U$, $\varphi_A(B) = \ln\left(A^{-1}B\right)$. The mappings φ_A are smooth homeomorphisms. Consequently, the transition functions $\varphi_B\varphi_A^{-1}$ are also smooth homeomorphisms.

Now with the help of the atlas of charts $\{V_A\}$ we construct an atlas of charts on the group $O(n)$. Let $P \subset \mathbb{R}^{n^2}$ be the subspace of skew-symmetric matrices in the space of all matrices of order n^2. Then, obviously, the matrix $\exp(X)$, $X \in P$, is an orthogonal matrix. And, conversely, if $A \in O(n) \cap V_E$, then $\ln A$ is a skew-symmetric matrix. Consequently, the set $U \cap P \subset P$ maps homeomorphically onto the open set $O(n) \cap V_E \subset O(n)$. We set

$$W_A = O(n) \cap V_A ,$$

$$\psi_A : W_A \to O(n) \cap P , \quad \psi_A(B) = \ln\left(A^{-1}B\right) = \varphi_A(B) .$$

We obtain an atlas of charts $\{W_A\}$ on the group $O(n)$, for which the transition functions $\varphi_B\varphi_A^{-1}$ are smooth transformations.

In an analogous way one can construct atlases of charts for other classical matrix groups as well: $GL(n, \mathbb{C})$, $SL(n, \mathbb{C})$, $U(n)$, etc. Using the exponential map, for each matrix group a corresponding subspace of the space of all matrices is constructed. Thus, for the group $SL(n, \mathbb{C})$ the corresponding subspace consists of all matrices whose trace equals zero. For the group

$U(n)$ the corresponding subspace consists of all skew-Hermitian complex matrices.

We show now that the homogenous spaces of left (right) cosets of a certain subgroup are also smooth manifolds. Let $H \subset GL(n, \mathbb{R})$ be a closed subgroup. One can show that there then exists a linear subspace $P \subset \mathbb{R}^{n^2}$, such that the neighborhood of zero $P \cap U$ is mapped diffeomorphically onto a neighborhood of the identity element in the subgroup H by the exponential map:

$$\exp : P \cap U \to H \cap V_E .$$

In the examples of subgroups considered below, the existence of such a subspace can be shown without difficulty and in explicit form. Thus, for example, for the case of the subgroups $SL(n, \mathbb{R})$, $O(n)$, $U(n)$ we have already indicated such subspaces above. Thus, a neighborhood of the identity element E, or, more precisely, of the conjugacy class $[E] = H$ of the homogenous space $GL(n, \mathbb{R})/H$ is homeomorphic to $(V_E \cdot H)/H$. Consequently one might expect that the space $(V_E \cdot H)/H$ is homeomorphically mapped onto a neighborhood of zero in the factor-space $\mathbb{R}^{n^2}/P$ by the map ln. More precisely, let $Q \subset \mathbb{R}^{n^2}$ be the algebraic complement of the subspace P, $\mathbb{R}^{n^2} = Q \oplus P$. We consider the composition

$$\psi : Q \cap U \xrightarrow{\exp} V_E \cdot H \to (V_E \cdot H)/H .$$

It is not hard to convince oneself that the mapping ψ is a homeomorphism, since each coset $A \cdot H$ of the subgroup H intersects the image $\exp(Q \cap U)$ in exactly one point. Thus we have constructed one chart $W_E = (V_E \cdot H)/H$, $\varphi_E : W_E \to Q \cap U$, $\varphi_E = \psi^{-1}$. To construct the entire atlas of charts $\{W_A\}$ we observe that the group $GL(n, \mathbb{R})$ acts on the factor space $GL(n, \mathbb{R})/H$ and maps the point $[E]$ to any other point. Consequently, setting $W_A = A \cdot W_E$, $\varphi_A(x) = \varphi_E(A^{-1}x)$, we obtain a complete atlas of charts on the factor space $GL(n, \mathbb{R})/H$.

In precisely the same way it can be shown that for any closed subgroups $GL(n, \mathbb{R}) \supset G \supset H$ the homogeneous space G/H is also a smooth manifold. Already beginning with dimension three the set of manifolds is so enormous that it does not yield to any kind of effective description, and we are forced to restrict ourselves only to a few classical examples of manifolds.

The classical examples of homogenous spaces have special names. Thus the set of all lines in Euclidean space $\mathbb{R}^{n+1}$ passing through the origin of the coordinate system is called real n-dimensional projective space, and is written $\mathbf{RP}_n$. It can be represented as the homogeneous space of cosets of the group $O(1) \times O(n-1)$ in the group $O(n) : \mathbf{RP}_n = O(n)/(O(1) \times O(n-1))$.

The spaces $G_{n,k}$ of k-dimensional subspaces of Euclidean space $\mathbb{R}^{n+k}$ have an important significance in various geometric investigations. This space is also a manifold, homeomorphic to $O(n+k)/(O(n)\times O(k))$, and is called the Grassmann manifold or the Grassmannian.

By anlogy with the scheme of definition of smooth manifolds, one can define other topological objects, which are called *bundles*. Suppose there are given topological spaces E, B, F and a continuous map $p : E \to B$ satisfying the following condition: there exists a covering $\{U_\alpha\}$ of the space B and a system of homeomorphisms

$$\varphi_\alpha : p^{-1}(U_\alpha) \to U_\alpha \times F ,$$

for which the following diagram is commutative:

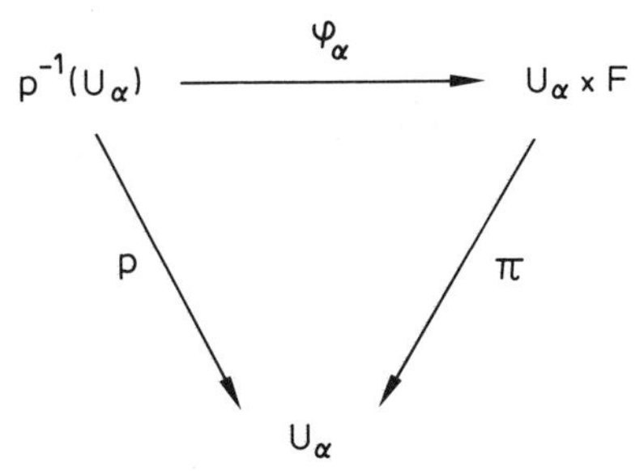

i.e. the equality

$$p = \pi \circ \varphi_\alpha$$

holds, where π is the projection onto the first factor. If we denote $U_{\alpha\beta} = U_\alpha \cap U_\beta$, as before, then the maps

$$\begin{gathered} \varphi_{\alpha\beta} = \left(\varphi_\beta | p^{-1}(U_{\alpha\beta})\right) \circ \left(\varphi_\alpha | p^{-1}(U_{\alpha\beta})\right)^{-1} , \\ \varphi_{\alpha\beta} : U_{\alpha\beta} \times F \to U_{\alpha\beta} \times F \end{gathered} \tag{1.2}$$

are homeomorphisms, which are identity maps on the first coordinate.

The collection $\xi = (E, B, p, F, \varphi_\alpha)$ is called a *locally trivial bundle*, E the *bundle space*, B the *base space*, F the *fiber*, and the maps $\{\varphi_{\alpha\beta}\}$ the *gluing functions*. If $\{V_\beta\}$ is a finer covering then the covering $\{U_\alpha\}$, then its own gluing functions of the original bundle are induced on it uniquely. The gluing functions $\varphi_{\alpha\beta}$ can be understood as mappings

$$\varphi_{\alpha\beta} : U_{\alpha\beta} \to \text{Homeo}\,(F, F) \tag{1.3}$$

into the group of homeomorphisms of the fiber F, for which such conditions of continuity are imposed on $\varphi_{\alpha\beta}$ that (1.2) becomes a homeomorphism. Moreover, the condition holds: for any point $x \in U_{\alpha\beta\gamma} = U_\alpha \cap U_\beta \cap U_\gamma$,

$$\varphi_{\beta\gamma}(x)\varphi_{\alpha\beta}(x) = \varphi_{\alpha\gamma}(x) \ . \tag{1.4}$$

If a covering $\{U_\alpha\}$ and functions (1.3) satisfying condition (1.4) are given on the space B, then there exists a locally trivial bundle $(E, B, p, F, \varphi_\alpha)$ such that (1.3) are its gluing functions.

This means that a locally trivial bundle can be expressed by means of its gluing functions, without considering the bundle space E.

Using the gluing functions (1.2) it is easy to define the *inverse image of a bundle* for a continuous map of bases. Indeed, if $f : B_1 \to B_2$ is a continuous map, $\xi = (E, B_2, p, F, \varphi_\alpha)$ is a locally trivial bundle, $\{\varphi_{\alpha\beta}\}$ are its gluing functions, then $f^*(\xi)$ denotes the locally trivial bundle over the space B_1 whose gluing functions are given by the formulas

$$\varphi^*_{\alpha\beta} = \varphi_{\alpha\beta} \circ f$$

for the atlas of charts $\{f^{-1}(U_\alpha)\}$.

The simplest example of a locally trivial bundle is the Cartesian product $E = B \times F$. In this case it is appropriate to take as the atlas of charts the single chart $U_\alpha = B$. On any other finer atlas of charts the gluing functions $\varphi_{\alpha\beta}$ induced will all have values equal to the identity homeomorphism. Thus, for example, if $B = S^1$ is the circle and $F = I$ is a segment of the real line, then $B \times F$ is simply a right circular cylinder. On the same base $B = S^1$ and with the same fiber $F = I$ it is possible to construct another, already nontrivial bundle. For this we separate the circle S^1 into two segments $U_1 = \{0 \leq \varphi \leq \pi\}$, $U_2 = \{\pi \leq \varphi \leq 2\pi\}$. We will give the gluing function φ_{12} on the charts U_1 and U_2 (we will for simplicity ignore the fact that U_1 and U_2 are not open sets). We set $\varphi_{12}(0) = \text{id}$, $\varphi_{12}(\pi) = -\text{id}$. Then the bundle E is the surface one obtains by gluing the opposite ends $(0 \times I)$ and $(2\pi \times I)$ of the strip $[0, 2\pi] \times I$ together with opposite orientations of the segment I (Fig. 4).

This surface is well known under the name of "the Möbius strip". Analogously it is possible to represent certain other surfaces in the form of locally trivial bundles: the torus and the "Klein bottle".

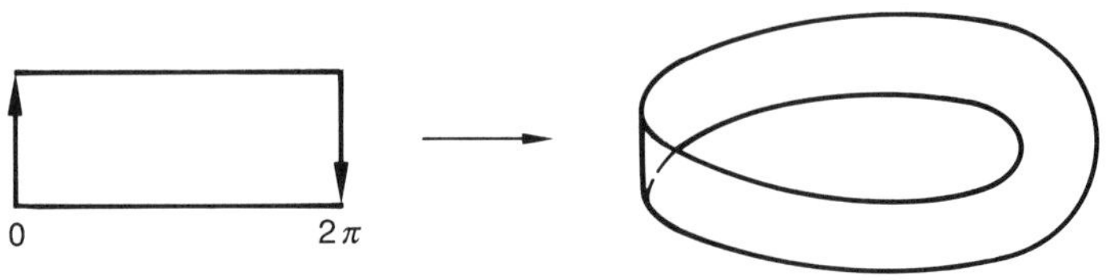

Fig. 4

A map $f : B \to E$ such that $pf(x) = x$, $x \in B$, is called a *section of the bundle.*

Two bundles $(E, B, p, F, \varphi_\alpha)$ and $\left(E', B, p'F, \varphi'_\beta\right)$ are considered *equivalent* if there exists a homeomorphism $h : E \to E'$ which forms a commutative diagram

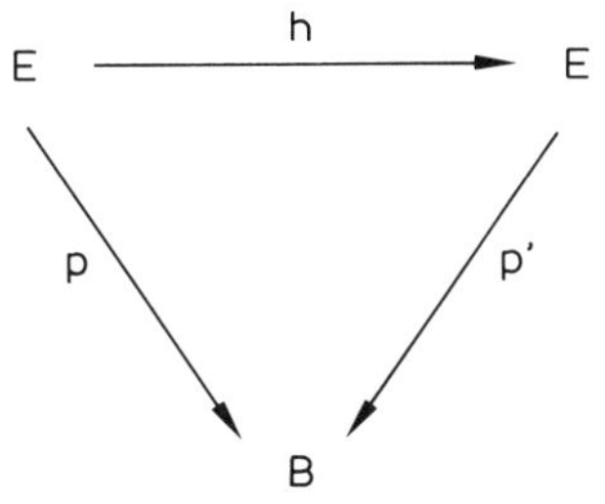

Replacing these two bundles by one finer one, we obtain that the gluing functions $\varphi_{\alpha\beta}$ and $\varphi'_{\alpha\beta}$ are associated by the relations

$$\varphi_{\alpha\beta}(x)h_\alpha(x) = h_\beta(x)\varphi'_{\alpha\beta}(x) \tag{1.5}$$

for certain maps

$$h_\alpha : U_\alpha \to \text{Homeo}\,(F, F)\ .$$

If $G \subset$ Homeo (F, F) is some subgroup of the group of homeomorphisms of the fiber F, and

$$\varphi_{\alpha\beta}\,(U_{\alpha\beta}) \subset G\ , \tag{1.6}$$

then we will say that we have given a locally trivial bundle with the structure group G. It might turn out that conditions (1.6) are not satisfied, but become true after applying Formula (1.5), i.e. replacing them by the gluing functions of an equivalent bundle. In this case it is said that the structure group of the bundle reduces to the group G. If $H \subset G$ is a subgroup, $\varphi_{\alpha\beta}(x)$, $h_\alpha(x) \in G$, $\varphi'_{\alpha\beta}(x) \in H$ in Formula (1.5), then it is said that the structure group G reduces to the subgroup H.

One of the most important classes of bundles is the class of real vector bundles, i.e. locally trivial bundles, whose fiber is the vector space $F = \mathbb{R}^n$ and whose structure group is the group of linear transformations $G = GL(n, \mathbb{R})$. If $F = \mathbb{C}^n$ and the structure group is $G = GL(n, \mathbb{C})$, then we speak of a complex vector bundle.

Proposition 1.2. *For a real vector bundle the structure group $GL(n, \mathbb{R})$ reduces to the subgroup $O(n)$ of orthogonal transformations of the fiber $\mathbb{R}^n$.*

For a complex vector bundle the group $GL(n, \mathbb{C})$ reduces to the subgroup of transformations $U(n)$ of the fiber $\mathbb{C}^n$.

The space $\Gamma(\xi)$ of all continuous sections of a real vector bundle naturally forms a real vector space, generally infinite-dimensional. The operation of addition of two sections f, g and the product of a section by a real number $\lambda \in \mathbb{R}^1$ are defined in every chart. The compositions

$$f_\alpha : U_\alpha \xrightarrow{f|U_\alpha} p^{-1}(U_\alpha) \xrightarrow{\varphi_\alpha} U_\alpha \times F \xrightarrow{\omega} F ,$$

$$g_\alpha : U_\alpha \xrightarrow{g|U_\alpha} p^{-1}(U_\alpha) \xrightarrow{\varphi_\alpha} U_\alpha \times F \xrightarrow{\omega} F$$

are vector functions on the chart U. The operations of addition and multiplication by a scalar are defined pointwise:

$$(f_\alpha + g_\alpha)(x) = f_\alpha(x) + g_\alpha(x) ,$$
$$(\lambda f_\alpha)(x) = \lambda f_\alpha(x) .$$

We obtain two new vector functions, according to which the maps

$$f\big|_{U_\alpha} + g\big|_{U_\alpha} : U_\alpha \to p^{-1}(U_\alpha) ,$$

$$\lambda f\big|_{U_\alpha} : U_\alpha \to p^{-1}(U_\alpha) ,$$

are constructed, and are sections on U.

It remains only to show that the maps obtained coincide on the intersections of charts. This fact follows from the fact that the gluing functions are linear transformations of the fiber at each point, i.e. they preserve the operations of addition and multiplication by a scalar in each fiber.

Moreover, the operation of scalar multiplication can be extended to the operation of multiplication by continuous functions, thus converting the space of sections $\Gamma(\xi)$ into a module over the ring $\mathbb{R}(B)$ of continuous functions on the base B.

For example, if ξ is the trivial bundle, i.e. $\xi = B \times \mathbb{R}^n$, then the space of sections $\Gamma(\xi)$ is simply the space of continuous vector functions defined on the base B with values in the space $\mathbb{R}^n$. If $n = 1$, then $\Gamma(\xi)$ coincides with the space of scalar functions on the base B.

Analogously, in the case of a complex vector bundle ξ the space of continuous sections is a complex vector space and a module over the ring $\mathbb{C}(B)$ of complex continuous functions on the base B. In case B is a finite polyhedron (for example, a compact manifold), $\Gamma(\xi)$ forms a projective $\mathbb{R}(B)$-module (a projective $\mathbb{C}(B)$-module respectively).

Let V_1, V_2 be finite-dimensional real (resp. complex) vector spaces. If

$$A_1 : V_1 \to V_1 , \quad A_2 : V_2 \to V_2$$

are linear isomorphisms, then we denote by $A_1 \oplus A_2$, $A_1 \otimes A_2$ the direct sum and tensor product of the isomorphisms respectively. Thus we obtain homeomorphisms of the groups

$$GL(n,\mathbb{R}) \times GL(m,\mathbb{R}) \to GL(n+m,\mathbb{R}) \ , \tag{1.7}$$

$$GL(n,\mathbb{R}) \times GL(m,\mathbb{R}) \to GL(nm,\mathbb{R}) \ , \tag{1.8}$$

associating to the pair of isomorphisms A_1, A_2 the isomorphisms $A_1 \oplus A_2$ and $A_1 \otimes A_2$ respectively. The homomorphisms (1.7) and (1.8) define operations on a vector bundle. If ξ_1, ξ_2 are two vector bundles over the space B with gluing functions $\varphi^1_{\alpha\beta}$ and $\varphi^2_{\alpha\beta}$, then by $\xi_1 \oplus \xi_2$ we denote the bundle whose gluing functions are $\varphi^1_{\alpha\beta} \oplus \varphi^2_{\alpha\beta}$, and by $\xi_1 \otimes \xi_2$ we denote the bundle with gluing functions $\varphi^1_{\alpha\beta} \otimes \varphi^2_{\alpha\beta}$.

Analogously, if $\wedge_i(V)$ is the i-th power of the vector space V under exterior multiplication, then the isomorphism $A : V \to V$ defines an isomorphism $\wedge_i(A) : \wedge_i(V) \to \wedge_i(V)$ of the exterior power of the space, so that we obtain a homomorphism of the groups of linear transformations of the spaces V and $\wedge_i(V)$. Then by $\wedge_i(\xi)$ we denote the vector bundle with gluing functions $\wedge_i(\varphi_{\alpha\beta})$. Furthermore, if V^* is the dual space to the space V, then this operation also induces an operation on vector bundles. The result is denoted ξ^*.

Let ξ_1, ξ_2 be two vector bundles:

$$\xi_1 = (E_1, p_1, B, F_1, \varphi^1_\alpha) \ , \quad \xi_2 = (E_2, p_2, B, F_2, \varphi^2_\alpha) \ .$$

Let $f : E_1 \to E_2$ be a fiber-to-fiber map, linear on each fiber. The condition of linearity, obviously, can be established only for the mapping $\varphi^2_\alpha f \left(\varphi^1_\alpha\right)^{-1}$, however, since the gluing functions are linear transformations this condition does not depend on the choice of index α. The map f is called a homomorphism of the bundle ξ_1 into the bundle ξ_2.

If the homomorphism $f : \xi_1 \to \xi_2$ is a monomorphism on each fiber, then the space consisting of the factor spaces of the fibers over each point is a locally trivial bundle and is called the *factor bundle.* A more precise definition is the following. Suppose $\{U_\alpha\}$ is an atlas of charts, and

$$f_\alpha : U_a \times F_1 \to U_a \times F_2$$

is determined by the formula

$$f_\alpha = \varphi^2_\alpha \circ f \circ \left(\varphi^1_\alpha\right)^{-1} \ .$$

If $\varphi^k_{\alpha\beta} = \varphi^k_\alpha \circ \left(\varphi^k_\beta\right)^{-1}$ are the gluing functions, then $f_\alpha \varphi^1_{\alpha\beta} = \varphi^2_{\alpha\beta} f_\beta$.

We define the gluing functions $\psi_{\alpha\beta}$ of a new bundle with fiber F_2/F_1. We denote by H_α the space of conjugacy classes of the subspace $f_\alpha(U_\alpha \times F_1)$

in the space $U_\alpha \times F_2$. One can show that for a small enough neighborhood U_α the space H_α is homeomorphic to $U_\alpha \times F_2/F_1$. Thus, let g_α be a homomorphism,

$$g_\alpha : U_\alpha \times F_2 \to U_\alpha \times F_2/F_1 \ ,$$

such that the kernel of the homomorphism g_α coincides with the image of f_α. Then there also exist unique homeomorphisms

$$\psi_{\alpha\beta} : U_{\alpha\beta} \times F_2/F_1 \to U_{\alpha\beta} \times F_2/F_1 \ ,$$

such that the diagram

$$\begin{array}{ccc} U_{\alpha\beta} \times F_2 & \xrightarrow{g_\alpha} & U_{\alpha\beta} \times F_2/F_1 \\ \uparrow \varphi^2_{\alpha\beta} & & \uparrow \psi_{\alpha\beta} \\ U_{\alpha\beta} \times F_2 & \xrightarrow{g_\beta} & U_{\alpha\beta} \times F_2/F_1 \end{array}$$

is commutative. The functions $\psi_{\alpha\beta}$ satisfy all the properties of gluing functions of a bundle. The bundle corresponding to the gluing functions $\psi_{\alpha\beta}$ is called the factor bundle.

If the base of a vector bundle ξ is a smooth manifold M, then one can require that the gluing functions

$$\psi_{kj} : U_{kj} \to GL(n, \mathbb{R})$$

be smooth functions, considering the group $GL(n, \mathbb{R})$ to be an open set in the n^2-dimensional Euclidean space of all matrices of order n.

It turns out that any vector bundle over a smooth manifold M admits smooth gluing functions for a given choice of charts and homeomorphisms φ_k. This assertion follows from standard theorems on uniform approximation of a continuous function by a sequence of smooth functions. If $\xi = (E, p, M, \mathbb{R}^n, \varphi_k)$ is a vector bundle over a smooth manifold M with smooth gluing functions, then the space of the bundle ξ is a smooth manifold, and all the transformations that appear in the definition of a bundle are smooth transformations.

A section $f : M \to E$ of the bundle ξ is called a *smooth section* if in any chart U_k the composition

$$\varphi_k \circ (f|U_k) : U_k \to U_k \times \mathbb{R}^n$$

is a smooth mapping.

The property of a section f of being smooth in a neighborhood of a point $x \in M$ does not depend on the choice of the chart U_k containing the point x.

An important example of a vector bundle on a manifold M is the so-called tangent bundle. Let $f : V_1 \to V_2$ be a smooth map of a domain

$V_1 \subset \mathbb{R}^n$ into a domain $V_2 \subset \mathbb{R}^n$. We denote by Df the matrix-valued function on V_1 composed of all the partial derivatives of the map f. If $g = f_1 \circ f_2$ is the composition of two maps, then

$$(Dg)(x) = Df_1\left(f_2(x)\right) \cdot Df_2(x) , \tag{1.9}$$

where the product on the right-hand side of (1.9) is the product of the matrices $Df_1\left(f_2(x)\right)$ and $Df_2(x)$.

Suppose M is a smooth manifold, $\{U_k\}$ is an atlas of local charts $\varphi_k : U_k \to V_k \subset \mathbb{R}^n$, $\varphi_{kj} : V_{kj} \to V_{jk}$ are the same as in the definition of a smooth manifold. Then the matrix-valued functions $\psi_{kj}(x) = D\varphi_{kj}\left(\varphi_k(x)\right)$ satisfy conditions (1.4). It is easy to check that

$$\det \psi_{kj}(x) \neq 0 .$$

Consequently the functions ψ_{kj} are gluing functions for some vector bundle. This bundle is called the *tangent bundle* of the manifold M and is denoted by TM.

We consider several examples of manifolds and tangent bundles. Let $M = S^1$ be the circle. Then as local coordinates on the circle we can take one of the values of the angular parameter φ. On the intersection of charts we obtain two local coordinates, which by definition will differ from one another by a constant function, a multiple of 2π. Consequently the matrix of partial derivatives (in this case simply the derivative) is identically equal to unity. Thus all the gluing functions of the tangent bundle TS^1 are identically equal to the unit homeomorphism of the fiber $\mathbb{R}^1$, and hence $TS^1 = S^1 \times \mathbb{R}^1$.

Analogously, one can show that on the two-dimensional torus the tangent bundle is trivial: $T\left(T^2\right) = T^2 \times \mathbb{R}^2$. But in the case of the two-dimensional sphere S^2 the tangent bundle TS^2 is nontrivial.

Since the space TM itself is a smooth manifold, then it makes sense to speak of a tangent bundle for it, too. Let

$$p : TM \to M$$

be the natural projection. Then

$$TTM = p^*(TM) \oplus p^*(TM) .$$

In the general case, for a vector bundle ξ over a smooth manifold M there is an isomorphism

$$T\xi = p^*(\xi) \oplus p^*TM .$$

The smooth sections of a tangent manifold TM are called *vector fields* on the manifold M. Every vector field X determines an operator of differentiation on the ring of smooth functions $C^\infty(M)$ on the manifold M.

This operator is defined in the following way. Let $U_k \subset M$ be a local chart, $x_k^1, \dots, x_k^n$ be a local system of coordinates in the chart U_k. Since the bundle TM over the chart U_k is homeomorphic to the direct product $U^k \times \mathbb{R}^n$, the section X can be written in terms of the basis $(e_1, \dots, e_n)$ of the space $\mathbb{R}^n$ in the form of a linear combination

$$X(x) = \sum_{j=1}^{n} X_k^j(x) e_j ,$$

where $X_k^j(x)$ are smooth functions on the chart U_k. Let $f \subset C^\infty(M)$ be an arbitrary smooth function on the manifold M. We set

$$X(f)(x) = \sum_{j=1}^{n} X_k^j(x) \frac{\partial f}{\partial x_k^j}(x) . \tag{1.10}$$

The value of the left-hand side of (1.10) is independent of the choice of a local system of coordinates $(x'_1, \dots, x'_n)$ by definition of the gluing functions of the bundle TM and the rules of differentiation of a composite function.

Incidentally, the converse assertion is also true. Any differential operator

$$A : C^\infty(m) \to C^\infty(M) ,$$

i.e. an operator A satisfying the equation

$$A(f \cdot g) = A(f)g + fA(g) ,$$

is determined by some smooth vector field.

A vector field X which takes the form $X(x) = e_j$ in a local system of coordinates $(x^1, \dots, x^n)$ is denoted $\partial/\partial x^j$. The value of a vector field X at a particular point $x \in M$ is called a tangent vector to the manifold M at x, and the entire fiber of the tangent bundle TM is called the tangent space. To each tangent vector X at the point x one can associate the functional of differentiation $X_x : C^\infty(M) \to \mathbb{C}$ defined by formula (1.10), and, conversely, any functional A which satisfies the equation

$$A(fg) = A(f)g + fA(g)$$

is defined by a certain tangent vector at the point $x \in M$.

Suppose $\varphi : [0,1] \to M$ is a smooth curve. Then at each point $\varphi(t_0)$, $0 < t_0 < 1$, one can define the tangent vector to the curve φ, denoted by $d\varphi/dt$, in a natural way. It is possible, for example, to represent this vector by the functional of differentiation

$$f \to \frac{d}{dt} f(\varphi(t)) \big|_{t=t_0} .$$

In a local system of coordinates a tangent vector to the curve φ is given by coordinates which are equal to the derivatives of the vector function φ.

Sometimes it is convenient to interpret the tangent vector X as an "infinitesimal curve" φ which passes through a given point x, i.e. as a bouquet of curves which pass through the point x and coincide up to second order with respect to a parameter. Thus, if the manifold M is smoothly imbedded in Euclidean space $\mathbb{R}^n$, then every tangent vector X_x, $x \in M$, can be realized as a directed segment in the space $\mathbb{R}^n$, whose origin is the point $x \in M \subset \mathbb{R}^n$. The set of all tangent vectors of a manifold at a point x generates a linear manifold in Euclidean space $\mathbb{R}^n$ which passes through the point x and is tangent to the manifold M. However, the entire tangent bundle, as a rule, cannot be realized in the space $\mathbb{R}^n$, since the linear manifolds tangent to M at distinct points may well have a nonempty intersection.

An example of the imbedding of a tangent bundle in Euclidean space is given by the ruled surface in $\mathbb{R}^3$ consisting of the lines tangent to a base curve (Fig. 5).

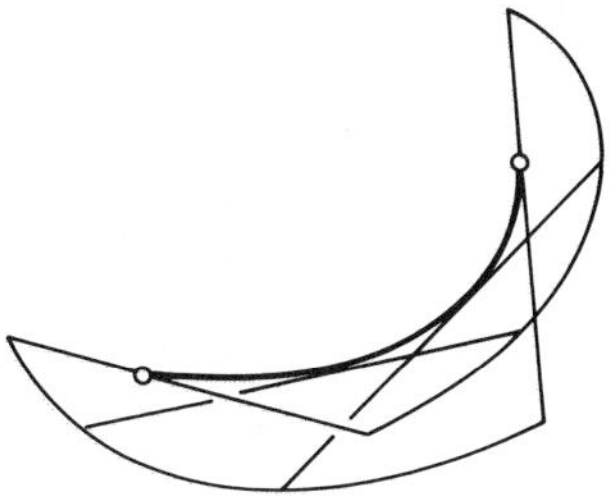

Fig. 5

Every smooth vector field X on a manifold M can be associated with a "system of ordinary differential equations", i.e. the problem of finding smooth curves φ such that

$$\frac{d\varphi}{dt}(t) = X(\varphi(t)) \ . \tag{1.11}$$

In each local system of coordinates Eq. (1.11) can be written in the form of an actual system of n ordinary differential equations. According to the theorems of existence and uniqueness for any point $x_0 \in M$ there exists a unique smooth curve $\varphi(t, x_0)$, $-\epsilon < t < \epsilon$, such that

$$\varphi(0, x_0) = x_0$$

and Eq. (1.11) holds. If M is a compact manifold, then ϵ can be taken as large as desired, so that the function $\varphi(t, x)$ forms a group of transformations of the manifold M, parametrized by the real numbers $t \in \mathbb{R}^1$:

$$\varphi(t_1, \varphi(t_2, x)) = \varphi(t_1 + t_2, x) \ .$$

We consider an example of a vector field on the two-dimensional torus, given in terms of the angular parameters by the system of equations

$$\frac{d\varphi_1}{dt} = a \ , \quad \frac{d\varphi_2}{dt} = b \ ,$$

where a, b are constants. The solutions of this system are the curves

$$\varphi_1 = at + \varphi_{10} \ , \quad \varphi_2 = bt + \varphi_{20} \ .$$

If the numbers a and b are rationally commensurable, i.e. if a/b is a rational number, then every trajectory is periodic, i.e. a closed curve on the torus. If a and b are incommensurable, i.e. a/b is irrational, then every trajectory forms a dense set on the torus T^2. In this case it is said that the motion given by the vector field is almost periodic.

Besides vector fields, another differential object on a smooth manifold M, called an *exterior differential form*, is also studied.

We define the exterior differential forms of degree k as smooth sections of the bundle $\wedge_k (T(M)^*)$. We denote the space of exterior differential forms of degree k by $\Omega_k(M)$. If $\omega \in \Omega_k(M)$ and $X_1, \ldots, X_k$ are vector fields, then the smooth function $\omega(X_1, \ldots, X_k)$ is defined as the value of the section ω on the vectors $X_1(x), \ldots, X_k(x)$ of the tangent space at the point x, since $\wedge_k (T(M)^*) = (\wedge_k T(M))^*$. We will denote by $\omega_1 \wedge \omega_2$ the exterior product of two forms induced by the exterior multiplication of vectors in the exterior algebra of a vector space.

Proposition 1.3. *There exists a unique operator*

$$d : \Omega_k(M) \to \Omega_{k+1}(M) \ ,$$

satisfying the following conditions:

a) $d(\omega_1 \wedge \omega_2) = d\omega_1 \wedge \omega_2 + (-1)^d \omega_1 \wedge d\omega_2$, *where* $d = \deg \omega_1$

b) $(df)(X) = X(f)$, *where* $f \in C^\infty(M) = \Omega_0(M)$ *and* X *is a vector field;*

c) $d^2 = 0$.

Proposition 1.4. *In a local system of coordinates* $(x^1, \ldots, x^n)$ *any exterior differential form* ω *of degree* k *has the form*

$$\omega(x) = \sum_{(i_1, \ldots, i_k)} a_{i_1 \ldots i_k} dx^{i_1} \wedge \ldots \wedge dx^{i_k} \ .$$

In particular, $\Omega_k = 0$ *for* $k > n$, *and any form* $\omega \in \Omega_n(M)$ *can be written as*

$$\omega(x) = f(x) dx^1 \wedge \ldots \wedge dx^n \ . \tag{1.12}$$

If $(y^1, \ldots, y^n)$ is another local system of coordinates, then

$$\omega(x) = f(x) \cdot \det \left\| \frac{\partial x^j}{\partial y^k} \right\| dy^1 \wedge \ldots \wedge dy^n \ .$$

The manifold M is called *orientable* if there exists a form $\omega \in \Omega_n(M)$ such that in a neighborhood of every point $x \in M$ there exists a local system of coordinates $(x^1, \ldots, x^n)$ for which in formula (1.12)

$$f(x) > 0 \ . \tag{1.13}$$

A collection of local coordinate systems which satisfy (1.13) is called an *orientation* of the manifold M.

For example, the circle S^1 is an orientable manifold, since there exists the differential form $d\varphi$, which has the same form for any angular parameter φ in a neighborhood of every point of the circle S^1.

In precisely the same way one can give a differential form on each of the six charts of the 2-dimensional sphere S^2, $x > 0$, $x < 0$, $y > 0$, $y < 0$, $z > 0$, $z < 0$, by the following equations:

for $z > 0$

$$\omega = \frac{1}{\sqrt{1 - x^2 - y^2}} dx \wedge dy \ ,$$

for $x > 0$

$$\omega = \frac{1}{\sqrt{1 - y^2 - z^2}} dy \wedge dz \ ,$$

for $y > 0$

$$\omega = \frac{1}{\sqrt{1 - x^2 - z^2}} dz \wedge dx \ ,$$

for $z < 0$

$$\omega = \frac{-1}{\sqrt{1 - x^2 - y^2}} dx \wedge dy \ ,$$

for $x < 0$

$$\omega = \frac{-1}{\sqrt{1 - y^2 - z^2}} dy \wedge dz \ ,$$

for $y < 0$

$$\omega = \frac{-1}{\sqrt{1 - x^2 - z^2}} dz \wedge dx \ ,$$

A trivial check shows that the definition of the form ω coincides on the intersection of charts and, hence, according to definition the manifold S^2 is orientable.

The definition of orientability can also be formulated in this way: there exists an atlas of charts such that the Jacobian of the change of coordinates is positive on the intersection of any pair of charts.

We show, for example, that the Möbius strip is a non-orientable manifold. One can imagine the Möbius strip as two rectangles U_1 and U_2 glued together. We introduce onto rectangle U_1 the coordinates (x_1, x_2), $0 < x_1 < 1$, $0 < x_2 < 1$, and the coordinates (y_1, y_2), $0 < y_1 < 1$, $0 < y_2 < 1$ onto rectangle U_2. Suppose the charts U_1 and U_2 are glued together by the mapping φ_{12}:

$$\psi_{12}(x_1, x_2) = \begin{cases} \left(x_1 + \frac{1}{2},\, x_2\right) , & 0 < x_1 < \frac{1}{2} , \\ \left(x_1 - \frac{1}{2},\, 1 - x_2\right) , & \frac{1}{2} < x_1 < 1 . \end{cases}$$

Then the Jacobian $\partial(y_1, y_2)/\partial(x_1, x_2)$ of the change of coordinates equals

$$\frac{\partial(y_1, y_2)}{\partial(x_1, x_2)} = \begin{cases} 1 , & 0 < x_1 < \frac{1}{2} , \\ -1 , & \frac{1}{2} < x_1 < 1 , \end{cases}$$

and, consequently, changes signs at different points on the intersection of the charts.

It is a more subtle problem to show that such a situation occurs in any atlas of charts on the Möbius strip. We will omit that proof.

If f is a smooth function on a smooth manifold M, then df is a differential form, that is a section of the tangent bundle T^*M. This section can be treated as a map

$$df : TM \to \mathbb{R}^1$$

from the tangent bundle TM into the space of real numbers $\mathbb{R}^1$ which is linear on each fiber. Then the diagonal map $DF = (df, f) : TM \to \mathbb{R}^1 \times \mathbb{R}^1 = TR^1$ is a map of tangent bundles which is linear on each fiber.

The indicated treatment of the differential of a function can be generalized to the case of an arbitrary smooth map of manifolds. A map $f : M_1 \to M_2$ of smooth manifolds is called a *smooth map* if f can be represented as a smooth vector function in any local system of coordinates of the manifolds M_1 and M_2. The condition of smoothness of a map f is independent both of the choice of a local system of coordinates on M_1 and on M_2.

Suppose $(x^1, \ldots, x^n)$, $(y^1, \ldots, y^m)$ are local systems of coordinates of the manifolds M_1 and M_2, in the charts U and V respectively. Then the mapping f has the form

$$y^j = f^j\left(x^1, \ldots, x^n\right) , \quad j = 1, \ldots, m .$$

We consider the matrix-valued function

$$Df(x) = \left(\frac{\partial f^j}{\partial x^k}(x) \right) . \tag{1.14}$$

The function (1.14) can be interpreted as a linear map

$$Df : U \times \mathbb{R}^n \to V \times \mathbb{R}^m . \tag{1.15}$$

Proposition 1.5. *The linear maps of the form* (1.15) *define a homomorphism of tangent bundles*

$$Df : TM_1 \to TM_2$$

such that the following diagram is commutative:

$$\begin{array}{ccc} TM_1 & \xrightarrow{Df} & TM_2 \\ \downarrow p_1 & & \downarrow p_2 \\ M_1 & \xrightarrow{f} & M_2 \end{array}$$

The proof is based on a formal application of the rules of differentiation of compositions of functions.

The mapping Df is called *the differential of the function* f. Since we have defined a map Df of the tangent vectors to the manifold M_1 to tangent vectors to the manifold M_2, we can use it to define a map of the spaces of differential forms

$$f^* : \Omega_k(M_2) \to \Omega_k(M_1) ,$$

setting

$$f^*(\omega)(X_1, \ldots, X_k)(x) = \omega(DfX_1(fx), \ldots, DfX_k(fx)) .$$

Then

$$f^*(d\omega) = df^*(\omega)$$

$$f^*(\omega_1 \wedge \omega_2) = f^*(\omega_1) \wedge f^*(\omega_2) . \tag{1.16}$$

The mapping $f : M_1 \to M_2$ is called an *imbedding* if f is a one-to-one mapping onto a closed subspace $f(M_1)$ and Df is a monomorphism on each fiber of the tangent bundle.

For example, an imbedding $f : [a, b] \to \mathbb{R}^3$ of a segment of the real line into three-dimensional Euclidean space is a regular smooth curve. The condition that the differential Df is a monomorphism on each fiber reduces to the following inequality: $df/dt(t) \neq 0$ for all $t \in [a, b]$, i.e. the velocity vector must be nonzero at each point of the curve.

Analogously, the condition of nondegeneracy of a surface in three-space $\mathbb{R}^3$ represented parametrically as the image of a rectangle, $f : [a, b] \times [c, d] \to$

$\mathbb{R}^3$, reduces to the condition that the partial derivatives $\partial f/\partial u(u,v)$ and $\partial f/\partial v(u,v)$ of the vector function $f(u,v)$ are linearly independent for all $u \in [a,b]$, $v \in [c,d]$. This condition is equivalent to saying that the Jacobian matrix

$$\begin{pmatrix} \frac{\partial f^1}{\partial u} & \frac{\partial f^1}{\partial v} \\ \frac{\partial f^2}{\partial u} & \frac{\partial f^2}{\partial v} \\ \frac{\partial f^3}{\partial u} & \frac{\partial f^3}{\partial v} \end{pmatrix}$$

of the map $f(u,v) = \left(f^1(u,v),\ f^2(u,v),\ f^3(u,v)\right)$ has maximal rank, i.e. the differential Df is a monomorphism.

Now we introduce the *de Rham cohomology* on a smooth manifold. The operator d of Theorem 1.3 forms a chain complex

$$\Omega_0(M) \xrightarrow{d} \Omega_1(M) \xrightarrow{d} \dots \xrightarrow{d} \Omega_n(M) ,$$

i.e. a sequence of homomorphisms such that

$$\mathrm{Ker}^k d \supset \mathrm{Im}^k d ,$$

where

$$\mathrm{Ker}^k d = \{\omega : \omega \in \Omega_k(M) \ \text{ and } d = 0\} ,$$
$$\mathrm{Im}^k d = d(\Omega_{k-1}(M)) .$$

We set

$$H_d^k(M) = \mathrm{Ker}^k d/\mathrm{Im}^k d .$$

The group $H_d^k(M)$ is called the *k-th de Rham cohomology group*. Condition (1.16) gives rise to the natural homomorphisms

$$f^* : H_d^k(M_2) \to H_d^k(M_1)$$

for a smooth map

$$f : M_1 \to M_2 .$$

There exists another definition of cohomology which holds for any topological space X. That is the so-called spectral cohomology. Suppose we are given an open covering $\mathfrak{U} = \{U_\alpha\}$ on the space X. We denote by $C^k(X,\mathfrak{U})$ the group whose elements are collections of constant functions $\{f_{\alpha_0 \dots \alpha_k}\}$, defined on all possible intersections of charts of the form $U_{\alpha_0 \dots \alpha_k} = U_{\alpha_0} \cap \dots \cap U_{\alpha_k} \neq \emptyset$. The group $C^k(X,\mathfrak{U})$ is called the *group of cochains* of the covering $\mathfrak{U}$, and the elements are *cochains* of the covering $\mathfrak{U}$. We define the homomorphism

$$\partial : C^k(X,\mathfrak{U}) \to C^{k+1}(X,\mathfrak{U})$$

by the equation

$$(\partial f)_{\alpha_0 \ldots \alpha_{n+1}} = \sum_{j=0}^{n+1} (-1)^j f_{\alpha_0 \ldots \widehat{\alpha}_j \ldots \alpha_{n+1}}$$

(the index $\widehat{\alpha}_j$ with a "hat" denotes that the given index is omitted). The equation

$$\partial^2 = 0$$

holds, so that, as in the case of de Rham cohomology, one can define the cohomology groups

$$H^k(X, \mathfrak{U}) = \mathrm{Ker}^k \partial / \mathrm{Im}^k \partial \ .$$

Let $\mathfrak{V} = \{V_\beta\}$ be another open covering, finer than $\mathfrak{U}$. This means that

$$V_\beta \subset U_{\alpha(\beta)} \tag{1.17}$$

and thus

$$V_{\beta_0 \ldots \beta_k} \subset U_{\alpha(\beta_0) \ldots \alpha(\beta_k)} \ . \tag{1.18}$$

Thus, relation (1.18) defines a map

$$\pi : C^k(X, \mathfrak{U}) \to C^k(X, \mathfrak{V}) \ ,$$

for which

$$\partial \pi = \pi \partial \ . \tag{1.19}$$

Condition (1.19) means that the homomorphisms π induce homomorphisms

$$\pi^* : H^k(X, \mathfrak{U}) \to H^k(X, \mathfrak{V}) \ . \tag{1.20}$$

Proposition 1.6. *The homomorphism* (1.20) *is independent of the choice of indices* $\alpha(\beta)$ *for the inclusion* (1.17).

In this way we obtain a well-defined sequence of groups $H_k(X, \mathfrak{U})$, directed by inclusion of one covering in another. We set

$$H^k(X) = \lim_{\rightarrow} H^k(X, \mathfrak{U})$$

and call the groups H^k *the spectral cohomology groups.*

Proposition 1.7. *Suppose the covering* $\mathfrak{U} = \{U_\alpha\}$ *is such that all the sets* $U_{\alpha_0 \ldots \alpha_k}$ *are contractible. Then the homomorphism*

$$H^k(X, \mathfrak{U}) \to H^k(X)$$

is an isomorphism.

An example of such a covering is the following. Let X be a simplicial complex, i.e. a space consisting of a finite union $X = \bigcup_{\alpha} X_\alpha$ of closed subspaces X_α homeomorphic to standard linear simplices of various dimensions, where the X_α intersect along certain boundaries. Then, if $e_1, \dots e_s$ are zero-dimensional simplices, then the sets

$$U_j = \bigcup (\mathrm{Int}\, X_\alpha : e_j \in X_\alpha\}$$

are open, cover the space X and satisfy the conditions of Proposition 1.7.

If the space X is a smooth manifold, then both definitions of cohomology coincide.

Theorem 1.8. *The de Rham cohomology $H_d^k(M)$ and the spectral cohomology $H^k(M)$ of a smooth manifold M are isomorphic.*

We show how it is possible to establish such an isomorphism in the case when the manifold M is represented as a complex made up of smooth simplices. First suppose M is an orientable manifold and $\omega \in \Omega_n(M)$, $\dim M = n$. If M is a region in Euclidean space $\mathbb{R}^n$ with coordinates $(x^1, \dots, x^n)$ then according to Formula (1.12)

$$\omega(x) = f(x) dx^1 \wedge \dots \wedge dx^n .$$

We set

$$\int_M \omega = \int_M \dots \int f(x) dx^1 dx^2 \dots dx^n . \tag{1.21}$$

If $(y^1, \dots, y^n)$ is another system of coordinates with the same orientation, then

$$\det \frac{\partial (x^1, \dots, x^n)}{\partial (y^1, \dots, y^n)} > 0$$

and, consequently, the definition (1.21) does not depend on the choice of a system of coordinates in the oriented region $M \subset \mathbb{R}^n$. Now it is already clear how to define the integral of the form ω for any oriented manifold M. One must cover M with local charts U_k, choose local systems of coordinates in them with the same orientation, and set

$$\int_M \omega = \sum_s (-1)^{s+1} \sum_{(\alpha_1, \dots, \alpha_s)} \int_{U_{\alpha_1 \dots \alpha_s}} \omega .$$

The addends corresponding to the various intersections are necessary to compensate for the repetition of the integration of the form over common sets.

Proposition 1.9. *If*

$$f : M_1 \to M_2$$

is an orientation-preserving diffeomorphism, then

$$\int_{M_1} f^*(\omega) = \int_{M_2} \omega \,.$$

Proposition 1.10. *Suppose M is a compact orientable smooth manifold with boundary ∂M, $\dim M = n$, $\omega \in \Omega_{n-1}(M)$. Then for an appropriate choice of orientation on ∂M Stokes' formula*

$$\int d\omega = \int_{\partial M} \omega \,. \tag{1.22}$$

is valid.

Proof. To prove Stokes' formula we note that formula (1.22) is linear with respect to the form ω. Hence if $\omega = \sum_i \omega_i$ and Formula (1.22) holds for each addend ω_i separately, then it also holds for the form ω. Since the manifold M is compact, there exists a finite atlas of charts $\{U_j\}$, and, according to Proposition (1.1), a partition of unity $\{\varphi_j\}$ subordinate to the covering $\{U_j\}$. Then the form ω can be represented in the form $\omega = \sum_j \omega_j$, where $\omega_j = \varphi_j \omega$. The support of the form ω_j is compact and lies in the chart U_j. Thus it is sufficient to to check Formula (1.22) for each addend ω_j separately. Since $\operatorname{supp} \omega_j \subset U_j$, then also $\operatorname{supp} d\omega_j \subset U_j$, and the integrals can be taken over the chart U_j:

$$\int_m d\omega_j = \int_{U_j} d\omega_j \,, \quad \int_{\partial M} \omega_j = \int_{\partial M \cap U_j} \omega_j \,.$$

Consequently, using the local coordinate chart U_j, we can suppose that the integration of the form ω_j takes place over the $(n-1)$-dimensional Euclidean space $\mathbb{R}^{n-1}$, and the integration of the form $d\omega_j$ takes place over the half-space $\mathbb{R}^n_+ = \{x_n \geq 0\}$.

Thus it is sufficient to prove Formula (1.22) for the special case of the manifold $\mathbb{R}^n_+$ with boundary $\mathbb{R}^{n-1}$ and a form ω on it with compact support.

In this special case the $(n-1)$-dimensional form ω can be represented in the form

$$\omega = \sum_{i=1}^{n} \lambda_i (x_1, \ldots, x_n) \, dx_1 \wedge \ldots \wedge \widehat{dx_i} \wedge \ldots \wedge dx_n \,,$$

where each function $\lambda_i(x_1, \ldots, x_n)$ has compact support. Using similar arguments, it is sufficient to check Formula (1.22) for the case of one addend

$$\omega = \lambda(x_1 \ldots x_n)\, dx_1 \wedge \ldots \wedge \widehat{dx_i} \wedge \ldots \wedge dx_n ,$$
$$d\omega = (-1)^{i-1} \frac{\partial \lambda}{\partial x_i}(x_1, \ldots, x_n)\, dx_1 \wedge \ldots \wedge dx_n .$$

There are two possible cases. Suppose $i \neq n$. Then

$$\int\limits_{R^n_+} d\omega = \int\limits_{\mathbb{R}^n_+} \!\!\ldots\! \int (-1)^{i-1} \frac{\partial \lambda}{\partial x^i}(x_1, \ldots, x_n)\, dx_1 \ldots dx_n$$
$$= \int\limits_{\mathbb{R}^n_+} \!\!\ldots\! \int dx_1 \ldots dx_{i-1} dx_{i+1} \ldots dx_n \left(\int\limits_{-\infty}^{\infty} (-1)^{i-1} \frac{\partial \lambda}{\partial x_i} dx_i \right) .$$

It is obvious that the interior integral is zero by virtue of the compactness of the support of the function λ. Hence

$$\int\limits_{\mathbb{R}^n_+} d\omega = 0 .$$

On the other hand,

$$\int\limits_{\mathbb{R}^{n-1}} \omega = \int\limits_{\mathbb{R}^{(n-1)}} \!\!\ldots\! \int \lambda(x_1, \ldots, x_{n-1}, 0)\, dx_1 dx_2 \ldots \widehat{dx_i} \ldots dx_n ,$$

where $x_n = 0$. Consequently

$$\int\limits_{\mathbb{R}^{n-1}} \omega = 0 .$$

Now suppose $i = n$. Then

$$\int\limits_{\mathbb{R}^n_+} d\omega = \int\limits_{\mathbb{R}^n_+} \!\!\ldots\! \int (-1)^{n-1} \frac{\partial \lambda}{\partial x_n}(x_1, \ldots, x_n)\, dx_1 \ldots dx_n$$
$$= \int\limits_{\mathbb{R}^n_+} \!\!\ldots\! \int dx_1 \ldots dx_{n-1} \left(\int\limits_{0}^{\infty} (-1)^{n-1} \frac{\partial \lambda}{\partial x_n} dx_n \right)$$
$$= \int\limits_{\mathbb{R}^n_+} \!\!\ldots\! \int dx_1 \ldots dx_n \left((-1)^n \lambda(x_1, \ldots, x_{n-1}, 0) \right) = (-1)^n \int\limits_{\mathbb{R}^{n-1}} \omega .$$

For an appropriate choice of orientation on $\mathbb{R}^{n-1}$ we obtain Formula (1.22). □

Stokes' formula (1.22) is a generalization of the classical formula for calculating line and surface integrals. If a curve Γ is situated in the plane $\mathbb{R}^2$ without self-intersections, then the corresponding formula is called Green's formula. Indeed, the line integral $\oint_\Gamma (Pdx + Qdy)$ along a closed curve Γ can be interpreted as the integral of the differential form $\omega = Pdx + Qdy$ on the one-dimensional manifold Γ.

Suppose Ω is a two-dimensional domain bounded by the curve Γ. Then Ω is a two-dimensional manifold with boundary, for which $\partial\Omega = \Gamma$. Then by Stokes' formula (1.22) we have $\int_{\partial\Omega} \omega = \int_\Omega d\omega$. We compute the differential $d\Omega$ in the standard way:

$$\begin{aligned} d\omega &= d(Pdx + Qdy) = dP \wedge dx + dQ \wedge dy \\ &= \left(\frac{\partial P}{\partial x}dx + \frac{\partial P}{\partial y}dy\right) \wedge dx + \left(\frac{\partial Q}{\partial x}dx + \frac{\partial Q}{\partial y}dy\right) \wedge \partial y \ . \end{aligned}$$

Since $dx \wedge dx = dy \wedge dy = 0$ by the skew-symmetry of differential forms,

$$d\omega = \frac{\partial P}{\partial y}dy \wedge dx + \frac{\partial Q}{\partial x}dx \wedge dy = \left(\frac{\partial Q}{\partial x} - \frac{\partial P}{\partial y}\right) dx \wedge dy \ .$$

Then

$$\oint_\Gamma (Pdx + Qdy) = \int\limits_{\partial\Omega} \omega = \int\limits_{\Omega} d\omega = \iint\limits_{\Omega} \left(\frac{\partial Q}{\partial x} - \frac{\partial P}{\partial y}\right) dx\, dy \ .$$

The last formula is, indeed, known as Green's formula.

Analogously, if a closed curve Γ is imbedded in three-dimensional space $\mathbb{R}^3$ and bounds a two-dimensional surface S, i.e. $\partial S = \Gamma$, then we interpret the line integral $\oint_\Gamma(Pdx+Qdy+Rdz)$ as the integral of the differential form $\omega = Pdx + Qdy + Rdz$ on the manifold Γ. Then, again applying Stokes' formula (1.22), we obtain the formula

$$\begin{aligned} \oint_\Gamma (Pdx + Qdy + Rdz) &= \int\limits_{\partial S = \Gamma} d\omega = \iint\limits_{S} \left\{\left(\frac{\partial Q}{\partial x} - \frac{\partial P}{\partial y}\right) dx\, dy \right. \\ &\left. + \left(\frac{\partial R}{\partial y} - \frac{\partial Q}{\partial z}\right) dy\, dz + \left(\frac{\partial P}{\partial z} - \frac{\partial R}{\partial x}\right) dz\, dx\right\} \ , \end{aligned}$$

which also bears the name of Stokes' formula.

Finally, if S is a closed surface in three-space $\mathbb{R}^3$, bounding a certain three-dimensional region V, $\partial V = S$, then we interpret the surface integral

$$\iint\limits_{S} (P dx\, dy + Q dy\, dz + R dz\, dx)$$

as the integral of the differential 2-form

$$\omega = P dx \wedge dy + Q dy \wedge dz + R dz \wedge dx$$

on the two-dimensional surface S. Applying Formula (1.22), we obtain the Gauss-Ostrogradski formula

$$\oint\!\!\oint_S (P dx\, dy + Q dy\, dz + R dz\, dx) = \int\limits_{S=\partial V} \omega = \int\limits_{V} d\omega$$
$$= \iiint\limits_{V} \left(\frac{\partial P}{\partial z} + \frac{\partial Q}{\partial x} + \frac{\partial R}{\partial y} \right) dx\, dy\, dz \ .$$

We now return to some simplicial decomposition of a manifold M. Let $e_1, \ldots, e_s$ be zero-dimensional simplices, U_j be the union of all the simplices which contain the point e_j, $\mathfrak{U} = \{U_j\}$. The intersection $U_{j_0 \ldots j_k} = U_{j_0} \cap \ldots \cap U_{j_k}$ is nonempty if and only if the points $e_{j_0}, \ldots, e_{j_k}$ are the vertices of a certain (unique) simplex. We denote this simplex by $\sigma_{j_0 \ldots j_k}$. Now we can construct the homomorphism $\alpha : \Omega_k(M) \to C^k(M, U)$. Let $\omega \in \Omega_k(M)$, $U_{j_0 \ldots j_k} \neq \emptyset$. We set

$$\alpha(\omega)\,(U_{j_0 \ldots j_k}) = \int\limits_{\sigma_{j_0 \ldots j_k}} \omega \ .$$

From (1.22) follows

Lemma 1.11. *For $\omega \in \Omega_k(M)$ the following identity holds:*

$$\partial a(\omega) = \alpha(d\omega) \ . \tag{1.23}$$

By (1.23) the homomorphism α induces a homomorphism

$$\alpha^* : H_d^k(M) \to H^k(M, \mathfrak{U}) \approx H^k(M) \ .$$

One can make Theorem 1.8 more precise by asserting that the homomorphism α^* is an isomorphism.

We illustrate how Theorem 1.8 is proven for one-dimensional cohomologies. Suppose $\mathfrak{U} = \{U_\alpha\}$ is some covering. Without loss of generality we will assume that the sets U_α and $U_\alpha \cap U_\beta$ are path-connected. Suppose $c \in C^1(M, \mathfrak{U})$, $c = \{c_{\alpha\beta}\}$, $\partial c = 0$, i.e. $c_{\alpha\beta} + c_{\beta\gamma} + c_{\gamma\alpha} = 0$ whenever ${}_{\alpha\beta\gamma} \neq 0$. Then there exist smooth functions f_α on the sets U_α which satisfy the condition

$$f_\alpha(x) - f_\beta(x) = c_{\alpha\beta}\ , \quad x \in U_{\alpha\beta}\ . \tag{1.24}$$

Indeed one can construct the functions f_α recursively, by giving a linear order to all the sets U_α of the covering $\mathfrak{U}$. If for $\alpha < \alpha_0$ functions f_α satisfying (1.24) have already been constructed, then, using (1.24) for $\beta = \alpha_0$, we obtain that the function f_{α_0} must be given uniquely on the subset $\cup_{\alpha<\alpha_0} U_{\alpha,\alpha_0} \subset U_{\alpha_0}$. Replacing, if necessary, each set U_α by a smaller set $V_\alpha \subset \overline{V}_\alpha \subset U_\alpha$, so that $\cup_\alpha V_\alpha = M$, we can extend the function f_{α_0} from the set $\cup_{\alpha<\alpha_0} U_{\alpha,\alpha_0}$ to the entire set U_{α_0}. Thus we have a collection of functions f_α which satisfy the condition (1.24). Supposing $\omega_\alpha = df_\alpha$, it then follows from (1.24) that $\omega_\alpha = \omega_\beta$ on $U_{\alpha\beta}$, i.e. there exists a form ω on the manifold M such that $\omega|U_\alpha = \omega_\alpha$. Clearly $d\omega = 0$. If f'_α is another collection of functions satisfying (1.24), then there exists a smooth function h on the manifold such that $f'_\alpha = f_\alpha + h$, $x \in U_\alpha$. Then $\omega' = \omega + dh$, i.e. ω' and ω define one and the same element in the group $H^1_d(M)$.

On the other hand, if $\omega \in \Omega_1(M)$ is a form such that $d\omega = 0$, then in each local chart U_α homeomorphic to Euclidean space, there exists a smooth function f_α such that $df_\alpha = \omega$. Indeed, the function f_α can be given by the curvilinear integral

$$f_\alpha(x) = \int_{x_0}^{x} \omega\ , \tag{1.25}$$

for which the curve $[x_0, x_1]$ lies entirely in the chart U_α. Since $U_\alpha \approx \mathbb{R}^n$, the integral (1.25) is independent of the path. Then the difference

$$f_\alpha - f_\beta$$

is a constant function on $U_{\alpha\beta}$ equal, say to $c_{\alpha\beta}$.

Thus we obtain a cochain $c = \{c_{\alpha\beta}\} \in C^1(M,U)$, $\partial c = 0$. We have constructed two mutually inverse mappings of the groups $H^1_d(M)$ and $H^1(M)$.

We have presented two different definitions of the cohomology of a smooth manifold M with real coefficients. There are other ways of defining the cohomology, for example, by partitioning a compact smooth manifold into simplices. The definition which is most convenient depends on the concrete problem being analyzed. In this presentation the spectral cohomology and the de Rham cohomology will arise naturally. Using the simplicial cohomology, on the other hand, one can effectively indicate a finite collection of integral conditions for the solvability of the equation $d\omega = \Omega$, where Ω is a differential k-form. Indeed, there exists a finite number of submanifolds $N_1, \ldots, N_s$, $\dim N_i = k$, $1 < i < s$, such that the equation $d\omega = \Omega$ can be solved if and only if

$$\int_{N_i} \Omega = 0\ , \quad 1 \le i \le s\ .$$

If the manifold M is decomposed into a finite set of smooth simplices, then the integral conditions can be rewritten in the form of a finite number of equations

$$\sum_i \lambda_i^j \int_{\sigma_i} \Omega = 0 ,$$

where σ_i are simplices of dimension k and λ_i^j are some numbers independent of the form Ω.

As an example we will compute the one-dimensional cohomology group of the circle S^1. To do this we introduce the angular coordinate φ on S^1 as a local coordinate. Then any one-dimensional form ω has the form $\omega = f(\varphi)d\varphi$, where $f(\varphi)$ must be a function of points on the circle, i.e. a periodic function, $f(\varphi) = f(\varphi + 2\pi)$. If $\omega = dg$, where g is some function on the circle S^1, i.e. g is a periodic function, $g(\varphi) = g(\varphi + 2\pi)$, then we arrive at the equation $f(\varphi) = d/d\varphi\,(g(\varphi))$ or $g(\varphi) = g(0) + \int_0^{\varphi} f(\gamma)d\gamma$. Since g is a periodic function, f must satisfy the condition

$$\int_0^{2\pi} f(\varphi)d\varphi = 0 .$$

Any periodic function $f(\varphi)$ can be written uniquely in the form of a sum $f(\varphi) = \lambda + h(\varphi)$, for which $\int_0^{2\pi} h(\varphi)d\varphi = 0$, $\lambda = \text{const}$. Then the form $h(\varphi)d\varphi$ belongs to the image of the differential d, $h(\varphi)d\varphi \in \operatorname{Im} d$. Thus every conjugacy class of $\operatorname{Ker} d/\operatorname{Im} d$ contains a unique form $\lambda d\varphi$ with a constant multiplier λ. Consequently, the first cohomology group of the circle S^1 is isomorphic to the group of real numbers $\mathbb{R}^1$.

We consider another example – the two-dimensional sphere S^2. It is convenient to represent the sphere S^2 as the extended complex plane. Then S^2 can be covered by two charts U_1 and U_2, each of which is homeomorphic to the complex plane $\mathbb{C}^1$ with complex coordinates z and w, which are associated by the relation $z = 1/w$. If $z = x + iy$, $w = u + iv$, then $x = u/\left(u^2 + v^2\right)$, $y = -v/\left(u^2 + v^2\right)$. We consider an arbitrary 1-form ω on S^2 for which $d\omega = 0$. Suppose that the form ω is given as $\omega = Pdx + Qdy$, $\omega = \overline{P}du + \overline{Q}dv$ on the charts U_1, U_2.

Since $d\omega = 0$, $\partial P/\partial y = \partial Q/\partial x$ and $\partial \overline{P}/\partial v = \partial \overline{Q}/\partial v$. Then in each chart separately ω can be represented as a complete differential

$$Pdx + Qdy = df_1(x,y) , \quad \overline{P}du + \overline{Q}dv = df_2(u,v) .$$

One sets $f_1(x,y) = \int_{(0,0)}^{(x,y)} (Pdx + Qdy)$, $f_2(u,v) = \int_{(0,0)}^{(u,v)} \left(\overline{P}du + \overline{Q}dv\right)$, where the integrals are taken along arbitrary smooth curves joining the point $(0,0)$ with the point (x,y) and the point $(0,0)$ with the point (u,v)

respectively. Let φ be a smooth function on the sphere S^2 which is identically equal to unity in a neighborhood of zero in the chart U_2 and has compact support in U_2. Then in the chart U_1 the function φ is equal to zero in a neighborhood of the point $(0,0)$. We set $g_2 = \varphi f_2$. Although the function f_2 is defined only in the chart U_2, since φ has compact support in the chart U_2 it is possible to extend $\varphi \cdot f_2$ to be zero at points which do not belong to the chart U_2.

Then the form $\omega' = \omega - dg_2$ is identically equal to zero in a neighborhood of the point $(0,0)$ of the chart U_2. Consequently, in the chart U_1 the form $\omega' = P'dx + Q'dy$ is identically equal to zero outside a certain circle, i.e. for $x^2 + y^2 > R^2$ we have the identities $P'(x,y) \equiv 0$, $Q'(x,y) \equiv 0$. We next consider a solution of the equation $dg_1 = \omega'$ in the chart U_1. We find a solution $g_1(x,y)$ of the form

$$g_1(x,y) = \int_{(0,0)}^{(x,y)} (P'(x,y)dx + Q'(x,y)dy) .$$

We show that for $x^2 + y^2 \geq R^2$ the function g_1 is constant. Indeed, the value of the function $g^1(x,y)$ can be calculated by integrating the form ω' along any path joining the origin to the point (x,y). We choose a path in the following way. Let γ be composed of three arcs, $\gamma = \gamma_1\gamma_2\gamma_3$. The path γ_1 is the ray from the point $(0,0)$ to the point $(R,0)$. The path γ_2 is the line segment joining $(R,0)$ and $\left((x^2+y^2)^{1/2}, 0\right)$. The path γ_3 is the arc of the circle of radius $(x^2+y^2)^{1/2}$ which joins the points $\left((x^2+y^2)^{1/2}, 0\right)$ and (x,y) (Fig. 6).

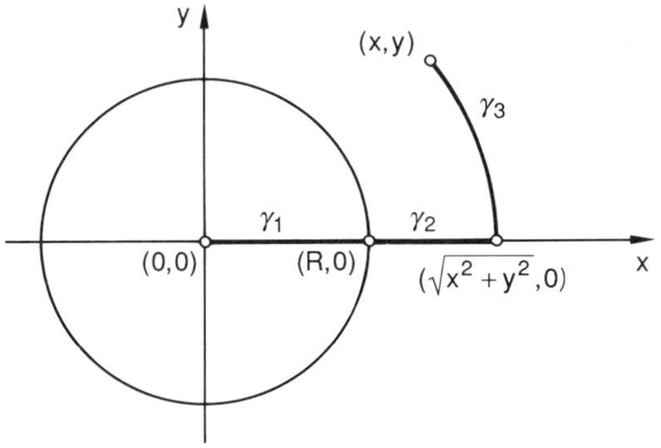

Fig. 6

Then all the points of the paths γ_2 and γ_3 lie outside the circle of radius R and therefore

$$\int_{\gamma_2\cup\gamma_3} \omega' = \int_{\gamma_2\cup\gamma_3} (P'dx + Q'dy) = 0 \ .$$

Consequently,

$$g_1(x,y) = \int_{\gamma} (P'dx + Q'dy) = \int_{\gamma_1} (P'dx + Q'dy)$$

$$+ \int_{\gamma_2\cup\gamma_3} (P'dx + Q'dy) = \int_{\gamma_1} (P'dx + Q'dy) = g_1\,(R,0) = \text{const.}$$

Since the function $g_1(x,y)$ is constant outside the circle of radius R in the chart U_1, it can be uniquely extended to have the value $g_1(R,0)$ at the infinitely distant point. Hence the function g_1 is constant in a neighborhood of the point $(0,0)$ in the chart U_2. Consequently, the identity

$$dg_1 \equiv \omega' \quad \text{or} \quad \omega = dg_1 + dg_2$$

holds on the entire sphere.

We have shown that for one-dimensional forms on the sphere S^2 the kernel of d coincides with the image of d, i.e. $H^1\left(S^2\right) = 0$.

We will now calculate the two-dimensional cohomology of the sphere S^2. For this it is sufficient to describe the subspace of all two-dimensional differential forms which are images of the differential d. We show that if $\int_{S^2}\omega = 0$, then the equation $d\Omega = \omega$ can be solved for some one-dimensional form Ω on the sphere S^2. On each of the charts U_1, U_2 such an equation is solvable, since the form ω can be written as $\omega = f(x,y)dx \wedge dy$ and one can set $\Omega = P(x,y)dy$, where $P(x,y) = \int_0^x f(x,y)dx$.

Thus there is some one-dimensional form Ω_1 on the chart U_1 such that $d\Omega_1 = \omega$; analogously there is some one-dimensional form Ω_2 on the chart U_2 such that $d\Omega_2 = \omega$. Then on the intersection of charts $U_1 \cap U_2$, which in (x,y) coordinates is the complex plane with the point $(0,0)$ removed, the equation $d(\Omega_1 - \Omega_2) = 0$ holds. We examine a closed curve Γ – the circle $x^2 + y^2 = R^2$. Then

$$\int_{\Gamma} \Omega_1 = \int_{x^2+y^2\le R^2} d\Omega_1 \ , \quad \int_{\Gamma} \Omega_2 = -\int_{x^2+y^2\ge R^2} d\Omega_2 \ .$$

Consequently,

$$\int_{\Gamma} (\Omega_1 - \Omega_2) = 2\int_{U_1} \omega = 2\int_{S^2} \omega = 0 \ .$$

Thus the integral of the form $\Omega_1 - \Omega_2$ along any closed curve lying in $U_1 \cap U_2$ is equal to zero. It follows that the equation $df = \Omega_1 - \Omega_2$ can be solved in the intersection $U_1 \cap U_2$.

We consider a smooth function g on the chart U_1 which is equal to zero on a neighborhood of the point $(0,0)$ and equal to one outside a circle of radius R. Then the function $h = gf$ can be extended at the point $(0,0)$ to become a smooth function on the chart U_1. We set $\Omega_1' = \Omega_1 - dh$. Then $d\Omega_1' = \omega$ in the chart U_1 and $\Omega_1' \equiv \Omega_2$ in the chart U_1 for $x^2 + y^2 > R^2$. Consequently the form Ω_1' can be extended to the entire sphere S^2 in such a way that $d\Omega_1' = \omega$ on the sphere S^2. Therefore every conjugacy class in the cohomology $H^2\left(S^2\right)$ is determined by a real number, equal to the integral $\int_{S^2} \omega$.

All that is left to show is that there exists a form ω on the two-dimensional sphere such that $\int_{S^2} \neq 0$. In the charts U_1 and U_2 we set

$$\omega' = \frac{dx \wedge dy}{\left(1 + x^2 + y^2\right)^2}\,, \qquad \omega'' = \frac{du \wedge dv}{\left(1 + u^2 + v^2\right)^2}\,.$$

It is not hard to convince oneself that the forms ω' and ω'' coincide on the intersection of charts $U_1 \cap U_2$. On the other hand, since the coefficient of $dx \wedge dy$ of the form ω' is nonnegative, $\int_{y_1} \omega' > 0$. Thus the form ω, $\omega|U_1 = \omega'$, $\omega|U_2 = \omega''$, satisfies all our requirements. Thus the two-dimensional cohomology group of the sphere S^2 is isomorphic to the group of real numbers: $H^2\left(S^2\right) = \mathbb{R}^1$.

1.2 Theorems on Transversal Regularity

Theorems on transversal regularity or, as they are still sometimes called, theorems on general position are based on the properties of vector spaces and their subspaces. For example, if V is a real vector space of dimension n and V_1, V_2 are two subspaces of dimensions p and q respectively, then when $p + q > n$ the intersections $V_1 \cap V_2$ is not zero, and the dimension $\dim\left(V_1 \cap V_2\right)$ of this intersection can be bounded from below by the number $p + q - n$:

$$\dim\left(V_1 \cap V_2\right) \geq p + q - n\,. \tag{2.1}$$

Moreover, for arbitrarily small motions of the linear subspaces V_1, V_2 the estimate (2.1) can be made into an equality. In this case the algebraic sum of the subspaces V_1 and V_2 coincides with the entire space V. Two linear subspaces V_1 and V_2 are said to be in general position if

$$V_1 + V_2 = V\,.$$

The term "general position" is justified by the circumstance that the imbeddings $V_1 \subset V$, $V_2 \subset V$ can be changed via an arbitrarily small motion into general position, and moreover the set of imbeddings of the spaces V_1 and V_2 into V for which the subspaces V_1 and V_2 lie in general position form a dense open set in the space of all imbeddings, as well as in the space of all linear maps. To put it another way, the "overwhelming majority" of imbeddings possess the property that the subspaces V_1 and V_2 lie in general position. It turns out that for nonlinear smooth mappings it is also possible to study questions analogous to the theory of the general position of linear maps. For this it is sufficient to consider the linear approximations to the smooth nonlinear maps. A classic example is the implicit function theorem. We will give this theorem, as well as well all the concepts used in it, in an invariant form, in order to be able to generalize it to the case of arbitrary smooth maps of smooth manifolds.

Suppose a smooth map

$$f : M_1 \to M_2 \tag{2.2}$$

of smooth manifolds, in general of different dimensions, is given. Then the point $x \in M_1$ is called a regular point for the map f if the differential of the map f

$$df : TM_1 \to TM_2$$

maps the tangent space $(TM_1)_x$ at the point $x \in M_1$ onto the tangent space $(TM_2)_{f(x)}$ at the point $f(x)$, i.e. the linear map

$$(df)_x : (TM_1)_x \to (TM_2)_{f(x)}$$

is an epimorphism. In local systems of coordinates in a neighborhood of the points x and $f(x)$ the map f can be represented as a collection of functions

$$y^k = f^k\left(x^1, \ldots, x^n\right) , \quad 1 < k < m ,$$

where $\left(x^1, \ldots, x^n\right)$ are local coordinates of the manifold M_1, and $\left(y^1, \ldots, y^m\right)$ are local coordinates of the manifold M_2. The fact that the map df is an epimorphism means that the rank of the Jacobian matrix

$$\left(\frac{\partial y^k}{\partial x^l}\right) = \left(\frac{\partial f^k\left(x^1, \ldots, x^n\right)}{\partial x^l}\right) \tag{2.3}$$

equals m. Consequently, $n \geq m$, and there must be some minor of order m of the matrix (2.3) which is different from zero, i.e. the coordinates $\left(x^1, \ldots, x^n\right)$ are divided into two groups $\left(x^{l_1}, \ldots, x^{l_m}\right)$ and $\left(x^{l_{m+1}}, \ldots, x^{l_n}\right)$. It is completely clear that the collection of implicit functions

$$f^k\left(x^1, \ldots, x^n\right) - y^k = 0$$

satisfies the conditions of solvability in terms of the coordinates $(x^{l_1}, \ldots, x^{l_m})$ given by the implicit function theorem. We call a point $y \in M_2$ a *regular* point for the mapping (2.2) if any point $x \in f^{-1}(y)$ is a regular point for the mapping (2.2). The implicit function theorem in our interpretation can be formulated in the following way.

Theorem 2.1. *Suppose $f : M_1 \to M_2$ is a smooth mapping of smooth manifolds,* $\dim M_1 = n$, $\dim M_2 = m$, *and $y \in M_2$ is a regular point for the map f. Then the preimage $f^{-1}(y)$ of the point $y \in M_2$ is a smooth manifold of dimension $(n - m)$, smoothly imbedded in the manifold M_1.*

A more general situation in which the implicit function theorem works is the following. Suppose three manifolds M_1, M_2, N are given, along with a smooth mapping

$$f : M_1 \to M_2$$

and a smooth imbedding

$$\varphi : N \to M_2 \ .$$

We recall that the differential $d\varphi$ is a monomorphism of the tangent bundles

$$d\varphi : TN \to TM_2 \ .$$

We will say that the mapping f is *transversally regular along the submanifold* N if at each point x of the preimage $f^{-1}(N)$ the composition of homomorphisms

$$(TM_1)_x \xrightarrow{df} (TM_2)_{f(x)} \to (TM_2)_{f(x)} \,/\, (TN)_{f(x)} \tag{2.4}$$

is an epimorphism. The definition of a regular point of a mapping f means precisely that the mapping f is transversally regular along the zero-dimensional manifold which consists of the single point y.

Condition (2.4) for the transversal regularity of a mapping f in a local system of coordinates, just as in the case of a regular point, implies the nondegeneracy of a certain minor of the Jacobi matrix of the mapping f. The implicit function theorem itself can be interpreted in the following way.

Theorem 2.2. *Let $f : M_1 \to M_2$ be a smooth mapping of smooth manifolds, $\varphi : N \to M_2$ be a submanifold,*

$$\dim M_1 = n \ , \quad \dim M_2 = m \ , \quad \dim N = k \ .$$

If the mapping f is transversally regular along the submanifold N, then the inverse image $f^{-1}(N)$ is a smooth manifold of dimension $n - m + k$, smoothly imbedded in the manifold M_1.

We will examine several examples of the application of Theorem 2.2.

1. Two submanifolds in general position. Suppose $\varphi_k : M_k \to M$, $k = 1, 2$, are two imbeddings. If the imbedding φ_1 is transversally regular along the submanifold M_2, then we will say that *the manifolds intersect transversally*, or that *the manifolds M_1 and M_2 are in general position.* Although the imbeddings φ_1 and φ_2 enter into this definition asymmetrically, in fact φ_1 and φ_2 can change places. Indeed, the preimages $\varphi_1^{-1}(M_2)$ and $\varphi_2^{-1}(M_1)$ coincide and are equal to the intersection $M_1 \cap M_2$. The condition of transversal regularity of one imbedding along the other manifold is equivalent to the following symmetric condition: for each point $x \in M_1 \cap M_2$ the identity

$$(TM_1)_x + (TM_2)_x = (TM)_x$$

holds. Then the intersection $M_1 \cap M_2$, according to Theorem 2.2, is a smooth submanifold,

$$\dim(M_1 \cap M_2) = \dim M_1 + \dim M_2 - \dim M \ .$$

For example, if the manifold M is compact and $\dim M = \dim M_1 + \dim M_2$, then the intersection $M_1 \cap M_2$ is a zero-dimensional compact manifold, i.e. consists of a finite set of points.

2. A tubular neighborhood of a submanifold. Suppose $\varphi : M_1 \to M_2$ is an imbedding, so that the differential $d\varphi$ is a monomorphism of bundles

$$d\varphi : TM_1 \to TM_2 \ ,$$

which can be decomposed into a composition

$$TM_1 \to \varphi^* TM_2 \to TM_2 \ .$$

The factor bundle $\varphi^* TM_2 / TM_1$ is called the *normal bundle* to the imbedding φ and is denoted by $\nu(M_1)$. The term "normal bundle" is justified by the following assertion. The space of the normal bundle $\nu(M_1)$ is diffeomorphic to a certain open neighborhood $V \subset M_2$ containing the submanifold M_1. Here the diffeomorphism

$$\psi : \nu(M_1) \to V \subset M_2$$

can be chosen such that the restriction $\psi | M_1$ of the diffeomorphism ψ to the null section of the normal bundle $\nu(M_1)$ coincides with the imbedding φ, the differential $d\psi$ maps the tangent bundle $T\nu(M_1)$ to the space of the normal bundle $\nu(M_1)$ diffeomorphically onto the tangent bundle $TV \subset TM_2$, and the restriction of $d\psi$ to the first term in the direct sum, $TM_1 \subset T\nu(M_1)$, coincides with the differential $d\varphi$ of the imbedding φ of M_1 into M_2.

In order to reduce the problem to Theorem 2.2, we proceed as follows. First we introduce onto the manifold M_2 a certain smooth Riemannian metric, i.e. a scalar product on each fiber of the tangent bundle TM_2. Then the restriction φ^*TM_2 of the tangent bundle TM_2 to the submanifold M_1 decomposes into an orthogonal direct sum of two bundles – TM_1 and its orthogonal complement $(TM_1)^\perp$:

$$\varphi^*TM_2 = TM_1 \oplus (TM_1)^\perp \ .$$

It follows directly from the definition that the normal bundle is isomorphic to the orthogonal complement $(TM_1)^\perp$:

$$\chi : \nu(M_1) \overset{\approx}{\to} (TM_1)^\perp \subset TM_2 \ .$$

Suppose $x \in M_2$ is an arbitrary point. The Riemannian metric in the manifold M_2 defines an exponential map of the tangent plane $(TM_2)_x$ into the manifold M_2:

$$\exp_x : (TM_2)_x \to M_2 \ ,$$

which maps each ray in the space $(TM_2)_x$ into a geodesic with its origin at the point $x \in M_2$, and the direction of the ray coincides with a tangent vector to the geodesic at the point $x \in M_2$. It is completely obvious that the differential $d(\exp_x)$ at the origin of the coordinates is the identity map. Now we construct a smooth map

$$\psi : \nu(M_1) \to M_2 \ .$$

Let $\xi \in \nu(M_1)$ be an arbitrary vector at the point $x \in M_1$. We set

$$\psi(\xi) = \exp_x(\chi(\xi)) \ .$$

It is easy to check in a local system of coordinates that the differential $d\psi$ of the map ψ is an isomorphism for all null vectors $\xi \in \nu(M_1)$. Consequently a certain neighborhood $U \subset \nu(M_1)$ of the null section $M_1 \subset \nu(M_1)$ is mapped diffeomorphically onto a neighborhood $V \subset M_2$ of the submanifold M_1.

In particular, if $M_1 \subset M_2$ is a smooth submanifold, $\dim M_1 = n$, $\dim M_2 = m$, then for each point $x_0 \in M_1$ there exists a local system of coordinates such that the manifold M is given by the system of equations

$$y_{n+1}(x) = 0, \ldots$$

$$y_m(x) = 0 \ .$$

We consider, for example, the circle S^1 in $\mathbb{R}^2$, which we can give by the equation $f(x,y) = (x^2 + y^2 - 1) = 0$. The function $f(x,y)$ is a mapping of two-space $\mathbb{R}^2$ into the real line $\mathbb{R}^1$. Here the point $0 \in \mathbb{R}^1$ is a regular point, since the Jacobian matrix of the function $f(x,y)$, which is equal to $\left(\frac{\partial f}{\partial x}, \frac{\partial f}{\partial y}\right) = (2x, 2y)$, has rank 1 at each point of the circle S^1. Then the normal bundle $\nu\left(S^1\right)$ is the trivial bundle equal to the preimage of the tangent bundle to the line $\mathbb{R}^1$ at the point 0. Obviously the space of the normal bundle $\nu\left(S^1\right)$ is mapped diffeomorphically onto the annular neighborhood V of the circle S^1:

$$V = \left\{(x,y) : 0 < x^2 + y^2 < 4\right\} ,$$

in such a way that the one-dimensional fiber of the normal bundle $\nu\left(S^1\right)$ over the point (x_0, y_0) on the circle is mapped diffeomorphically onto a segment of the ray passing through the point (x_0, y_0):

$$\psi : \mathbb{R}^1 \to \mathbb{R}^2 , \quad \psi(t) = \left(\frac{2}{\pi} x_0 \tan^{-1} t \, , \ \frac{2}{\pi} y_0 \tan^{-1} t\right) .$$

Analogously, the normal bundle to the two-dimensional sphere $S^2 = \left\{(x,y,z) :\ x^2 + y^2 + z^2 - 1 = 0\right\}$ is diffeomorphic to the neighborhood $V = \left\{(x,y,z) :\ 0 < x^2 + y^2 + z^2 < 4\right\}$ and is diffeomorphic to the direct product $S^2 \times \mathbb{R}^1$.

We now proceed to formulate the theorems which we whill use to find maps which are transversally regular along a certain submanifold. For simplicity we will consider all the manifolds compact, although this restriction is not essential.

First of all we introduce a metric on the space of all maps of the manifold M_1 into the manifold M_2:

$$\rho(f,g) = \sup_{x \in M_1} \left[\rho_1\left(f(x), g(x)\right) + \rho_2\left(df(x), dg(x)\right)\right] , \tag{2.5}$$

where ρ_1 is some metric on the manifold M_2 which agrees with the topology on that manifold, and ρ_2 is a metric on the cotangent space T^*M_2.

If $N \subset M_2$ is a submanifold in M_2 and the mapping f is transversally regular along the submanifold N, then the property of being transversally regular along the manifold M holds for all smooth maps which are sufficiently close to f in terms of the metric (2.5). Indeed, in a local system of coordinates transversal regularity means the nondegeneracy of a certain matrix composed of partial derivatives of the map f. Consequently, for a sufficiently close map g in the sense of (2.5), the partial derivatives of g are close to the corresponding partial derivatives of f and, therefore, the nondegeneracy of the corresponding matrix is preserved. The compactness of all the manifolds allows one to escape from the constraint of locality.

Thus the set of all mappings which are transversally regular along a submanifold N is open in the space of all smooth mappings. As it turns out, the overwhelming majority of mappings have the property of transversal regularity. A precise formulation is given in the next theorem.

Theorem 2.3 (Thom). *The set of maps which are transversally regular along a submanifold N is everywhere dense in the set of all smooth mappings of a manifold M_1 into a manifold M_2.*

The proof of Theorem 2.3 contains a substantially different, more particular assertion.

Lemma 2.4 (Sard). *Lef $f : M_1 \to M_2$ be a smooth mapping. The set of all regular points of the manifold M_2 with respect to the function f is an open, everywhere dense set.*

Theorem 2.3 follows from Lemma 2.4. Indeed, it is sufficient to verify the assertion of Theorem 2.3 for mappings for which the property of transversal regularity holds only in a neighborhood of one point $y \in N \subset M_2$. But then both the manifold M_2 and its submanifold N can be considered to be Euclidean spaces, for which N is linearly imbedded in M_2. In this case, if $f : M_1 \to M_2$ is a smooth mapping, then the transversal regularity of the mapping f along the submanifold N is equivalent to the assertion that the origin of the coordinates in the factor space M_2/N is a regular point for the composition of maps

$$g = \pi \circ f : M_1 \to M_2 \to M_2/N ,$$

where $\pi : M_2 \to M_2/N$ is the projection.

If f is an arbitrary smooth map then, generally speaking, the origin $0 \in M_2/N$ is not a regular point. Let $f(x) = (g(x), h(x))\, , x \in M_1$. According to Sard's lemma, in an arbitrarily small neighborhood of the point $0 \in M_2/N$ there exists a regular point $y \in M_2/N$, $\|0 - y\| < \epsilon$, for the mapping g. We consider some diffeomorphism

$$\varphi : M_2/N \to M_2/N$$

for which

$$\varphi(y) = 0 \ , \quad \rho(\varphi, \mathrm{id}) \leq 2 \ .$$

Then for the composition $g' = \varphi \circ g$ the point $0 \in M_2/N$ is a regular point. On the other hand, the mapping g' splits into a composition

$$g' = \pi \circ f'$$

where

$$f'(x) = (g'(x), h(x)) \ , \quad x \in M_1 \ . \tag{2.6}$$

Thus the mapping $f'(x)$ is transversally regular along the submanifold N, and from (2.6) it follows that

$$\rho(f, f') < 4\epsilon \ .$$

□

It is possible to strengthen Thom's Theorem 2.3. Since the property "the mapping f is transversally regular along a submanifold" is stable with respect to small deformations of the map f, one can restrict the class of mappings f by imposing several supplementary conditions on it. For example, one can suppose *a priori* that the mapping f is already transversally regular along some other submanifold. Thus, suppose $f : M_1 \to M_2$ is a smooth mapping of manifolds with boundary, $f(\partial M_1) \subset \partial M_2$, φ is a smooth imbedding of a manifold N with boundary into M_2, for which $\varphi(\partial N) \subset \partial M_2$, and ∂M_2 and N lie in general position. Suppose, moreover, that the mapping $f|\partial M_1$ is transversally regular along the submanifold $\partial N \subset \partial M_2$. Then for any $\epsilon > 0$ there exists a map $g : M_1 \to M_2$ such that

a) $g(x) \equiv f(x)$ if x belongs to a certain neighborhood V_1 of the boundary ∂M_1;

b) $\rho(g, f) < \epsilon$;

c) g is transversally regular along the submanifold N.

Indeed, a neighborhood V_2 of the boundary ∂M_2 can be represented in the form of a direct product $V_2 = \partial M_2 \times [0,1)$ and some neighborhood V of the boundary ∂N in the form $V = \partial N \times [0,1)$. Since ∂M_2 and N are in general position, a tangent vector d/dt, $t \in [0,1]$, to the manifold N is mapped to a nonzero tangent vector to the manifold M_2. Consequently, at the point $y \in \partial M_2 \cap \partial N$ the identity

$$(T\partial M_2)_y \,/\, (T\partial N)_y = (TM_2)_y \,/\, (TN)_y$$

holds. Consequently, from the transversal regularity of the function $f|\partial M$ along ∂N follows the transversal regularity of the mapping $f|V_1$ along the submanifold N. Next one must apply Thom's theorem to change the mapping f in those local charts of the manifold M_1 which do not intersect the boundary ∂M_1. As a consequence of Thom's theorem we cite the following assertion. We will define a *homotopy of a mapping* $f : X \to Y$ to be a continuous mapping

$$F : X \times [0,1] \to Y$$

such that $F(x, 0) = f(x)$, $x \in X$. Then the mappings $F(x,t)$ for fixed t are called *homotopic maps*.

From Thom's theorem it follows that every smooth map

$$f : M_1 \to M_2$$

is homotopic to a map g which is transversally regular to a submanifold $N \subset M_2$ given in advance. Moreover, one can consider the homotopy between f and g also to be a smooth map.

In Theorem 2.3 of Thom one seeks transversally regular maps in the class of all smooth maps of one manifold into another. However there are sometimes cases in which we cannot change the map $f : M_1 \to M_2$ in an arbitrary fashion, but must remain the entire time in a certain, more limited class of maps.

We present an example of this. Suppose a smooth function f on a smooth manifold M is given. The point $x \in M$ is called a critical point if $(df)_x = 0$. In a local system of coordinates this condition means that $(\operatorname{grad} f)_x = 0$ or that all the partial derivatives $\partial f/\partial x^k$ become zero at the point x. A critical point $x \in M_2$ of a function f is called a nondegenerate critical point if the matrix Hess f composed of the second partial derivatives is nondegenerate. This property of a critical point is independent of the choice of a local system of coordinates. A smooth function f on a smooth manifold is called a Morse function if all of its critical points are nondegenerate. We will investigate Morse functions when we study integrals of rapidly oscillating functions.

Let us attempt to determine, using Thom's theorem, how great the stock of Morse functions on a smooth manifold M is. First of all, we need to express the property of a function f being a Morse function in terms of the property of transversal regularity. It turns out that, if one interprets the differential df as a section in the cotangent bundle,

$$df : M \to (TM)^* \;, \tag{2.7}$$

then the function f is a Morse function if and only if the mapping (2.7) is transversally regular along the null section $M_0 \subset (TM)^*$. Thus, if we wished to apply Thom's theorem, all we could obtain from this theorem is that there exists a map

$$g : M \to (TM)^* \;, \tag{2.8}$$

arbitrarily close to df which is transversally regular along the null section M_0. However, the map g, generally speaking, need not be the differential of any function on the manifold M. Consequently, we have to find transversally regular maps not in the class of all smooth maps (2.8), but in the narrower class of maps of the form (2.7).

We will pose the general problem in the following way. Suppose some set F of smooth mappings of the manifold M_1 into the manifold M_2 is given. It is necessary to find a condition on the set F such that the maps in F

which are transversally regular along a fixed submanifold $N \subset M_2$ is an everywhere dense set in F.

We organize the set of mappings F in the form of a single mapping

$$A : F \times M_1 \to M_2 \ ,$$

in such a way that the particular map f is recovered as the restriction of the map A to $f \times M_1$:

$$f(x) = A(f, x) \ .$$

In making this definition we do not exclude any parametrization of the set F.

Sufficient conditions are known for the set $F^t \subset F$ of mappings which are transversally regular along a submanifold $N \subset M_2$ to be an everywhere dense set in F (Abraham's theorem). We will give the particular case of Abraham's theorem when the set F is a finite-dimensional smooth manifold.

Theorem 2.5 (Abraham). *Suppose F, M_1, M_2 are smooth manifolds and $N \subset M_2$ is a submanifold, and*

$$A : F \times M_1 \to M_2$$

is a smooth mapping which is transversally regular along the submanifold N. We denote by $F^t \subset F$ the subset of points for which the mapping

$$f(x) = A(f, x) \ , \quad f \in F^t \ ,$$

is transversally regular along the submanifold N. Then the set F^t is dense in F.

We will derive Theorem 2.5 from Sard's lemma. Since the mapping A is transversally regular along the submanifold $N \subset M_2$, the preimage $W = A^{-1}(N)$ is a smooth manifold. Suppose $f \in F^t$. The condition of transversal regularity of the mapping f along the submanifold N means that at each point $(f, x) \in W \cap (f \times M_1)$ the differential dA of the mapping A maps the tangent space $(T(f \times M_1))_{(f,x)}$ into the tangent space $(TM_2)_{A(f,x)}$ in such a way that

$$DA((T(f \times M_1)))_{(f,x)} + (TN)_{A(f,x)} = (TM_2)_{A(f,x)} \ . \tag{2.9}$$

The condition of transversal regularity of the entire map A at the same point (f, x) means that

$$DA\Big((T(F \times M_1))_{(f,x)}\Big) + (TN)_{A(f,x)} = (TM_2)_{A(f,x)} \ .$$

The tangent space $T(F \times M_1)_{(f,x)}$ decomposes into a direct sum of two subspaces:

$$T(F \times M_1)_{(f,x)} = (TW)_{(f,x)} \oplus \nu(W)_{(f,x)} ,$$

where DA maps the subspace $(\nu(W))_{(f,x)}$ isomorphically onto $(\nu(N))_{A(f,x)}$, and the subspace $(TW)_{(f,x)}$ into $(TN)_{A(f,x)}$. We prove that

$$T(f \times M_1)_{(f,x)} + (TW)_{(f,x)} = T(F \times M_1)_{(f,x)} .$$

Indeed, if $\xi \in T(F \times M_1)_{(f,x)}$, then

$$\xi = \xi_1 + \xi_2 , \quad \xi_1 \in TW_{(f,x)} , \quad \xi \in (\nu(M))_{(f,x)} .$$

Then $DA(\xi_2) \in (\nu(N))_{A(f,x)}$ and, according to (2.9), can be represented in the form

$$DA(\xi_2) = D(A)(\eta_1) + \eta_2 ,$$

$$\eta_1 \in (T(f \times M_1))_{(f,x)} , \quad \eta_2 \in (TN)_{A(f,x)} .$$

We consider the vector $\xi' = \xi_1 + \eta_1$. We have $DA(\xi - \xi') = DA(\xi_2 - \eta_1) = \eta_2 \in TN_{A(f,x)}$. Then $\xi'' = \xi - \xi' \in (TW)_{(f,x)}$. Thus $\xi = \xi_1 + \eta_1 + \xi''$, where $\eta_1 \in (T(f \times M_1))_{(f,x)}$, $\xi_1 + \xi'' \in (TW)_{(f,x)}$. So we have established that $f \in F^t$ if and only if the submanifolds W and $f \times M_1$ are in general position. This condition is precisely equivalent to the following condition: $f \in F^t$ if and only if the point $f \in F$ is a regular point for the mapping

$$\pi|W : W \to F , \tag{2.10}$$

where $\pi : F \times M_1 \to F$ is the projection. According to Sard's lemma the set of regular points F^t for the map (2.10) is everywhere dense in the space F. This completes the proof of Theorem 2.5. □

Theorem 2.5 is a special case of Abraham's theorem, whose formulation differs from the formulation of Theorem 2.5 only in that the space of parameters F is taken to be an infinite-dimensional Banach space. But in practice, as a rule, it is sufficient to consider only finite-dimensional families of mappings which satisfy the conditions of Abraham's theorem.

For example, in the case of seeking Morse functions it is sufficient to include the original smooth function f on the manifold M in a finite-dimensional family of functions f_t, $t \in T$, such that the differential of the function $F(x,t) = f_t(x)$ is transversally regular along the null section. In a local system of coordinates one should choose as such a family the functions

$$f_t(x) = f(x) + \sum_{k=0}^{n} t_k x^k ,$$

$$x = \left(x^1, \ldots, x^n\right) \ , \quad t = (t_1, \ldots, t_n) \in \mathbb{R}_n \ .$$

It is completely obvious that the differential of the mapping

$$Df_t : M \times \mathbb{R}^n \to (TM)^*$$

is transversally regular along the null section $M_0 \subset (TM)^*$, since at stationary points $x \in M$ the mapping Df_t is linear on $(x \times \mathbb{R}^n)$.

We now present one noteworthy property of Morse functions on a manifold M.

Proposition 2.6. *Suppose the point $x_0 \in M$ of a smooth n-dimensional manifold M is a nondegenerate critical point of a Morse function f. Then in a neighborhood of the point $x_0 \in M$ there exist local coordinates $\left(x^1, \ldots, x^n\right)$ such that in these local coordinates the function f has the form*

$$f\left(x^1, \ldots, x^n\right) = f(x_0) + \left(x^1\right)^2 + \ldots + \left(x^k\right)^2 - \left(x^{k+1}\right)^2 - \ldots - \left(x^n\right)^n \ .$$

Proof. It is sufficient to assume immediately that the function f is a function of n independent variables $\left(x^1, \ldots, x^n\right)$, where the point $(0, \ldots, 0)$ is a nondegenerate critical point of the function f. This means that the partial derivatives $\frac{\partial f}{\partial x^k}(0, \ldots, 0)$ are equal to zero.

We consider the matrix Hess f of second partial derivatives of f at the point $(0, \ldots, 0)$. Since the matrix Hess f is nondegenerate and changes under linear coordinate transformations as the matrix of a quadratic form, there exists some linear transformation of the coordinates such that in the new coordinates Hess f is a diagonal matrix whose diagonal elements are all nonzero. We will denote the new coordinates by $\left(y^1, \ldots, y^n\right)$. We consider the Taylor expansion of the function $f\left(y^1, \ldots, y^n\right)$,

$$f\left(y^1, \ldots y^2\right) = f(0) + \sum y^i y^j h_{ij}\left(y^1, \ldots, y^n\right) \ , \quad h_{ij} = h_{ji} \ .$$

Then $h_{11}(0, \ldots, 0) \neq 0$. Consequently, if $h_{11}(0, \ldots, 0) > 0$, one can introduce new coordinates

$$z^1 = y^1 \sqrt{h_{11}\left(y^1, \ldots, y^2\right)} + \sum_{k \geq 2} y^k \frac{h_{1k}\left(y^1, \ldots, y^n\right)}{\sqrt{h_{11}\left(y^1, \ldots, y^n\right)}} \ ,$$

$$z^2 = y^2 \ , \ \ldots \ , \ z^n = y^n \ .$$

Then

$$f\left(z^1, \ldots, z^n\right) = f(0) + \left(z^1\right)^2 + \sum_{i,j \geq 2} z^i z^j h_{ij}\left(z^1, \ldots, z^n\right) \ .$$

If $h_{11}(0, \ldots, 0) < 0$, then we set

$$z^1 = y^1\sqrt{-h_{11}\left(y^1,\ldots,y^n\right)} - \sum_{k\geq 2} y^k \frac{h_{1k}\left(y^1,\ldots,y^2\right)}{\sqrt{h_{11}\left(y^1,\ldots,y^n\right)}} ,$$

$$z^2 = y^2 , \ \ldots \ , \ z^n = y^n .$$

Then

$$f\left(z^1,\ldots z^n\right) = f(0) - \left(z^1\right)^2 + \sum_{i,j\geq 2} z^i z^j h_{ij}\left(z^1,\ldots,z^n\right) .$$

In both cases the process of changing coordinates can be carried out by induction. This proves Proposition 2.6. □

As a consequence of Proposition 2.6, we obtain that any nondegenerate critical point of the function f is isolated. Indeed, since in a neighborhood of a nondegenerate critical point x_0 the function f has the form $f\left(x^1,\ldots,x^n\right) = \sum_{i=1}^n \pm\left(x^i\right)^2$ in some system of coordinates, the partial derivatives can only be zero at the point $x^1 = x^2 = \ldots = x^n = 0$. Thus in the entire neighborhood of the point x_0 there are no other critical points. Consequently a Morse function f can only have finitely many critical points on a compact manifold M.

1.3 The Index of Intersection of Submanifolds

In the previous section the situation of two submanifolds in general position was considered as an example. Suppose $\varphi_k : M_k \to M$, $k = 1,2$, are two manifolds in general position, i.e. each imbedding φ_k is transversally regular along the other manifold. In this case the intersection $M_3 = M_1 \cap M_2$ is a submanifold, whose dimension is equal to $\dim M_3 = \dim M_1 + \dim M_2 - \dim M$. If $\dim M_1 + \dim M_2 = \dim M$, then the dimension of the intersection is zero, i.e. the manifold M_3 consists of a finite number of isolated points (we assume that one of the manifolds M_k is compact).

The number of points of intersection of M_1 and M_2 can serve as a certain invariant of the imbeddings. More precisely, we denote by $\operatorname{ind}_2\left[(M_1,\varphi_1):(M_2,\varphi_2)\right] \in \mathbb{Z}_2$ the number of points of the intersection $M_1 \cap M_2$ modulo 2. Then $\operatorname{ind}_2\left[(M_1,\varphi_1):(M_2,\varphi_2)\right]$ depends only on the homotopy classes of the imbeddings, i.e. if φ'_k are other imbeddings of the manifold M_k which are homotopic to the imbeddings φ_k, $k = 1,2$, then

$$\operatorname{ind}_2\left[(M_1,\varphi_1):(M_2,\varphi_2)\right] = \operatorname{ind}_2\left[(M_1,\varphi'_1):(M_2,\varphi'_2)\right] .$$

Consequently, in the notation $\operatorname{ind}_2\left[(M_1,\varphi_1):(M_2,\varphi_2)\right]$ one need not indicate the imbeddings φ_1,φ_2, but simply write $\operatorname{ind}_2\left[M_1 : M_2\right]$.

The number of intersections itself is not a homotopy invariant.

We consider the example of the two-dimensional torus $T^2 = S^1 \times S^1$ with the angular parameters φ_1 and φ_2 as coordinates. We will call the first angular parameter φ_1 the latitude, and the second – φ_2 – the longitude. Then a net of two mutually orthogonal curves is formed on the torus: the curves $\varphi_1 =$ const. are circles, which are called parallels, and the curves $\varphi_2 =$ const. are circles, which are called meridians (Fig. 7).

Every parallel intersects every meridian in precisely one point, at which the intersection is transversal. Consequently, the index of intersection of a parallel and a meridian is equal to unity. If we undertake a deformation, say, of a meridian, then the number of points of intersection of the meridian (rather, of its deformed image) with the parallel is changed by an even number of points (Fig. 8).

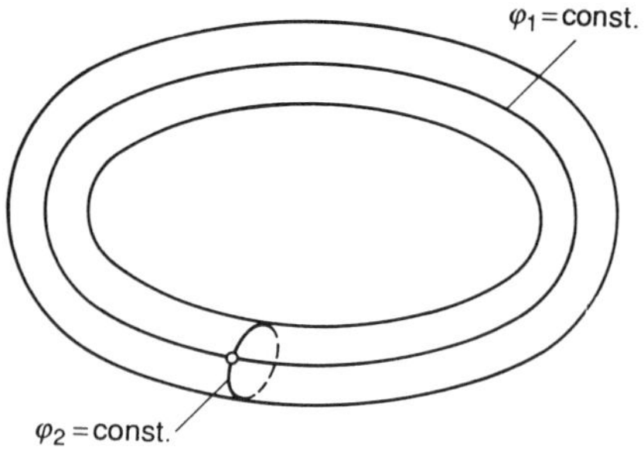

Fig. 7

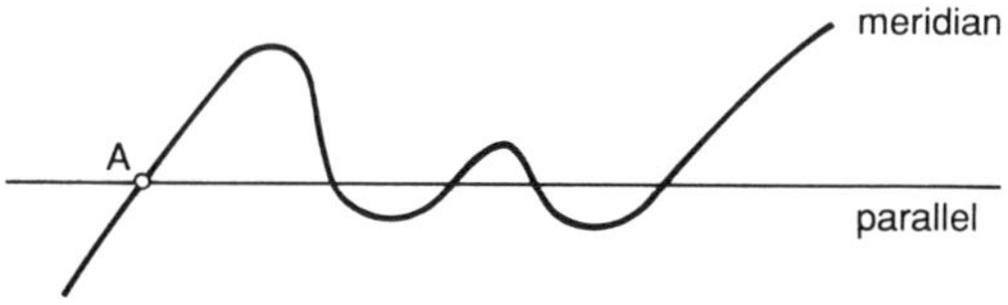

Fig. 8

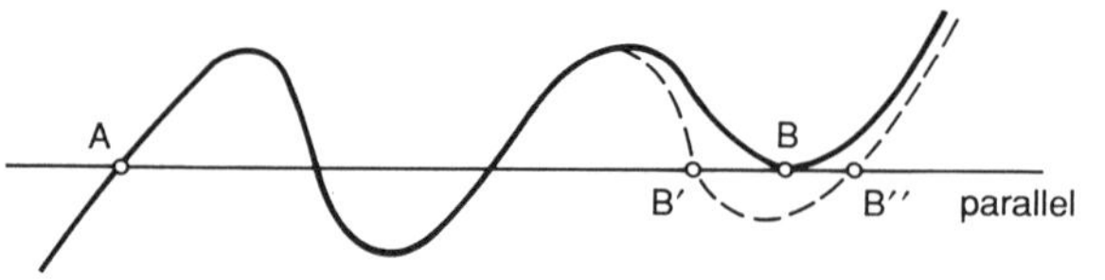

Fig. 9

In case the meridian and the parallel are not in general position, the number of new points may be odd, since some pair of points of the intersection may "merge" into one under deformation. In Fig. 9 we show how under a deformation of the meridian the pair of points B' and B'' combine into the single point B. It turns out that, in order to construct an integral index of intersection, it is necessary to make several of the properties of manifolds and their intersections more precise.

As was said in § 1.1, a smooth manifold M is called an orientable manifold if it is possible to choose an atlas of charts $\{U_\alpha\}$ on M with local coordinates $x_\alpha^1, \ldots, x_\alpha^n$ on them such that at any point $x \in U_\alpha \cap U_\beta$ the Jacobian matrix has a positive determinant

$$\det \left(\frac{\partial x_\alpha^k}{\partial x_\beta^j} \right) > 0 . \tag{3.1}$$

An atlas of charts $\{U_\alpha\}$ satisfying conditions (3.1) is called an oriented atlas of the smooth manifold M. Two oriented atlases $\{U_\alpha, x_\alpha^k\}$ and $\left\{U'_\beta x'^k_\beta\right\}$ are called equivalent if the Jacobian matrix of the change of coordinates at any point $x \in U_\alpha \cap U'_\beta$ has a positive determinant

$$\det \left(\frac{\partial x_\alpha^k}{\partial x'^j_\beta} \right) > 0 ,$$

i.e. if the union of charts $\{U_\alpha, x_\alpha^k\} \cup \left\{U'_\beta, x'^k_\beta\right\}$ is an oriented atlas. Then all atlases of charts which give an orientation to a smooth manifold M can be broken into two classes, in each of which that atlases of charts are pairwise equivalent. Thus, on a connected orientable manifold M there exist two distinct orientations in all.

If a manifold M is connected and orientable, its orientation can also be given in the following way. We fix a vector space V of dimension n. The set of (ordered) bases can be divided into two classes: two bases belong to the same class if the matrix of transformation from one basis to the other has a positive determinant. We then say that these two bases determine the same orientation on the space V. On the other hand, if the matrix of transformation from one basis to the other has negative determinant, then we assign the two bases to different classes and say that the two bases determine distinct (opposite) orientations of the space V.

We consider a local system of coordinates $(x_\alpha^1, \ldots, x_\alpha^2)$ in the chart U_α. Then this local system of coordinates assigns to each tangent space $(TM)_x$, $x \in U_\alpha$, a basis $\{\partial/\partial x_\alpha^1, \ldots, \partial/\partial x_\alpha^n\}$ and hence a certain orientation on the space $(TM)_x$. The oriented atlas $\{U_\alpha, x_\alpha^k\}$ of the manifold M determines a certain orientation on every tangent space $(TM)_x$, $x \in M$, which is independent of the chart U_α which contains the point x.

Conversely, if $x \in U_\alpha$ and an orientation is given on the space $(TM)_x$, then one can choose a local system of coordinates $\{x_\alpha^k\}$ such that the basis $\{\partial/\partial x_\alpha^1, \ldots, \partial/\partial x_\alpha^n\}$ determines the original orientation on the space $(TM)_x$. Suppose two points $x, y \in M$ are given, along with a continuous curve

$$\varphi : [0,1] \to M , \quad \varphi(0) = x , \quad \varphi(1) = y ,$$

and two bases $(e_1, \dots, e_n) \in (TM)_x$, $(f_1, \dots, f_n) \in (TM)_y$. We will say that these bases *determine compatible orientations along the curve* φ if there exists a continuous family of bases

$$\{e_1(t), \dots, e_n(t)\} \in (TM)_{\varphi(t)} , \quad 0 \le t \le 1 ,$$

for which

$$\begin{aligned} e_k(0) &= e_k , \quad 1 \le k \le n , \\ e_k(1) &= f_k , \quad 1 \le k \le n . \end{aligned}$$

It is easy to show that for two bases $\{e_1, \dots, e_n\} \in (TM)_x$, $\{f_1, \dots, f_n\} \in (TM)_y$ and any curve φ there exists a continuous family $\{e_1(t), \dots, e_n(t)\} \in (TM)_{\varphi(t)}$, $0 \le t \le 1$, such that $e_k(0) = e_k$, $1 \le k \le n$, and for $t = 1$ the basis $\{e_k(1)\}$ either coincides with $\{f_1, \dots, f_k\}$ or else determines the opposite orientation on the space $(TM)_y$.

Suppose that a basis $\{e_1(x), \dots, e_n(x)\}$ is given on each tangent space $(TM)_x$, $x \in M$ (we do not assume a continuous dependence on $x \in M$!). We say that these bases determine compatible orientations if they give compatible orientations for any pair of points $x, y \in M$ and any continuous curve joining x to y. If a compatible set of bases of the tangent spaces is given at every point $x \in M$, then the manifold M is orientable; moreover, the bases $\{\partial/\partial x^1_\alpha, \dots, \partial/\partial x^n_\alpha\}$, which are given by the local system of coordinates in the atlas $\{U_\alpha, x^k_\alpha\}$ oriented in the appropriate manner, determine the original orientation on each tangent space $(TM)_x$.

We proceed now to study the index of intersection of two orientable submanifolds M_1, M_2 lying in general position in an orientable manifold M. Since the manifolds M_1 and M_2 lie in general position, their intersection $M_3 = M_1 \cap M_2$ is a smooth manifold. We will show that the manifold M_3 is orientable. For this we must present a compatible collection of bases on each tangent space $(TM)_x$, $x \in M_3$. We have the following indentities:

$$\begin{aligned} (TM_1)_x + (TM_2)_x &= (TM)_x , \\ (TM_1)_x \cap (TM_2)_x &= (TM_3)_x . \end{aligned}$$

We are already given compatible orientations on the spaces $(TM)_x$, $(TM_1)_x$, $(TM_2)_x$. We fix a certain basis

$$\{a_1, \dots, a_k\} \in (TM_3)_x . \tag{3.2}$$

We extend it to the bases

$$\{a_1, \dots, a_k, b_1, \dots, b_s\} \in (TM_1)_x , \tag{3.3}$$

$$\{a_1, \dots, a_k, c_1, \dots, c_l\} \in (TM_2)_x . \tag{3.4}$$

Then the vectors

$$(c_1, \ldots c_l, a_1, \ldots, a_k, b_1, \ldots, b_s\} \tag{3.5}$$

form a basis of the space $(TM)_x$. We will assume that $k \neq 0$, $l \neq 0$, $s \neq 0$.

In general position, when the manifolds M_1, M_2 are not zero-dimensional, it is sufficient to assume that $k \neq 0$, and the inequalities $s \neq 0$, $l \neq 0$ hold automatically. Changing, if necessary, the sign of vector b_1, we can assume that basis (3.3) is compatible with the orientation of the manifold M_1. Analogously, possibly changing the sign of c_1, we consider basis (3.4) to be compatible with the orientation of the manifold M_2.

We consider the basis (3.5). If this basis does not agree with the orientation of the manifold M, then we change the sign of the vectors a_1, b_1, c_1. Thus, as a result, we obtain vectors such that the bases (3.3), (3.4) and (3.5) have orientations compatible with the orientations of the manifolds M_1, M_2 and M. We give an orientation of the space $(TM_3)_x$ by means of the basis (3.2). It is trivial to check the consistency of the bases (3.2) at different points of the manifold M_3.

We consider now the case when $k = 0$, i.e. $\dim M_3 = 0$. Then the manifold M_3 consists of separate isolated points. Although a zero-dimensional manifold formally has no orientation, we can construct a function $\epsilon(x), x \in M_3$, on M_3 which takes the value ± 1. We consider the bases (3.3) and (3.4), which are compatible with the orientations of the manifolds M_1 and M_2 (the vectors $\{a_j\}$ are absent). Then we set $\epsilon(x) = 1$ if basis (3.5) is compatible with the orientation of the manifold M, $\epsilon(x) = -1$ if the orientation of the basis is opposite to that of M.

Definition 3.1. *Suppose M_1, $M_2 \subset M$ are two oriented submanifolds in general position,*

$$M_3 = M_1 \cap M_2 \ , \quad \dim M_3 = 0 \ .$$

We set

$$\operatorname{ind}[M_2 : M_1] = \sum_{x \in M_3} \epsilon(x)$$

and call this number the index of intersection.

It is simple to check the formula

$$\operatorname{ind}[M_2 : M_1] = (-1)^{\dim M_1 \dim M_2} \operatorname{ind}[M_1 : M_2] \ .$$

The index of intersection for two submanifolds M_1 and M_2 depends *a priori* on the way the submanifolds M_1 and M_2 are imbedded. In fact it is possible to avoid a series of limitations on the imbedding of the manifolds. For example, if the imbeddings

$$\varphi_k : M_k \to M \ , \quad k = 1, 2 \ ,$$

are not in general position, then it is possible to solve the problem of approximation of the imbeddings φ_k, $k = 1, 2$, by other imbeddings $\widetilde{\varphi}_k$ which are in general position, define the index of intersection for the altered imbeddings $\widetilde{\varphi}_k$ and study the dependence of the index of intersection on the method of approximation. The first step can be taken care of with the help of Abraham's theorem. Indeed, we have to show that the set of imbeddings of M_1 into M which lie in general position with respect to the submanifold $M_2 \subset M$ forms a dense subset of the class of all imbeddings. In order to apply Abraham's theorem in the form in which it was presented in the previous section (Theorem 2.5), one must include the given imbedding $\varphi_1 : M_1 \to M$ in a family of imbeddings, parametrized by a smooth manifold F, for which the hypotheses of Theorem 2.5 hold. For example, one can take F to be the family of imbeddings constructed in the following way. Let G be a certain finite dimensional group of diffeomorphisms of the manifold M. Suppose that this condition holds for that group: for any point $x \in M_2 \subset M$ the map

$$f_x : G \to M \ , \quad f_x(g) = g(x) \ , \quad g \in G \ , \tag{3.6}$$

is a transversally regular map along the submanifold $M_2 \subset M$. We consider the map

$$A : M_1 \times G \to M \ ,$$

$$A(x, g) = g\left(\varphi_1(x)\right) \ , \quad x \in M_1 \ , \quad g \in G \ .$$

We prove that the mapping A is transversally regular along the submanifold $M_2 \subset M$ in a neighborhood of $M_1 \times 1 \subset M_1 \times G$. In fact, if $A(x, 1) = y \in M_2$, then $y = \varphi_1(x)$. Furthermore,

$$DA\left(T(M_1 \times G)_{(x,1)}\right) \supset DA\left(T(x \times G)_{(x,1)}\right) = Df_y\,(TG)_1 \ .$$

Consequently, from the transversal regularity (3.6) it follows that

$$DA\left(T(M_1 \times G)_{(x,1)}\right) + (TM_2)_y = (TM)_x \ .$$

All that remains is to construct the group of diffeomorphisms G.

We consider the normal bundle $\nu(M_2)$ of the submanifold $M_2 \subset M$. This is a finite-dimensional vector space, and therefore there exists a finite number of sections $s_1, \ldots, s_N$ of the bundle $\nu(M_2)$ such that at each point $x \in M_2$ the vectors $\{s_1(x), \ldots, s_N(x)\}$ generate the fiber of the vector bundle $\nu(M_2)$. Suppose $\{X_1, \ldots, X_N\}$ is a collection of vector fields on the manifold M such that the restriction of the field X_k to the manifold M_2 coincides with the section $s_k(x)$. Every vector field X_k determines a one-parameter group G_k of diffeomorphisms of the manifold M_2, for which, if t_k is the parameter of the group G_k, then the tangent vector to the curve $y = t_k(x)$, $-\epsilon \leq t_k \leq \epsilon$, at the point x coincides with the vector $X_k(x)$.

We consider the manifold $G = \prod_{k=1}^{N} G_k$, each of whose points determines a certain diffeomorphism of the manifold M_2:

$$(t_1, \dots, t_n(x)) = t_1 (t_2 \dots (t_n(x)) \dots) \ .$$

It is then simple to establish that the map of the form (3.6) is transversally regular along the manifold M_2.

Thus we have established that every imbedding of a manifold M_1 into M can be approximated arbitrarily closely by a transversally regular imbedding along the manifold M_2. If we show that the index of intersection does not depend on the method of approximation, then the index of intersection of two manifolds which are not in general position will, by this result, be well defined.

It will be more convenient for us to define the index of intersection not for imbeddings, but for arbitrary smooth maps φ_1 of the manifold M_1 into M which are transversally regular along M_2. If $\dim M = \dim M_1 + \dim M_2$, then, as in the case of imbeddings, the preimage $\varphi_1^{-1}(M_2)$ consists of a finite set of points. In the case of oriented manifolds M, M_1, M_2, each point x of the preimage $\varphi_1^{-1}(M_2)$ can be assigned a sign $\epsilon(x)$. If $\dim M < \dim M_1 + \dim M_2$, then the preimage $\varphi_1^{-1}(M_2)$ is provided with an orientation compatible with the orientations of the manifolds M, M_1, M_2. Thus we can assign a number

$$\operatorname{ind}[\varphi_1 : M_2] = \sum_{x \in \varphi_1^{-1}(M_2)} \epsilon(x)$$

to each transversally regular mapping along M_2. This number depends only upon the homotopy class of the map φ_1 of the manifold M_1. Indeed, if φ_1' is another transversally regular mapping of M_1 along the submanifold M_2, and the mapping φ_1, φ_1' are homotopic, then there exists a smooth map

$$F : M_1 \times I \to M_2 \ ,$$

$$F(x, 0) = \varphi_1(x) \ , \quad x \in M_1 \ , \tag{3.7}$$

$$F(x, 1) = \varphi_1'(x) \ , \quad x \in M_1 \ . \tag{3.8}$$

According to Thom's theorem it is possible to choose the mapping F, without changing conditions (3.7), (3.8), to be transversally regular along the submanifold M_2. We consider the preimage $W = F^{-1}(M_2)$. The manifold W is one-dimensional and therefore consists of a finite number of segments and closed curves (Fig. 10).

The boundary of the manifold W lies on the boundary of $M_1 \times I$, $\partial(M_1 \times I) = (M_1 \times 0) \cup (M_1 \times 1)$, and is equal to the union of the preimages $\varphi_1^{-1}(M_2) \cup \varphi_1'^{-1}(M_2)$. All the points of $\varphi_1^{-1}(M_2) \cup \varphi_1'^{-1}(M_2)$ can be

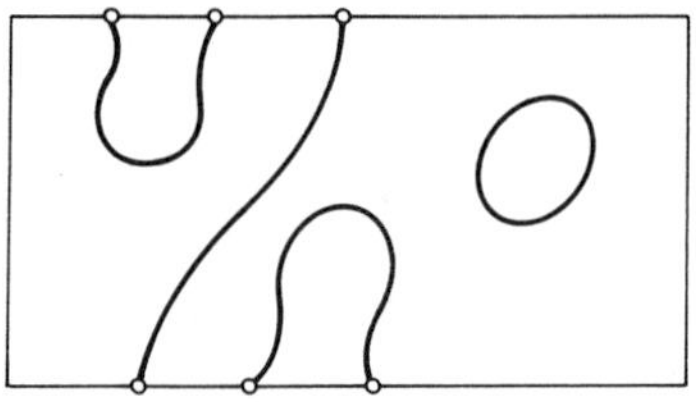

Fig. 10

broken up into pairs – the boundaries of the connected components of the manifold W. If a pair of points x_1, x_2 lie on one edge of the manifold $M_1 \times I$, then $\epsilon(x_1) = -\epsilon(x_2)$, and if they are on separate edges, then $\epsilon(x_1) = \epsilon(x_2)$. Hence it follows that

$$\operatorname{ind}[\varphi_1 : M_2] = \operatorname{ind}[\varphi_1' : M_2] \ .$$

In the future we will only be interested in the case when M_1 is a one-dimensional connected compact manifold, i.e. a circle, and M_2 is a manifold of codimension 1. We note that the manifold M_2 need not be a compact manifold; it is sufficient that its topological boundary $\overline{M}_2 \setminus M_2$ have codimension ≥ 3.

1.4 Homotopy Groups

Here we will briefly formulate several concepts and theorems which are necessary for concrete calculations. A detailed presentation of the concepts introduced here can be found in the book of Fuchs, Fomenko and Gutenmacher [1]. In further reading one may omit these concrete calculations, leaving only the results of the calculations.

It is assumed that all our spaces are homeomorphic to finite simplicial complexes. A space X is called *pointed* if a point x_0 is singled out in X. A map f of pointed spaces

$$f : (X, x_0) \to (Y, y_0)$$

must map the distinguished point x_0 to the distinguished point y_0, $f(x_0) = y_0$. A homotopy f_t of mappings of pointed spaces must preserve the distinguished points, $f_t(x_0) = y_0$.

The set of all homotopy classes of mappings of the k-dimensional sphere (S^k, s_0) into a space (X, x_0) is denoted by $\pi_k(X, x_0)$. A group structure can be introduced naturally onto the set $\pi_k(X, x_0)$, and $\pi_k(X, x_0)$ is called the k-dimensional homotopy group of the space (X, x_0). For $k > 1$ the groups $\pi_k(X, x_0)$ are abelian. If $f : (X, x_0) \to (Y, y_0)$ is a map of pointed spaces, it induces a homomorphism of homotopy groups

$$f_k : \pi_k(X, x_0) \to \pi_k(Y, y_0) \ .$$

Suppose $p : E \to Y$ is a locally trivial bundle with fiber F, $j : F \to E$ is an imbedding, $x_0 \in F$, $p(x_0) = y_0$. Then there is a homomorphism

$$\partial : \pi_k(Y, y_0) \to \pi_{k-1}(F, x_0)$$

such that the sequence

$$\begin{aligned} \ldots \xrightarrow{\partial} \pi_k(F, x_0) \xrightarrow{j_*} \pi_k(E, x_0) \xrightarrow{p_*} \pi_k(Y, y_0) \\ \xrightarrow{\partial} \pi_{k-1}(F, x_0) \xrightarrow{j_*} \pi_{k-1}(E, x_0) \xrightarrow{p_*} \ldots \end{aligned}$$

is exact, i.e. that the composition of two successive homomorphisms is zero, and the kernel of the second homomorphism coincides with the image of the first homomorphism.

Now suppose X is a simplicial complex, $C_k(X)$ is the free abelian group generated by all the k-dimensional simplices. If σ is a k-dimensional simplex with vertices $e_0, \ldots, e_k$, we will write $\sigma = (e_0, \ldots, e_k)$. We define the homomorphism

$$d : C_k(X) \to C_{k-1}(X)$$

by setting

$$d(e_0, \ldots, e_k) = \sum (-1)^j (e_0, \ldots, \widehat{e}_j, \ldots, e_k) \ .$$

Then $d^2 = 0$ and we can define the group

$$H_k(X) = \left(\operatorname{Ker}^k d\right) / \left(\operatorname{Im}^k d\right) \ ,$$

where

$$\begin{aligned} \operatorname{Ker}^k d &= \{x \in C_k(X) : dx = 0\} \ , \\ \operatorname{Im}^k d &= d(C_{k+1}(X)) \ . \end{aligned}$$

The following assertion holds: if $\pi_j(X, x_0) = 0$ for $j < k$, $k > 1$, then

$$\pi_k(X, x_0) = H_k(X)$$

(the Hurewicz theorem). If $\pi_1(X, x_0)$ is abelian, then $\pi_1(X, x_0) = H_1(X)$ for a connected space X. In the general case

$$H_1(X) = \pi_1(X, x_0) / [\pi_1(X, x_0), \pi_1(X, x_0)] \ ,$$

where $[\pi, \pi]$ denotes the commutator of the group π, i.e. the normal subgroup generated by elements of the form

$$aba^{-1}b^{-1}, a, b \in \pi \ .$$

The spectral cohomologies $H^k(X)$ of the space X are isomorphic to the groups $\mathrm{Hom}\,(H_k(X), \mathbb{R})$. The last assertion allows one to find out if a differential form $\omega \in \Omega_k(M)$ on a smooth manifold determines a null or non-null class of the cohomology. For this it is necessary to consider a simplicial decomposition of the manifold M and choose a (finite) basis $(a_1, \ldots, a_r)$ in the group $H_k(M)$. Let $(b_1, \ldots, b_r)$ be representatives of the elements $(a_1, \ldots, a_r)$. Then every element b_j is a linear combination of k-dimensional simplices of the manifold M, so that the number

$$\gamma_j = \int\limits_{b_j} \omega$$

is well-defined. In order for the form to determine the null cohomology class, it is necessary and sufficient that

$$\gamma_j = 0 \ , \quad 1 \leq j \leq r \ .$$

Chapter 2
The Geometry of Real Lagrangian Manifolds

The basic geometric concept for the construction of asymptotic solutions of equations with a small parameter by Maslov's canonical operator is the concept of a Lagrangian manifold in phase space. The present chapter is devoted to a systematic study of the topological properties of Lagrangian manifolds which are necessary for the construction of the canonical operator.

2.1 Lagrangian Manifolds in Hamiltonian Space

Let $\Phi_{\mathbb{R}}(2n)$ be a real $2n$-dimensional vector space with a specified basis $(e, f) = \left(e_1, \ldots, e_n, f^1, \ldots, f^n\right)$. We will call the space $\Phi_{\mathbb{R}}(2n)$ the phase space. We will separate the coordinates of an arbitrary vector $\xi \in \Phi_{\mathbb{R}}(2n)$ into two groups: the x-coordinates $\left(x^1, \ldots, x^n\right)$ and the p-coordinates $(p_1, \ldots, p_n)$:

$$\xi = \sum_{k=1}^{n} x^k e_k + \sum_{k=1}^{n} p_k f^k \ .$$

We consider the external differential form

$$dp \wedge dx = \sum_{k=1}^{n} dp_k \wedge dx^k \tag{1.1}$$

in the phase space $\Phi_{\mathbb{R}}(2n)$. Since the coefficients of the form (1.1) are constant, this form induces at every point $\xi \in \Phi_{\mathbb{R}}(2n)$ an identical skew-symmetric form $<,>$ on the tangent space, which we can identify with the phase space $\Phi_{\mathbb{R}}(2n)$ itself. This bilinear form can be represented in terms of the basis (e, f) by the skew-symmetric matrix I_n, equal to

$$I_n = \begin{pmatrix} 0 & E \\ -E & 0 \end{pmatrix} \ . \tag{1.2}$$

Definition 1.1. Let $\varphi : M \to \Phi_{\mathbb{R}}(2n)$ be an n-dimensional submanifold in phase space. If the restriction of the form (1.1) to the submanifold M is identically equal to zero:

$$\varphi^*(dp \wedge dx) \equiv 0 \ ,$$

then the manifold M is called a *Lagrangian manifold in the phase space* $\Phi_{\mathbb{R}}(2n)$.

A particular example of a Lagrangian manifold is a linear subspace $\varphi : L \to \Phi_{\mathbb{R}}(2n)$ on which the restriction of the form (1.1) is identically zero. Such linear subspaces, which are simultaneously Lagrangian manifolds, will be called Lagrangian planes in the phase space $\Phi_{\mathbb{R}}(2n)$.

It is completely obvious that an n-dimensional submanifold $\varphi : M \to \Phi_{\mathbb{R}}(2n)$ is a Lagrangian manifold if and only if all its tangent spaces $T_m(M)$, $m \subset M$, are Lagrangian planes under the imbedding $d\varphi : T_m(M) \to \Phi_{\mathbb{R}}(2n)$.

A typical example of a Lagrangian manifold is the graph of a gradient of a smooth function in the variables $(x^1, \dots, x^n)$. To be precise, let $S(x^1, \dots, x^n)$ be a smooth function defined in some domain V of the n-dimensional space $\mathbb{R}^n$ with basis $e = (e_1, \dots, e_n)$ and coordinates $x = (x^1, \dots, x^n)$. We consider the imbedding

$$\varphi : V \to \Phi_{\mathbb{R}}(2n)$$

given by the equations

$$\begin{aligned} x^k &= x^k \ , \quad 1 \le k \le n \ , \\ p^k &= \frac{\partial S}{\partial x^k}(x^1, \dots, x^n) \ , \quad 1 \le k \le n \ . \end{aligned} \tag{1.3}$$

Then the submanifold V is a Lagrangian manifold. Indeed, the restriction $\varphi^*(dp \wedge dx)$ of the form (1.1) has the following appearance in a local system of coordinates $x = (x^1, \dots, x^n)$:

$$\begin{aligned} \varphi^*(dp \wedge dx) &= \sum_{k=1}^{n} \left(d\frac{\partial S}{\partial x^k} \wedge dx^k \right) = \sum_{k=1}^{n} \left(\sum_{l=1}^{n} \frac{\partial^2 S}{\partial x^k \partial x^l} dx^l \wedge dx^k \right) \\ &= \sum_{k=1}^{n} \frac{\partial^2 S}{(dx^k)^2} dx^k \wedge dx^k + \sum_{k \neq l} \frac{\partial^2 S}{\partial x^k \partial x^l} (dx^l \wedge dx^k + dx^k \wedge dx^l) \equiv 0 \ . \end{aligned}$$

In a certain sense any Lagrangian manifold can be represented in the form (1.3). A precise formulation will be given below under the name of the lemma on local canonical coordinates.

Suppose I, J denote some subsets of the indices from 1 to n, $\overline{I}$ denotes the complement of I, $\overline{I} = \{1, \ldots, n\} - I$. We denote by $\mathbb{R}^I \times \mathbb{R}_J \subset \Phi_{\mathbb{R}}(2n)$ the subspace generated by the vectors $\{e_k : k \in I; f^l, l \in J\}$. Correspondingly we will denote the coordinates of vectors of $\mathbb{R}^I \times \mathbb{R}_J$ by (x^I, p_J). Let

$$P_J^I : \varphi_{\mathbb{R}}(2n) \to \mathbb{R}^I \times \mathbb{R}_J$$

denote the projection of phase space $\Phi_{\mathbb{R}}(2n)$ along the complementary subspace $\mathbb{R}^{\overline{I}} \times R_{\overline{J}}$. It follows from the implicit function theorem that the coordinates (x^I, p_J) serve as a local coordinate system for the submanifold M in a neighborhood of a point $m \in M$ if an only if the projection

$$P_J^I : T_m(M) \to \mathbb{R}^I \times R_J$$

is an isomorphism. These same coordinates (x^I, p_J) are a system of (linear) coordinates also for the tangent space $T_m(M)$.

Lemma 1.1. (Lemma on a Local Canonical System of Coordinates) (Maslov [1]). *For any Lagrangian plane $\varphi : L \to \Phi_{\mathbb{R}}(2n)$ there exists a collection of indices I such that the projection*

$$P_{\overline{I}}^I : L \to \mathbb{R}^I \times \mathbb{R}_{\overline{I}}$$

is an isomorphism, i.e. the coordinates $(x^I, p_{\overline{I}})$ form a local system of coordinates on the plane L.

Proof. We consider the projection

$$P^n : \Phi_{\mathbb{R}}(2n) \to \mathbb{R}^n ;$$

here n denotes the set consisting of all the indices $(1, \ldots, n)$. Let $L' = P^n(L)$ be a linear subspace in $\mathbb{R}^n$, $\dim L' \leq \dim L$. Quite obviously,

$$L = L' \oplus L'' , \quad L'' = \operatorname{Ker} P^n , \quad \dim L' \leq \dim L .$$

Then the projection $P^n : L' \to P^n(L')$ is an isomorphism.

Let I be a collection of indices such that the projection

$$P^n : L' \to \mathbb{R}^I$$

is an isomorphism. But in this case the projection $P^I : L' \to R^I$ is also an isomorphism. We show that the projection

$$P_{\overline{I}}^I : L \to \mathbb{R}^I \times R_{\overline{I}}$$

is then an isomorphism as well. For this it is sufficient to show that $\operatorname{Ker}\left(P_{\overline{I}}^I\right) | L = 0$, since

$$n = \dim L = \dim \mathbb{R}^I + \dim R_{\overline{I}} \ .$$

Let $\xi \in L$ be a vector such that

$$P^I_{\overline{I}}(\xi) = 0 \ .$$

Then the coordinates of the vector ξ have the form $\xi = \left(0, x^{\overline{I}}, p_I, 0\right)$ in the factorization $\Phi_{\mathbb{R}}(2n) = \mathbb{R}^I \times \mathbb{R}^{\overline{I}} \times \mathbb{R}_I \times \mathbb{R}_{\overline{I}}$. In particular, $P^I(\xi) = 0$, hence $P^n(\xi) = 0$. Consequently, $\xi \in L'' \subset \mathbb{R}_I$, i.e. $x^{\overline{I}} = 0$. Thus $\xi = (0, 0, p_I, 0)$. Since the projection $P^I : L \to \mathbb{R}^I$ is an epimorphism, for any set of coordinates x^I there is a vector η such that $P^I(\eta) = x^I$. Then $\eta = \left(x^I, x^{\overline{I}}, p_I, p_{\overline{I}}\right)$. We calculate the value of the bilinear form $<,>$ on the pair of vectors ξ, η (see Eq. (1.2)):

$$\langle \xi, \eta \rangle = \sum_{k \in I} p_k x^k \ .$$

Since for a Lagrangian plane L the value $<,>$ is identically zero, we obtain the identity

$$\sum_{k \in I} p_k x^k \equiv 0$$

in the variables x^k, from which it follows that $p_k = 0$, $k \in I$. Thus $\xi = 0$, i.e. $\operatorname{Ker} P^I_{\overline{I}} = 0$. This proves Lemma 1.1. □

It follows from Lemma 1.1 that for any Lagrangian manifold $\varphi : M \to \Phi_{\mathbb{R}}(2n)$ and any point $m \in M$ there exists some neighborhood $U \ni m$ and a collection of indices I such that the functions $(x^I, p_{\overline{I}})$ can serve as local coordinates – the coordinates of the point m in phase space $\Phi_{\mathbb{R}}(2n)$. The remaining coordinates $\left(x^{\overline{I}}, p_I\right)$, consequently, are functions in the variables $(x^I, p_{\overline{I}})$:

$$\begin{aligned} x^{\overline{I}} &= x^{\overline{I}}\left(x^I, p_{\overline{I}}\right) \ , \\ p_I &= p_I\left(x^I, p_{\overline{I}}\right) \ . \end{aligned} \tag{1.4}$$

We will call a chart U on which one can take the coordinates $(x^I, p_{\overline{I}})$ as a local coordinate system a *canonical chart* and denote it by U_I. Then by Lemma 1.1 it follows that on any Lagrangian manifold M there exists an atlas $\{U_I\}$ of *canonical charts.*

Analogously to formulas (1.3) let us try to find a function $S_I\left(x^I, p_{\overline{I}}\right)$ which satisfies the equations

$$\frac{\partial S_I}{\partial x^I} = p_I \ , \quad \frac{\partial S_I}{\partial p_{\overline{I}}} = -x^{\overline{I}} \ . \tag{1.5}$$

The system (1.5) is equivalent to the equation in terms of differential forms

$$dS_I = p_I dx^I - x^{\overline{I}} dp_{\overline{I}} \; . \tag{1.6}$$

Here

$$p_I dx^I = \sum_{k \in I} p_k x^k \; ,$$
$$x^{\overline{I}} dp_{\overline{I}} = \sum_{l \in I} x^l p_l \; .$$

The left and right-hand sides of (1.6) are functions in the variables $(x_I, p_{\overline{I}})$. Since the functions $p_I, x^{\overline{I}}$ on the right-hand side are taken from (1.4) as the coordinates of the point m of the Lagrangian manifold M, we can, by considering the function S as being defined in the neighborhood $U \ni m$ of the manifold M, view (1.6) as an equation on differential forms on the manifold M in the neighborhood U:

$$dS_I = \varphi^* \left(p_I dx^I - x^{\overline{I}} dp_{\overline{I}} \right) \; . \tag{1.7}$$

The form

$$p_I dx^I - x^{\overline{I}} dp_{\overline{I}} \tag{1.8}$$

is already defined on the entire phase space $\Phi_{\mathbb{R}}(2n)$. Forms of the type (1.8) are associated by the relationship

$$p_I dx^I - x^{\overline{I}} dp_{\overline{I}} = p\, dx - d\left(p_{\overline{I}} x^{\overline{I}} \right) \; .$$

Therefore on a Lagrangian manifold the right-hand side of (1.7) is a closed differential form and, consequently, Eq. (1.7) is solvable in a sufficiently small neighborhood U of the point $m \in M$. Thus every Lagrangian manifold can locally by represented in the form (1.5).

We consider now, besides a Lagrangian manifold M in the phase space $\Phi_{\mathbb{R}}(2n)$, a smooth function $H(x,p)$ on $\Phi_{\mathbb{R}}(2n)$. With each such function H one can associate a vector field $V(H)$ on $\Phi_{\mathbb{R}}(2n)$, the dual of the differential dH relative to the form $dp \wedge dx$:

$$(dp \wedge dx)(X, V(H)) = dH(X) \; . \tag{1.9}$$

The function H is called a Hamiltonian function, and the vector field $V(H)$ is the Hamiltonian vector field. If

$$V(H) = \sum_k \left(\alpha^k \frac{\partial}{\partial x^k} + \beta_k \frac{\partial}{\partial p_k} \right) \; ,$$

and X is one of the vector fields $\partial/\partial x^l$, $\partial/\partial p_l$, then we obtain from (1.9)

$$\beta_l = -\frac{\partial H}{\partial x^l} , \quad \alpha^l = \frac{\partial H}{\partial p_l} ,$$

i.e.

$$V(H) = \sum \frac{\partial H}{\partial p_l}\frac{\partial}{\partial x^l} - \frac{\partial H}{\partial x^l}\frac{\partial}{\partial p_l} .$$

Suppose that the manifold M lies on a level surface of the function H:

$$H|M \equiv \text{const.}$$

Then for any vector field on the surface M

$$dH(X) \equiv 0 ,$$

i.e. at any point $m \in M$ the property

$$\langle V(H)_m, X_m \rangle = 0$$

holds on the tangent space $L = T_m(M)$, for any vector $X_m \in L$. If the vector $V(H)_m$ did not belong to the tangent space L, then in a suitable system of coordinates the bilinear form $\langle \, , \rangle$ would define a matrix for which some $(n+1) \times (n+1)$ minor would consist of null elements, i.e. the matrix of the bilinear form $\langle \, , \rangle$ would be degenerate, which contradicts (1.2). Hence

$$V(H)_m \in T_m(M) ,$$

i.e. the vector field $V(H)$ is tangent to the manifold M or, which is the same, the manifold M is invariant relative to the vector field $V(H)$.

Conversely, if the vector field $V(H)$ is tangent to a manifold M, then for any vector field X on the Lagrangian manifold M we obtain from (1.9) the equation

$$dH(X) \equiv 0 ,$$

i.e. $(dH)_M \equiv 0$, $H|M \equiv \text{const.}$

We have shown the following assertion.

Lemma 1.2. *The function H is constant on a Lagrangian manifold M if and only if the Hamiltonian vector field $V(H)$ is tangent to M.*

2.2 The Cohomology of the Lagrangian Grassmannian

The Lagrangian planes in phase space $\Phi_{\mathbb{R}}(2n)$ can be organized into a special manifold – the Lagrangian Grassmannian $G_{2n}^{\mathbb{R}}(I_n)$. The usefulnes of such a manifold is due to the fact that any Lagrangian manifold $\varphi : M \to \Phi_{\mathbb{R}}(2n)$ can be mapped naturally into the Lagrangian Grassmannian

$G_{2n}^{\mathbb{R}}(I_n)$. Indeed, each point $m \in M$ can be placed into correspondence with its tangent space $T_m(M) \in G_{2n}^{\mathbb{R}}(I_n)$. The map $M \to G_{2n}^{\mathbb{R}}(I_n)$ obtained in this way helps us to explain more clearly the geometric nature of the various characteristic classes on the manifold, one of which is the so-called Maslov index of a Lagrangian manifold M. For convenience we will first study the Lagrangian Grassmannian of the complex phase space $\Phi_{\mathbb{C}}(2n)$.

Suppose $\Phi_{\mathbb{C}}$ is a $2n$-dimensional complex space, on which we fix a complex basis of vectors

$$\{e, f\} = \{e_1, \ldots, e_n, f^1, \ldots, f^n\} .$$

We call the space $\Phi_{\mathbb{C}}$ a *complex phase space*. Any vector $a \in \Phi_{\mathbb{C}}$ can be uniquely represented in the form of a linear combination of the basis vectors $\{e, f\}$ with complex coefficients

$$a = \sum_k z^k e_k + \sum_j \zeta_j f^j .$$

We will call the numbers z^k, $k = 1, \ldots, n$, the *z-coordinates*, and the numbers ζ_j, $j = 1, \ldots, n$ the *ζ-coordinates* of the vector a of the phase space $\Phi_{\mathbb{C}}$.

We set

$$z^k = x^k + iy^k , \quad \zeta_j = p_j + i\eta_j , \quad j, k = 1, \ldots, n ,$$

where x^k, y^k, p_j, η_j are real numbers.

The real linear subspace generated by the vectors $\{e, f\}$ forms a $2n$-dimensional real subspace of $\Phi_{\mathbb{C}}$. We will denote it by $\Phi_{\mathbb{R}}$. Thus if $a \in \Phi_{\mathbb{R}}$, then

$$a = \sum_k x^k e_k + \sum_J p_j f^j .$$

In this case the numbers x^k will be called *x-coordinates*, and p_j the *p-coordinates* of the vector $a \in \Phi_{\mathbb{R}}$. In the complex phase space $\Phi_{\mathbb{C}}$ we consider the skew-symmetric complex-valued form $\langle a, b\rangle$ given in the basis $\{e, f\}$ by the skew-symmetric matrix

$$I_n = \begin{pmatrix} 0 & E \\ -E & 0 \end{pmatrix} ,$$

where E is the identity matrix of dimension $(n \times n)$. Furthermore, we write

$$\langle a, b\rangle = a^t I_n b ,$$

where by a, b we will also denote the column vectors composed of the coordinates of the vectors a, b in the basis $\{e, f\}$; the operation $X \to X^t$ is the transposition of a complex matrix. Thus, if

$$a^t = (z^1, \ldots, z^n, \zeta_1, \ldots, \zeta_n) ,$$
$$b^t = \left(z'^1, \ldots, z'^n, \zeta'_1, \ldots, \zeta'_n\right) ,$$

then

$$\langle a, b\rangle = \sum_k z^k \zeta'_k - \sum_j \zeta_j z'^j .$$

It is obvious that the identity

$$\langle a, b\rangle = -\langle b, a\rangle$$

holds.

Definition 2.1. A complex subspace $L \subset \Phi_{\mathbb{C}}$ is called a *Lagrangian plane* if:

a) $\dim_{\mathbb{C}} L = n$;
b) $\langle a, b\rangle = 0$ for any pair of vectors $a, b \in L$.

Proposition 2.1. *$L \subset \Phi_{\mathbb{C}}$ be a Lagrangian plane, $\{a_1, \ldots, a_n\}$ be an arbitrary complex basis of the space L, X be the $(n \times 2n)$ matrix consisting of the coordinates of the vectors $a_1, \ldots, a_n$ in terms of the basis $\{e, f\}$ of phase space $\Phi_{\mathbb{C}}$, i.e. $X = (a_1, \ldots, a_n)$. Then*

$$X^t I_n X = 0 ,$$

where 0 is the null matrix of dimension $(n \times n)$.

The proof is obvious.

An example of a Lagrangian plane is the subspace $\mathbb{C}^n \subset \Phi_{\mathbb{C}}$ generated by the vectors $e = \{e_1, \ldots, e_n\}$. Indeed, the vectors $e_1, \ldots, e_n$ form a basis in the subspaces $\mathbb{C}_n$, and the corresponding matrix X of their coordinates has the form

$$X = \begin{pmatrix} E \\ 0 \end{pmatrix} .$$

Then

$$X^t I_n X = (E, 0) \begin{pmatrix} 0 & E \\ -E & 0 \end{pmatrix} \begin{pmatrix} E \\ 0 \end{pmatrix} = (0, E) \begin{pmatrix} E \\ 0 \end{pmatrix} = 0 .$$

Let $C : \Phi_{\mathbb{C}} \to \Phi_{\mathbb{C}}$ be an invertible linear (complex) transformation of the phase space $\Phi_{\mathbb{C}}$. The corresponding $(2n \times 2n)$ matrix in the basis $\{e, f\}$ will also be denoted by C.

Definition 2.2. The transformation C is called *symplectic* or a *Hamiltonian transformation* if it preserves the antisymmetric form, i.e. if

$$\langle a, b\rangle = \langle Ca, Cb\rangle \ ,$$

or, equivalently, if

$$C^t I_n C = I_n \ .$$

It is easy to see that if $L \subset \Phi_{\mathbb{C}}$ is a Lagrangian plane, and C is a Hamiltonian transformation, then the space $C(L)$ is also a Lagrangian plane.

The set of all Hamiltonian transformations forms a group relative to the operation of composition of functions. We will denote this group by $GL(2n, I_n)$.

Proposition 2.2. *Suppose $L \subset \Phi_{\mathbb{C}}$ is a Lagrangian plane. Then there exists a Hamiltonian map $C \in GL(2n, I_n)$ such that*

$$L = C(\mathbb{C}_n) \ .$$

Proof. We introduce a Hermitian metric on the complex phase space, considering the basis $\{e, f\}$ to be orthonormal. Thus, if the vectors $a, b \in \Phi_{\mathbb{C}}$ have coordinates

$$\left(z^1, \ldots, z^n, \zeta_1, \ldots, \zeta_n\right) \ , \quad \left(z'^1, \ldots, z'^n, \zeta'_1, \ldots, \zeta'_n\right)$$

respectively, then their scalar product is defined by the formula

$$(a, b) = \sum_k \overline{z}^k z'^k + \sum_j \overline{\zeta}_j \zeta'_j \ .$$

Now suppose $L \subset \Phi_{\mathbb{C}}$ is an arbitrary Lagrangian plane. Then there exists an orthonormal system of vectors $\{a_1, \ldots, a_n\} \subset L$ which form a complex basis of the space L.

First, we formulate the following obvious result.

Lemma 2.3. *Suppose the complex-valued matrix*

$$X = \begin{pmatrix} A \\ B \end{pmatrix}$$

of dimension $(n \times 2n)$ is such that its columns define an orthonormal system of vectors $\{h_1, \ldots, h_n\}$ of the space $\Phi_{\mathbb{C}}$. Then

$$A^* A + B^* B = E \ , \tag{2.1}$$

where $A^* = \overline{A}^t$.

Now we conclude the proof of Proposition 2.2. We choose an orthonormal basis of the space L; suppose the coordinates of this basis form a matrix

$$X = \begin{pmatrix} A \\ B \end{pmatrix} .$$

We set

$$C = \begin{pmatrix} A & -\overline{B} \\ B & \overline{A} \end{pmatrix} .$$

Then

$$C^t I C = \begin{pmatrix} A^t & B^t \\ -\overline{B}^t & \overline{A}^t \end{pmatrix} \begin{pmatrix} 0 & E \\ -E & 0 \end{pmatrix} \begin{pmatrix} A & -\overline{B} \\ B & \overline{A} \end{pmatrix}$$

$$= \begin{pmatrix} -B^t & A^t \\ -A^* & -B^* \end{pmatrix} \begin{pmatrix} A & -\overline{B} \\ B & \overline{A} \end{pmatrix} = \begin{pmatrix} A^t B - B^t A & B^t \overline{B} + A^t \overline{A} \\ -(A^* A + B^* B) & A^* \overline{B} - B^* \overline{A} \end{pmatrix} .$$

Since the space L is a Lagrangian plane, according to Proposition 2.1,

$$X^t I_n X = 0$$

or

$$A^t B - B^t A = 0 , \tag{2.2}$$

from which we obtain

$$A^* \overline{B} = B^* \overline{A} = 0 . \tag{2.3}$$

From (2.1) we obtain

$$B^t \overline{B} + A^t \overline{A} = E . \tag{2.4}$$

Thus, using Eqs. (2.1), (2.2), (2.3), (2.4), we obtain

$$C^t I_n C = \begin{pmatrix} 0 & E \\ -E & 0 \end{pmatrix} = I_n .$$

We have established that the matrix C determines a Hamiltonian transformation of the phase space $\Phi_{\mathbb{C}}$.

The equation $L = C(\mathbb{C}^n)$ can be seen readily from the definition of the matrix C. This proves Proposition 2.2. □

Proposition 2.2 can be interpreted in the following way, which will be useful to us. Let $G_{2n}(I_n)$ denote the set of all Lagrangian planes in phase space $\Phi_{\mathbb{C}}$. Then there is a natural map

$$\Theta : GL(2n, I_n) \to G_{2n}(I_n) ,$$

which assigns to each Hamiltonian transformation $C \in GL(2n, I_n)$ the Lagrangian plane

$$\Theta(C) = C(\mathbb{C}^n) .$$

Proposition 2.2 asserts that the map Θ is an epimorphism. Now let us study the preimages of individual points under the map Θ. We denote the subgroup of $GL(2n, I_n)$ consisting of all Hamiltonian transformations C which map the subspace $\mathbb{C}^n \subset \Phi_{\mathbb{C}}$ into itself by H.

The subgroup H consists of matrices of the form

$$C = \begin{pmatrix} A & B \\ 0 & (A^t)^{-1} \end{pmatrix} , \quad A \in GL(n) , \quad (A^{-1}B) = (A^{-1}B)^t .$$

The preimage of every point of the map Θ coincides with some left conjugacy class of the subgroup H in the group $GL(2n, I_n)$, i.e. the points of the set $G_{2n}(I_n)$ are in one-to-one correspondence with the points of the homogeneous space of all left conjugacy classes of

$$GL(2n, I_n)/H .$$

We note that the proof of Proposition 2.2 actually contains a more powerful assertion. We denote by $U(2n, I_n)$ the subgroup of $GL(2n, I_n)$ consisting of all Hamiltonian transformation which are also unitary transformations. If the matrix

$$C = \begin{pmatrix} A & -\overline{B} \\ B & \overline{A} \end{pmatrix}$$

determines a Hamiltonian transformation and $A^*A + B^*B = E$, then C is a unitary matrix.

Proposition 2.4. *The space $G_{2n}(I_n)$ is isomorphic to the homogeneous space of left cosets of the subgroup H' in the group $U(2n, I_n)$. The subgroup H' is isomorphic to the group $U(n)$ of unitary matrices and consits of matrices of the form*

$$C = \begin{pmatrix} A & 0 \\ 0 & \overline{A} \end{pmatrix} , \quad A \in U(n) .$$

Proof. The first assertion is obvious. Suppose a unitary matrix C has the form

$$C = \begin{pmatrix} A & B \\ 0 & (A^t)^{-1} \end{pmatrix} .$$

using the condition $C^*C = \begin{pmatrix} E & 0 \\ 0 & E \end{pmatrix}$, we obtain $A^*A = E$, $B = 0$. This proves Proposition 2.4. □

In order to proceed to a real phase space $\Phi_{\mathbb{R}}$, one must consider the space $\Phi_{\mathbb{R}}$ to be a subspace of $\Phi_{\mathbb{C}}$, generated by the basis $\{e, f\}$ over the field of real numbers. The form $\langle\,,\rangle$ takes real values on real vectors, i.e. vectors in the subspace $\Phi_{\mathbb{R}}$.

Definition 2.3. Let $\tau : \Phi_{\mathbb{C}} \to \Phi_{\mathbb{C}}$ be the transformation which satisfies the conditions $\tau(e_k) = e_k$, $\tau(f^j) = f^j$, $\tau(\lambda x) = \overline{\lambda}\tau(x)$.

Lemma 2.5. *If $L \subset \Phi_{\mathbb{C}}$ is a Lagrangian plane, then $\tau(L)$ is also a Lagrangian plane.*

Proof. Let $x, y \in \Phi_{\mathbb{C}}$; then $\langle \tau(x), \tau(y)\rangle = \overline{\langle x, y\rangle}$. Consequently, if $x, y \in \tau(L)$, then $\langle x, y\rangle = \langle \tau^2 x, \tau^2 y\rangle = \overline{\langle \tau x, \tau y\rangle} = 0$, since $\tau x \in L$. □

Definition 2.4. A Lagrangian plane $L \subset \Phi_{\mathbb{C}}$ is called a *real Lagrangian plane* if $\tau(L) = L$.

Proposition 2.6. *$L \subset \Phi_{\mathbb{C}}$ is a real Lagrangian plane, then:*

a) *$L \cap \Phi_{\mathbb{R}} = L^\tau$, where L^τ denotes the subspace of vectors fixed under the transformation τ;*

b) *$L \cap \Phi_{\mathbb{R}}$ is a Lagrangian plane in $\Phi_{\mathbb{R}}$;*

c) *$L = (L \cap \Phi_{\mathbb{R}}) \oplus i(L \cap \Phi_{\mathbb{R}})$.*

Proof. Assertion (a) is obvious. To show assertion (b) it is sufficient to show that $\dim_{\mathbb{R}}(L \cap \Phi_{\mathbb{R}}) = n$. If $x \in L^\tau$, then $\tau(ix) = -i\tau x = -ix$. Consequently, $L^\tau \cap iL^\tau = 0$.

On the other hand, if $x \in L$, then

$$y = \frac{x + \tau x}{2} \in L^\tau , \quad z = \frac{x - \tau x}{2i} \in L^\tau ,$$

i.e. $x = y + iz, y, z \in L^\tau$.

Thus $L = L^\tau + iL^\tau$. We have shown assertion (c). Since multiplication by the number i is invertible,

$$\dim_{\mathbb{R}} L^\tau = \dim_{\mathbb{R}} iL^\tau ,$$

i.e.

$$\dim_{\mathbb{R}} L = 2\dim_{\mathbb{R}} L^\tau = 2\dim_{\mathbb{C}} L = 2n .$$

Hence $\dim_{\mathbb{R}} L^\tau = n$. This proves Proposition 2.6 completely. □

Corollary. *The set of Lagrangian planes in the real phase space $\Phi_{\mathbb{R}}$ is in one-to-one correpondence with the set of real Lagrangian planes in the space $\Phi_{\mathbb{C}}$.*

We denote the set of real Lagrangian planes by $G_{2n}^{\mathbb{R}}(I_n)$. As we established in Proposition 2.4, it is possible to represent the manifold of Lagrangian planes in the form of a homogeneous space $U(2n, I_n)/U(n)$.

We denote by $O(2n, I_n)$ the subgroup $O(2n, I_n) = U(2n, I_n) \cap O(2n)$, where $O(2n)$ is the group of orthogonal matrices. As before, let $\Theta : U(2n, I_n) \to G_{2n}(I_n)$ be the projection.

Lemma 2.7.
(a) $\Theta(O(2n, I_n)) = G_{2n}^{\mathbb{R}}(I_n)$;
(b) $O(2n, I_n) \cap U(n) = O(n)$;
(c) $G_{2n}^{\mathbb{R}}(I_n) = O(2n, I_n)/O(n)$;
(d) *the group* $O(2n, I_n)$ *is isomorphic to the group* $U(n)$, *and the imbedding* $O(n) \subset O(2n, I_n)$ *corresponds under the imbedding to the natural imbedding of the real matrices into the complex.*

Proof. Assertions (a), (b) and (c) follow instantly from Proposition 2.6. We will show assertion (d). Let C be a real Hamiltonian transformation, $C \in O(2n)$. The condition of being Hamiltonian means that

$$C^t I_n C = I_n \ .$$

Since $C \in O(2n)$, $C^t = C^{-1}$, i.e. the relation $C^t I_n C = I_n$ implies the following identity:

$$I_n C = C I_n \ .$$

Thus any matrix $C \in O(2n, I_n)$ gives an orthogonal transformation of the space $\Phi_{\mathbb{R}}$ which commutes with the transformation I_n. Since the transformation I_n is orthogonal and $I_n^2 = -E$, it is possible to define a structure of an n-dimensional complex space on the space $\Phi_{\mathbb{R}}$, setting

$$(\lambda_1 + i\lambda_2)x = \lambda_1 x + \lambda_2 I_n(x) \ ,$$

under which definition the Euclidean metric on $\Phi_{\mathbb{R}}$ becomes a Hermitian metric on this complex space. Consequently, any transformation $C \in O(2n, I_n)$ is a unitary transformation of the complex space $(\Phi_{\mathbb{R}}, I_n)$. Conversely, if C is some unitary transformation of the space $(\Phi_{\mathbb{R}}, I_n)$, then
a) C is an orthogonal transformation;
b) $CI_n = I_n C$.

Let C, I_n denote the matrices of the transformation with respect to the basis (e, f) in the space $\Phi_{\mathbb{R}}$. Then the condition $CI_n = I_n C$ gives the identity

$$C^t I_n C = I_n \ .$$

Consequently, $C \in O(2n, I_n)$. In order to show the second part of assertion (d), one must convince oneself that if $C = A + iB$ is the matrix of a complex transformation in the basis $(e_1, \ldots, e_n)$, then the matrix of the same transformation in the basis$(e_1, \ldots, e_n, ie_1, \ldots, ie_n)$ has the form

$$\begin{pmatrix} A & -B \\ B & A \end{pmatrix} .$$

This proves Lemma 2.7 in entirety. □

In the future we will have use for the set of real Lagrangian planes with a assigned orientation. Thus the points of this new set will be pairs (L, ϵ), where $L \subset \Phi_{\mathbb{R}}$ is a Lagrangian plane in a real phase space, and ϵ is some orientation on the space, i.e. a class of bases, for any two of which the change of basis matrix has positive determinant. We denote this set by $G_{2n}^{SO}(I_n)$. By ignoring orientation we obtain a natural mapping

$$\pi : G_{2n}^{SO}(I_n) \to G_{2n}^{\mathbb{R}}(I_n) \ ,$$

which is a two-sheeted covering.

Lemma 2.8. *The space $G_{2n}^{SO}(I_n)$ is homeomorphic to the homogeneous space $U(n)/SO(n)$, where $SO(n) \subset O(n)$ is the subgroup of orthogonal matrices with determinant equal to* 1. *Under the homeomorphism the following diagram of bundles commutes:*

$$\begin{array}{ccccc} O(n) & \to & U(n) & \to & G_{2n}^{\mathbb{R}}(I_n) \\ \uparrow & & \uparrow & & \uparrow \\ SO(n) & \to & U(n) & \to & G_{2n}^{SO}(I_n) \end{array}$$

Lemma 2.9. $H_1\left(G_{2n}^{SO}(I_n)\right) = \mathbb{Z}$.

The calculation of the one-dimensional cohomology group $H^1\left(G_{2n}^{SO}(I_n)\right)$ is based on a regular application of exact homotopy sequences for bundles. First of all, it is necessary to convert the problem to homology groups, by using the equality

$$H^1\left(G_{2n}^{SO}(I_n)\right) = \operatorname{Hom}\left(H_1\left(G_{2n}^{SO}(I_n)\right), \mathbb{Z}\right) \ ,$$

and, by the Hurewicz theorem, to the fundamental group $\pi_1\left(G_{2n}^{SO}(I_n)\right)$:

$$H_1\left(G_{2n}^{SO}(I_n)\right) = \frac{\pi_1\left(G_{2n}^{SO}(I_n)\right)}{[\pi_1, \pi_1]} \ ,$$

where $[\,,]$ is the commutator of the given fundamental group. Then, applying the exact homotopy sequence to the lower line of the diagram of Lemma 2.8, we obtain the following sequence:

$$\pi_1(SO(n)) \to \pi_1(U(n)) \to \pi_1\left(G_{2n}^{SO}(I_n)\right) \to \pi_0(SO(n)) \ .$$

Thus it is necessary as a preliminary to compute the fundamental groups of the spaces $SO(n)$, $U(n)$, and also observe that $\pi_0(SO(n)) = 0$.

For the group $U(n)$ we again apply the exact homotopy sequence for the bundle

$$U(n-1) \to U(n) \to S^{2n-1} \ .$$

Here the map $U(n) \to S^{2n-1}$ assigns to each unitary matrix its first column vector. Analogously, for the group $SO(n)$ we use the fibration

$$SO(n-1) \to SO(n) \to S^{n-1} \ .$$

By simple arguments one can show that

$$\pi_1 (U(1)) = \ldots = \pi_1 (U(n)) \ ,$$
$$\pi_1 (SO(3)) = \ldots = \pi_1 (SO(n)) \ .$$

The groups $U(1)$ and $SO(3)$ are easy to describe. The first is the circle S^1, and the second is homeomorphic to 3-dimensional projective space. Consequently,

$$\pi_1 (U(n)) = \mathbb{Z} \ , \quad n \geq 1 \ ; \qquad \pi_1 (SO(n)) = \mathbb{Z}_2 \ , \quad n \geq 3 \ .$$

Moreover, $SO(2) = S^1$, $\pi_1 (SO(2)) = \mathbb{Z}$; $SO(1) = 1$, $\pi_1 (SO(1)) = 0$.

From here it follows instantly that

$$\pi_1 \left(G_{2n}^{SO} (I_n)\right) = \mathbb{Z} \ ,$$

and that the natural imbedding

$$G_{2n-2}^{SO} (I_n) \to G_{2n}^{SO} (I_n)$$

induces an isomorphism of fundamental groups. In particular, the space $G_2^{SO} (I_1)$ is homeomorphic to the circle S^1.

It is useful to describe the generator of the group $H^1 \left(G_{2n}^{SO} (I_n)\right)$ in some universal way. For this purpose we consider the mapping $\det : U(n) \to S^1$. This map decomposes into a composition

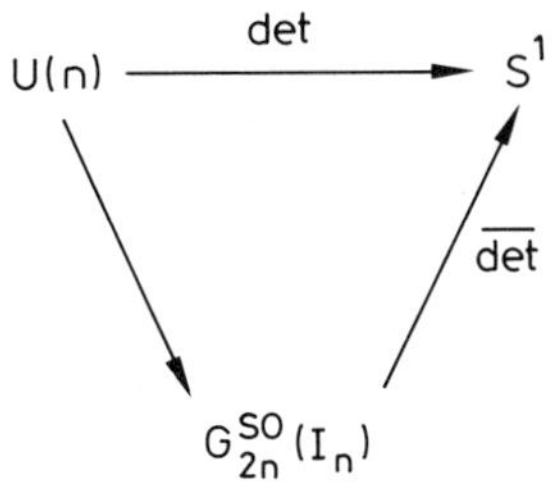

When $n = 1$, the map det is a homeomorphism, hence

$$\overline{\det}^* : H^1 (S^1) \to H^1 \left(G_{2n}^{SO} (I_n)\right)$$

is an isomorphism. Thus the generating element $U_n \in H^1\left(G_{2n}^{SO}(I_n)\right)$ can be represented as the preimage of the differential form $d\varphi \in H^1\left(S^1\right) : U_n = \overline{\det}^*(d\varphi)$.

2.3 Characteristic Classes of Lagrangian Manifolds

During the construction of Maslov's canonical operator on a Lagrangian manifold $\varphi : M \to \Phi_{\mathrm{IR}}(2n)$, which will be carried out in Chap. 3, several restrictions will be imposed on the manifold.

Definition 3.1. Suppose $\varphi : M \to \Phi_{\mathrm{IR}}(2n)$ is a Lagrangian manifold in phase space $\varphi_{\mathrm{IR}}(2n)$, and $\{U_i\}$ is an atlas of canonical charts on the manifold M. We introduce the following notation for two collections of indices I and J:

$$I_1 = I \cap J\ , \quad I_2 = I \setminus J\ , \quad I_3 = J \setminus I\ ,$$
$$I_4 = \{1,\ \ldots,\ n\} \setminus (I \cup J)\ .$$

A Lagrangian manifold M is said to be quantized if the following two conditions are satisfied (the Maslov conditions of quantization):

(a) In each chart U_I there exists a function S_I satisfying the equation

$$dS_I = p_I dx^I - x^{\overline{I}} dp_{\overline{I}}\ ,$$

and on the intersection of two charts U_I and V_J the equation

$$S_I - S_J = p_{I_2} x^{I_2} - p_{I_3} x^{I_3} \tag{3.1}$$

holds.

(b) There exists a collection of whole numbers τ_I, indexed by the canonical charts, such that for every intersection of U_I and V_J the relation

$$\tau_I - \tau_J = \operatorname{sign} \frac{\partial\left(x^{I_2}, -p_{I_3}\right)}{\partial\left(p_{I_2}, x^{I_3}\right)} \pmod 8\ . \tag{3.2}$$

holds. We will show the Maslov conditions of quantization can be interpreted as relations on certain cohomology classes of the Lagrangian manifold M.

First we consider the first Maslov condition of quantization (3.1). If condition (3.1) holds on the manifold M, then the functions

$$\Phi_J = S_J + x^{\overline{J}} p_{\overline{J}}\ , \tag{3.3}$$

defined on each of the respective charts V_J, assume identical values on pairwise intersections. Indeed, for two charts U_I and V_J we have

$$\Phi_J = S_J + x^{I_2} p_{I_2} + x^{I_4} p_{I_4} ,$$
$$\Phi_I = S_I + x^{I_3} p_{I_3} + x^{I_4} p_{I_4} .$$

Then

$$\Phi_J - \Phi_I = S_J - S_I + p_{I_2} x^{I_2} - p_{I_3} x^{I_3} \equiv 0 .$$

Consequently there exists a general smooth function S on the entire Lagrangian manifold M which coincides with Φ_I on each chart U_I:

$$S|U_I = \Phi_I .$$

Moreover,

$$dS = d\Phi_I = p\,dx \tag{3.4}$$

Thus, if condition (3.1) holds, then the cohomology class $[p\,dx]$ determined by the one-dimensional closed form $p\,dx$ is trivial:

$$[p\,dx] = 0 \in H^1(M) . \tag{3.5}$$

Conversely, if condition (3.5) holds, then there exists a smooth function S satisfying condition (3.4). Then we find the functions S_I by formulas (3.3). Hence the first Maslov condition of quantization is equivalent to the triviality of a certain one-dimensional cohomology class $[p\,dx] \in H^1(M)$ on the Lagrangian manifold M.

We now study the second Maslov condition of quantization (3.2). First of all we note that condition (3.2), translated into the language of cohomology, means that the one-dimensional integer-valued cochain

$$C_{I,J} = \operatorname{sign} \frac{\partial \left(x^{I_2}, -p_{I_3}\right)}{\partial \left(p_{I_2}, x^{I_3}\right)} , \tag{3.6}$$

defined on coverings by canonical charts $\{U_I\}$ is a cocycle (mod 8) and is cohomologous to zero (mod 8). Thus the second Maslov condition of quantization of a Lagrangian manifold can be interpreted as the triviality of a certain one-dimensional cohomology class in the group $H^1(M, \mathbb{Z}_8)$, called the Maslov index of the Lagrangian manifold.

The value of the cochain (3.6) depends only upon the positioning in the phase space $\Phi_{\mathbb{R}}(2n)$ of the Lagrangian tangent plane $T_m(M) \subset \Phi_{\mathbb{R}}(2n)$ at the point $m \in M$. Consequently, cochain (3.6) is the preimage of some universal one-dimensional integral cochain of the Lagrangian Grassmannian $G_{2n}^{SO}(I_n)$, given by a certain covering of the Grassmannian $G_{2n}^{SO}(I_n)$. For this we consider open sets $W_J \subset G_{2n}^{SO}(I_n)$ consisting of those Lagrangian planes $L \in G_{2n}^{SO}(I_n)$ for which the projection

$$P_{\overline{J}}^{J} : L \to \mathbb{R}^J \times \mathbb{R}_{\overline{J}}$$

is an isomorphism. According to Lemma 1.1 the family of open sets $\mathfrak{W} = \{W_J\}$ covers the Grassmannian $G_{2n}^{SO}(I_n)$:

$$\bigcup_J W_J = G_{2n}^{SO}(I_n) \ .$$

We define a one-dimensional integral cochain on $\mathfrak{W}$,

$$a_{I,J} = \operatorname{sign} \frac{\partial\left(x^{I_2}, -p_{I_3}\right)}{\partial\left(p_{I_2}, x^{I_3}\right)} \ , \tag{3.7}$$

which is a locally constant function on each intersection $W_I \cap W_J$. In order for the functions (3.7) to be constant, it is sufficient to replace the covering $\mathfrak{W}$ by another, finer covering. Therefore we can assume without loss of generality that (3.7) defines an integral cochain with constant functions $a_{I,J}$. Thus

$$C_{I,J} = \varphi^*\left(a_{I,J}\right) \ .$$

It turns out that the cochain $a_{I,J}$ is an integral cocycle, and the one-dimensional cohomology class $[a_{I,J}] \in H^1\left(G_{2n}^{SO}(I_n)\right)$ defined by the cochain $a_{I,J}$ is equal to the fourth power of the generator of the group $H^1\left(G_{2n}^{SO}(I_n)\right) = \mathbb{Z}$. Thus the one-dimensional cohomology class $[C_{I,J}] \in H^1(M, \mathbb{Z}) \subset H^1(M, \mathbb{R})$ can be represented by a certain closed differential 1-form $\omega \in \Omega^1(M)$. One can even write out the form explicitly in any arbitrary local system of coordinates $\left(\alpha^1, \ldots, \alpha^n\right)$ of the manifold M. To do this one must construct a map

$$M \xrightarrow{d\varphi} G_{2n}^{SO}(I_n) \xrightarrow{\overline{\det}} S^1 \subset \mathbb{C} \tag{3.8}$$

from the imbedding functions $x = x(\alpha)$, $p = p(\alpha)$.

If $\alpha = \left(\alpha^1, \ldots, \alpha^n\right)$ is a local system of coordinates, the columns of the matrix

$$\begin{pmatrix} \frac{\partial x}{\partial \alpha} \\ \frac{\partial p}{\partial \alpha} \end{pmatrix}$$

form a basis of the tangent space $T_m(M)$ to the manifold M. By a linear change of coordinates (defined by some matrix C) we replace this basis by an orthonormal basis, whose matrix has the form

$$\begin{pmatrix} \frac{\partial x}{\partial \alpha} C \\ \frac{\partial p}{\partial \alpha} C \end{pmatrix} .$$

Then the matrix $\left(\frac{\partial x}{\partial \alpha} + i \frac{\partial p}{\partial \alpha}\right) \cdot C$ is unitary, and the image of the point $m \in M$ under the map (3.8) equals

$$\det\left(\left(\frac{\partial(x+ip)}{\partial \alpha}\right) \cdot C\right) \in S^1 \subset \mathbb{C} \ .$$

The generator of the group $H^1(S^1)$ can be represented by a 1-form $(2\pi)^{-1}d\varphi$, where φ is the argument of a complex number in $S^1 \subset \mathbb{C}$. Consequently,

$$\omega = \frac{4}{2\pi i} d \ln \left(\det \left(\frac{\partial(x+ip)}{\partial\alpha} \right) \cdot C \right) . \tag{3.9}$$

To get rid of the indefinite matrix C, we note that

$$\left| \det \left(\frac{\partial(x+ip)}{\partial\alpha} \right) \cdot C \right| = 1 ,$$

i.e.

$$\left| \det C \cdot \det \left(\frac{\partial(x+ip)}{\partial\alpha} \right) \right| = 1 .$$

Since $\det C > 0$, we have

$$\det C = \left| \det \left(\frac{\partial(x+ip)}{\partial\alpha} \right) \right|^{-1} . \tag{3.10}$$

Substituting (3.10) into (3.9), we finally obtain

$$\omega = \frac{2}{\pi i} d \ln \left\{ \det \left(\frac{\partial(x+ip)}{\partial\alpha} \right) \cdot \left| \det \left(\frac{\partial(x+ip)}{\partial\alpha} \right) \right|^{-1} \right\} . \tag{3.11}$$

Thus the second Maslov condition of quantization can be written as a series of conditions

$$\oint_{\nu} \omega \equiv 0 \,(\mathrm{mod}\, 8)$$

for any closed contour γ in the manifold M.

We now prove that the integral cochain (3.7) is an integral cocyle. We have to show that for three charts W_I, W_J, W_K the equation

$$a_{I,J} + a_{J,K} + a_{K,I} = 0 \tag{3.12}$$

holds. In particular, if $K = \overline{n}$, it is necessary that

$$a_{I,J} + a_{J,\overline{n}} + a_{\overline{n},I} = 0 . \tag{3.13}$$

In fact, if Eq. (3.13) hold for any pair W_I, W_J, then all equations of the form (3.12) hold. To see this we note that the open set W_n is everywhere dense in the space $G_{2n}^{SO}(I_n)$. Consequently, using (3.13) and substituting the values of the cochains $a_{I,J}$, $a_{J,K}$, $a_{K,I}$ in (3.12), we obtain

$$-(a_{J,\overline{n}} + a_{\overline{n},I}) - (a_{K,\overline{n}} + a_{\overline{n},J}) - (a_{I,\overline{n}} + a_{\overline{n},K}) = 0 .$$

Thus it is sufficient to prove (3.13). We have

$$a_{I,J} = \operatorname{sign} \frac{\partial\left(x^{I_2}, -p_{I_3}\right)}{\partial\left(p_{I_2}, x^{I_3}\right)} ,$$

$$a_{J,\overline{n}} = \operatorname{sign} \frac{\partial x_J}{\partial p_J} , \qquad a_{\overline{n},I} = -\operatorname{sign} \frac{\partial p_I}{\partial x^I} .$$

We choose some Lagrangian plane $L \subset \Phi_{\mathbb{R}}(2n)$ in the intersection $W_I \cap W_J \cap W_n$. Since the coordinates x are a system of coordinates for vectors of L, we obtain a linear function

$$\begin{pmatrix} p_{I_1} \\ p_{I_2} \\ p_{I_3} \\ p_{I_4} \end{pmatrix} = \begin{pmatrix} p_{11} & p_{12} & p_{13} & p_{14} \\ p_{21} & p_{22} & p_{23} & p_{24} \\ p_{31} & p_{32} & p_{33} & p_{34} \\ p_{41} & p_{42} & p_{43} & p_{44} \end{pmatrix} \begin{pmatrix} x^{I_1} \\ x^{I_2} \\ x^{I_3} \\ x^{I_4} \end{pmatrix} . \tag{3.14}$$

In order to obtain the values of $a_{I,J}$, $a_{J,\overline{n}}$ and $a_{\overline{n},I}$, it is necessary to carry out the changes of coordinates explicitly. Since the values of $a_{I,J}$ are locally constant, the plane L can be chosen arbitrarily in the intersection of the three charts, so without loss of generality we can assume the nondegeneracy of various minors of the matrices (p_{ij}) in (3.14). Now, to calculate $a_{I,J}$ it is necessary to express the coordinates $\left(x^{I_1}, x^{I_2}, p_{I_3}, p_{I_4}\right)$ in terms of the coordinates $\left(p_{I_1}, p_{I_2}, x^{I_3}, x^{I_4}\right)$. After some simple cancellations we obtain

$$a_{I,J} = \operatorname{sign}$$

$$\times \begin{pmatrix} Q^{-1} & Q^{-1}\left(p_{21}p_{11}^{-1}p_{13} - p_{23}\right) \\ \left(p_{31}p_{11}^{-1}p_{12} - p_{32}\right)Q^{-1} & \left(p_{32} - p_{31}p_{11}^{-1}p_{12}\right)Q^{-1}\left(p_{23} - p_{21}p_{11}^{-1}p_{13}\right) + \\ & +p_{31}p_{11}^{-1}p_{13} - p_{33} \end{pmatrix}$$

$$= \operatorname{sign}\begin{pmatrix} Q^{-1} & 0 \\ 0 & p_{31}p_{11}^{-1}p_{13} - p_{33} \end{pmatrix} = \operatorname{sign}\begin{pmatrix} Q & 0 \\ 0 & p_{32}p_{11}^{-1}p_{13} - p_{33} \end{pmatrix} ,$$

where $Q = p_{22} - p_{21}p_{11}^{-1}p_{12}$. To calculate $a_{J,\overline{n}}$ it is necessary to express the coordinates $\left(x^{I_1}, x^{I_3}, p_{I_2}, p_{I_4}\right)$ in terms of the coordinates $\left(p_{I_1}, p_{I_3}, x^{I_2}, x^{I_4}\right)$. We have

$$\begin{pmatrix} p_{I_1} \\ p_{I_3} \end{pmatrix} = \begin{pmatrix} p_{11} & p_{13} \\ p_{31} & p_{33} \end{pmatrix}\begin{pmatrix} x^{I_1} \\ x^{I_3} \end{pmatrix} + \begin{pmatrix} p_{12} & p_{14} \\ p_{32} & p_{34} \end{pmatrix}\begin{pmatrix} x^{I_2} \\ x^{I_4} \end{pmatrix} ,$$

i.e.

$$a_{J,\overline{n}} = \operatorname{sign}\begin{pmatrix} p_{11} & p_{13} \\ p_{31} & p_{33} \end{pmatrix}^{-1} = \operatorname{sign}\begin{pmatrix} p_{11} & p_{13} \\ p_{31} & p_{33} \end{pmatrix} . \tag{3.15}$$

For $a_{\overline{n},I}$ we obtain

$$a_{n,I} = -\operatorname{sign}\begin{pmatrix} p_{11} & p_{12} \\ p_{21} & p_{22} \end{pmatrix} . \tag{3.16}$$

Reducing (3.15) and (3.16) to diagonal form, we obtain

$$a_{J,\overline{n}} = \operatorname{sign}\left(p_{33} - p_{31}p_{11}^{-1}p_{13}\right) + \operatorname{sign} p_{11} \ ,$$
$$a_{\overline{n},I} = -\operatorname{sign}\left(p_{22} - p_{21}p_{11}^{-1}p_{12}\right) - \operatorname{sign} p_{11} \ ,$$

i.e. $a_{I,J} + a_{J,\overline{n}} + a_{\overline{n},I} = 0$.

Since in the definition of the cocycle $a_{I,J}$ on each intersection $W_I \cap W_J$ only those indices l appear for which the coordinates p_l, x^l change places under the change of coordinates, the value of the cocycle $a_{I,J}$ does not depend on the dimension of the Grassmanian $G_{2n}^{SO}(I_n)$.

Thus, in order to calculate the cohomology class of $[a_{I,J}] \in H^1\left(G_{2n}^{SO}(I_n)\right)$ it is sufficient to compute its cohomology class when $n = 1$. For $n = 1$ the Lagrangian Grassmannian $G_2^{SO}(I_1)$ is homeomorphic to the circle S^1, parametrized by the angle between the ray of abscissas in $\Phi_{\mathbb{R}}(2n) = \mathbb{R}^2$ and a Lagrangian plane passing through the origin of the system of coordinates – a line (Fig. 11).

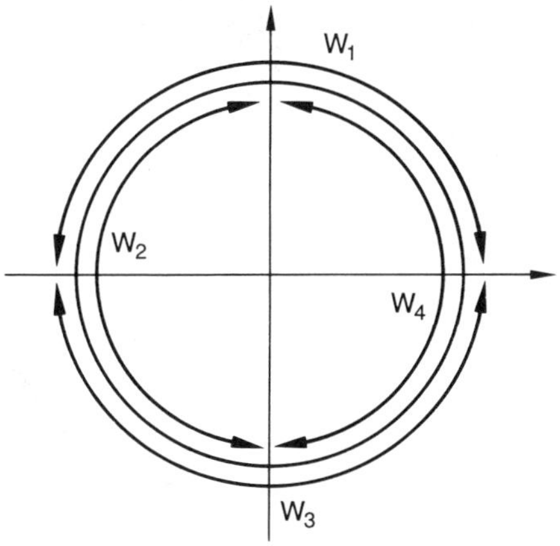

Fig. 11

Then the manifold $G_2^{SO}(I_1)$ can be covered by the four charts W_k, $1 \leq k < 4$:

$$W_1 = \{\varphi;\ 0 < \varphi < \pi\} \ , \quad W_2 = \left\{\varphi; \tfrac{\pi}{2} < \varphi \leq \tfrac{3\pi}{2}\right\} \ ,$$
$$W_3 = \{\varphi;\ \pi < \varphi < 2\pi\} \ , \quad W_4 = \left\{\varphi;\ -\tfrac{\pi}{2} < \varphi < \tfrac{\pi}{2}\right\} \ .$$

We calculate the value of the cocycle a_{kl} according to Formula (3.7). For example, in the intersection $W_1 \cap W_2$ we have $I = \{1\}$, $J = \emptyset$, $I_2 = \{1\}$, $I_3 = \emptyset$. Thus, since the function $x = x(p)$ is increasing in the interval $W_1 \cap W_2$.

$$a_{12} = \operatorname{sign} \frac{\partial x}{\partial p} = 1 \ .$$

Analogously it can be shown that the cocycle a_{kl} takes on the following values:

$$a_{12} = 1 \,, \quad a_{23} = 1 \,, \quad a_{34} = 1 \,, \quad a_{41} = 1 \,.$$

Consequently, $[a_{kl}] = 4u$, where u is the generator of $H^1\left(S^1\right)$.

We consider as an example the one-dimensional Lagrangian manifold in two-dimensional phase space $\Phi_{\mathbb{R}}(2)$ given by the equation $x^2 + p^2 = R^2$. The first Maslov condition of quantization is equivalent to the triviality of the cohomology class defined by the form $\Omega = p\,dx$. This form defines a non-null cohomology class, since the integral of the form Ω over the manifold M is nonzero. Indeed,

$$\int\limits_M \Omega = \oint\limits_{x^2+y^2=R^2} p\,dx = \iint\limits_{x^2+p^2\leq R^2} dp \wedge dx = \pi R^2 \neq 0 \,.$$

Thus the first Maslov condition of quantization is not satisfied.

The second Maslov condition of quantization defines a form ω according to Formula (3.11).

We have

$$\omega = \frac{2}{\pi i} d \ln \left\{ \frac{\partial(x+ip)}{\partial\alpha} \left| \frac{\partial(x+ip)}{\partial\alpha} \right|^{-1} \right\} ,$$

where α is the angular parameter of the circle

$$x^2 + p^2 = R^2 \,, \quad x = R\cos\alpha \,, \quad p = R\sin\alpha \,,$$

$$x + ip = Re^{i\alpha} \,, \quad \frac{\partial(x+ip)}{\partial\alpha} = iRe^{i\alpha} \,.$$

Therefore

$$\omega = 2(\pi i)^{-1} d \ln\left(iRe^{i\alpha} \cdot R^{-1}\right) = 2\,(\pi i)^{-1}\, d \ln e^{i\alpha} = 2\pi^{-1} d\alpha \,.$$

Consequently

$$\oint\limits_M \omega = \oint\limits_M \frac{2}{\pi} d\alpha = \int\limits_0^{2\pi} \frac{2}{\pi} d\alpha = 4 \not\equiv 0 \,(\mathrm{mod}\, 8) \,.$$

Hence the second Maslov condition of quantization is also not satisfied.

2.4 Lagrangian Manifolds in General Position

We will introduce another of calculating the Maslov index of a Lagrangian manifold $\varphi : M \to \Phi_{\mathbb{R}}(2n)$ in phase space by using the index of intersection. The one-dimensional cohomology class $[\omega] \in H^1(M)$, whose triviality is guaranteed by the second Maslov condition of quantization, is entirely determined by its values on one-dimensional cycles. It turns out that for "almost" any Lagrangian manifold these values can be defined as the index of intersection of a closed curve with a certain submanifold $\Gamma \subset M$ of codimension 1. We already observed in § 1.3 that to calculate the index of intersection $\operatorname{ind}[\gamma : \Gamma]$ of a closed curve γ with a submanifold Γ it is sufficient for Γ to be smooth (not necessarily compact) and for its topological boundary $\overline{\Gamma} \setminus \Gamma$ to have codimension 3 in the manifold M.

We denote by $\left[G_{2n}^{SO}(I_n)\right]^k$ the set of all Lagrangian planes $L \subset \Phi_{\mathbb{R}}(2n)$ for which the dimension of the projection $P^n(L) \subset \mathbb{R}^n$ equals $n-k$. In particular, for $k=0$ we obtain all the Lagrangian planes for which the x-coordinates serve as a system of coordinates. The projection P^n assigns to each Lagrangian plane $L \in \left[G_{2n}^{SO}(I_n)\right]^k$ its image $P^n(L)$ – an $(n-k)$-dimensional subspace in $\mathbb{R}^n$. Thus we obtain a map

$$q : \left[G_{2n}^{SO}(I_n)\right]^k \to G(n,k) ,$$

where $G(n,k)$ is the Grassmannian manifold of $(n-k)$-dimensional subspaces in $\mathbb{R}^n$. The preimage $q^{-1}(\xi)$ of a arbitrary point $\xi \in G(n,k)$ consists of all the Lagrangian planes L such that $q(L)=\xi$. The set $q^{-1}(\xi)$ can be described as follows. Let $\varphi : \xi \to \mathbb{R}^n$ be an arbitrary linear map, and $\xi^\perp \subset \mathbb{R}^n$ consist of all vectors $x \in \mathbb{R}^n$ such that $\langle x, \xi\rangle = 0$; $\dim \xi^\perp = k$. We consider the set Ω_ξ of maps φ such that $\xi^\perp \perp \varphi(\xi)$. Then the map q is a locally trivial bundle with fiber Ω_ξ. Consequently, $\left[G_{2n}^{SO}(I_n)\right]^k$ is a manifold, whose dimension is determined by the equation

$$\dim \left[G_{2n}^{SO}(I_n)\right]^k = \dim G(n,k) + \dim \Omega_\xi = \frac{(n-k)(n+k+1)}{2} .$$

In particular,

$$\dim \left[G_{2n}^{SO}(I_n)\right] = \frac{(n+1)n}{2} ,$$
$$\dim \left[G_{2n}^{SO}(I_n)\right]^1 = \frac{(n-1)(n+2)}{2} = \frac{(n+1)n}{2} - 1 .$$

$$\dim \left[G_{2n}^{SO}(I_n)\right] - \dim \left[G_{2n}^{SO}(I_n)\right]^k = \frac{k(k+1)}{2} \geq 3 \quad \text{for } k \geq 2 .$$

The manifold $G_{2n}^{SO}(I_n)$ is an orientable manifold. Indeed, since $G_{2n}^{SO}(I_n) = U(n)/SO(n)$, it is sufficient to show that the homogeneous space G/H of left conjugacy classes of a Lie group G by a connected subgroup H is an orientable manifold. Indeed, suppose $p : G \to G/H$ is the projection. Let $g \in G$, $[Hg] = p(g) \in G/H$, $D_p : T_g G \to T_{p(g)}(G/H)$ be the differential of the map p.

One can represent the tangent space $T_p(G)$ as the image of $T_e(G)$ under a right translation, i.e. $T_g(G) = R_g\left(T_e(G)\right)$. Then the kernel $\operatorname{Ker} D_p$ of the mapping p at the point $g \in G$ consists of the vectors $R_g\left(T_e H\right)$. Let $V \subset T_e G$ be the complementary subspace of the space $T_e H$, $(T_e H) \oplus V = T_e G$. Then $\operatorname{Ker} p = R_g\left(T_e H\right)$ is transversal to $R_g(V)$. Consequently D_p maps $R_g(V)$ isomorphically onto $T_{p(g)}(G/H)$. One need only show that if $p(g) = p(g')$ then the orientations of $D_p\left(R_g(V)\right)$ and $D_p\left(R_{g'}(V)\right)$ coincide.

We consider the element $h = g' g^{-1} \in H$. Then $D_p : T_{g'} G \to T_{p(g)}(G/H)$ decomposes into a composition

$$T_{g'} G \xrightarrow{L_h} T_g G \xrightarrow{D_p} T_{p(g)}\left(G/H\right) .$$

Hence $D_p\left(R_{g'}(V)\right) = D_p\left(L_h R_g V\right) = D_p\left(R_{g'} L_h R_h^{-1}(V)\right)$. Thus it is sufficient to verify that the subspaces V and $L_h R_h^{-1}(V) \subset T_e G$ have the same orientation after projection along $T_e H$. Since the group H is connected, it is possible to join any element h to the unity e by a path h_t in such a way that the orientation of $L_{h_t} R_{h_t^{-1}}(V)$ under projection onto V along $T_e(H)$ does not change under a change of the parameter t. Since $h_0 = e$, $L_e = R_e = id$, the orientations of the spaces V and $L_h R_h^{-1}$ coincide, as was required.

The submanifold $\Gamma' = \left[G_{2n}^{SO}\left(I_n\right)\right]^1$ is also an orientable manifold. For this it is sufficient to note that the submanifold $\left[G_{2n}^{SO}\left(I_n\right)\right]^1$ can be represented as the set of regular points of the null level surface of a certain function $f : G_{2n}^{SO}\left(I_n\right) \to \mathbb{R}^1$. The function f is defined as follows. As was already shown, every oriented Lagrangian plane L can be defined by an orthonormal basis $\{a_1, \ldots, a_n\} \subset L$ and this basis, in turn, by a matrix of coordinates in phase space $\Phi_{\mathbb{R}}(2n)$:

$$\{a_1, \ldots, a_n\} \to X = \begin{pmatrix} A \\ B \end{pmatrix} .$$

If $\{a_1', \ldots, a_n'\}$ is another orthonormal basis in the plane L (with the same orientation), and $X' = \begin{pmatrix} A' \\ B' \end{pmatrix}$ is the matrix of coordinates of the new basis, then the matrices X and X' are related by the equation $X' = XC$, where C is an orthogonal matrix, $\det C = 1$. Then we set

$$f(L) = \det A .$$

The function f is well-defined, i.e. is independent of the choice of basis in the plane L. Then the submanifold $\left[G_{2n}^{SO}\left(I_n\right)\right]^1$ consists precisely of the regular points of the function f which lie on the null level surface, i.e. $f(L) = 0$, $df(L) \neq 0$. We denote the corresponding orientation on Γ' by Γ'_{det}. However, we can introduce an orientation on the manifold $\Gamma' = \left[G_{2n}^{SO}(I_n)\right]^1$ in another way. More precisely, we orient the one-dimensional normal bundle $\nu(\Gamma')$ to the submanifold Γ' in $G_{2n}^{SO}(I_n)$. To do this we will give a direction in the manifold $G_{2n}^{SO}(I_n)$ transversal to Γ' at each point $L \in \Gamma'$. Let $L \in \Gamma'$ be an oriented Lagrangian plane, $\dim P^n(L) = (n-1)$. Let $f \in L \cap \mathbb{R}_n^p$ be a basis vector (note that $\dim L \cap \mathbb{R}_n^p = 1$). Let $e \in \mathbb{R}_x^n$ be a vector orthogonal to $P^n(L)$ such that $\langle f, e\rangle > 0$. If $L^\perp$ is the orthogonal complement to the vector f in the plane L, then the spaces $L(t) = L^\perp \oplus \{f + te\}$, $t > 0$, are Lagrangian planes, for which $P^n(L(t)) = \mathbb{R}_x^n$. It is quite obvious that the family $L(t)$ does not depend on the choice of the vector f and depends continuously on the point of Γ'. Thus we have obtained another orientation, say Γ_{ind}, on the manifold Γ', which does not in general coincide with the orientation given by the function f. Indeed, even in the case $n = 1$ the manifold Γ' is zero-dimensional and consists of two isolated points. One can identify the Lagrangian Grassmannian $G_2^{SO}(I_1)$ with the points of the circle $x^2 + p^2 = 1$, considering a one-dimensional oriented Lagrangian plane to be represented by the vector with coordinates (x, p) whose endpoint lies on the circle. Then the submanifold Γ' consists of two points: $A = (0, 1)$ and $B = (0, -1)$. If α is the angular parameter on the circle, the function f has the form $f(\alpha) = \cos\alpha$. Consequently f is positive for $x > 0$ and negative for $x < 0$. This means that point A has the orientation $(-)$ and point B has the orientation $(+)$ for the orientation Γ'_{det}. In the case of the orientation Γ'_{ind} both points A and B have the orientation $(-)$.

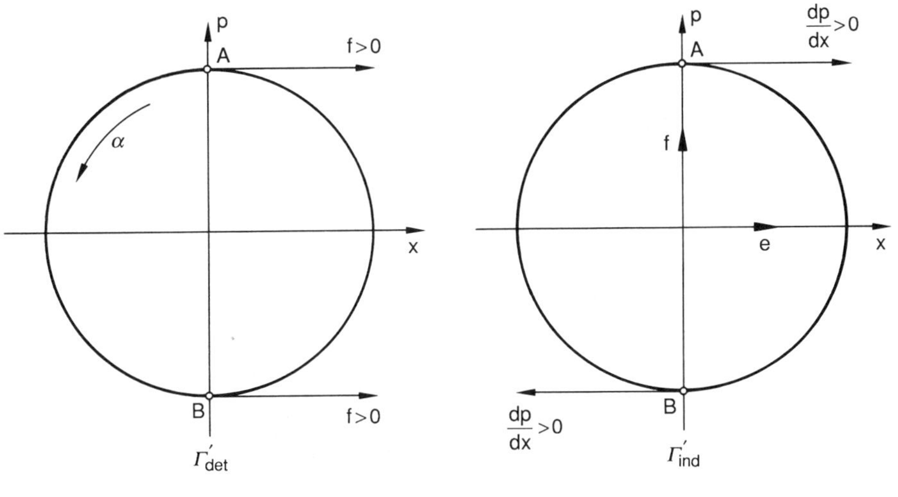

Fig. 12

The directions of orientation of the normal bundles at the points A and B under the different methods of orientation are illustrated in Fig. 12. The condition $\langle f, e\rangle > 0$ is equivalent to $dp/dx > 0$. In the case of $\Gamma' = \left[G_4^{SO}(I_2)\right]^1 \subset G_4^{SO}(I_2)$ the manifold Γ' is two-dimensional and also consists of two connected components, on which the orientations Γ'_{det} and Γ'_{ind} do not agree. Each component has the form of the surface of rotation of a circle around a line which is tangent to it at some point C (a degenerate torus). The central point C actually belongs to $\left[G_4^{SO}(I_2)\right]^2$ and is the common boundary of both connected components of the manifold Γ' (Fig. 13).

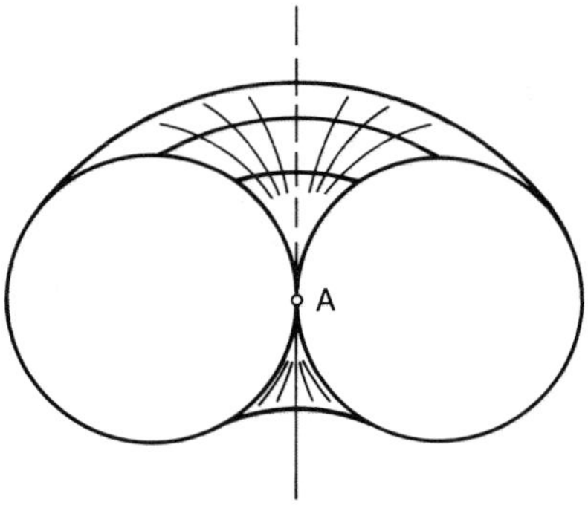

Fig. 13

Thus the theory of intersections with a closed curve is applicable to the manifold $\Gamma' = \left[G_{2n}^{SO}(I_n)\right]^1$.

The function $\operatorname{ind}[\gamma : \Gamma']$ determines a certain one-dimensional cohomology class of the manifold $G_{2n}^{SO}(I_n)$, whose construction is independent of the dimension n. Thus it is sufficient to calculate this class for $n = 1$, i.e. for $G_2^{SO}(I_1)$. In this special case the manifold Γ' consists of two points, so that if γ is the generating cycle, $\operatorname{ind}[\gamma : \Gamma'] = 2$.

Thus for any Lagrangian Grassmannian $G_{2n}^{SO}(I_n)$ and closed cycle γ the identity

$$\int_{\nu} \omega = 2 \operatorname{ind}[\gamma : \Gamma'] \ .$$

holds, as it does in the one-dimensional case.

In order to pass to the theory of intersections directly on a Lagrangian manifold $\varphi : M \to \Phi_{\mathbb{R}}(2n)$, it is sufficient to assume that the map

$$d\varphi : M \to G_{2n}^{SO}(I_n) \tag{4.1}$$

is transversally regular along all submanifolds $\left[G_{2n}^{SO}(I_n)\right]^k$. Then the preimage $\Gamma = (d\varphi)^{-1}(\Gamma')$ will be a smooth manifold of codimension 1, and its topological boundary $\overline{\Gamma} \setminus \Gamma$ will be equal to the union $\bigcup_{n\geq 2}(d\varphi)^{-1}\left[G_{2n}^{SO}(I_n)\right]^k$, and will have codimension ≥ 3. We will show that for "almost" all imbeddings

$$\varphi : M \to \Phi_{\mathbb{R}}(2n)$$

the mapping (4.1) satisfies the condition of transversal regularity along any finite collection of submanifolds.

Lemma 4.1. *There exists a canonical transformation ψ of the phase space $\Phi_{\mathbb{R}}(2n)$ arbitrarily close to the identity such that the composition $\psi\varphi : M \to \Phi_{\mathbb{R}}(2n)$ induces a transversally regular map $d(\psi\varphi) : M \to G_{2n}^{SO}(I_n)$ along a finite collection of submanifolds.*

Indeed, the composition of the map φ and a canonical transformation can be viewed as a general map $A : M \times U(n) \to \Phi_{\mathbb{R}}(2n)$.

We consider the "differential" of the map A,

$$dA : M \times U(n) \to G_{2n}^{SO}(I_n) \ .$$

Since $G_{2n}^{SO}(I_n) = U(n)/O(n)$, dA is transversally regular along any submanifold in $G_{2n}^{SO}(I_n)$. One can thus apply Abraham's theorem. □

Thus the second Maslov condition of quantization can be written in terms of the index of intersection of closed curves γ with a particular submanifold Γ.

Theorem 4.2. *Suppose $M \subset \Phi_{\mathbb{R}}(2n)$ is a Lagrangian manifold which is in general position in the sense that the differential* (4.1)

$$d\varphi : M \to G_{2n}^{SO}(I_n)$$

is transversally regular along the submanifolds $\left[G_{2n}^{SO}(I_n)\right]^k$, and let $\Gamma = (d\varphi)^{-1}\left(\left[G_{2n}^{SO}(I_n)\right]^1\right)$ be the special manifold of codimension 1 lying in M. Then the second Maslov condition of quantization is equivalent to the congruences

$$\operatorname{ind}[\gamma : \Gamma] \equiv 0 \pmod 4$$

for any closed curve $\gamma \subset M$.

Chapter 3
Complex Lagrangian Manifolds

The present chapter contains a systematic study of all the properties of complex Lagrangian manifolds which are necessary for constructing Maslov's canonical operator in the complex case. The first two sections study questions which do not require the application of s-analytic analysis. The third and fourth sections are devoted to setting up the analysis of s-analytic functions on s-analytic manifolds and the application of this analysis to the special case of Lagrangian manifolds.

We will study here in detail the second Maslov quantization condition in the complex case, since the study of this condition leads to effects which are not met with in the real case. On the other hand, these investigations shed new light on the second Maslov quantization conidtion for real Lagrangian manifolds as well.

3.1 The Grassmannian of Positive Lagrangian Planes

Suppose, as before, I denotes some subset of the set of natural numbers from 1 to n, and $\overline{I}$ is the complement of the set I in the interval of natural numbers $[1, n]$. We will denote by $\mathbb{C}^I$ the subspace of the phase space $\Phi_{\mathbb{C}}$ generated by the vectors $\{e_k, k \in I\}$. $\mathbb{C}_{\overline{I}}$ will denote the subspace generated by the vectors $\{f^k, k \in \overline{I}\}$. It is easy to see that the subspace $\mathbb{C}^I \times \mathbb{C}_{\overline{I}} \subset \Phi_{\mathbb{C}}$ is a Lagrangian plane. Let

$$P_I : \Phi_{\mathbb{C}} \to \mathbb{C}^I \times \mathbb{C}_{\overline{I}}$$

denote the projection of the phase space $\Phi_{\mathbb{C}}$ along the subspace $\mathbb{C}^{\overline{I}} \times \mathbb{C}_I$.

The proof of the next lemma does not differ from the proof of Lemma 1.1 of § 2.1 in the real case.

Lemma 1.1. *Suppose $L \subset \Phi_{\mathbb{C}}$ is an arbitrary Lagrangian plane. There exists some set I such that*

$$P_I|L : L \to \mathbb{C}^I \times \mathbb{C}_{\overline{I}}$$

is an isomorphism.

Suppose $L \subset \Phi_{\mathbb{C}}$ is a Lagrangian plane and $C \in GL(2n, I_n)$ is a Hamiltonian transformation such that $L = C(\mathbb{C}^n)$. Let I be a collection of indices such that $P_I : L \to \mathbb{C}^I \times C_{\overline{I}}$ is an isomorphism. Then, in terms of the decomposition $\Phi_{\mathbb{C}} = \mathbb{C}^I \times \mathbb{C}^{\overline{I}} \times \mathbb{C}_I \times \mathbb{C}_{\overline{I}}$, the transformation C is given by a matrix of the form

$$C = \begin{pmatrix} A_1 & A_2 & * & * \\ A_3 & A_4 & * & * \\ B_1 & B_2 & * & * \\ B_3 & B_4 & * & * \end{pmatrix} ,$$

where the matrix

$$\begin{pmatrix} A_1 & A_2 \\ B_3 & B_4 \end{pmatrix}$$

is invertible. □

Definition 1.1. A Lagrangian plane is called *positive* if the matrix

$$\operatorname{Im} \begin{pmatrix} B_1 & B_2 \\ -A_3 & A_4 \end{pmatrix} \begin{pmatrix} A_1 & A_2 \\ B_3 & B_4 \end{pmatrix}^{-1}$$

is nonnegative definite.

We observe that by virtue of the fact that L is Lagrangian, the latter matrix is symmetric. Indeed, suppose first that $I = \{1, 2, \ldots, n\}$. Then

$$C = \begin{pmatrix} A & * \\ B & * \end{pmatrix}$$

and from the fact that L is Lagrangian it follows (see Definition 2.2 of § 2.2) that

$$A^t B = B^t A ,$$

from which, taking into account the nonsingularity of the matrix A, the symmetry of the matrix BA^{-1} follows. The general case reduces to the one considered by using the Hamiltonian transformation

$$P_I = \begin{pmatrix} E & 0 & 0 & 0 \\ 0 & 0 & 0 & E \\ 0 & 0 & E & 0 \\ 0 & -E & 0 & 0 \end{pmatrix} .$$

It is necessary to show that Definition 1.1 is consistent, i.e. that it is independent of the choice of indices I.

Lemma 1.2. *Suppose the complex matrix*

$$C = \begin{pmatrix} C_1 & C_2 \\ C_3 & C_4 \end{pmatrix}$$

satisfies the conditions:

a) $C^t = C$;

b) $\operatorname{Im} C$ *is positive definite;*

c) $\operatorname{Im} C_4$ *is positive definite.*

Then the matrix

$$\operatorname{Im} \begin{pmatrix} C_1 & C_2 \\ 0 & -E \end{pmatrix} \begin{pmatrix} E & 0 \\ C_3 & C_4 \end{pmatrix}^{-1}$$

is positive definite.

Since the matrix

$$\begin{pmatrix} E \\ C \end{pmatrix}$$

defines a Lagrangian plane by virtue of condition a), by the remark after Definition 1.1 the latter matrix is symmetric.

Proof. Let $C_k = A_k + iB_k$. From condition c) of the lemma it follows that the matrices C_4 and $B_4 = \operatorname{Im} C_4$ are invertible. From the equations

$$(A_4 + iB_4)\left(B_4^{-1}A_4 - iE\right) = A_4 B_4^{-1} A_4 + B_4 \ ,$$

$$B_4^{-1}A_4 - iE = B_4^{-1}\left(\overline{A_4 + iB_4}\right)$$

it follows that the matrix $A_4 B_4^{-1} A_4 + B_4$ is invertible and

$$C_4^{-1} = \left(B_4^{-1}A_4 - iE\right) Z \ ,$$

where

$$Z^{-1} = \left(A_4 B_4^{-1} A_4 + B_4\right) \ .$$

We compute the matrix

$$D = \operatorname{Im} \begin{pmatrix} C_1 & C_2 \\ 0 & -E \end{pmatrix} \begin{pmatrix} E & 0 \\ C_3 & C_4 \end{pmatrix}^{-1} .$$

We have

$$\begin{pmatrix} E & 0 \\ C_3 & C_4 \end{pmatrix}^{-1} = \begin{pmatrix} E & 0 \\ -C_4^{-1}C_3 & C_4^{-1} \end{pmatrix} ,$$

$$D = \mathrm{Im}\begin{pmatrix} C_1 - C_2C_4^{-1}C_3 & C_2C_4^{-1} \\ C_4^{-1}C_3 & -C_4^{-1} \end{pmatrix}$$

$$= \begin{pmatrix} B_1 - A_2B_4^{-1}A_4ZB_3 - B_2B^{-1}A_4ZA_3 + A_2ZA_3 - B_2ZB_3 & B_2B_4^{-1}A_4Z - A_2Z \\ B_4^{-1}A_4ZB_3 - ZA_3 & Z \end{pmatrix}$$

We simplify the symmetric matrix D by using the matrix

$$K = \begin{pmatrix} E & 0 \\ A_3 & E \end{pmatrix} .$$

Then

$$K^tDK = \begin{pmatrix} B_1 - B_2ZB_3 & B_2B_4^{-1}A_4Z \\ B_1^{-1}A_1ZB_3 & Z \end{pmatrix} .$$

The latter matrix is symmetric because of the symmetry of D, that is $B_2B_4^{-1}A_4Z = \left(B_4^{-1}A_4ZB_3\right)^t = B_2ZA_4B_4^{-1}$.

Now suppose

$$K_1 = \begin{pmatrix} E & 0 \\ -Z^{-1}B_4^{-1}A_4ZB_3 & E \end{pmatrix} .$$

Then

$$K_1^tK^tDKK_1 = \begin{pmatrix} B_1 - B_2ZB_3 - B_2B_4^{-1}A_4B_4^{-1}A_4ZB_3 & 0 \\ 0 & Z \end{pmatrix} .$$

All that is left is to show that the matrix

$$H = B_1 - B_2ZB_3 - B_2B_4^{-1}A_4B_4^{-1}A_4ZB_3$$

is positive definite. We have

$$\begin{aligned} H &= B_1 - B_2\left(1 + B_4^{-1}A_4B_4^{-1}A_4\right)ZB_3 \\ &= B_1 - B_2B_4^{-1}\left(B_4 + A_4B_4^{-1}A_4\right)ZB_3 = B_1 - B_2B_4^{-1}B_3 . \end{aligned}$$

It follows from condition b) of the lemma that the matrix

$$\mathrm{Im}\, C = \begin{pmatrix} B_1 & B_2 \\ B_3 & B_4 \end{pmatrix}$$

is positive definite. Since the matrix B_4 is invertible, the matrix $\mathrm{Im}\, C$ is conjugate to[1] the matrix

[1] By using the matrix

$$\begin{pmatrix} E & 0 \\ -B_4^{-1}B_3 & B_4^{-1} \end{pmatrix} .$$

(We call the matrices D and K^tDK conjugate.)

$$\begin{pmatrix} B_1 - B_2 B_4^{-1} B_3 & 0 \\ 0 & B_4 \end{pmatrix} ,$$

i.e. the matrix H is positive definite. This proves Lemma 1.2. □

Using Lemma 1.2 we can establish the consistency of Definition 1.1. Indeed, if $C \in GL(2n, I_n)$,

$$C = \begin{pmatrix} E & * \\ B & * \end{pmatrix} ,$$

then $B^t = B$, and, conversely, if $B^t = B$ then there exists a matrix C of the form

$$C = \begin{pmatrix} E & * \\ B & * \end{pmatrix} ,$$

which belongs to the group $GL(2n, I_n)$. Let

$$P_I = \begin{pmatrix} E & 0 & 0 & 0 \\ 0 & 0 & 0 & E \\ 0 & 0 & -E & 0 \\ 0 & -E & 0 & 0 \end{pmatrix}$$

be a Hamiltonian transformation expressed in terms of the decomposition $\Phi_{\mathbb{C}} = \mathbb{C}^I \times \mathbb{C}^{\bar{I}} \times \mathbb{C}_I \times \mathbb{C}_{\bar{I}}$. If $C \in GL(2n, I_n)$ determines a positive Lagrangian plane, then by Definition 1.1 there exists some collection of indices I such that the matrix $D = P_I C$ has the form

$$D = \begin{pmatrix} D_1 & * \\ D_2 & * \end{pmatrix} ,$$

where D_1 is an invertible matrix and $\operatorname{Im} D_2 D_1^{-1}$ is nonnegative definite. Let J be another collection of indices,

$$H = P_J C = \begin{pmatrix} H_1 & * \\ H_2 & * \end{pmatrix} ,$$

where H_1 is an invertible matrix. Then

$$H = P_J P_I^{-1} D = P_K D ,$$

where

$$K = (I \cup J) \setminus (I \cap J) .$$

We now observe that from Lemma 1.2 follows

Lemma 1.3. *Suppose the complex matrix*

$$C = \begin{pmatrix} C_1 & C_2 \\ C_3 & C_4 \end{pmatrix}$$

satisfies the following conditions:

a) $C^t = C$;

b) $\operatorname{Im} C$ *is nonnegative definite;*

c) *the matrix* C_4 *is invertible.*

Then the matrix

$$D = \operatorname{Im} \begin{pmatrix} C_1 & C_2 \\ 0 & E \end{pmatrix} \begin{pmatrix} E & 0 \\ C_3 & C_4 \end{pmatrix}^{-1}$$

is nonnegative definite.

Proof. Suppose C' is another complex matrix, for which conditions a), b) and c) hold, and let $D' = \operatorname{Im} \begin{pmatrix} C_1' & C_2' \\ 0 & E \end{pmatrix} \begin{pmatrix} E & 0 \\ C_3' & C_4 \end{pmatrix}^{-1}$. Then for any $\epsilon > 0$ there exists $\delta > 0$ such that if (in any norm of a finite-dimensional space) $\|C - C'\| < \delta$, then $\|D - D'\| < \epsilon$. If the matrix D has any negative eigenvalues, then there exists some number $\epsilon_0 > 0$ such that any symmetric matrix D', $\|D - D'\| < \epsilon_0$, will also have negative eigenvalues. On the other hand, for any $\delta > 0$ there exists a matrix C' such that $\|C - C'\| < \delta$, $C'^t = C'$, $\operatorname{Im} C$, is positive definite. Then it follows from Lemma 1.2 that the matrix D' is positive definite, yet $\|D - D'\| < \epsilon_0$. Lemma 1.3 follows by contradiction. □

Applying Lemma 1.3 to the matrix $D_2 D_1^{-1}$, we obtain that the matrix $H_2 H_1^{-1}$ is nonnegative definite. This completes the proof of the consistency of Definition 1.1.

Definition 1.3. We will denote the set of positive Lagrangian planes by $G_{2n}^+(I_n) \subset G_{2n}(I_n)$.

We will describe the space $G_{2n}^+(I_n) \subset G_{2n}(I_n)$ for $n = 1$. In this case the phase space is two dimensional, $\Phi_{\mathbb{C}}(2) = \mathbb{C}^1 \oplus \mathbb{C}_1$, and the Lagrangian Grassmannian $G_{2n}(I_n)$ consists of all complex lines $L \subset \mathbb{C}^1 + \mathbb{C}_1$ which pass through the origin. If (z, ζ) is a linear system of coordinates in the phase space $\Phi_{\mathbb{C}}(2)$, every Lagrangian plane L is given by the equation $\lambda z + \mu\zeta = 0$; $\lambda, \mu \in \mathbb{C}$. Thus $G_2(I_1)$ is isomorphic to the complex projective space $\mathbb{C}P^1$ or, equivalently, to the extended complex plane $\mathbb{C}P^1 = \mathbb{C}^1 \cup \{\infty\}$. Then the subspace $G_2^+(I_1)$ is given by the inequality $\operatorname{Im} \lambda/\mu \geq 0$. Hence $G_2^+(I_1)$ coincides with the upper half-plane in the extended complex plane, i.e. is homeomorphic to a two-dimensional disk. Thus the space $G_2^+(I_1)$ is a contractible space.

There is a canonical complex n-dimensional bundle ξ_n over the manifold $G_{2n}(I_n)$, which consists of the Lagrangian planes themselves. We shall describe the bundle ξ_n more precisely. We consider the direct product $G_{2n}(I_n) \times \Phi_{\mathbb{C}}$. Let $E \subset G_{2n}(I_n) \times \Phi_{\mathbb{C}}$ be the subspace consisting of all points of the form

$$(L, x) \in G_{2n}(I_n) \times \Phi_{\mathbb{C}} , \qquad x \in L .$$

The projection onto the first coordinate

$$G_{2n}(I_n) \times \Phi_{\mathbb{C}} \to G_{2n}(I_n)$$

induces a map

$$\pi : E \to G_{2n}(I_n) .$$

It is obvious that the preimage of each point $L \in G_{2n}(I_n)$ is a subspace homeomorphic to the n-dimensional complex space L. One can show that $\{E, G_{2n}(I_n), \pi\}$ is a locally trivial bundle with structure group $U(n)$.

It is this complex n-dimensional bundle $\{E, G_{2n}(I_n), \pi\}$ which we will denote by ξ_n.

Theorem 1.4. *Let η denote the restriction of the bundle $[\wedge^n(\xi_n)]^*$ to the subspace $G_{2n}^+(I_n)$. Then there exists an analytic section $\sigma : G_{2n}^+(I_n) \to \eta$ which is nonzero at every point.*

Proof. In the case $n = 1$ the assertion is trivial, since $G_2^+(I_1)$ is a contractible space and, consequently, any bundle with the base $G_2^+(I_1)$ is trivial. We present the proof in the general case. We recall that $\wedge^n(\xi_n)$ denotes the n-th exterior power of the bundle ξ_n. Since $\dim \xi_n = n$, $\dim \wedge^n(\xi_n) = \dim \eta = 1$. Thus in order to construct a nonzero section $\sigma : G_{2n}^+(I_n) \to \eta$ at each point it is sufficient to assign to each point $x \in G_{2n}^+(I_n)$ a nonzero vector $\sigma(x)$ of the n-th external product $\wedge^n(L)$ of the fiber L of ξ_n. Suppose $(a_1, \ldots, a_n)$ is a basis of the vector space L. Then the vector $a_1 \wedge a_2 \wedge \ldots \wedge a_n$ will be the basis vector of the subspace $\wedge^n L$. If $(a'_1, \ldots, a'_n)$ is another basis of the space L, $a_i = \sum_j c_i^j a'_j$, then

$$a_1 \wedge \ldots \wedge a_n = \det\left(c_i^j\right) a'_1 \wedge \ldots \wedge a'_n .$$

Consequently, to construct a section $\sigma : G_{2n}^+(I_n) \to \eta$, it is sufficient to assign a complex number

$$f(a_1, \ldots, a_n)$$

to each basis $(a_1, \ldots, a_n)$ of the Lagrangian plane L in such a way that

$$f(a_1, \ldots, a_n) \det\left(c_i^j\right)^{-1} = f(a'_1, \ldots, a'_n) ,$$

where $(a'_1, \ldots, a'_n)$ is another basis of the space L, related by the equation

$$a_i \sum_j c_i^j a'_j \ .$$

Lemma 1.5. *Suppose C is a complex-valued matrix, $C^t = C$, $\operatorname{Im} C$ is a nonnegative definite matrix. If λ is an eigenvalue of the matrix C, then* $\operatorname{Im} \lambda \geq 0$.

The proof is obvious.

Lemma 1.6. *Suppose $C \in GL(2n, I_n)$, $C(\mathbb{C}^n) \in G_{2n}^{+}(I_n)$, $C = \begin{pmatrix} A & * \\ B & * \end{pmatrix}$. Then*

$$\det(A - iB) \neq 0 \ .$$

Proof. In the case when $\det A \neq 0$, we have

$$\det(A - iB) = \det A \det\left(E - iBA^{-1}\right) \ .$$

But for the matrix $E - iBA^{-1}$ any eigenvalue has a real part larger than one. The general case reduces to the above case by means of the symplectic transformation (1.1.) □

Now suppose $(a_1, \ldots, a_n)$ is an arbitrary basis of the positive Lagrangian plane $L \in G_{2n}^{+}(I_n)$, and

$$X = \begin{pmatrix} A \\ B \end{pmatrix}$$

is the coordinate matrix of the vectors $(a_1, \ldots, a_n)$. We set

$$f(a_1, \ldots, a_n) = \det(A - iB) \ .$$

If $(a'_1, \ldots, a'_n)$ is another basis and

$$X' = \begin{pmatrix} A' \\ B' \end{pmatrix}$$

is its matrix of coordinates,

$$a_i = \sum_j c_i^j a'_j \ ,$$

then

$$X = X'C^t \ ,$$

where $C = \left(c_i^j\right)$. Thus

$$\begin{pmatrix} A \\ B \end{pmatrix} = \begin{pmatrix} A' \\ B' \end{pmatrix} C^t ,$$

$$A = A'C^t , \quad B = B'C^t , \quad (A - iB) = (A' - iB')\,C^t .$$

Consequently

$$f\,(a_1, \ldots, a_n) = f\,(a_1', \ldots, a_n') \det \left(c_i^j\right) .$$

Thus the function $f\,(a_1, \ldots, a_n)$ gives a well-defined section

$$\sigma : G_{2n}^{+}\,(I_n) \to \eta .$$

The analyticity of the section σ follows from the facts that

a) the group $U\,(2n, I_n)$ is an analytic manifold, and the operation of multiplication in the group $U\,(2n, I_n)$ is an analytic map;

b) the linear action of the group $U(n)$ on the space $\mathbb{C}^n$ is an analytic action;

c) the fibration $U(n) \to U\,(2n, I_n) \to G_{2n}^{+}\,(I_n)$ is an analytic fibration of analytic manifolds;

d) the basis $(a_1, \ldots, a_n)$ of the Lagrangian plane L can be chosen so as to depend analytically on the point $L \in G_{2n}^{+}\,(I_n)$;

e) the function $\det\,(A - iB)$ is an analytic function of the elements of the matrices A and B.

This completes the proof of Theorem 1.4. □

3.2 The Maslov Index of Complex Lagrangian Manifolds

Analogously to the second quantization condition for real Lagrangian manifolds, it is possible to formulate a Maslov quantization condition on the section σ of the bundle η in the case of positive Lagrangian planes.

Suppose $\{U_I\}$ is the set of Lagrangian planes $L \in G_{2n}^{+}\,(I_n)$ such that the projection

$$P_{\bar{I}}^{I} : L \to C^I \times C_{\bar{I}}$$

is an isomorphism. Then

$$\bigcup U_I = G_{2n}^{+}\,(I_n) .$$

Let $X \subset G_{2n}^{+}\,(I_n)$ be an open set and $\eta|_X$ be the restriction of the bundle η to X.

If μ is a nonzero section of the bundle $\eta|_X$, then in the chart U_I the function

$$\mu_I = \frac{\partial \mu}{\partial \left(z^I, \zeta_{\overline{I}}\right)}$$

is defined. Then the second quantization condition takes the following form: it is possible to choose the argument $\arg \mu_I$ in each chart $X \cap U_I$ in such a way that in the intersection $U_I \cap V_J$ the equation

$$\arg \mu_I - \arg \mu_J = \sum_k \arg \lambda_k + |I_2| \, \pi \ , \tag{2.1}$$

holds, where

$$I_1 = I \cap J \ , \quad I_2 = I \setminus J \ , \quad I_3 = J \setminus I \ , \quad I_4 = \overline{I} \cap \overline{J} \ ,$$

and $\{\lambda_k\}$ are the eigenvalues of the matrix

$$-\frac{\partial \left(\zeta_{I_2}, -z^{I_3}\right)}{\partial \left(z^{I_2}, \zeta_{I_3}\right)} \ . \tag{2.2}$$

In order to show that condition (2.1) is well-defined, it is sufficient to show that the eigenvalues of the matrix (2.2) lie in the lower half-plane. Indeed, suppose $L \subset \Phi_{\mathbb{C}}$ is a positive Lagrangian plane, for which both $\left(z^I, \zeta_{\overline{I}}\right)$ and $\left(z^J, \zeta_{\overline{J}}\right)$ serve as coordinates. The positivity condition on a Lagrangian plane implies that the matrix

$$-\mathrm{Im} \frac{\partial \left(\zeta_J, -z^{\overline{J}}\right)}{\partial \left(z^J, \zeta_{\overline{J}}\right)} \tag{2.3}$$

is non-positive definite or that the matrix

$$-\mathrm{Im} \frac{\partial \left(\zeta_I, -z^{\overline{I}}\right)}{\partial \left(z^I, \zeta_{\overline{I}}\right)}$$

is non-positive definite. Since the matrix (2.2) is a principal minor of the matrix (2.3), its imaginary part is also non-positive definite.

Suppose the arguments of the functions μ_I are chosen arbitrarily. We define a cochain $C(U, V)$ of the canonical covering with coefficients in the sheaf of germs of smooth functions by the equation

$$C(U, V) = \arg \mu_I - \arg \mu_J - \sum_k \arg \lambda_k + |I_2| \, \pi \ ,$$

where $\{\lambda_K\}$ have the same meaning as in (2.1).

Lemma 2.1.
a) $C(U,V) \equiv 0 \bmod 2\pi$.
b) *The cochain* $C(U,V)$ *is a cocycle.*
c) *Condition* (2.1) *holds if and only if the cocycle* $C(U,V)$ *is cohomologous to zero in the group* $H^1(X,\mathbb{Z})$.

Proof. a) Obviously, the equation

$$\mu_I = \mu_J \det \frac{\partial\,(\zeta_{I_2}, z^{I_3})}{\partial\,(z^{I_2}, \zeta_{I_3})} \,.$$

is true. Therefore (2.1) holds modulo 2π.

b) It suffices to show that the cochain

$$d(U,V) = \sum_k \arg \lambda_k + |I_2|\,\pi$$

is a cocycle, where $\{\lambda_k\}$ are the eigenvalues of the matrix (2.2), since this cochain differs from $C(U,V)$ by a coboundary. Without loss of generality[2], we can assume that one of the charts is a chart of the form U_n.

Thus we need to show that

$$d(U,V) + d(V,W) + d(W,U) = 0 \,,$$

or, equivalently,

$$\begin{aligned} &\arg\det \frac{\partial\,(z^{I_3}, -\zeta_{I_2})}{\partial\,(\zeta_{I_3}, z^{I_2})} + |I_2|\,\pi + \arg\det \frac{\partial\,(z^{I_2}, z^{I_4})}{(\zeta_{I_3}, \zeta_{I_4})} \\ &+ \arg\det \frac{\partial(-\zeta_{I_3}, -\zeta_{I_4})}{\partial\,(z^{I_3}, z^{I_4})} + |I_3 + I_4|\,\pi = 0 \,, \end{aligned} \tag{2.4}$$

where by $\arg\det$ of a matrix we mean the sum of the arguments of its eigenvalues, each of which is taken to be in the interval $(-3\pi/2, \pi/2)$. We observe that this is well-defined because all of the matrices which appear in expression (2.4) are non-positive definite, being principal minors of the following matrix, taken with a negative sign:

$$\operatorname{Im} \begin{pmatrix} \dfrac{\partial \zeta_I}{\partial z^I} & -\dfrac{\partial z^{\bar{I}}}{\partial z^I} \\ \dfrac{\partial \zeta_I}{\partial \zeta_{\bar{I}}} & \dfrac{\partial z^{\bar{I}}}{\partial \zeta_{\bar{I}}} \end{pmatrix} \,.$$

[2] Indeed, it is sufficient to check condition (b) at any point. Furthermore, this condition depends only on the Lagrangian plane at the given point, and consequently can be verified in the case of a Lagrangian Grassmannian where a nonsingular chart exists in a neighborhood of every point.

We choose a basis for the Lagrangian plane in such a way that its coordinate matrix has the form

$$X = \begin{pmatrix} 1 & 0 & 0 & 0 \\ 0 & 1 & 0 & 0 \\ 0 & 0 & 1 & 0 \\ 0 & 0 & 0 & 1 \\ B_{11} & B_{12} & B_{13} & B_{14} \\ \vdots & \vdots & \vdots & \vdots \\ B_{41} & B_{42} & B_{43} & B_{44} \end{pmatrix}$$

in the decomposition of phase space given by

$$\Phi_{\mathbb{C}} = \mathbb{C}^{I_1} \times \mathbb{C}^{I_2} \times \mathbb{C}^{I_3} \times \mathbb{C}^{I_4} \times \mathbb{C}_{I_1} \times \mathbb{C}_{I_2} \times \mathbb{C}_{I_3} \times \mathbb{C}_{I_4} \ .$$

The conditions

$$B^t_{kj} = B_{jk}$$

hold for the matrices B_{kj}. Thus

$$\zeta_{I_s} = \sum_{k=1}^{4} B_{sk} z^{I_k} \ .$$

Replacing the Lagrangian plane by a nearby one, one can assume that all the matrices B_{kk} are invertible. Then, setting

$$Q_{ks} = B_{ks} - B_{k4} B_{44}^{-1} B_{4k} \ ,$$

we obtain

$$\frac{\partial\left(z^{I_3}, -\zeta_{I_2}\right)}{\partial\left(\zeta_{I_3}, z^{I_2}\right)} = \begin{pmatrix} -Q_{22}^{-1} & Q_{22}^{-1} Q_{23} \\ Q_{32} Q_{22}^{-1} & Q_{33} - Q_{32} Q_{22}^{-1} Q_{23} \end{pmatrix} .$$

This matrix is equivalent to the matrix

$$\begin{pmatrix} -Q_{22}^{-1} & 0 \\ 0 & Q_{33} \end{pmatrix} .$$

On the other hand,

$$\frac{\partial\left(-\zeta_{I_3}, -\zeta_{I_4}\right)}{\partial\left(z^{I_3}, z^{I_4}\right)} = \begin{pmatrix} B_{33} & B_{34} \\ B_{43} & B_{44} \end{pmatrix} . \tag{2.5}$$

The matrix (2.5) is equivalent to

$$\begin{pmatrix} Q_{33} & 0 \\ 0 & B_{44} \end{pmatrix} . \tag{2.6}$$

Finally,

$$\frac{\partial(-\zeta_{I_2}, -\zeta_{I_4})}{\partial(z^{I_2}, z^{I_4})} = \begin{pmatrix} B_{22} & B_{24} \\ B_{42} & B_{44} \end{pmatrix} . \tag{2.7}$$

and the matrix (2.7) is equivalent to

$$\begin{pmatrix} Q_{22} & 0 \\ 0 & B_{44} \end{pmatrix} . \tag{2.8}$$

Therefore we must show that

$$\arg\det \begin{pmatrix} -Q_{22}^{-1} & 0 \\ 0 & Q_{33} \end{pmatrix} + |I_2|\,\pi + \arg\det \begin{pmatrix} -Q_{33}^{-1} & 0 \\ 0 & -B_{44}^{-1} \end{pmatrix} + \arg\det \begin{pmatrix} Q_{22} & 0 \\ 0 & B_{44} \end{pmatrix} + |I_3 + I_4|\,\pi = 0 .$$

Indeed, it is not hard to see that

1) $\arg\det Q_{22} + \arg\det\left(-Q_{22}^{-1}\right) = -\,|I_2|\,\pi$
2) $\arg\det Q_{33} + \arg\det\left(-Q_{33}^{-1}\right) = -\,|I_3|\,\pi$
3) $\arg\det Q_{44} + \arg\det\left(-Q_{44}^{-1}\right) = -\,|I_4|\,\pi$.

Adding these together, we obtain assertion (b) of Lemma 2.1. Assertion (c) of the lemma is trivial. □

Theorem 2.2. *Let $X \subset G_{2n}^{+}(I_n)$ be an open set, μ a section of the bundle $\eta|_X$. Condition* (2.1) *holds if and only if μ/σ defines a cocycle which is cohomologous to zero in the first homology group $H^1(X, \mathbb{Z})$ of the manifold X.*

Proof. It is enough to establish that condition (2.1) holds if $\mu = \sigma$. Hence, suppose $\mu = \sigma$. We must choose the arguments J_I in such a way that condition (2.1) is fulfilled. It is sufficient to do this when $J = [n]$. (See footnote on p. 126.) In this case we choose a basis of the Lagrangian plane in such a way that its coordinate matrix is

$$X = \begin{pmatrix} 1 & 0 \\ 0 & 1 \\ B_1 & B_2 \\ B_3 & B_4 \end{pmatrix} .$$

Then the matrices

$$\operatorname{Im}\begin{pmatrix} B_1 & B_2 \\ B_3 & B_4 \end{pmatrix} , \qquad \operatorname{Im} B_4$$

are nonnegative definite. Using the form of the matrix X, we obtain

$$\sigma_I^{-1} = \det\left(E - i\begin{pmatrix} B_1 & B_2 \\ B_3 & B_4 \end{pmatrix}\right) \cdot \det\begin{pmatrix} 1 & 0 \\ -B_4^{-1}B_3 & B_4^{-1} \end{pmatrix}$$
$$= \det\left(E - i\begin{pmatrix} B_1 & B_2 \\ B_3 & B_4 \end{pmatrix}\right) \cdot \det B_4^{-1} , \tag{2.9}$$

$$\sigma_J^{-1} = \det\left(E - i\begin{pmatrix} B_1 & B_2 \\ B_3 & B_4 \end{pmatrix}\right) , \tag{2.10}$$

and the matrix (2.2) is just

$$-\frac{\partial\left(\zeta_{I_2}, -z^{I_3}\right)}{\partial\left(z^{I_2}, \zeta_{I_3}\right)} = \frac{\partial\left(z^{\bar{I}}\right)}{\partial\left(\zeta_{\bar{I}}\right)} = B_4^{-1} .$$

We choose the arguments of the eigenvalues of the matrix

$$E - i\begin{pmatrix} B_1 & B_2 \\ B_3 & B_4 \end{pmatrix}$$

in the right half-plane, and for the matrix B_4^{-1} we choose them in the lower half-plane. Under such a choice of the arguments we obtain a uniquely defined argument for the numbers (2.9), (2.10), and (2.1) holds. This follows from formulas (2.9), (2.10) in view of the equation $|I_2| = 0$, which holds when $J = [n]$. □

Now we will clarify how the Maslov quantization conditions for the real Lagrangian Grassmannians $G_{2n}^{SO}(I_n)$ are related to those for $G_{2n}^{+}(I_n)$. There exists a natural map

$$\pi : G_{2n}^{SO}(I_n) \to G_{2n}^{+}(I_n) .$$

Since there existed a nonzero section $\sigma : G_{2n}^{+}(I_n) \to \eta$ over the space $G_{2n}^{+}(I_n)$, there also exists a nonzero section which is the preimage of the section σ:

$$\pi^*(\sigma) : G_{2n}^{SO}(I_n) \to \pi^*(\eta) .$$

On the other hand, it is possible to construct another section on the bundle $\pi^*(\eta) = \eta_{SO}$, using the orientation of the Lagrangian planes $(L, \epsilon) \in G_{2n}^{SO}(I_n)$. As we did for the section σ, we will construct a function $g(a_1, \ldots, a_n)$ on the orthonormal bases $(a_1 \ldots a_n)$ of the plane L, in such a way that

$$g(a_1, \ldots, a_n) \det(c_{ij})^{-1} = g(a_1', \ldots, a_n') , \tag{2.11}$$

where $a_k = \sum_j c_{kj} a_j'$.

For any real orthonormal basis $(a_1, \ldots, a_n)$ which gives the orientation ϵ on the real Lagrangian plane $L \subset \Phi_{\mathbb{R}}$, we set

$$g\left(a_1, \ldots, a_n\right) = 1 \ . \tag{2.12}$$

It is obvious that condition (2.11) holds for any other basis which gives the same orientation ϵ on L, since $\det\left(c_{ij}\right) = 1$.

We will denote the section constructed by Formula (2.12) by $\sigma_{\mathbb{R}}$. Thus we have two nonzero sections:

$$\pi^*\sigma : G_{2n}^{SO}\left(I_n\right) \to \eta_{SO}$$

and

$$\sigma_{\mathbb{R}} : G_{2n}^{SO}\left(I_n\right) \to \eta_{SO}$$

of the bundle η_{SO}. Let

$$f : \mathbb{C} \times G_{2n}^{SO}\left(I_n\right) \to \eta_{SO}$$

be the isomorphism of bundles defined by

$$f\left(\lambda, x\right) = \lambda \pi^* \sigma(x) \ .$$

Then the second section $\sigma_{\mathbb{R}}$ gives us a mapping of the space $G_{2n}^{SO}\left(I_n\right)$ into S^1:

$$g : G_{2n}^{SO}\left(I_n\right) \to S^1 \ ,$$

which is defined in the following way. The composition of maps $f^{-1} \circ \sigma_{\mathbb{R}}$ has the form

$$\left(f^{-1} \circ \sigma_{\mathbb{R}}\right)(x) = \left(h(x), x\right), h(x) \in \mathbb{C} \ ,$$

where $h(x) \neq 0$. We set

$$g(x) = h(x)/|h(x)| \in S^1 \subset \mathbb{C} \ .$$

Theorem 2.3.

a) $H^1\left(G_{2n}^{SO}\left(I_n\right)\right) = \mathbb{Z}$.

b) *Let u_n be a generator of $H^1\left(G_{2n}^{SO}\left(I_n\right)\right)$. Then it is possible to choose a generator $s \in H^1\left(S^1\right)$ of the group $H^1\left(S^1\right) = \mathbb{Z}$ such that $g^*(s) = u_n$.*

Theorem 2.3 can be formulated in the following way as well. Let

$$F : \mathbb{C} \times G_{2n}^{+}\left(I_n\right) \to \eta$$

be the isomorphism of bundles defined by

$$F(\lambda, x) = \lambda \sigma(x) \ .$$

We denote by η_0 the image

$$\eta_0 = F\left(\mathbb{C}^* \times G_{2n}^{+}\left(I_n\right)\right) \ ,$$

where $\mathbb{C}^* = \mathbb{C} \setminus \{0\}$. The space η_0 is homeomorphic to $\mathbb{C}^* \times G_{2n}^+(I_n)$.

Let

$$a_n = \left(F^{-1}\right)^* (s) \in H^1(\eta_0) \ , \tag{2.13}$$

where s is a generator of $H^1(C^*) = \mathbb{Z}$.

Theorem 2.4. *Let*

$$\omega : \pi^*(\eta) \to \eta$$

be a mapping of the above bundles. Then

a) $\omega\sigma_{\mathbb{R}}\left(G_{2n}^{SO}(I_n)\right) \subset \eta_0$;

b) $(\omega\sigma_{\mathbb{R}})^*(a_n) = u_n$.

To prove Theorems 2.3 and 2.4 we carry out several ancillary topological constructions.

We define imbeddings

$$\begin{aligned} i_n &: U(2n, I_n) \to U(2n+2, I_{n+1}) \ , \\ i_n^U &: U(n) \to U(n+1) \ , \\ i_n^G &: G_{2n}(I_n) \to G_{2n+2}(I_{n+1}) \end{aligned}$$

by the formulas:

a) if $C \in U(2n, I_n)$,

$$C = \begin{pmatrix} C_1 & C_2 \\ C_3 & C_4 \end{pmatrix} \ ,$$

then

$$i_n(C) = \begin{pmatrix} C_1 & 0 & C_2 & 0 \\ 0 & 1 & 0 & 0 \\ C_3 & 0 & C_4 & 0 \\ 0 & 0 & 0 & 1 \end{pmatrix} \ ;$$

b) if $C \in U(n)$, then

$$i_n^U(C) = \begin{pmatrix} C & 0 \\ 0 & 1 \end{pmatrix} \ ; \tag{2.14}$$

c) if $L \in G_{2n}$, then

$$i_n^G(L) = I_n^{\Phi}(L) + \{e_{n+1}\} \ .$$

Let

$$j_n : U(n) \to U(2n, I_n)$$

be given by the formula

$$j_n(C) = \begin{pmatrix} C & 0 \\ 0 & \overline{C} \end{pmatrix} ,$$

and let

$$\theta : U(2n, I_n) \to G_{2n}(I_n)$$

be the projection described earlier,

$$\theta_n(C) = C(\mathbb{C}^n) .$$

Then

a) The diagram

$$\begin{array}{ccccc} U(n) & \xrightarrow{i_n} & U(2n, I_n) & \xrightarrow{\theta_n} & G_{2n}(I_n) \\ \downarrow i_n^U & & \downarrow i_n & & \downarrow i_n^G \\ U(n+1) & \xrightarrow[j_{n+1}]{} & U(2n+2, I_{n+1}) & \xrightarrow[\theta_{n+1}]{} & G_{2n+2}(I_{n+1}) \end{array}$$

commutes.

b) $i_n^G(G_{2n}^+(I_n)) \subset G_{2n+2}^+(I_{n+1})$.

c) There exists an isomorphism

$$\varphi_n : \zeta_n \oplus 1 \to (i_n^G)^* (\zeta_{n+1}) ,$$

under which, if $x \in L$, $L \in G_{2n}(I_n)$, then

$$\varphi_n(x) = i_n^{\Phi}(x) \oplus 0 \in i_n^G(L) ;$$

if $\lambda \in \mathbb{C}$, then

$$\varphi_n(\lambda) = 0 \oplus \lambda e_{n+1} \in i_n^G(L) = i_n^{\Phi}(L) \oplus \{e_{n+1}\} .$$

d) The isomorphism φ_n induces an isomorphism of bundles

$$\psi : \eta_n \to (i_n^G)^* \eta_{n+1} ,$$

where $\eta_n = (\wedge^n \zeta_n)^*$, by the formula

$$\psi_n = \left\{ \left(\wedge^{n+1} \varphi_n \circ \chi_n\right)^{-1} \right\}^* , \tag{2.15}$$

where $\chi_n : \wedge^n \zeta_n \to \wedge^{n+1} (\zeta_n \oplus 1)$ is the natural isomorphism.

e) Let

$$\sigma_n : G_{2n}^+(I_n) \to \eta_n$$

be the section constructed by Theorem 1.4. Then

$$\sigma_n = \psi_n^* (i_n^G)^* \sigma_{n+1} .$$

Next, let $\eta_{0n} \subset \eta_n$ be the space of nonzero vectors of the bundle η_n, and let $a_n \in H^1(\eta_{0n})$ be defined by Eq. (2.13).

Let the imbeddings

$$i_n^o : O(n) \to O(n+1) ,$$
$$i_n^{SO} : SO(n) \to SO(n+1)$$

be defined by Eq. (2.14). Let

$$\alpha_n : SO(n) \to O(n) ,$$
$$\beta_n : O(n) \to U(n)$$

be the natural imbeddings.

Then the following diagram commutes:

$$\begin{array}{ccccc} SO(n) & \xrightarrow{\alpha_n} & O(n) & \xrightarrow{\beta_n} & U(n) \\ \downarrow i_n^{SO} & & \downarrow i_n^O & & \downarrow i_n^U \\ SO(n+1) & \xrightarrow[\alpha_{n+1}]{} & O(n+1) & \xrightarrow[\beta_{n+1}]{} & U(n+1) \end{array}$$

and we obtain

$$\pi_0(U(n)) = 0 , \quad \pi_1(U(n)) = \mathbb{Z} .$$

The homomorphism

$$\left(i_n^U\right)_* : \pi_1(U(n)) \to \pi_1(U(n+1))$$

is an isomorphism,

$$\pi_0(O(n)) = \mathbb{Z}_2 , \quad \pi_1(O(1)) = 0 ,$$
$$\pi_1(O(2)) = \mathbb{Z} , \quad \pi_1(O(n)) = \mathbb{Z}_2 \quad \text{for } n \geq 3 ,$$

and the homomorphism

$$\left(i_n^O\right)_* : \pi_1(O(n)) \to \pi_1(O(n+1))$$

is an isomorphism for $n \geq 3$.

Next, $\pi_0(SO(n)) = 0$, the homomorphism

$$(\alpha_n)_* : \pi_1(SO(n)) \to \pi_1(O(n))$$

is an isomorphism, and the homomorphism

$$(\beta_n)_* : \pi_1(O(n)) \to \pi_1(U(n))$$

is trivial.

Proceeding to the spaces $G_{2n}^{SO}(I_n)$, $G_{2n}^O(I_n)$, we obtain the following commutative diagram:

$$\begin{array}{ccc}
\pi_1(SO(n)) & \overset{(\alpha_n)_*}{\longrightarrow} & \pi_1(O(n)) \\
(\beta_n\alpha_n)_* \downarrow & & \downarrow (\beta_n)_* \\
\pi_1(U(n)) & \overset{\cong}{\longrightarrow} & \pi_1(U(n)) \\
(\widetilde{\theta_n})_* \downarrow & & \downarrow (\theta_n)_* \\
\pi_1\left(G^{SO}_{2n}(I_n)\right) & \overset{\pi_*}{\longrightarrow} & \pi_1\left(G^{O}_{2n}(I_n)\right) \\
\partial_1 \downarrow & & \downarrow \partial \\
\pi_0(SO(n)) & \overset{(\alpha_n)_*}{\longrightarrow} & \pi_0(O(n)) \\
(\beta_n\alpha_n)_* \downarrow & & \downarrow (\beta_n)_* \\
\pi_0(U(n)) & \overset{\cong}{\longrightarrow} & \pi_0(U(n))
\end{array} \tag{2.16}$$

Since $(\beta_n)_* = 0$, this diagram reduces to the following:

$$\begin{array}{ccc}
0 & & 0 \\
\downarrow & & \downarrow \\
\mathbb{Z} & \overset{\cong}{\longrightarrow} & \mathbb{Z} \\
\downarrow & & \downarrow \\
\pi_1\left(G^{SO}_{2n}(I_n)\right) & \overset{\pi_0}{\longrightarrow} & \pi_1\left(G^{O}_{2n}(I_n)\right) \\
\downarrow & & \downarrow \\
0 & & \mathbb{Z}_2 \\
 & & \downarrow \\
 & & 0
\end{array}$$

Diagram (2.16) induces a mapping

$$\begin{array}{ccc}
i^G_n : U(n)/O(n) & \longrightarrow & U(n+1)/O(n+1) \\
\| & & \| \\
G^O_{2n}(I_n) & \longrightarrow & G^O_{2n+2}(I_{n+1})
\end{array}$$

so that the following diagram commutes:

$$\begin{array}{ccc}
O(n) & \overset{i^O_n}{\longrightarrow} & O(n+1) \\
\beta_n \downarrow & & \downarrow \beta_{n+1} \\
U(n) & \overset{i^U_n}{\longrightarrow} & U(n+1) \\
\theta_n \downarrow & & \downarrow \theta_{n+1} \\
G^O_{2n}(I_n) & \overset{i^G_n}{\longrightarrow} & G^O_{2n+2}(I_{n+1})
\end{array}$$

Thus we obtain the following commutative diagram of homotopy groups:

$$
\begin{array}{ccc}
0 & & 0 \\
\downarrow & & \downarrow \\
\pi_1(U(n)) & \xrightarrow{\left(i_n^U\right)_*} & \pi_1(U(n+1)) \\
\downarrow & & \downarrow \\
\pi_1\left(G_{2n}^O(I_n)\right) & \xrightarrow{\left(i_n^G\right)_*} & \pi_1\left(G_{2n+2}^O(I_{n+1})\right) \\
\downarrow & & \downarrow \\
\pi_0(O(n)) & \xrightarrow{\left(i_n^O\right)_*} & \pi_0(O(n+1)) \\
\downarrow & & \downarrow \\
0 & & 0
\end{array}
$$

Since $\left(i_n^U\right)_*$ and $\left(i_n^O\right)_*$ are isomorphisms, so is $\left(i_n^G\right)_*$. Thus the group $\pi_1\left(G_{2n}^O(I_n)\right)$ is isomorphic to $\pi_1\left(G_n^O(I_n)\right)$. It is easy to describe the space $G_2^O(I_1)$ in explicit form: it is the manifold of all lines in $\mathbb{R}^2$ passing through the origin, and, consequently, is homeomorphic to S^1. Thus,

$$\pi_1\left(G_{2n}^O(I_n)\right) = \pi_1\left(S^1\right) = \mathbb{Z} .$$

Consequently, the homomorphism π_* in Diagram (2.16) maps a generator of the group $\pi_1\left(G_{2n}^{SO}(I_n)\right)$ into the square of a generator of the group $\pi_1\left(G_{2n}^O(I_n)\right)$. We can now formulate the following assertion.

Lemma 2.5.

a) $H^1\left(G_{2n}^{SO}(I_n)\right) = \mathbb{Z}$.

b) $H^1\left(G_{2n}^O(I_n)\right) = \mathbb{Z}$.

c) *It is possible to choose generators* $u_n \in H^1\left(G_{2n}^{SO}(I_n)\right)$, $v_n \in H^1\left(G_{2n}^O(I_n)\right)$ *such that*

$$\pi_*(v_n) = 2u_n ,$$

$$\left(i_n^G\right)_*(v_{n+1}) = v_n ,$$

$$\left(i_n^G\right)_*(u_{n+1}) = u_n .$$

d) Let

$$\psi_n : \eta_n^{SO} \to \left(i_n^G\right)^* \eta_{n+1}^{SO}$$

be the isomorphism induced by Eq. (2.15) *and*

$$\sigma_{\mathbb{R},n} : G_{2n}^{SO}(I_n) \to \eta_n^{SO}$$

be the section constructed by Formula (2.12). *Then*

$$\sigma_{\mathbb{R},n} = \psi_n^* \left(i_n^G\right)^* \left(\sigma_{\mathbb{R},n+1}\right) .$$

Assertion (a), (b) and (c) have already been proven, and proposition (d) is trivial. □

Proof of Theorems 2.3 *and* 2.4. We observe that the assertions of Theorems 2.3 and 2.4 are equivalent. It is sufficient to prove the assertion only for $n = 1$. Indeed, the maps

$$g_n : G_{2n}^{SO}\left(I_n\right) \to S^1$$

satisfy the equation

$$g_n = g_{n+1} \circ i_n^G .$$

Therefore, if

$$u_n = g_n^*(s) ,$$

then

$$\left(i_n^G\right)^* u_{n+1} = u_n = \left(g_{n+1} \circ i_n^G\right)^* (s) = \left(i_n^G\right)^* g_{n+1}^*(s)$$

Since $\left(i_n^G\right)^*$ is an isomorphism, $u_{n+1} = g_{n+1}^*(s)$ and Theorem 2.3 will be proven by induction if we can show that

$$u_1 = g_1^*(s) .$$

Hence we construct a map g_1 of the space $G_2^{SO}\left(I_1\right)$ into the circle S^1. A point $L \in G_2^{SO}\left(I_1\right)$ is parametrized by a basis vector $a \in L$, $|a| = 1$. The section $\sigma_1 : G_2^{SO}\left(I_1\right) \to \eta_1^{SO}$ is determined by its value on the basis vector $a \in \eta_1^{SO}$, namely $\sigma_1(a) = a_1 + ia_2$, where

$$a = \begin{pmatrix} a_1 \\ a_2 \end{pmatrix}$$

are the coordinates of the vector $a \in \mathbb{R}^2$.

Meanwhile, the section $\sigma_{\mathbb{R},1} : G_2^{SO}\left(I_1\right) \to \eta_1^{SO}$ takes the value

$$\sigma_{\mathbb{R},1}(a) = 1$$

on the basis vector a. Therefore the function $g_1 : G_2^{SO}\left(I_1\right) \to S^1$ takes the following value:

$$g_1(a) = \left(a_1 + ia_2\right)^{-1} .$$

Then the mapping

$$g_1 : G_2^{SO}\left(I_1\right) = S^1 \to S^1$$

has degree 1, i.e. it is possible to choose a generator of the group $H^1\left(G_2^{SO}(I_1)\right) = \mathbb{Z}$ such that

$$g_1^*(s) = u_1 .$$

This proves Theorem 2.3. □

3.3 Analysis on s-Analytic Manifolds

3.3.1 Fundamental Definitions. We consider a smooth (class C^∞) manifold M of even dimension $2n$. Assume that a nonnegative function ρ_M such that $\rho_M^2 \in C^\infty(M)$ is given on M. We will call such a function a weight function. We denote by $\Omega(M, \rho_M)$ the zero set of the function ρ_M:

$$\Omega(M, \rho_M) = \{\alpha \in M | \rho_M(\alpha) = 0\} \ . \tag{3.1}$$

Note. In fact, as will be seen below, we will only need a certain neighborhood of the set $\Omega(M, \rho_M)$ in the manifold M. Thus we will always assume that all our considerations take place in a certain sufficiently small neighborhood of the set $\Omega(M, \rho_M)$.

Definition 3.1. We denote by ${}^sI(M, \rho_M)$ the space of all complex-valued functions $f \in C^\infty(M, \mathbb{C})$ for which the following condition holds: each point x_0 is contained in some local chart $U \subset M$ with local coordinates $(x^1, \ldots, x^{2n})$ such that for any multi-index $\gamma = (\gamma_1, \ldots, \gamma_{2n})$, $|\gamma| = \gamma_1 + \ldots + \gamma_{2n} < s$ there exists a constant $c > 0$ such that

$$|D_x^\gamma f(x)| \le C\,[\rho(x)]^{s-|\gamma|} \ . \tag{3.2}$$

It is not difficult to see that condition (3.2) is independent of the choice of a local system of coordinates. More precisely, the following lemma is true.

Lemma 3.1. *Suppose $f \in {}^sI(M, \rho_M)$. For any point $x_0 \in M$ and any local coordinate system $(y^1, \ldots, y^{2n})$ in a neighborhood of this point there exists a neighborhood V such that $x_0 \in V$ and*

$$\left|D_y^\gamma f(y)\right| \le C\,|\rho(y)|^{s-|\gamma|}$$

for any multi-index γ, $|\gamma| < s$.

Proof. Let (U, x) be the local coordinate system whose existence is guaranteed by Definition 3.1, and suppose V is a neighborhood of the point x_0, compactly imbedded in U and contained in the range of the coordinates y. Then

$$D_y^\gamma f(y) = \sum_{|\delta| \le |\gamma|} a_\delta(x) D_x^\delta f(x)\big|_{x=x(y)} \ , \tag{3.3}$$

where the functions $a_\delta(x)$ are bounded in the neighborhood V. Estimating each term of the right-hand side of (3.3) using the inequalitiies (3.2), we obtain the assertion of the lemma.

We point out the following property of the spaces ${}^sI(M, \rho_M)$.

1. If $f \in {}^sI(M, \rho_M)$, M' is a submanifold of M, and $\rho_{M'} = \rho_M|M'$, then $f|M' \in {}^sI(M', \rho_M)$.

2. If $\{U_j\}$ is a locally finite covering of the manifold M by open sets, then the inclusion $f \in^s I(M, \rho_M)$ is equivalent to the system of inclusions $f|U_i \in {}^sI(U_i, \rho_M|U_i)$. In other words, belonging to the space ${}^sI(M, \rho_M)$ is a local property.

Lemma 3.2. *The space ${}^sI(M, \rho_M)$ is an ideal of the ring $C^\infty(M, \mathbb{C})$ with respect to pointwise multiplication of smooth functions. Moreover,*

a) ${}^tI(M, \rho_M) \subseteq {}^sI(M, \rho_M)$, $s < t$.

b) ${}^0I(M, \rho_M) = C^\infty(M, \mathbb{C})$.

The proof of the lemma follows immediately from Leibniz's formulas for differentiating the product

$$D_x^\gamma(f(x)g(x)) = \sum_{\delta \leq \gamma} \frac{\gamma!}{\delta!(\gamma - \delta)!} D_x^\delta f(x) D_x^{\gamma-\delta} g(x) , \tag{3.4}$$

where $\delta \leq \gamma$ means that $\delta_i \leq \gamma_i$, $i \in 1, \ldots, 2n$; $\gamma - \delta = (\gamma_1 - \delta_1, \ldots, \gamma_{2n} - \delta_{2n})$, $\gamma! = \gamma_1!, \ldots, \gamma_{2n}!$. Assertions a) and b) are now obvious.

Now suppose that $U \subset \mathbb{C}^n$ is an open set and ρ is a weight function on it. We will denote the coordinates in $\mathbb{C}^n$ by $\alpha^1, \ldots, \alpha^n$; $\alpha^j = a^j + ib^j$. We introduce the operators

$$\frac{\partial}{\partial \alpha^k} = \frac{1}{2}\left(\frac{\partial}{\partial a^k} - i\frac{\partial}{\partial b^k}\right) , \quad \frac{\partial}{\partial \overline{\alpha}^k} = \frac{1}{2}\left(\frac{\partial}{\partial a^k} + i\frac{\partial}{\partial b^k}\right) . \tag{3.5}$$

Definition 3.2. A smooth function $f \in C^\infty(U, \mathbb{C})$ is called s-analytic with respect to the weight function ρ if

$$\frac{\partial f}{\partial \overline{\alpha}^k} \in {}^sI(U, \rho) , \qquad k = 1, \ldots, n .$$

It is obvious that the set of s-analytic functions in the domain U is a *ring* with respect to pointwise multiplication of functions.

We now return to the manifold M. Suppose a covering $M = \cup_k U_k$ of M by open sets is given, as well as a system of homeomorphims φ_k: $U_k \to V_k \subset \mathbb{C}^n$. We set

$$U_{kl} = U_k \cap U_l , \quad V_{kl} = \varphi_k(U_{kl}) \subset V_k ;$$

$$\varphi_{kl} : V_{kl} \to V_{lk} , \quad \varphi_{kl} = (\varphi_l|U_{kl}) \circ (\varphi_k|U_{kl})^{-1} .$$

Definition 3.3. We will say that the system (U_k, φ_k) defines an *s-analytic manifold structure* on (M, ρ_M) if the mappings φ_{kl} are given by s-analytic functions on V_{kl} with respect to the weight functions

$$\rho_{kl}(\alpha) = \rho\left(\varphi_k^{-1}(\alpha)\right) , \qquad \alpha \in V_{kl} \subset \mathbb{C}^n .$$

A manifold M along with an s-analytic manifold structure will be called an *s-analytic manifold.*

A smooth function $f : M \to \mathbb{C}$ will be called *s-analytic* if the functions

$$f_k(\alpha) = f\left(\varphi_k^{-1}(\alpha)\right) , \qquad \alpha \in V_k \subset \mathbb{C}^n \tag{3.6}$$

are s-analytic on the domain V_k with respect to the weight functions

$$\rho_k(\alpha) = \rho\left(\varphi_k^{-1}(\alpha)\right) , \qquad \alpha \in V_k \subset \mathbb{C}^n . \tag{3.7}$$

Two s-analytic manifold structures $\{U_k, \varphi_k\}$ and $\{U_l', \varphi_l'\}$ are called *equivalent* if their union is also an s-analytic manifold structure. A local coordinate system consisting of a domain $U \subset M$ and a homeomorphism $\varphi : U \to V \subset \mathbb{C}^n$ is called an s-analytic system of coordinates if adding it to the original atlas of charts defining the s-analytic manifold structure results in a new atlas of charts which also defines an s-analytic manifold structure.

Lemma 3.3. *Suppose f is an s-analytic function, and (U, φ) is an s-analytic system of coordinates. Then the function $\widetilde{f} = f\left(\varphi^{-1}(x)\right)$ is s-analytic on the domain V with respect to the weight function $\widetilde{\rho} = \rho\left(\varphi^{-1}(x)\right)$.*

Proof. We need to verify that the functions $\partial f/\partial \overline{\alpha}^k$ belong to the space ${}^sI(V, \rho)$. Since this is a local inclusion, we can check it in a neighborhood of any point x_0 in V. Suppose (U_k, φ_k) is a system of coordinates in the original atlas, containing the point $\varphi^{-1}(x_o)$. Then in a neighborhood of the point x_0

$$\widetilde{f}(\alpha) = f_k\left(\psi(\alpha)\right) , \quad \psi_k \circ \varphi^{-1} ,$$

where f_k is given by formula (3.6).

We note that by s-analyticity of the chart, ψ is given by s-analytic functions. Therefore

$$\frac{\partial \widetilde{f}(\alpha)}{\partial \overline{\alpha}^k} = \sum_{j=1}^{n} \left[\frac{\partial f_k}{\partial \overline{\beta}^j} \frac{\partial \overline{\psi}^j}{\partial \overline{\alpha}^k} + \frac{\partial f_k}{\partial \beta^j} \frac{\partial \psi^j}{\partial \overline{\alpha}^k} \right] ,$$

where $\{\beta^j\}$ are coordinates on V_k. Furthermore, $\frac{\partial f_k}{\partial \overline{\beta}^j} \in {}^sI\left(V_k, \rho_k\right)$, and hence, by virtue of Lemma 3.1, $\frac{\partial f_k}{\partial \overline{\beta}^j} \in {}^sI(V, \widetilde{\rho})$. The functions ψ^j are also

s-analytic, i.e. $\frac{\partial \psi^j}{\partial \alpha^k} \in {}^sI(V, \widetilde{\rho})$. The necessary inclusion now follows from Lemma 3.2. □

From Lemma 3.3 it follows that if M is an s-analytic manifold, then a function f which is s-analytic in one structure is s-analytic in any other equivalent structure.

Definition 3.4. We will denote by ${}^s\mathcal{O}'(M, \rho_M)$ the ring of s-analytic functions on M.

Definition 3.5. Suppose (M, ρ_M) and (N, ρ_N) are s-analytic manifolds. The map $f : M \to N$ is called s-analytic if it is given by s-analytic functions in s-analytic coordinates and if

$$\rho_N(f(x)) \leq C\rho_M(x) . \tag{3.8}$$

Lemma 3.4. *If f is an s-analytic mapping of the manifold (M, ρ_M) into the manifold (N, ρ_N), then for any t-analytic function F on $N, t \leq s$, the function $F \circ f(x)$ is t-analytic.*

Proof. Let $(\alpha^1, \ldots, \alpha^n)$ be an s-analytic system of coordinates in a neighborhood of the point $x_0 \in M$, $(\beta^1, \ldots, \beta^n)$ s-analytic coordinates in a neighborhood of $f(x_0) \in N$. Then

$$F \circ f(\alpha) = F\left(f^1\left(\alpha^1, \ldots, \alpha^n\right), \ldots, f^m\left(\alpha^1, \ldots, \alpha^n\right)\right)$$

where $f^1, \ldots, f^m$ represent the mapping f in the coordinates α, β.

Differentiating the latter equation and using the t-analyticity of the functions F, f^j, $j = 1, \ldots, m$, as well as condition (3.8), we obtain the desired assertion. □

Definition 3.6. An s-analytic mapping $f : (M, \rho_M) \to (N, \rho_N)$ is called an *imbedding* if for any s-analytic systems of coordinates $(\alpha^1, \ldots, \alpha^n)$ on M and $(\beta^1, \ldots, \beta^n)$ on N the rank of the matrix

$$Df = \left(\frac{\partial \beta^k}{\partial \alpha^j} \right)$$

is n on the set $\Omega(M, \rho_M)$ and the mapping f is one-to-one.

3.3.2 The Operators of Restriction and Continuation. Partitions of unity. Suppose M is an s-analytic manifold with weight function ρ_M and atlas of s-analytic charts $\{U_k, \varphi_k\}$. From here on we will always assume that the following condition on the atlas of s-analytic charts is satisfied.

Condition 3.1. *For any chart* (U_k, ρ_k) *there exists a constant* $C > 0$ *such that*

$$\rho_M(x) \geq C \sum_{l=1}^{n} \left|\operatorname{Im} \varphi_k^l(x)\right| , \tag{3.9}$$

where $\varphi_k(x) = \left(\varphi_k^1(x), \ldots, \varphi_k^n(x)\right) \in \mathbb{C}^n$.

In the future we will consider only s-analytic charts which satisfy Condition 3.1.

We note that it follows from Condition 3.1, in particular, that the functions $\varphi_k(x)$ are real on the set $\Omega(M, \rho_M)$ for any s-analytic chart U, since $\Omega \cap U$ is contained in the submanifold U^0 given by the equation $\operatorname{Im} \alpha = 0$. Let (U, α) be a chart having a nonempty intersection with Ω. Since we are only interested in a neighborhood of the set Ω, we may assume without loss of generality that

$$U = U^0 \times W ,$$

where W is the image of the chart U under projection onto the subspace of imaginary coordinates.

Let R^0 be the operator of restriction of the ring $C^\infty(U, \mathbb{C})$ to the ring $C^\infty(U^0, \mathbb{C})$:

$$R^0 : C^\infty(U, \mathbb{C}) \to C^\infty\left(U^0, \mathbb{C}\right) .$$

We prove the following assertion.

Proposition 3.5. *For any* $t \leq s$ *and any function* $f \in C^\infty\left(U^0, \mathbb{C}\right)$ *there exists a* t*-analytic function* $f' \in {}^t\mathcal{O}'(U, \rho_M|U)$ *such that* $R^0 f' = f$. *Furthermore, if for some* t*-analytic function* g *in the chart* U *we have* $R^0 g = 0$, *then* $g \in {}^{t+1}I(U, \rho_M|U)$.

Proof. We define an operator ${}^tA^0 : C^\infty\left(U^0, \mathbb{C}\right) \to C^\infty(U, \mathbb{C})$ by the formula

$$\left({}^tA^0 f\right)(\alpha) = \sum_{|k| \leq t} (ib)^k \frac{\partial^{|k|} f(a)}{\partial a^k} , \tag{3.10}$$

where $k = (k_1, \ldots, k_n)$ is a multi-index and $\alpha^j = a^j + ib^j$. We check the s-analyticity of the function (3.10). A direct calculation shows that

$$\frac{\partial}{\partial \bar{\alpha}^j} \left({}^tA^0 f\right) = \frac{1}{2} \sum_{|k|=t} \frac{1}{k!} (ib)^k \frac{\partial^{|k|}}{\partial a^k} \frac{\partial f}{\partial a^j} . \tag{3.11}$$

By condition 3.1 we have $|b| \leq C\rho_M|U$. From this it is clear that the right-hand side of formula (3.11) lies in the ideal ${}^tI(U, \rho_M|U)$. Thus there is a well-defined operator

$$ {}^{t}A^0 : C^{\infty}\left(U^0, \mathbb{C}\right) \to {}^{t}\mathcal{O}'\left(U, \rho_M|_U\right) . \tag{3.12}$$

From the definition (3.10) of the operator ${}^{t}A^0$ it is obvious that $R^0 \circ {}^{t}A^0$ is the identity map. This proves the first half of the proposition. Now suppose $R^0 g = 0$. By Taylor's formula we have

$$ g(a,b) = \sum_{|k|\le t} \frac{1}{k!} b^k \left[\left(\frac{\partial}{\partial b}\right)^k g(a,b)|_{b=0} \right] + \sum_{|k|=t+1} b^k G_k(a,b) . \tag{3.13}$$

But the function $g(a,b)$ is t-analytic, and hence

$$ \frac{1}{2}\left[\frac{\partial}{\partial \alpha^j} + i\frac{\partial}{\partial \beta^j}\right] g \in {}^{t}I\left(U, \rho_M|U\right)$$

from which

$$ \frac{\partial g}{\partial \beta^j} = i\frac{\partial g}{\partial \alpha^j} + h_j , \qquad h_j \in {}^{t}I\left(U, \rho_M|U\right) .$$

Differentiating the latter equation with respect to the coordinates a, b and using the fact that $\frac{\partial h_l}{\partial a^j}, \frac{\partial h_j}{\partial b^l} \in {}^{t-1}I\left(U, \rho_M|U\right)$ if $h_j, h_l \in {}^{t}I\left(U, \rho_M|U\right)$, we obtain by induction for any multi-index k, $|k| \le t$, that

$$ \frac{\partial^{|k|} g}{\partial b^k} = i^{|k|}\frac{\partial^{|k|} g}{\partial a^k} + h_k , \qquad h_k \in {}^{t-|k|+1}I\left(U, \rho_M|_U\right) . \tag{3.14}$$

We note now that by Condition 3.1 the last term in (3.13) lies in the ideal ${}^{t+1}I\left(U, \rho_M|U\right)$. The equation $R^0 g = 0$ gives us $\partial^{|k|} g/\partial \alpha^k\big|_{b=0} = 0$, and therefore it is sufficient to verify the inclusion

$$ b^k h_k(a,0) \in {}^{t+1}I\left(U, \rho_M|U\right) .$$

Using Taylor's formula, we have $\left(\text{since } \rho_M^2 \in C^{\infty}(M)\right)$

$$\begin{aligned} \rho^2(a,0) = \rho^2(a, b-b) &= \rho^2(a,b) - \sum_{j=1}^{n} b^j \frac{\partial}{\partial b^j}\rho^2(a,b) + \sum_{i,j=1}^{n} b_i b_j F_{ij}(a,b) \\ &\le \rho^2(a,b) + C_1|b|\rho(a,b) + C_2|b|^2 \le C_3\rho^2(a,b) , \end{aligned}$$

and, consequently,

$$ \rho^2(a,0) + |b|^2 \le C\rho^2(a,b) , \tag{3.15}$$

where we have used the inequality $|\operatorname{grad}\rho^2| \le C\rho$, which follows from the nonnegativity of the function ρ^2, and Condition 3.1. The desired conclusion is now obvious. □

Remark 3.2. Suppose the function $f(a)$ belongs to the ideal ${}^{t}I\left(U^0, \rho_M|U_0\right)$. Then the function $\left({}^{r}A^0 f\right)(\alpha)$ belongs to the ideal ${}^{t}I\left(U, \rho_M|U\right)$ for any

$r \geq 0$. The proof of this fact follows from the last estimate for $\rho^2(a,0)$ and Condition 3.1, analogously to the proof of the second part of Proposition 3.5.

Corollary 3.6.

$$^{t}A^0 \circ R^0 f - f \in {}^{t+1}I(U, \rho_M|U) \ . \tag{3.16}$$

Proof. Since $R^0\left({}^{t}A^0 \circ R^0 f - f\right) = R^0 f - R^0 f = 0$, the inclusion (3.16) follows from the second assertion of Proposition 3.5. □

Proposition 3.7.

$$\left(\frac{\partial}{\partial \alpha^j}{}^{t}A^0 - {}^{t-1}A^0\frac{\partial}{\partial a^j}\right) f \in {}^{t}I(U, \rho_M|_U) \ , \tag{3.17}$$

$$\left(\frac{\partial}{\partial a^j}R^0 - R^0\frac{\partial}{\partial \alpha^j}\right) f \in {}^{t}I\left(U^0, \rho_M\big|_{U^0}\right) \ . \tag{3.18}$$

Proof. Since $f \in {}^{t}\mathcal{O}'(U, \rho_M|U)$,

$$\begin{aligned}\frac{\partial}{\partial \alpha^j} f &= \frac{1}{2}\left[\frac{\partial}{\partial a^j} - i\frac{\partial}{\partial b^j}\right] f = \frac{\partial f}{\partial a^j} - \frac{1}{2}\left[\frac{\partial}{\partial \alpha^j} + \frac{\partial}{\partial b^j}\right] f \\ &= \frac{\partial f}{\partial a^j} + h \ , \qquad h \in {}^{t}I(U, \rho_M|U) \ .\end{aligned}$$

Restricting this relation to U^0, we obtain Eq. (3.18).

Furthermore, in Eq. (3.18) we can replace f by the function ${}^{t}A^0 f$ for some $f \in C^\infty\left(U^0, \mathbb{C}\right)$. Using the identity $R^0 \circ {}^{t}A^0 f = f$, we obtain

$$\left[\frac{\partial}{\partial a^j} - R^0\frac{\partial}{\partial a^j}{}^{t}A^0\right] f = h \in {}^{t}I\left(U^0, \rho_M|U^0\right) \ .$$

We apply the operator ${}^{t-1}A^0$ to the latter relation and use the inclusion (3.16) with t replaced by $t-1$. We obtain

$$\left[{}^{t-1}A^0\frac{\partial}{\partial a^j} - \frac{\partial}{\partial \alpha^j}{}^{t}A^0\right] f = {}^{t-1}A^0 h + h' \ , \qquad h' \in {}^{t}I(U, \rho_M|U) \ .$$

But ${}^{t-1}A^0 h \in {}^{t}I(U, \rho_M|U)$ by virtue of Remark 3.2. This proves the proposition. □

Now we prove that an s-analytic partition of unity exists on an s-analytic manifold.

Proposition 3.8. *Suppose $\{V_k\}$ is a locally finite covering of the manifold M by open sets. Then there exists a sytem of s-analytic functions $\{\varphi_k\}$ such that*

a) $\operatorname{supp} \varphi_k \subset V_k$,

b) $\sum \varphi_k \equiv 1$.

Proof. Without loss of generality it is possible to assume that the sets V_k lie in a sufficiently small neighborhood of the set $\Omega(M, \rho_M)$. We consider a locally finite atlas $\{U_j^0 \times W_j\}$ such that each chart of this atlas lies completely in one of the sets V_k. Suppose $\widetilde{U}_j^0$ are open sets, compactly imbedded in U_j^0, such that $\{U_j^0 \times W_j\}$ form a covering of the manifold M (more precisely, some neighborhood of the set $\Omega(M, \rho_M)$). In each chart $\{U_j^0 \times W_j\}$ we consider a function ψ_j of compact support on U_j^0 which is identically equal to one on the set U_j^0. The functions

$$\widetilde{\psi}_j = ({}^s A^0 \psi_j) \cdot \chi_j ,$$

where χ_j are functions of compact support on W_j which are equal to 1 on neighborhood of zero, clearly have the following properties:

a) $\widetilde{\psi}_j$ are s-analytic (by Proposition 3.5);

b) $\widetilde{\psi}_j$ are real and nonnegative on the set $\Omega(M, \rho_M)$;

c) $\widetilde{\psi}_j$ form a locally finite family and $\widetilde{\psi}_j = 1$ on $\widetilde{U}_j^0$.

Now we consider the functions

$$\widetilde{\varphi}_k = \sum_{U_j^0 \times W_j \subset V_j} \widetilde{\psi}_j .$$

These are analytic functions and $\operatorname{supp} \widetilde{\varphi}_k \subset V_k$. Furthermore the sum $\widetilde{\varphi} = \sum_k \widetilde{\varphi}_k$ is real and ≥ 1 on the set $\Omega(M, \rho_M)$, as follows from property c) of the functions $\widetilde{\psi}_j$. Therefore in some neighborhood of this set $\widetilde{\varphi} \geq \frac{1}{2}$. Restricting our attention to this neighborhood, we define the functions

$$\varphi_k = \frac{\widetilde{\varphi}_j}{\widetilde{\varphi}} .$$

It is obvious that the functions thus constructed satisfy all the requirements of Proposition 5.8. □

Corollary 3.9. *Suppose F is a closed set in M and V is some neighborhood of F. There exists an s-analytic function φ such that* $\operatorname{supp} \varphi \subset V$, $\varphi | F \equiv 1$.

Proof. It is sufficient to apply Proposition 3.8 to the sets V and $M \setminus F$. □

Corollary 3.10. *Suppose (M, ρ_M) is an s-analytic manifold, $V \subset M$ is an open set, f is an s-analytic function on M, and g is an s-analytic function on the set V. Suppose*

$$f|_V - g \in {}^sI(V, \rho_M|_V) \ . \tag{3.19}$$

Then for any closed set $F \subset V$ there exists an s-analytic fuction $\widetilde{f}$ on the manifold M such that

$$\widetilde{f}|_F = g|_{F'} \ , \qquad \widetilde{f} - f \in {}^sI(M, \rho_M) \ . \tag{3.20}$$

Proof. Let φ be the function whose existence was asserted in Corollary 3.9. We set

$$\widetilde{f} = (1 - \varphi)f + \varphi g$$

(extending the function $\varphi \cdot g$ by the zero function on M). Since $\varphi \equiv 1$ on F, $\widetilde{f}|F = (1-\varphi)f|F + \varphi g|F = 0 + g|F = g|F$. Furthermore,

$$\widetilde{f} - f = (1-\varphi)f + \varphi g - f = \varphi(g - f) \in {}^sI(M, \rho_M)$$

by virtue of Eq. (3.19) and the fact that $\operatorname{supp}\varphi \subset V$. □

3.3.3 Structure Rings. Vector Fields and Forms. In this subheading we will show that it is possible to study not only s-analytic functions, but also s-analytic vector fields and differential forms on s-analytic manifolds, and that it is, moreover, possible to interpret them in a certain sense as the sections of certain vector bundles over the manifold (M, ρ_M).

We consider the subring ${}^t\mathcal{O}'(M, \rho_M)$ of the ring $C^\infty(M)$ of all smooth functions on the manifold M. The ideal ${}^{t+1}I(M, \rho_M)$ belongs to the ring ${}^t\mathcal{O}'(M, \rho_M)$.

We denote by ${}^t\mathcal{O}(M, \rho_M)$ the factor ring

$${}^t\mathcal{O}(M, \rho_M) = {}^t\mathcal{O}'(M, \rho_M) \,/\, {}^{t+1}I(M, \rho_M) \ . \tag{3.21}$$

Definition 3.7. The *support* $\operatorname{supp} f$ of an element $f \in {}^t\mathcal{O}(M, \rho_M)$ is the intersection of the supports of all the functions in the equivalence class of f.

It is obvious that the inclusion

$$\operatorname{supp} f \subset \Omega(M, \rho_M) \tag{3.22}$$

holds for any element $f \in {}^t\mathcal{O}(M, \rho_M)$.

Proposition 3.11. *Suppose* $f \in {}^t\mathcal{O}(M, \rho_M)$ *is an element such that*

$$\operatorname{supp} f = \emptyset \ .$$

Then

$$f = 0 \ .$$

Proof. Suppose $\operatorname{supp} f = \emptyset$. By the definition of the support of an element in ${}^t\mathcal{O}(M, \rho_M)$, for each point $\alpha \in \Omega(M, \rho_M)$ there exists a neighborhood U_α of this point and a function f_α such that

$$f' - f_\alpha \in {}^{t+1}I(M, \rho_M) \ , \qquad f_\alpha | U_\alpha = 0 \ ,$$

where f' is some representative of the equivalence class of f. We choose from the sets U_α a locally finite system $\{U_{\alpha_i}\}$ which covers the set $\Omega(M, \rho_M)$. Let $\{e_i\}$ be an s-analytic partition of unity subordinate to this covering (Proposition 3.8). By construction $\sum_i f_{\alpha_i} e_i = 0$. Furthermore, we have

$$f' = f' - \sum_i f_{\alpha_i} e_i = \sum_i (f' - f_{\alpha_i}) e_i \in {}^{t+1}I(M, \rho_M) \ ,$$

since $f' - f_{\alpha_i} \in {}^{t+1}I(M, \rho_M)$ and the sum is locally finite (see property 2 of the ideals ${}^sI(M, \rho_M)$, formulated at the beginning of this subheading). The latter inclusion means precisely that $f = 0$. □

Lemma 3.12. *Whenever* $t_1 \leq t_2$ *the inclusion*

$$ {}^{t_2}\mathcal{O}'(M, \rho_M) \subseteq {}^{t_1}\mathcal{O}'(M, \rho_M) \ , \tag{3.23}$$

holds, which, together with the inclusion (a) *of Lemma* 3.2

$$ {}^{t_2+1}I(M, \rho_M) \subseteq {}^{t_1+1}I(M, \rho_M) \tag{3.24}$$

induces a homomorphism

$$ {}^{t_2}\mathcal{O}(M, \rho_M) \to {}^{t_1}\mathcal{O}(M, \rho_M) \ . \tag{3.25}$$

If the element $f \in {}^{t_2}\mathcal{O}(M, \rho_M)$ *is mapped by the homomorphism* (3.25) *into the element* $g \in {}^{t_1}\mathcal{O}(M, \rho_M)$, *then*

$$\operatorname{supp} g \subset \operatorname{supp} f \ .$$

Proof. The first assertion follows from Lemma 3.2. The second assertion follows from the inclusions (3.23), (3.24).

The assertion about the supports follows from the inclusion

$$f' + {}^{t_2+1}I(M, \rho_M) \subset f' + {}^{t_1+1}I(M, \rho_M)$$

for any $f' \in f + {}^{t_2}\mathcal{O}(M, \rho_M)$. This proves the lemma. □

Suppose $f : M \to N$ is an s-analytic mapping of s-analytic manifolds. According to Lemma 3.4 the composition of a t-analytic function φ on the manifold N with the mapping f will be a t-analytic function on the manifold M. Thus the mapping f induces a ring homomorphism

$$f^* : {}^t\mathcal{O}(N, \rho_n) \to {}^t\mathcal{O}(M, \rho_M) .$$

Lemma 3.13. *Suppose M is an s-analytic manifold, $\{U_\alpha\}$ is a covering by open sets; $\rho_\alpha = \rho_M|U_\alpha$, $\rho_{\alpha\beta} = \rho_M|U_\alpha \cap U_\beta$. Suppose φ_α are t-analytic functions on U_α, subject to*

$$\varphi_\alpha|_{U_\alpha \cap U_\beta} - \varphi_\beta|_{U_\alpha \cap U_\beta} \in {}^{t+1}I(U_\alpha \cap U_\beta, \rho_{\alpha\beta}) . \qquad (3.26)$$

Then there exists a t-analytic function $\varphi \in {}^t\mathcal{O}'(M, \rho_M)$ such that

$$\varphi|_{U_\alpha} - \varphi_\alpha \in {}^{t+1}I(U_\alpha, \rho_\alpha) . \qquad (3.27)$$

Proof. Suppose ψ_α are s-analytic functions on the manifold M such that

$$\sum \varphi_\alpha \equiv 1 , \qquad \operatorname{supp} \psi_\alpha \subset U_\alpha .$$

Such functions ψ_α exist by Proposition 3.8. We set

$$\varphi = \sum_\alpha \psi_\alpha \varphi_\alpha \in {}^t\mathcal{O}'(M, \rho_M) .$$

Then

$$\varphi|U_\alpha = \sum_\beta (\psi_\beta \varphi_\beta)|U_\alpha .$$

The support $\operatorname{supp}(\psi_\beta \varphi_\beta)|U_\sigma$ lies in the set $U_\alpha \cap U_\beta$. Therefore

$$\psi_\beta \varphi_\beta = \psi_\beta \varphi_\alpha + r_{\alpha\beta} , \qquad r_{\alpha\beta} \in {}^{t+1}I(U_\alpha, \rho_\alpha) .$$

Hence

$$\varphi|U_\alpha = \sum_\beta \psi_\beta \varphi_\alpha + r_{\alpha\beta} = \varphi_\alpha + \sum_\beta r_{\alpha\beta} ,$$

in other words, condition (3.27) holds. This proves Lemma 3.13. □

Definition 3.8. Suppose $t \leq s-1$. A smooth vector field X on an s-analytic manifold (M, ρ_M) is called t-analytic if in each s-analytic chart with coordinates $(\alpha^1, \dots, \alpha^n)$ the field X can be represented in the form

$$X = \sum_{k=1}^{n} a^k(\alpha) \frac{\partial}{\partial a^k} + Y \ , \tag{3.28}$$

where the functions $a^k(\alpha)$ are t-analytic functions:

$$a^k(\alpha) \in {}^t\mathcal{O}'(M, \rho_M) \ ,$$

and the field Y has coefficients in the ideal ${}^{t+1}I(M, \rho_M)$:

$$\begin{gathered} Y = \sum_{k=1}^{n} \left(y_1^k(\alpha) \frac{\partial}{\partial a^k} + y_2^k(\alpha) \frac{\partial}{\partial \overline{\alpha}^k} \right) \ , \\ y_1^k, y_2^k \in {}^{t+1}I(M, \rho_M) \ , \quad k = 1, \dots, n \ . \end{gathered} \tag{3.29}$$

Lemma 3.14. *If the representation* (3.28) *holds for some atlas of s-analytic charts, then for any s-analytic chart (not in the fixed atlas) the representation* (3.28) *is possible. The condition* (3.29) *is invariant under s-analytic changes of coordinates.*

Proof. In essence it is necessary to show that if the functions

$$\beta^k = \beta^k (\alpha^1, \dots, \alpha^n)$$

define an s-analytic change of coordinates, then in the coordinates $\beta = (\beta^1, \dots, \beta^n)$ the field X can also be represented in the form (3.28). Indeed, we have

$$\begin{aligned} X = & \sum_{j,k=1}^{n} a^k(\alpha) \left[\frac{\partial \beta^j}{\partial \alpha^k} \frac{\partial}{\partial \beta^j} + \frac{\partial \overline{\beta}^j}{\partial \alpha^k} \frac{\partial}{\partial \overline{\beta}^j} \right] \\ & + \sum_{j,k=1}^{n} \left[y_1^k(\alpha) \left[\frac{\partial \beta^j}{\partial \alpha^k} \frac{\partial}{\partial \beta^j} + \frac{\partial \overline{\beta}^j}{\partial \alpha^k} \frac{\partial}{\partial \overline{\beta}^j} \right] + y_2^k(\alpha) \left[\frac{\partial \beta^j}{\partial \overline{\alpha}^k} \frac{\partial}{\partial \beta^j} + \frac{\partial \overline{\beta}^j}{\partial \overline{\alpha}^k} \frac{\partial}{\partial \overline{\beta}^j} \right] \right] \\ = & \sum_{j=1}^{n} \left[\sum_{k=1}^{n} a^k(\alpha) \frac{\partial \beta^j}{\partial \alpha^k} \right] \frac{\partial}{\partial \beta^j} + Z \ , \end{aligned}$$

where the field Z is equal to

$$Z = \sum_{j=1}^{n} \left[\sum_{k=1}^{n} \left[\frac{\partial \beta^j}{\partial \alpha^k} y_1^k(\alpha) + \frac{\partial \beta^j}{\partial \overline{\alpha}^k} y_2^k(\alpha) \right] \frac{\partial}{\partial \beta^j} \right.$$

$$\left. + \sum_{k=1}^{n} \left[\frac{\partial \overline{\beta}^j}{\partial \alpha^k} y_1^k(\alpha) + \frac{\partial \overline{\beta}^j}{\partial \overline{\alpha}^k} y_2^k(\alpha) + a^k(\alpha) \left[\frac{\partial \overline{\beta}^j}{\partial \overline{\alpha}^k} \right] \right] \frac{\partial}{\partial \overline{\beta}^j} \right] .$$

It is clear that Z satisfies condition (3.29), if $t \leq s - 1$. On the other hand, the functions $\sum_{k=1}^{n} a^k(\alpha) \frac{\partial \beta^j}{\partial \alpha^k}$ are t-analytic for $t \leq s-1$. This proves the lemma. □

Lemma 3.15. *Suppose X is a t-analytic field, $f \in {}^{t+1}\mathcal{O}'(M, \rho_M)$. Then $X(f) \in {}^{t}\mathcal{O}'(M, \rho_M)$.*

Proof. It is sufficient to show the assertion for some local s-analytic chart. Then the field X, by definition, can be represented in the form (3.28), where the field Y satisfies condition (3.29).

Thus

$$X(f) = \sum_{k=1}^{n} a^k(\alpha) \frac{\partial f}{\partial \alpha^k} + Y(f) . \tag{3.30}$$

If $f \in {}^{(t+1)}\mathcal{O}'(M, \rho_M)$ then the functions $\partial f / \partial \alpha^k$ are t-analytic functions, and therefore the first term in the sum in (3.30) is a t-analytic function. Since condition (3.29) holds,

$$Y(f) \in {}^{t+1}I(M, \rho_M) \subset {}^{t}\mathcal{O}'(M, \rho_M) .$$

Thus the second summand in (3.30) is also a t-analytic function. This proves Lemma 3.15. □

Corollary 3.16. *A t-analytic vector field X induces a ring homomorphism*

$$X : {}^{(t+1)}\mathcal{O}(M, \rho_M) \to {}^{t}\mathcal{O}(M, \rho_M) ,$$

under which the following relation holds:

$$X(f \cdot g) = g f(X) + f X(g) ,$$

$$f, g \in {}^{(t+1)}\mathcal{O}(M, \rho_M) .$$

We will denote by ${}^{t}T'(M, \rho_M)$ the space of all t-analytic vector fields on an s-analytic manifold M. The space ${}^{t}T'(M, \rho_M)$ is a ${}^{t}\mathcal{O}'(M, \rho_M)$-module. Suppose, furthermore, that ${}^{t+1}(IT)'(M, \rho_M)$ is the subspace of all vector fields satisfying condition (3.29). We set

$$^{t}T\left(M,\rho_{M}\right)={}^{t}T'\left(M,\rho_{M}\right)/\,{}^{t+1}(IT)'\left(M,\rho_{M}\right) \tag{3.31}$$

It is clear that the space ${}^{t}T\left(M,\rho_{M}\right)$ is a ${}^{t}\mathcal{O}\left(M,\rho_{M}\right)$-module.

Lemma 3.17. *Suppose U is an s-analytic chart. Then the ${}^{t}\mathcal{O}\left(U,\rho_{N}\right)$-module ${}^{t}T\left(U,\rho_{U}\right)$ is a free module with n independent generators.*

Proof. Let $\alpha=\left(\alpha^{1},\ldots,\alpha^{n}\right)$ be a local system of s-analytic coordinates. The vector fields

$$X_{k}=\frac{\partial}{\partial\alpha^{k}}$$

are t-analytic vector fields. Let ξ_{k} be their representatives in the factor space (3.31) Let Γ be the free ${}^{t}\mathcal{O}\left(U,\rho_{U}\right)$-module with generators X_{k}.

We construct a mapping of ${}^{t}\mathcal{O}\left(U,\rho_{U}\right)$-modules

$$h:\Gamma\to{}^{t}T\left(U,\rho_{U}\right)\ ,$$

setting

$$h\left(X_{k}\right)=\xi_{k}\ .$$

We will prove that h is an isomorphism. Suppose $\gamma=\sum_{k=1}^{n}a^{k}X_{k}\in\Gamma$ and $h(\gamma)=0$. This means that if f^{k} are the representatives of the elements a^{k} in the ring ${}^{t}\mathcal{O}'\left(U,\rho_{U}\right)$, then

$$\sum_{k=1}^{n}f^{k}\frac{\partial}{\partial\alpha^{k}}\in{}^{t+1}(IT)'\left(U,\rho_{U}\right)\ . \tag{3.32}$$

It is completely obvious that condition (3.32) means that $f^{k}\in{}^{t+1}I\left(U,\rho_{U}\right)$, i.e. $a^{k}=0$ for all k.

Suppose, conversely, that $\gamma\in{}^{t}T\left(U,\rho_{U}\right)$ and X is the vector field representing the element γ. Then, by Definition 3.8,

$$X=\sum_{k=1}^{n}f^{k}(\alpha)\frac{\partial}{\partial\alpha^{k}}+Y\ ,\qquad f^{k}(\alpha)\in{}^{t}\mathcal{O}'\left(U,\rho_{U}\right)\ ,$$

$$Y\in{}^{t+1}(IT)'\left(U,\rho_{U}\right)\ .$$

Consequently we obtain that

$$\gamma=h\left(\sum_{k=1}^{n}a^{k}X_{k}\right)\ ,$$

where a^{k} are the elements of the ring ${}^{t}\mathcal{O}\left(U,\rho_{U}\right)$ defined by the t-analytic functions f^{k}. This proves the lemma. □

Definition 3.9. Suppose

$$f : M \to N$$

is an s-analytic imbedding of manifolds, $X \in {}^tT(N, \rho_N)$. If there exists an element $Y \in {}^tT(M, \rho_M)$ such that the equation

$$f^*(X(\varphi)) = Y(f^*(\varphi)) \in {}^{t-1}\mathcal{O}(M, \rho_M) \tag{3.33}$$

holds for all $\varphi \in {}^t\mathcal{O}(N, \rho_N)$, then we will say that the submanifold M is invariant with respect to the vector field X.

We denote by $\wedge^k(M)$ the space of k-dimensional smooth forms on the manifold M.

Definition 3.10. Suppose M is an s-analytic manifold and ω is a smooth differential k-form on the manifold M. The form ω is called t-analytic if in each local s-analytic coordinate system $\alpha = (\alpha^1, \ldots, \alpha^n)$ ω can be expressed in the form

$$\begin{aligned} \omega = \sum_{j_1 < \ldots < j_k} a_{j_1 \ldots j_k}(\alpha) d\alpha^{j_1} \wedge \ldots \wedge d\alpha^{j_k} + \omega' \,, \\ a_{j_1 \ldots j_k}(\alpha) \in {}^t\mathcal{O}'(M, \rho_M) \,, \end{aligned} \tag{3.34}$$

where

$$\omega' \text{ is a form with coefficients in the ideal } {}^{t+1}I(M, \rho_M). \tag{3.35}$$

Lemma 3.18. *If the representation* (3.34) *is satisfied for some atlas of s-analytic charts, then for any s-analytic chart (not in the fixed atlas) the representation* (3.34) *is possible. Condition* (3.35) *is invariant with respect to s-analytic changes of coordinates.*

The proof is analogous to the proof of Lemma 3.14.

Lemma 3.19. *Suppose ω is a t-analytic k-form and $X_1, \ldots, X_k$ are t-analytic vector fields. Then*

$$\omega(X_1, \ldots, X_k) \in {}^t\mathcal{O}'(M, \rho_M) \,.$$

Moreover, if ω satisfies condition (3.35) *or if one of the vector fields satisfies condition* (3.29), *then*

$$\omega(X_1, \ldots, X_k) \in {}^{t+1}I(M, \rho_M) \,.$$

The proof is obvious.

We denote by ${}^t\Lambda'(M,\rho_M)$ the space of t-analytic forms, and by ${}^{t+1}(I\Lambda)'(M,\rho_M)$ the space of all forms satisfying condition (3.35). We set

$$ {}^t\Lambda(M,\rho_M) = {}^t\Lambda'(M,\rho_M) \,/\, {}^{t+1}(I\Lambda)'(M,\rho_M)\ . \tag{3.36} $$

The spaces ${}^t\Lambda_k(M,\rho_M)$ are ${}^t\mathcal{O}(M,\rho_M)$-modules, and by Lemma 3.19 the elements of the ${}^t\mathcal{O}(M,\rho_M)$-module ${}^t\Lambda_k(M,\rho_M)$ are induced as skew-symmetric ${}^t\mathcal{O}(M,\rho_M)$-linear functionals in the k variables $X_j \in {}^tT(M,\rho_M)$, $j = 1,\dots k$.

Proposition 3.20. *If ω_1, ω_2 are t-analytic forms, then $\omega_1 \wedge \omega_2$ is also a t-analytic form. If $\omega_1 \in {}^{t+1}(I\Lambda)'(M,\rho_M)$, then*

$$ \omega_1 \wedge \omega_2 \in {}^{t+1}(I\Lambda)'(M,\rho_M)\ . $$

The operation of exterior multiplication of forms induces a binary skew-symmetric operation on the spaces ${}^t\Lambda_k(M,\rho_M)$. If $\omega \in {}^t\Lambda'_k(M,\rho_M)$, then

$$ d\omega \in {}^{t-1}\Lambda'_{k+1}(M,\rho_M)\ . $$

If $\omega \in {}^{t+1}(I\Lambda)'_k(M,\rho_M)$, then

$$ d\omega \in {}^t(I\Lambda)'_{k+1}(M,\rho_M)\ . $$

The operator d induces a linear operator

$$ d : {}^t\Lambda_k(M,\rho_M) \to {}^{t-1}\Lambda_{k+1}(M,\rho_M)\ , $$

for which, if $f \in {}^t\mathcal{O}(M,\rho_M) = {}^t\Lambda_0(M,\rho_M)$ and $X \in {}^{t-1}T(M,\rho_M)$, then

$$ df(X) = X(f)\ ; $$

if $\omega_1 \in {}^t\Lambda_k(M,\rho_M)$, $\omega_2 \in {}^t\Lambda_s(M,\rho_M)$, then

$$ d(\omega_1 \wedge \omega_2) = d\omega_1 \wedge \omega_2 + (-1)^k \omega_1 \wedge d\omega_2\ . $$

The proof is trivial.

Proposition 3.21. *Suppose*

$$ f : N \to M $$

is an s-analytic mapping of s-analytic manifolds. If $\omega \in {}^t\Lambda'_k(M,\rho_M)$ is a t-analytic form, then

$$ f^*\omega \in {}^t\Lambda'_k(N,\rho_N)\ . \tag{3.37} $$

If $\omega \in {}^{t+1}(I\Lambda)'_k(M,\rho_M)$, then

$$f^*\omega \in {}^{t+1}(I\Lambda)'_k(N, \rho_n) ,$$

Proof. It is sufficient to verify condition (3.37) locally.

Suppose $\alpha = \left(\alpha^1, \ldots, \alpha^n\right)$ is a local s-analytic coordinate system on the manifold N, $\beta = \left(\beta^1 \ldots \beta^n\right)$ is a local s-analytic coordinate system on the manifold M, and $h_k(\alpha)$ are s-analytic functions such that

$$\beta^k = h_k(\alpha) .$$

Without loss of generality we may assume that

$$\omega = a(\beta) d\beta^1 \wedge \ldots \wedge d\beta^k$$

or

$$\omega = a(\beta) d\beta^1 \wedge \ldots \wedge d\beta^r \wedge d\overline{\beta}^{i_1} \wedge \ldots \wedge d\overline{\beta}^{i_s} ,$$

$$a(\beta) \in {}^{t+1}I(M, \rho_M) .$$

In the first case we have

$$f^* = a(h(\alpha))\, dh_1(\alpha) \wedge \ldots \wedge dh_k(\alpha) . \tag{3.38}$$

We compute the differentials which appear on the right-hand side of the latter expression:

$$dh_j(\alpha) = \sum_{i=1}^{n} \left(\frac{\partial h_j}{\partial \alpha^i} d\alpha^i + \frac{\partial h_j}{\partial \overline{\alpha}^i} d\overline{\alpha}^i \right) , \tag{3.39}$$

where

$$\frac{\partial h_j}{\partial \overline{\alpha}^i} \in {}^{s}I(M, \rho_M) \subset {}^{t+1}I(M, \rho_M) . \tag{3.40}$$

Thus it follows from formulas (3.38)–(3.40) that conditions (3.34), (3.35) of Definition 3.10 are satisfied.

The second case is trivial in view of condition (3.8) of Definition 3.5. □

Corollary 3.22. *The homomorphism f^* induces a homomorphism*

$$f^* : {}^{t}\Lambda_k(M, \rho_M) \to {}^{t}\Lambda_k(N, \rho_n) .$$

3.3.4 The Tangent Bundle. In this section we will construct on an s-analytic manifold M an n-dimensional complex bundle which is isomorphic to the real ($2n$-dimensional) tangent bundle, and for which the complex structure will in a certain sense be uniquely determined by the s-analytic structure on the manifold M.

Suppose $\{U_j\}$ is an atlas of s-analytic charts on the s-analytic manifold M with weight function ρ_M, and $\alpha = \{\alpha_j^1, \ldots, \alpha_j^n\}$ are local s-analytic coordinates on the chart U_j. We set $U_{jk} = U_j \cap U_k$, $U_{jkl} = U_j \cap U_k \cap U_l$. We define matrix functions $A_{jk}(x)$, $x \in U_{jk}$, by setting

$$A_{jk}(x) = \begin{bmatrix} \frac{\partial \alpha_j^1}{\partial \alpha_k^1} & \cdots & \frac{\partial \alpha_j^1}{\partial \alpha_k^n} \\ \cdots & \cdots & \cdots \\ \frac{\partial \alpha_j^n}{\partial \alpha_k^1} & \cdots & \frac{\partial \alpha_j^n}{\partial \alpha_k^n} \end{bmatrix} \tag{3.41}$$

Lemma 3.23. *Suppose $x \in U_{jkl}$. Then*

$$A_{jk}(x)A_{kl}(x) - A_{jl}(x) \in {}^sI(U_{jkl}, \rho) \ . \tag{3.42}$$

The proof follows trivially from the formulas for differentiating a complex function and from the s-analyticity of the transition functions $\alpha_j(\alpha_k)$.

Corollary 3.24. *If $x \in \Omega(M) \cap U_{jkl}$, then*

$$A_{jk}(x)A_{kl}(x) - A_{jl}(x) = 0 \ . \tag{3.43}$$

Corollary 3.25. *There exists a neighborhood V of the set Ω such that if $x \in V \cap U_{jk}$, then*

$$\det A_{jk}(x) \neq 0 \ . \tag{3.44}$$

Hence, for the manifold V and its covering $\{U_j\}$ there are given functions $A_{jk}(x)$, $x \in U_{jk}$, $\det A_{jk}(x) \neq 0$. Suppose $\|A_{jk}(x)\| \leq C$ for all x, j, k.

Lemma 3.26. *There exist s-analytic functions B_{jk} on the sets U_{jk} such that:*

a) $B_{jk}B_{kl} - B_{jl}|U_{jkl} = 0$,

b) $B_{jk} - A_{jk} \in {}^{s+1}I(U_{jk}, \rho)$.

Proof. We order all the sets U_k. We fix a number k_0. Suppse the s-analytic functions B_{kl} are constructed for all l, k, Min $\{l, k\} < k_0$, and suppose conditions a) and b) hold when Min $\{j, k\} < k_0$, Min $\{k, l\} < k_0$, Min $\{l, j\} < k_0$.

We will construct the functions B_{kl} satisfying a) and b) for Min $\{k, l\} < k_0 + 1$. We consider all the sets of the form $U_{k_0,l}$, $l > k_0$. We choose the minimum index l such that $U_{k_0 l} \neq \emptyset$. Condition b) defines the function $B_{k_0 l}(x)$ on $\Omega \cap U_{k_0 l}$. Condition a) uniquely defines the function $B_{k_0 l}(x)$

on the submanifold $U_{ik_0l} \in U_{k_0l}$, $i < k_0$. It is not hard to see that these conditions agree at common points.

Thus the function $B_{k_0l}(x)$ is defined on the set

$$W_{k_0l} = (\Omega \cap U_{k_0l}) \cup \bigcup_{i<k_0} U_{ik_0l} \,. \tag{3.45}$$

We continue the function $B_{k_0l}(x)$ to the whole set U_{k_0l} in such a way that conditions a) and b) continue to hold. This continuation is possible if the neighborhood V is chosen in such a way that the matrix B_{k_0l} remains nonsingular on V. By induction, we proceed further to construct all the functions $B_{k_0l'}$, where $l' > k_0$. This proves Lemma 3.26. □

Definition 3.11. We will denote the complex bundle defined by the gluing functions $B_{kj}(x)$, satisfying the conditions of Lemma 3.26, on some neighborhood V of the set $\Omega(M, \rho_M)$, by $\tau(M)$.

Proposition 3.27. *The bundle in Definition* 3.11 *is well-defined, i.e. it does not depend on the choice of the gluing functions* $B_{jk}(x)$.

Proof. The main arbitrariness in the choice of the functions $B_{kl}(x)$ occurs in continuing the function $B_{kl}(x)$ from the closed set (3.45) to the entire set U_{kl} in such a way that conditions a) and b) of Lemma 3.26 hold. Since we consider the estimate in condition b) to be sufficiently small uniformly in $x \in U_{kl}$, there exists a homotopy between any two distinct continuations of the functions B_{kl} in the class of functions satisfying the conditions of Lemma 3.26. Consequently it suffices to apply Lemma 3.26 to the manifold $M \times [0,1]$. Thus a different choice of the gluing functions $B_{kl}(x)$ defines a homotopy of these functions, while the space of homotopies is contractible, by virtue of condition b). This proves Proposition 3.27. □

Proposition 3.28. *Suppose* $T(M)$ *is the real* $2n$*-dimensional tangent bundle. The bundles* $T(M)$ *and* $r\tau(M)$ *have identical gluing functions at the points* $x \in \Omega(M, \rho_M)$, *in other words there exists a canonical isomorphism*

$$\varphi_\Omega : T(M) \to r\tau(M)$$

on the set Ω *(here* r *is the operator of making a complex bundle real). Moreover, there exists an isomorphism*

$$\varphi : T(M) \to r\tau(M) \tag{3.46}$$

on a certain neighborhood $V \supset \Omega(M, \rho_M)$ *which continues the isomorphism* φ_Ω. *Any two isomorphisms of the form* (3.46) *are homotopic in the class of isomorphisms continuing* φ_Ω.

Proposition 3.28 is proved in analogous manner to Proposition 3.27.

A vector bundle $\xi \to M$ over an s-analytic manifold (M, ρ_M) is called an s-analytic bundle if the gluing functions $B_{jk}(x)$, $x \in U_{jk}$ are s-analytic functions on the manifolds $U_{jk} = U_j \cap U_k$. Thus in Lemma 3.26 we constructed the gluing functions of the s-analytic vector bundle $\tau(M)$. Suppose $\xi \to M$ is an s-analytic vector bundle over the s-analytic manifold (M, ρ_M), and $\sigma : M \to \xi$ is a smooth section. The section σ will be called s-analytic if in each local chart it is an s-analytic vector function. It is completely obvious that the definition of s-analyticity of the section σ does not depend on the choice of a local chart, since the gluing functions are s-analytic matrix functions.

We will denote by ${}^s\Gamma'(\xi)$ the space of all s-analytic sections of the bundle ξ, and by ${}^{s+1}I\Gamma'(\xi)$ the space of all those sections which in each local chart U are vector functions whose coordinates lie in the ideal ${}^{s+1}I(U, \rho_U)$. We set

$$ {}^s\Gamma(\xi) = {}^s\Gamma'(\xi)/\, {}^{s+1}I\Gamma'(\xi) \,. \tag{3.47} $$

It is not hard to see that the space ${}^s\Gamma(\xi)$ is an ${}^s\mathcal{O}(M, \rho_M)$-module.

Proposition 3.29. *The ${}^s\mathcal{O}(M, \rho_M)$-modules ${}^s\Gamma(\tau(M))$ and ${}^sT(M, \rho_M)$ (see (3.31)) are isomorphic. The ${}^s\mathcal{O}(M, \rho_M)$-modules $\Lambda^k[{}^s\Gamma(\tau^*(M))]$ and ${}^s\Lambda_k(M, \rho_M)$ (see 3.36)) are isomorphic. ($\tau^*(M)$ is the dual bundle to $\tau(M)$.)*

Proof. We construct a homomorphism

$$ Q : {}^sT(M, \rho_M) \to {}^s\Gamma(\tau(M)) \,. $$

Let X be an s-analytic vector field on the manifold M. Then, according to Definition 3.8, in each local coordinate system the field X has the form

$$ X = \sum_{k=1}^{n} a^k(\alpha)\frac{\partial}{\partial \alpha^k} + Y \,. $$

Suppose $\{U_j\}$ is an atlas of s-analytic charts on the manifold (M, ρ_M), φ_j is an s-analytic partition of unity on the manifold (M, ρ_M).

We will associate to the vector field X a section $\sigma : M \to \tau(M)$ of the bundle $\tau(M)$ which is a sum of sections $\sigma_j : M \to \tau(M)$, where σ_j is defined in each chart U_j as a vector function $\sigma_j = \{\varphi_j a^k(\alpha)\}$. We set

$$ Q(x) = [\sigma] \in {}^s\Gamma(\tau(M)) \,, $$

where the symbol $[\sigma]$ denotes the conjugacy class of the element σ. The verification that this is an isomorphism and the proof of the second half of the conclusion of Proposition 3.29 are carried out analogously. □

Thus we have established a link between s-analytic vector fields, s-analytic differential forms and the sections of certain bundles over the manifold (M, ρ_M).

3.3.5 Nonsingular Real Submanifolds of S-Analytic Manifolds. In the second section we defined a submanifold U^0 for each s-analytic chart U of the manifold (M, ρ_M). Moreover, it is clear that the submanifold U^0 in the α coordinates will not also be the submanifold U^0 in terms of other s-analytic coordinates β. The goal of this section is to describe submanifolds of the chart (U, α) with respect to different s-analytic coordinates.

Definition 3.12. A smooth mapping

$$g : U^0 \to \mathbb{C}^n ,$$

given by functions $g^i\left(a^1, \ldots, a^n\right)$; $U^0 \to \mathbb{C}$, $j = 1, \ldots, n$ is called a nonsingular germ if the conditions:

1) $g^j\left(a^1, \ldots, a^n\right) \in {}^1I^0\left(U^0, \rho\right),\ j = 1, \ldots, n$;

2) $\det\left(\frac{\partial\left(a^j + ig^j\right)\left(a^1, \ldots, a^n\right)}{\partial \alpha^k}\right)\Bigg|_{\Omega(U, \rho_U)} \neq 0 .$

We will denote the set of all nonsingular germs in the chart U by $G(U)$.

Definition 3.13. Suppose $g \in G(U)$ is a nonsingular germ. An n-dimensional submanifold in the chart U, given in parametric form by the equations

$$\alpha^j = a^j + ig^j\left(a^1, \ldots, a^n\right) , \qquad j = 1, \ldots, n \tag{3.48}$$

is called the nonsingular manifold corresponding to the germ g, and is denoted by U^g.

From condition 2) of Definition 3.12 it follows that equation (3.48) correctly defines a submanifold U^g imbedded in the manifold U.

Indeed,

$$\alpha^j\left(a^1, \ldots, a^n\right) = a^j + ig^j\left(a^1, \ldots, a^n\right) = f_1^j\left(a^1, \ldots, a^n\right) + if_2^j\left(a^1, \ldots, a^n\right) ,$$

where f_1^j, f_2^j are real-valued functions. To prove that U^g is a submanifold it is sufficient to prove that the rank of the matrix

$$\begin{pmatrix} \frac{\partial f_1^1}{\partial a^1} \cdots \frac{\partial f_1^n}{\partial a^1} & \frac{\partial f_2^1}{\partial a^1} \cdots \frac{\partial f_2^n}{\partial a^1} \\ \cdots\cdots\cdots & \cdots\cdots\cdots \\ \frac{\partial f_1^1}{\partial a^n} \cdots \frac{\partial f_1^n}{\partial a^n} & \frac{\partial f_2^1}{\partial a^n} \cdots \frac{\partial f_2^n}{\partial a^n} \end{pmatrix}$$

equals n. If this is not so, then the rows of the matrix are linearly dependent, and therefore for any λ the determinant

$$\det \begin{pmatrix} \frac{\partial f_1^1}{\partial a^1} + \lambda \frac{\partial f_2^1}{\partial a^1} & \dots & \frac{\partial f_1^n}{\partial a^1} + \lambda \frac{\partial f_2^n}{\partial a^1} \\ \dots\dots\dots\dots & & \dots\dots\dots\dots \\ \frac{\partial f_1^1}{\partial a^n} + \lambda \frac{\partial f_2^1}{\partial a^n} & \dots & \frac{\partial f_1^n}{\partial a^n} + \lambda \frac{\partial f_2^n}{\partial a^n} \end{pmatrix}$$

is equal to zero. Setting $\lambda = i$, we obtain a contradiction with requirement 2) of Definition 3.12.

We note that the concepts we have introduced depend on the *s-analytic chart* U on the manifold M, in other words not only on the set U but also on the system of coordinates α.

Proposition 3.30. *Suppose* $g \in G(U)$ *and* U^g *is the nonsingular manifold corresponding to the germ* g. *For each* $t \leq s$ *there exists an operator*

$$^tT^g : C^\infty(U^g) \to C^\infty(U^0) , \tag{3.49}$$

such that

1) *for any* t*-analytic function* $\mathcal{F}$ *on* (U, ρ_U)

$$^tT^g(R^g\mathcal{F}) - R^0\mathcal{F} \in {}^{t+1}I^0(U^0, \rho_{U^0}) , \tag{3.50}$$

where $R^g : C^\infty(U) \to C^\infty(U^g)$ *is the restriction operator;*

2) *the inclusions*

$$\begin{aligned} {}^tT^g \circ \left[R^g \circ {}^tA^0\right] f - f \in {}^{t+1}I^0(U^0, \rho_{U^0}) , \qquad & f \in C^\infty(U^0) ; \\ \left[R^g \circ {}^tA^0\right] \circ {}^tT^g f - f \in {}^{t+1}I^g(U^g) , \qquad & f \in C^\infty(U^g) , \end{aligned} \tag{3.51}$$

hold, where ${}^{t+1}I^g\{U^g)$ *is the restriction of the set* ${}^{t+1}I^g(U)$ *to* U^g:

$${}^{t+1}I^g(U^g) = R^g\left({}^{t+1}I(U)\right) .$$

Proof. We will break the proof of this proposition up into a series of lemmas.

We consider the functions

$$\alpha^j = \alpha^j(a^1, \dots, a^n) = a^j + ig^j(a^1, \dots, a^n)$$

and define operators $\frac{d}{d\alpha^j}$ acting on the space of smooth functions in the variables $a^1, \dots, a^n$ by the relations

$$\frac{\partial}{\partial a^l} = \sum_{j=1}^n \frac{\partial \alpha^j(a^1, \dots, a^n)}{\partial a^l} \frac{d}{d\alpha^j} . \tag{3.52}$$

The operators $\frac{d}{d\alpha^j}$ are defined uniquely by conditions (3.52), because by Definition 3.12

$$\left(\frac{\partial\alpha^j\left(a^1,\ldots,a^n\right)}{\partial a^k}\right)\neq 0$$

in some neighborhood V of the set

$$\Omega\left(U,\rho_U\right)=\Omega\left(U^0,\rho_{U^0}\right) .$$

Lemma 3.31. *For the operators* $\frac{d}{d\alpha^j}$ *the identities*

$$\frac{d}{d\alpha^j}\frac{d}{d\alpha^k}=\frac{d}{d\alpha^k}\frac{d}{d\alpha^j} \tag{3.53}$$

hold.

Proof. Restricting our attention to the neighborhood V we denote by $\partial a^l/\partial\alpha^r$ the matrix inverse to the matrix $\partial\alpha^r/\partial a^l$, in other words

$$\sum_{r=1}^{n}\frac{\partial\alpha^r}{\partial a^l}\frac{\partial a^m}{\partial\alpha^r}=\delta_l^m . \tag{3.54}$$

Now we compute the left-hand side of Eq. (3.53)

$$\begin{aligned}\frac{d}{d\alpha^j}\frac{d}{d\alpha^k}f=\sum_{l,m}\frac{\partial a^l}{\partial\alpha^j}\frac{\partial}{\partial a^l}\left[\frac{\partial a^m}{\partial\alpha^k}\frac{\partial f}{\partial a^m}\right]=\sum_{l,m}\frac{\partial a^l}{\partial\alpha^j}\\ \times\frac{\partial}{\partial a^l}\left[\frac{\partial a^m}{\partial\alpha^k}\right]\frac{\partial f}{\partial a^m}+\sum_{l,m}\frac{\partial a^l}{\partial\alpha^j}\frac{\partial a^m}{\partial\alpha^k}\frac{\partial^2 f}{\partial a^m\partial a^l} .\end{aligned} \tag{3.55}$$

Now we use Eq. (3.54), taking the partial derivatives of both sides with respect to a^l:

$$0=\sum_r\frac{\partial}{\partial a^l}\left[\frac{\partial\alpha^r}{\partial a^s}\frac{\partial a^m}{\partial\alpha^r}\right]=\sum_r\frac{\partial^2\alpha^r}{\partial a^l\partial a^s}\frac{\partial a^m}{\partial\alpha^r}+\sum_r\frac{\partial\alpha^r}{\partial a^s}\frac{\partial}{\partial a^l}\left[\frac{\partial a^m}{\partial\alpha^r}\right] .$$

Multiplying the latter equation by $\frac{\partial a^s}{\partial\alpha^j}\frac{\partial a^l}{\partial\alpha^j}$ we obtain

$$\sum_l\frac{\partial a^l}{\partial\alpha^j}\frac{\partial}{\partial a^l}\left[\frac{\partial a^m}{\partial\alpha^k}\right]=-\sum_{l,r,s}\frac{\partial a^s}{\partial\alpha^k}\frac{\partial a^l}{\partial\alpha^j}\frac{\partial a^m}{\partial\alpha^r}\frac{\partial^2\alpha^r}{\partial a^l\partial a^s} . \tag{3.56}$$

The latter expression is symmetric with respect to the indices j and k. □

Now we define the operator ${}^tT^g$ by the equation

$$\left({}^{t}T^{g} f\right)(a)=\sum_{|k| \leq t} \frac{(-1)^{|k|}}{k!}(i g(a))^{k} \frac{d^{|k|}}{d \alpha^{k}} f(a) . \tag{3.57}$$

Here $k=(k_1,\ldots,k_n)$ is a multi-index, $k!=k_1!,\ldots,k_n!$,

$$(i g(a))^{k}=\left(i g^{1}(a)\right)^{k_{1}} \ldots\left(i g^{n}(a)\right)^{k_{n}} ;$$

$$\frac{d^{|k|}}{d \alpha^{k}}=\frac{d^{k_{1}}}{\left(d \alpha^{1}\right)^{k_{1}}} \cdots \frac{d^{k_{n}}}{\left(d \alpha^{n}\right)^{k_{n}}} .$$

Lemma 3.32. *For the operator ${}^tT^g$ defined by formula* (3.57) *the relations*

$$\frac{\partial}{\partial a^{j}} \circ {}^{t}T^{g} f-{}^{t-1}T^{g} \circ \frac{d}{d \alpha^{j}} f \in {}^{t}I^{0}\left(U^{0}, \rho_{U^{0}}\right) \tag{3.58}$$

hold for any function $f \in C^{\infty}(U^g)$.

Proof. We calculate

$$\begin{aligned}
\left(\frac{\partial}{\partial a^{j}} \circ {}^{t}T^{g}\right) f(a) &=\frac{\partial}{\partial a^{j}}\left[\sum_{|k| \leq t} \frac{(-1)^{|k|}}{k!}(i g(a))^{k} \frac{d^{|k|}}{d \alpha^{k}} f(a)\right] \\
&=\sum_{|k| \leq t-1} \frac{(-1)^{|k|}}{k!}(i g(a))^{k}\left[\frac{\partial}{\partial a^{j}}-i \frac{\partial g^{l}(a)}{\partial a^{j}} \frac{d}{d \alpha^{l}}\right] \frac{d^{|k|}}{d \alpha^{k}} f(a) \\
&+\sum_{|k|=t} \frac{(-1)^{k}}{k!}(i g(a))^{k} \frac{\partial}{\partial a^{j}} \frac{d^{|k|}}{d \alpha^{k}} f(a) .
\end{aligned} \tag{3.59}$$

By requirement 1) of Definition 3.12 the last term on the right-hand side of Eq. (3.59) belongs to the ideal ${}^tI^0\left(U^0,\rho_{U^0}\right)$. Furthermore, from definition (3.53) of the operators $\frac{d}{d\alpha^j}$ we have

$$\begin{aligned}
\frac{\partial}{\partial a^{l}} &=\frac{\partial \alpha^{j}\left(a^{1}, \ldots, a^{n}\right)}{\partial a^{l}} \frac{d}{d \alpha^{j}}=\frac{\partial\left(a^{j}+i g^{j}\left(a^{1}, \ldots, a^{n}\right)\right)}{\partial a^{l}} \frac{d}{d \alpha^{j}} \\
&=\left[\delta_{l}^{j}+i \frac{\partial g^{j}(a)}{\partial a^{l}}\right] \frac{d}{d \alpha^{j}}=\frac{d}{d \alpha^{l}}+i \frac{\partial g^{j}(a)}{\partial a^{l}} \frac{d}{d \alpha^{j}} ,
\end{aligned}$$

from which it follows that

$$\frac{d}{d \alpha^{j}}=\frac{\partial}{\partial a^{j}}-i \frac{\partial g^{l}(a)}{\partial a^{j}} \frac{d}{d \alpha^{l}} . \tag{3.60}$$

Substituting (3.60) into (3.59), we finally have

$$\left(\frac{\partial}{\partial a^j}\circ {}^tT^g\right) f(a) = \sum_{|k|\le t-1} \frac{(-1)^{|k|}}{k!} (ig(a))^k \frac{d}{d\alpha^j}\frac{d^{|k|}}{d\alpha^k} f(a) + F(a) \,,$$

where $F(a)\in {}^tI^0\left(U^0, \rho_{U^0}\right)$. Using the commutativity of the operators $\frac{d}{d\alpha^k}$ and $\frac{d}{d\alpha^l}$ (Lemma 3.31), we have

$$\begin{aligned}\left(\frac{\partial}{\partial a^j}\circ {}^tT^g\right) f(a) &= \sum_{|k|\le t-1} \frac{(-1)^{|k|}}{k!} (ig(a))^k \frac{d^{|k|}}{d\alpha^k}\left[\frac{df}{d\alpha^j}(a)\right] + F(a) \\ &= \left({}^{t-1}T^g\circ \frac{d}{d\alpha^j}\right) f(a) + F(a) \,.\end{aligned}$$

Since $F(a)\in {}^tI^0\left(U^0, \rho_{U^0}\right)$, this proves Lemma 3.32. □

Now we write out an analytic expression for the operator $R^g\circ {}^tA^0$;

$$\begin{aligned}\left(R^g\circ {}^tA^0\right) f(a) &= \sum_{k=1}^{t} \frac{1}{k!}\left(\sum_{j=1}^{n} ib^j \frac{\partial}{\partial a^j}\right)^k f(a)\Bigg|_{b^j=g^j(a)} \qquad (3.61)\\ &= \sum_{|k|\le t} \frac{1}{k!}(ig(a))^k \frac{\partial^{|k|}}{\partial a^k} f(a) \,.\end{aligned}$$

Lemma 3.33. *The inclusions*

$$\frac{d}{d\alpha^j}\circ\left(R^g\circ {}^tA^0\right) f - \left(R^g\circ {}^{t-1}A^0\right)\frac{\partial}{\partial a^j} f \in {}^tI^g\left(U^g\right) \qquad (3.62)$$

hold for any function $f\in C^\infty\left(U^0\right)$.

Proof. First we show that the inclusion

$$\left(R^g\circ\frac{\alpha}{\partial\alpha^j}\right)\mathcal{F} - \left(\frac{d}{d\alpha^j}\circ R^g\right\}\mathcal{F}\in {}^tI^g\left(U^g\right) \qquad (3.63)$$

holds for any t-analytic function $\mathcal{F}$ on U. Indeed, denoting

$$\varphi(a) = R^g\mathcal{F}(\alpha) = \mathcal{F}\left(a + ig(a)\right) \,,$$

by differentiating with respect to a^l we obtain the equation

$$\frac{\partial\varphi(a)}{\partial a^l} = \frac{\partial\mathcal{F}}{\partial\alpha^j}\left(a+ig(a)\right)\frac{\partial\alpha^j}{\partial a^l} + \frac{\partial\mathcal{F}}{\partial\overline{\alpha}^j}\left(a+ig(a)\right)\frac{\partial\overline{\alpha}^j}{\partial a^l} \,.$$

The second summand in this formula is, obviously, an element of $R^g\left({}^tI\left(U,\rho_U\right)\right) = {}^tI^g\left(U^g\right)$. Therefore, multiplying by the matrix $\partial a/\partial\alpha$ inverse to $\partial\alpha/\partial a$, we have

$$\frac{\partial\mathcal{F}}{\partial\alpha^j}(a+ig(a)) = \frac{\partial a^l}{\partial\alpha^j}\frac{\partial\varphi(a)}{\partial a^l} + \mathcal{F}_1(a) = \frac{d\varphi(a)}{d\alpha^j} + \mathcal{F}_1(a)\ , \tag{3.64}$$

where $\mathcal{F}_1(a) \in {}^tI^g\left(U^g\right)$. Noting that

$$\frac{\partial\mathcal{F}}{\partial\alpha^j}(a+ig(a)) = \left(R^g \circ \frac{\partial}{\partial\alpha^j}\right)\mathcal{F}$$

and recalling that $\varphi(a) = R^g\mathcal{F}(a)$, we see that Eq. (3.64) coincides with the inclusion (3.63).

Now we substitute into (3.63) $\mathcal{F} = {}^tA^0 f$, $f \in C^\infty\left(U^0\right)$. By Proposition 3.5, $\mathcal{F}$ is a t-analytic function, and therefore (3.63) is valid for it. We obtain

$$R^g \circ \frac{\partial}{\partial\alpha^j} \circ {}^tA^0 f - \frac{d}{d\alpha^j}\left(R^g \circ {}^tA^0\right) f \in {}^tI^g\left(U^g\right)\ . \tag{3.65}$$

By virtue of the inclusion (3.17),

$$\frac{\partial}{\partial\alpha^j} \circ {}^tA^0 f - {}^{t-1}A^0\frac{\partial}{\partial a^j} f \in {}^tI\left(U,\rho_U\right)\ ,$$

hence

$$\begin{aligned} R^g \circ \tfrac{\partial}{\partial\alpha^j} \circ {}^tA^0 f - \left(R^g \circ {}^{t-1}A^0\right) \circ \tfrac{\partial}{\partial a^j} f &\in R^g\left[{}^tI\left(U,\rho_U\right)\right] \\ &= {}^tI^g\left(U^g\right)\ . \end{aligned} \tag{3.66}$$

Substituting (3.66) into (3.65), we obtain (3.62). This proves the lemma. □

Proof of Proposition 3.30. First we prove the inclusion (3.51). Suppose $f \in C^\infty\left(U^0\right)$. By Eq. (3.57) we have

$${}^tT^g \circ \left[R^g \circ {}^tA^0\right] f = \sum_{|k|\le t} \frac{(-1)^{|k|}}{k!}(ig(a))^k \frac{d^{|k|}}{d\alpha^k}\left[R^g \circ {}^tA^0\right] f\ .$$

Furthermore, from formula (3.62) it follows that

$$\begin{aligned} {}^tT^g \circ \left[R^g \circ {}^tA^0\right] f = &\sum_{|k|\le t} \frac{(-1)^{|k|}}{k!}(ig(a))^k \left[R^g \circ {}^{t-|k|}A^0\right] \\ &\times \frac{\partial^{|k|}}{\partial a^k} f + F\ , \end{aligned} \tag{3.67}$$

where $F \in {}^{t+1}I^g\left(U^g\right)$. Since, by requirement 1) of Definition 3.12, $g(a) \in {}^1I^0\left(U^0,\rho_{U^0}\right)$, any function in ${}^tI^g\left(U^g\right)$, considered as a function on U^0, belongs to ${}^tI^0\left(U^0,\rho_{U^0}\right)$. Now using Eq. (3.61), we bring (3.67) to the form

$$
\begin{aligned}
{}^tT^g \circ \left[R^g \circ {}^tA^0\right] f &= \sum_{|k|\le t} \frac{(-1)^{|k|}}{k!} (ig(a))^k \\
&\quad \cdot \sum_{|r|\le t-|k|} \frac{1}{r!} \{ig(a))^r \frac{\partial^{|k+r|}}{\partial a^{k+r}} f + F \\
&= \sum_{|q|\le t} \left\{ \sum_{k+r\le q} \frac{1}{k!r!} (-ig(a))^k (ig(a))^r \right\} \frac{\partial^{|q|}}{\partial a^q} f + F \,.
\end{aligned}
$$

Since for $|q| > 0$

$$
\sum_{k+r=q} \frac{1}{k!r!} (-ig(a))^k ((ig(a))^r = \frac{1}{q!} (ig(a) - ig(a))^q = 0 \,,
$$

the first of the inclusions (3.51) is proven.

To prove the second of the inclusions (3.51) we take a function $f \in C^\infty(U^g)$. Using formulas (3.57), (3.58) and (3.61) we obtain

$$
\begin{aligned}
\left[R^g \circ {}^tA^0\right] \circ {}^tT^g f &= \sum_{|k|\le t} \frac{1}{k!} (ig(a))^k \frac{\partial}{\partial a^k} {}^tT^g f \\
&= \sum_{|k|\le t} \frac{1}{k!} (ig(a))^k \, {}^{t-|k|}T^g \frac{d^{|k|}}{d\alpha^k} f + F_1 \\
&= \sum_{|k|\le t} \frac{1}{k!} (ig(a))^k \sum_{|r|\le t-|k|} \frac{1}{r!} (-ig(a))^r \frac{d^{|k+r|}}{d\alpha^{k+r}} f + F_1 \,,
\end{aligned}
$$

where $F_1 \in {}^{t+1}I^g(U^g)$. The proof of the second of the inclusions (3.51) now finishes in exactly the same way as the proof of the first.

In order to finish the proof of the theorem we have to prove relation (3.50). However, for any t-analytic function $\mathcal{F}$ the function $R^0\mathcal{F} \in C^\infty\left(U^0\right)$. Applying to $f = R^0\mathcal{F}$ the first of inclusions (3.51), we have

$$
{}^tT^g \circ R^g \circ {}^tA^0 \circ R^0\mathcal{F} - R^0\mathcal{F} \in {}^tI^0\left(U^0, \rho_{U^0}\right) \,.
$$

Furthermore, from (3.16) and the requirement 1) of Definition 3.12 we finally obtain the inclusion (3.50). This proves the proposition. □

Now we define the factor spaces

$$
{}^tC^\infty\left(U^0\right) = C^\infty\left(U^0\right) / {}^{t+1}I\left(U^0, \rho_{U^0}\right) \,, \tag{3.68}
$$

$$
{}^tC^\infty\left(U^g\right) = C^\infty\left(U^g\right) / {}^{t+1}I^g\left(U^g\right) \,. \tag{3.69}
$$

Corollary 3.34. *The operators* ${}^tT^g$, R^g *and* R^0 *induce operators on the factor spaces*

$$
\begin{aligned}
{}^tT^g &: {}^tC^\infty(U^g) \to {}^tC^\infty(U^0) \ , \\
R^g &: {}^t\mathcal{O}(U, \rho_U) \to {}^tC^\infty(U^g) \ , \\
R^0 &: {}^t\mathcal{O}(U, \rho_U) \to {}^tC^\infty(U^0) \ ;
\end{aligned}
$$

all of these operators are isomorphisms, and the diagram

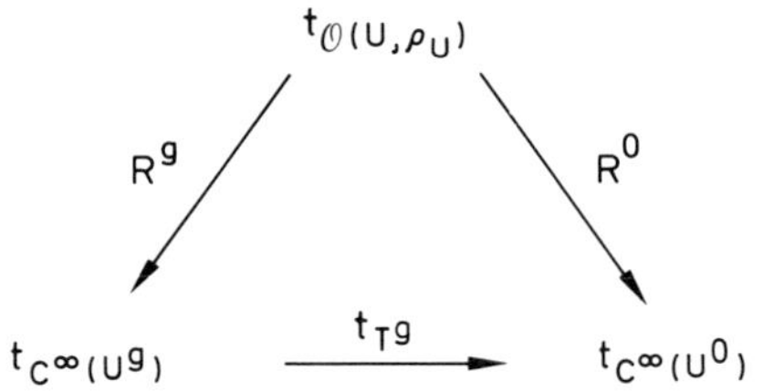

(3.70)

commutes.

Proof. The commutativity of the diagram (3.70) follows from the relation (3.50). The fact that R^0 is an isomorphism is a consequence of Proposition 3.5. The fact that ${}^tT^g$ is an isomorphism follows from relations (3.51). Finally, the fact that R^g is an isomorphism follows from the commutative diagram (3.70) and the fact that the operators R^0 and ${}^tT^g$ are isomorphisms. This proves the corollary. □

Condition 2) of Definition 3.12 of a nonsingular germ is equivalent to the condition that the operator

$$R^g : {}^t\mathcal{O}(U, \rho_U) \to {}^tC^\infty(U^g)$$

is an isomorphism. More precisely, the following assertion holds.

Proposition 3.35. *Suppose the manifold* U^g *is given by the germ* $g : U^0 \to \mathbb{C}^n$ *in the sense of Eq. (3.48) (the germ* g *is not assumed to be nonsingular!). Suppose*

$$g(a) \in {}^1I^0(U, \rho_U)$$

and the operator

$$R^g : {}^t\mathcal{O}(U, \rho_U) \to {}^tC^\infty(U^g) = C^\infty(U^g) / R^g\left({}^{t+1}I(U, \rho_U)\right)$$

is an isomorphism for some $t \geq 2$. *Then the germ is nonsingular.*

Proof. Obviously it is necessary to verify that condition 2) of Definition 3.12 holds. By the hypothesis of the theorem there exists an operator

$$(R^g)^{-1} : {}^t C^\infty(U^g) \to {}^t \mathcal{O}(U, \rho_U) \ .$$

Let $f(a) \in {}^t C^\infty(U^g)$, $\mathcal{F} = (R^g)^{-1} f$. Then

$$f(a) = \mathcal{F}(a + ig(a)) \ .$$

Differentiating this equation by a^j, we obtain

$$\frac{\partial f(a)}{\partial a^j} = \frac{\partial \mathcal{F}}{\partial \alpha^l} \frac{\partial \alpha^l(a)}{\partial a^j} \quad \text{on } \Omega(U) \ .$$

Choosing the coordinate functions a^k as f, we obtain $\left(\mathcal{F}^k = (R^g)^{-1} a^k\right)$

$$\delta_j^k = \frac{\partial \mathcal{F}^k}{\partial \alpha^l}(\alpha(a)) \frac{\partial \alpha^l(a)}{\partial a^j} \quad \text{on } \Omega(U) \ ,$$

which shows the nondegeneracy of the matrix $\partial \alpha^l(a)/\partial a^j$ on $\Omega(U)$. This proves the lemma. □

Now we establish a link between nonsingular germs and submanifolds of the type U^0.

Proposition 3.36. *Suppose U^0 is a submanifold in the chart (U, α), given by the equations*

$$\operatorname{Im} \alpha^j = 0 \ , \qquad j = 1, \ldots, n \ .$$

Then for any s-analytic coordinates α' acting on the region U, there exists a nonsingular germ $g(a')$ such that the manifold U^0 is given by the equations

$$a'^j = a'^j + ig^j(a') \ .$$

Conversely, for any nonsingular germ $g(a)$ there exist s-analytic coordinates ξ such that the submanifold U^g is given by the equations

$$\operatorname{Im} \xi^j = 0 \ , \qquad j = 1, \ldots, n \ .$$

Proof. Suppose the formula for the change of coordinates from α' to α is

$$\alpha^j = \alpha^j\left(\alpha'^1, \ldots, \alpha'^n\right) \ , \qquad j = 1, \ldots, n \ . \tag{3.71}$$

The functions $\alpha^j(\alpha')$ are s-analytic functions in the chart U. The formulas (3.71) can be written in more detail as follows:

$$a^j = a^j\left(a'^1, \ldots, a'^n; b'^1, \ldots, b'^n\right) ,$$
$$b^j = b^j\left(a'^1, \ldots, a'^n; b'^1, \ldots, b'^n\right) .$$

At a certain point in the set $\Omega(U)$ we have

$$\frac{\partial \alpha^j}{\partial \alpha'^k} = \frac{\partial a^j(a', 0)}{\partial a'^k} + i\frac{\partial b^j(a', 0)}{\partial a'^k} \neq 0 . \tag{3.72}$$

We prove that there exists a number $\lambda \in \mathbb{R}$ such that

$$\det\left(\frac{\partial a^j(a', 0)}{\partial a'^k} + \lambda\frac{\partial b^j(a', 0)}{\partial a'^k}\right) \neq 0 . \tag{3.73}$$

Indeed, the left-hand side of the inequality (3.73) is a polynomial of degree n in λ. If this polynomial is identically equal to zero on $\mathbb{R}^1$, then it is identically zero on $\mathbb{C}^1$. Hence, in particular, it follows that (3.72) fails to hold (setting $\lambda = i$).

Now we consider a certain germ $g(a')$ in the coordinates α'. The conditions which guarantee that $U^{g'} \subset U^0$ have the form

$$b^j(a' - g_2(a'), g_1(a')) = 0 ,$$

where $g(a') = g_1(a') + ig_2(a')$. Thus, to find the germ $g(a')$ we must solve a system of n equations in $2n$ unknowns. We add on the n additional equations

$$\begin{cases} a^j(a' - g_2(a'),\ g_1(a')) = a^j(a') , \\ b^j(a' - g_2(a'),\ g_1(a')) = 0 , \end{cases} \tag{3.74}$$

where the functions $a^j(a')$ are determined by the formulas

$$a^j(a') = a^j(a', 0) + \lambda b^j(a', 0) . \tag{3.75}$$

Equation (3.75), by virtue of inequality (3.73), defines a one-to-one correspondence between the coordinates a and a'. Furthermore, we define g_1, g_2 from Eq. (3.74). To prove that system (3.74) can be solved, we rewrite it in more detail, using the s-analyticity of the functions $\sigma^j(\alpha')$:

$$\begin{aligned} a^j(a' - g_2(a'), g_1(a')) &= a^j(a', 0) + \sum \frac{\partial a^j}{\partial b'^k} g_1^k(a') \\ &\quad - \sum \frac{\partial a^j}{\partial a'^k} g_2^k(a') + O\left(|g(a')|^2\right) \\ &= a^j(a', 0) - \sum \frac{\partial b^j}{\partial a'^k} g_1^k(a') \\ &\quad - \sum \frac{\partial a^j}{\partial a'^k} g_2^k(a') + O\left(|g(a')|^2\right) \\ &= a^j(a', 0) + \lambda b^j(a', 0) , \end{aligned} \tag{3.76}$$

$$\begin{aligned} b^j\left(a' - g_2(a'), g_1(a')\right) &= b^j(a',0) + \sum \frac{\partial b^j}{\partial b'^k} g_1^k(a') \\ &\quad - \sum \frac{\partial b^j}{\partial b'^k} g_2^k(a') + O\left(|g'(a')|^2\right) \\ &= b^j(a',0) + \sum \frac{\partial a^j}{\partial a'^k} g_1^k(a) \\ &\quad - \sum \frac{\partial b^j}{\partial a'^k} g_2^k(a') + O\left(|g'(a')|^2\right) = 0 \,. \end{aligned}$$

We prove that the system (3.76) is solvable. For this we observe that the inequality

$$\det\left(\frac{\partial \alpha^j}{\partial \alpha'^k}\right) = \det\left(\frac{\partial a^j(a',0)}{\partial a'^k} + i\frac{\partial b^j(a',0)}{\partial a'^k}\right) \neq 0$$

holds on the set $\Omega(M)$. Therefore for any values of f^j the system of linear equations

$$\begin{aligned} f^j &= f_1^j + i f_2^j \\ &= \left(\frac{\partial a^j(a',0)}{\partial a'^k} + i\frac{\partial b^j(a',0)}{\partial a'^k}\right)\left(x^k + iy^k\right) \\ &= \frac{\partial a^j(a',0)}{\partial a'^k} x^k - \frac{\partial b^j(a',0)}{\partial a'^k} y^k + i\left(\frac{\partial a^j}{a'^k} y^k + \frac{\partial b^j}{\partial a'^k} x^k\right) \,. \end{aligned} \tag{3.77}$$

is uniquely solvable with respect to the variables $z^k = x^k + iy^k$. The system of equations resulting from dividing (3.77) into real and imaginary parts:

$$\left\{ \begin{aligned} &\frac{\partial a^j(a',0)}{\partial a'^k} x^k - \frac{\partial b^j(a',0)}{\partial a'^k} y^k = f_1^j \,, \\ &\frac{\partial b^j(a',0)}{\partial a'^k} x^k + \frac{\partial a^j(a',0)}{\partial a'^k} y^k = f_2^j \,, \end{aligned} \right. \tag{3.78}$$

is, consequently, also uniquely solvable for any right-hand side f_1^j, f_2^j. Thus the determinant Δ of this system is nonzero on the set $\Omega\left(U, \rho_U\right)$, and therefore also in some neighborhood of $\Omega\left(U, \rho_U\right)$. We restrict our attention to this neighborhood. We observe that the determinant of the linear system (3.78) coincides with the Jacobian of the system up to a change of sign. Taking into account the fact that manifolds of type U^0 (in different coordinates) intersect in the set $\Omega\left(U, \rho_U\right)$, we obtain that the system (3.76) has a solution.

Let us estimate the solution $g_1(a'), g_2(a')$ of the system (3.76). To do this we consider it as a linear system of equations

$$\sum \frac{\partial b^j(a'_j 0)}{\partial a'^k} g_1^k(a') + \sum \frac{\partial a^j(a',0)}{\partial a'^k} g_2^k(a') = -\lambda b^j(a',0) + O\left(|g(a')|^2\right) ,$$

$$\sum \frac{\partial a^j(a'_j 0)}{\partial a'^k} g_1^k(a') - \sum \frac{\partial b^j(a',0)}{\partial a'^k} g_2^k(a') = -b^j(a',0) + O\left(|g(a')|^2\right) .$$

From this system we obtain the estimate

$$|g(a')| \leq C\left[\max_j \left|b^j(a',0)\right| + |g(a')|^2\right] .$$

Restricting to a small enough neighborhood of the set $\Omega(U, \rho_U)$ that $|g(a')| < (2C)^{-1}$, we arrive at the estimate

$$|g(a')| \leq 2C \max_j \left|b^j(a',0)\right| \leq 2C\rho|_{U^0(\alpha')} , \tag{3.79}$$

where $U^0(\alpha')$ is the manifold given by the equations $\operatorname{Im}\alpha' = 0$ in the coordinates α'. The last inequality follows from Condition 3.1.

Thus the manifold U^g coincides with the manifold $U^0(\alpha)$ and is determined by a germ $g(\alpha') \in {}^1I\left(U^0(\alpha'), \rho_{U^0(\alpha')}\right)$. Now Lemma 3.35 proves that the germ $g(a')$ is nonsingular.

Now suppose $g(a)$ is a nonsingular germ. $\left(a^1, \dots, a^n\right)$ serve as coordinates on the manifold U^g. We note at the same time that $a^j \neq R^g \alpha^j$. By Corollary 3.34 there exist s-analytic functions $A^j(\alpha)$ such that $R^g A^j = a^j$. According to Corollary 3.34 these functions are

$$A^j = \left({}^sA^0 \circ {}^sT^g\right) a^j .$$

Let us compute ${}^sT^g a^j$. By the definition of the operator ${}^sT^g$ we have

$${}^sT^g a^j = a^j - \left(ig^k(a)\right) \frac{\partial a^j}{\partial \alpha^k} + \dots .$$

By this formula and the definition of the operators ${}^sA^0$ we have the estimate

$$\left|\operatorname{Im} A^j(\alpha)\right| \leq C\left(\rho_0 + |b|\right) , \tag{3.80}$$

where $\rho_0(\alpha) = \rho(a)$ $(\alpha = a + ib)$.

In view of the estimate (3.15), the inequality (3.80) can be rewritten in the form

$$\left|\operatorname{Im} A^j(\alpha)\right| \leq C\rho(\alpha) ,$$

therefore it remains only to verify the solvability of the system of equations

$$A^j(\alpha) = \xi^j ,$$

which follows from the equations

$$\left.\frac{\partial A^j(\alpha)}{\partial \alpha^k}\right|_\Omega = \left.\frac{\partial a^j}{\partial \alpha^k}\right|_\Omega \quad \text{and} \quad \frac{\partial \alpha^k}{\partial a^l}\frac{\partial a^j}{\partial \alpha^k} = \delta^j_l \; .$$

This proves the proposition. □

3.4 Positive Lagrangian s-Analytic Manifolds

Suppose, as in § 2.2, $\Phi_{\mathbb{C}}$ denotes a $2n$-dimensional complex phase space, in which a complex basis of vectors

$$\{e, f\} = \{e_1, \dots, e_n, f^1, \dots, f^n\}$$

is fixed. We introduce on the space $\Phi_{\mathbb{C}}$ a weight function ρ_Φ, setting

$$\rho_\Phi(a) = \sum_k \left(\left|\operatorname{Im} z^k\right|^2 + \left|\operatorname{Im}\zeta_k\right|^2\right)^{1/2} , \tag{4.1}$$

where

$$a = \sum_k z^k l_k + \sum_j \zeta_j f^j \; . \tag{4.2}$$

Then the set $\Omega(\Phi_{\mathbb{C}}, \rho_\Phi)$ (see § 3.3) coincides with the subspace $\Phi_{\mathbb{R}}$ (see § 2.2). Thus the weight function ρ_Φ and the coordinates (z^k, ζ_j), $k, j = 1, \dots, n$, at a point $a \in \mathbb{C}$ determine an s-analytic manifold structure (Definition 3.3), satisfying Condition 3.1, on the space $\Phi_{\mathbb{C}}$.

It is easy to see that any analytic function on the space $\Phi_{\mathbb{C}}$, as well as any analytic form, is automatically an s-analytic function (respectively s-analytic form) on the space $\Phi_{\mathbb{C}}$. In particular, the form

$$\omega = \sum_j d\zeta_j \wedge dz^j \tag{4.3}$$

is an s-analytic form.

Suppose (M, ρ_M) is an s-analytic n-dimensional manifold and

$$f : (M, \rho_M) \to (\Phi_{\mathbb{C}}, \rho_\Phi) \tag{4.4}$$

is an s-analytic imbedding of s-analytic manifolds.

Definition 4.1. A manifold (M, ρ_M) is called a Lagrangian s-analytic manifold if the form $f^*(\omega)$ defines the null element in the ${}^s\mathcal{O}(M, \rho_M)$-module ${}^s\Lambda^2(M, \rho_M)$:

$$f^*(\omega) = 0 \in {}^s\Lambda^2(M, \rho_M) \; , \tag{4.5}$$

i.e. by definition, in each local s-analytic coordinates system the form $f^*(\omega)$ has all its coefficients in the ideal ${}^{s+1}I(M,\rho_M)$.

Lemma 4.1. *Suppose (M,ρ_M) is a Lagrangian manifold in the phase space $\Phi_{\mathbf{C}}$, T_x is the tangent space to the manifold M at the point $x \in M$. If $x \in \Omega(M,\rho_M)$, then the space T_x is a complex subspace of $\Phi_{\mathbf{C}}$, and is a Lagrangian plane.*

Proof. Let $\alpha = (\alpha^1,\dots,\alpha^n)$ be local s-analytic coordinates in a neighborhood of a point $x \in \Omega(M,\rho_M)$ of the manifold M. The imbedding (4.4) defines $2n$ s-analytic functions – the coordinates of a point of the manifold M in the standard basis of phase space:

$$z^k = z^k(\alpha^1,\dots,\alpha^n)\ , \qquad k = 1,\dots,n\ ; \tag{4.6}$$

$$\zeta_k = \zeta_k(\alpha^1,\dots,\alpha^n)\ , \qquad k = 1,\dots,n\ . \tag{4.7}$$

Then in the space T_x, $x \in M$, one may choose as a real basis the $2n$ vectors r_k, r'_k, $k = 1,\dots,n$, with the following coordinates:

$$r_k = \begin{bmatrix} \partial z^1/\partial a^k \\ \vdots \\ \partial z^n/\partial a^k \\ \partial\zeta_1/\partial a^k \\ \vdots \\ \partial\zeta_n/\partial a^k \end{bmatrix}\ , \quad r'_k = \begin{bmatrix} \partial z^1/\partial b^k \\ \vdots \\ \partial z^n/\partial b^k \\ \partial\zeta_1/\partial b^k \\ \vdots \\ \partial\zeta_n/\partial b^k \end{bmatrix}\ , \tag{4.8}$$

where $\alpha^k = a^k + ib^k$.

By virtue of the s-analyticity of the functions (4.6), (4.7), we have

$$\frac{\partial z^l}{\partial a^k} = -i\frac{\partial z^l}{\partial b^k}\ , \quad \frac{\partial \zeta_l}{\partial a^k} = -i\frac{\partial \zeta_l}{\partial b^k} \tag{4.9}$$

for $x \in \Omega(M,\rho_M)$.

Thus it follows from (4.8) and (4.9) that

$$ir_k = r'_k\ ,$$

i.e. if $x \in \Omega(M,\rho_M)$, then T_x is a complex subspace.

Furthermore, since by Definition 4.1 all the coefficients of the form $f^*(\omega)$ belong to the ideal ${}^{s+1}I(M,\rho_M)$, we have

$$f^*(\omega) = 0 \quad \text{on} \quad \Omega(M,\rho_M)\ . \tag{4.10}$$

The last equation means that the plane T_x is Lagrangian, since for any vectors a, b, $\omega(a,b) = \langle a, b\rangle$. This proves the lemma. □

Lemma 4.2. *Suppose (M, ρ_M) is a Lagrangian manifold in the phase space $\Phi_{\mathbb{C}}$, $c \in \Omega(M, \rho_M)$. There exists a choice of indices I such that the functions*

$$\{z^k(x)\ , \qquad k \in I\} \tag{4.11}$$

$$\{\zeta_j(x)\ , \qquad j \in \overline{I}\} \tag{4.12}$$

form an local s-analytic coordinate system in a neighborhood of the point x.

Proof. According to Lemma 4.1, the tangent space T_x at the point $x \in \Omega(M, \rho_M)$ is a Lagrangian plane. Then by Lemma 1.1 there exists a collection of indices I such that the projection

$$p_I|T_x : T_x \to \mathbb{C}^I \times \mathbb{C}_{\overline{I}}$$

is an isomorphism. Consequently, by the implicit function theorem there exists a neighborhood $U \subset M$ of the point x such that the projection

$$p_I|U : U \to \mathbb{C}^I \times \mathbb{C}_{\overline{I}}$$

is a diffeomorphism, hence the functions (4.11) and (4.12) form an s-analytic local coordinate system in the neighborhood U of the manifold (M, ρ_M). This proves Lemma 4.2. □

Proposition 4.3. *Suppose (M, ρ_M) is a Lagrangian s-analytic manifold in the phase space $\Phi_{\mathbb{C}}, f : M \to \Phi_{\mathbb{C}}$ is an s-analytic imbedding. Then there exists a map*[3]

$$f : \tau(M) \to \Phi_{\mathbb{C}}\ , \tag{4.13}$$

such that

a) $F|M = f$, (4.14)

b) the mapping F is linear on each fiber, and the image of each fiber is a Lagrangian plane,

c) if $x \in \Omega(M, \rho_M)$, then

$$F|_{\tau_x} = df_x \circ \varphi_\Omega^{-1}\ , \tag{4.15}$$

where τ_x is the fiber over x of the bundle $\tau(M)$, and $df_x : T_x \to \Phi_{\mathbb{C}}$ is an imbedding. Here $\tau(M)$ is the bundle constructed in § 3.3 (Definition 3.11), and φ_Ω is defined in Proposition 3.28.

[3] As usual we restrict ourselves to some neighborhood of Ω.

Proof. Let $\{U_j\}$ be an atlas of s-analytic charts on the manifold (M, ρ_M) with local s-analytic coordinates $(\alpha_j^1, \dots, \alpha_j^n)$. Then, by Definition 3.5, the coordinates of a point $x \in M$ in the phase space $\Phi_{\mathbb{C}}$ are s-analytic functions in the local coordinates $(\alpha_j^1, \dots, \alpha_j^n)$ on the manifold U_j with the weight function $\rho_{U_j} = \rho_M | U_j$:

$$z^k = z^k\left(\alpha_j^1, \dots, \alpha_j^n\right) , \qquad k = 1, \dots, n ; \tag{4.16}$$

$$\zeta_k = \zeta_k\left(\alpha_j^1, \dots, \alpha_j^n\right) , \qquad k = 1, \dots, n . \tag{4.17}$$

We consider the matrix function

$$\begin{bmatrix} \dfrac{\partial z^1}{\partial \alpha_j^1} & \cdots & \dfrac{\partial z^1}{\partial \alpha_j^n} \\ \cdots & \cdots & \cdots \\ \dfrac{\partial z^n}{\partial \alpha_j^1} & \cdots & \dfrac{\partial z^n}{\partial \alpha_j^n} \\ \dfrac{\partial \zeta_1}{\partial \alpha_j^1} & \cdots & \dfrac{\partial \zeta_1}{\partial \alpha_j^n} \\ \cdots & \cdots & \cdots \\ \dfrac{\partial \zeta_n}{\partial \alpha_j^1} & \cdots & \dfrac{\partial \zeta_n}{\partial \alpha_j^n} \end{bmatrix} = X_j(x) . \tag{4.18}$$

If $x \in \Omega(M, \rho_M)$, then the elements of the matrix (4.18) define a Lagrangian plane in phase space $\Phi_{\mathbb{C}}$. If $x \in U_{jk} = U_j \cap U_k$, then by s-analyticity of the functions (4.16) and (4.17) we obtain the inclusion

$$X_j(x) - X_k(x) A_{kj}(x) \in {}^{s+1}I\left(U_{jk}, \rho_{U_{jk}}\right) ,$$

which becomes an equality

$$X_j(x) = X_k(x) A_{kj}(x)$$

in case $x \in \Omega(M, \rho_M) \cap U_{jk}$. From Lemma 3.26 it follows that

$$X_j(x) - X_k(x) B_{kj}(x) \in {}^{s+1}I\left(U_{jk}, \rho_{U_{jk}}\right) .$$

Condition (4.5) can be written in the following form in a local s-analytic system of coordinates:

$$X_j^t(x) I_n X_j(x) \in {}^{s}I\left(U_j, \rho_{U_j}\right) ,$$

where I_n is the matrix (1.2) defined in § 2.1.

Lemma 4.4. *There exist s-analytic functions*

$$Y_j(x) , \quad x \in U_j ,$$

such that

a)

$$X_j(x) - Y_j(x) \in {}^sI\left(U_j, \rho_{U_j}\right)\ , \qquad x \in U_j\ , \tag{4.19}$$

b)

$$Y_j^t(x) I_n Y_j(x) = 0\ , \qquad x \in U_j\ , \tag{4.20}$$

c)

$$Y_j(x) - Y_k(x) B_{kj}(x) = 0\ , \qquad x \in U_{jk}\ . \tag{4.21}$$

Proof. By Lemma 1.1, for every point $x \in U_j$ there exists a collection of indices I such that the matrix

$$\begin{bmatrix} \dfrac{\partial z^I}{\partial \alpha^j} \\ \dfrac{\partial \zeta_{\overline{I}}}{\partial \alpha^j} \end{bmatrix}$$

is nondegenerate. Without loss of generality we assume that $I = \{1, \ldots, n\}$, i.e. the matrix $X_j'(x) = \partial z / \partial \alpha^j$ is nondegenerate. Let the matrix X_j'' be defined by

$$X_j(x) = \begin{bmatrix} X_j'(x) \\ X_j''(x) \end{bmatrix}\ .$$

The the matrix

$$Z_j(x) = X_j''(x) \left(X_j'(x)\right)^{-1}$$

satisfies the condition

$$[Z_j(x)]^t - Z_j(x) \in {}^sI\left(U_j, \rho_{U_j}\right)\ .$$

We set

$$\widetilde{Z}_j(x) = \frac{1}{2}\left(Z_j(x) + Z_j(x)^t\right)\ .$$

Then

$$Z_j(x) - \widetilde{Z}_j(x) \in {}^sI\left(U_j, \rho_{U_j}\right)\ .$$

Furthermore, we set

$$Y_j(x) = \begin{bmatrix} X_j'(x) \\ \widetilde{Z}_j(x) X_j'(x) \end{bmatrix}\ . \tag{4.22}$$

It is clear that the inclusions (4.19) and (4.20) hold for the matrix (4.22), and if for some point $x \in U_j$ the equation

$$X_j^t(x) I_n X_j(x) = 0$$

holds, then for this point x

$$X_j(x) = Y_j(x)\ .$$

Suppose, further, that the matrix function $Y_j(x)$ satisfies conditions (4.19) and (4.20), and that the matrix function $Y_j^0(x)$ is defined on some closed set $F \subset U_j$ and also satisfies conditions (4.19) and (4.20) for points $x \in F$. Then there exists a continuation $Y_j^0(x)$ to the entire set U_j, for which conditions (4.19) and (4.20) hold.

Indeed, suppose

$$Y_j(x) = \begin{bmatrix} Y_j'(x) \\ Y_j''(x) \end{bmatrix} , \quad Y_j^0(x) = \begin{bmatrix} Y_j^{0'}(x) \\ Y_j^{0''}(x) \end{bmatrix} .$$

Then the matrices $Y_j''(x)\left(Y_j'(x)\right)^{-1}$ and $Y_j^{0''}(x)\left(Y_j^{0''}(x)\right)^{-1}$ are symmetric:

$$\begin{gathered} \left[Y_j''(x)\left(Y_j'(x)\right)^{-1}\right]^t = \left(Y_j''(x)\left(Y_j'(x)\right)^{-1}\right) , \\ \left[Y_j^{0''}(x)\left(Y_j^{0'}(x)\right)^{-1}\right]^t = \left(Y_j^{0''}(x)\left(Y_j^{0'}(x)\right)^{-1}\right) , \end{gathered} \tag{4.23}$$

and their difference belongs to the ideal ${}^{s+1}I\left(F, \rho_M|F\right)$. Since the matrices $Y_j'(x)$ and $Y_j^{0'}(x)$ are invertible, and

$$Y_j'(x) - Y_j^{0'}(x) \in {}^{s+1}I\left(F, \rho_M|_F\right) , \tag{4.24}$$

then the matrix $Y_j^{0'}(x)$ can be continued by Corollary 3.10 to an invertible matrix function on some neighborhood of the set $\Omega\left(M, \rho_M\right) \cap U_j$, with property (4.24) remaining true. Analogously we continue the matrix function $Y_j^{0''}(x)\left(Y_j^{0''}(x)\right)^{-1}$ on a neighborhood of the set $\Omega\left(M, \rho_M\right) \cap U_j$, so that condition (4.23) and the condition

$$Y_j''(x)\left(Y_j'(x)\right)^{-1} - Y_j^{0''}(x)\left(Y_j^{0'}(x)\right)^{-1} \in {}^{s}I\left(U_j, \rho_{U_j}\right)$$

are fulfilled. This completes the continuation of the function $Y_j^0(x)$ to a neighborhood of the set $\Omega\left(M, \rho_M\right) \cap U_j$.

Now we finish the proof of Lemma 4.4. We place an ordering on the sets U_j. If the functions $Y_k(x)$ are already constructed for $k < j_0$ so that (4.19), (4.20) and (4.21) are satisfied for $k, j < j_0$, then by (4.21) the function $Y_{j_0}(x)$ will be uniquely defined on the set $\cup_{k<j_0} U_{kj_0} \subset U_{j_0}$. Then the function $Y_{j_0}(x)$ can be continued to some neighborhood of the set $\Omega\left(M, \rho_M\right) \cap U_{j_0}$ in such a way that (4.19), (4.20) and (4.21) hold for $k, j \leq j_0$. Lemma 4.4 follows by induction. □

Now we return to the proof of Proposition 4.3. We set

$$F(\xi) = Y_j(x)\xi + f(x) \tag{4.25}$$

in a local coordinate system of the chart U_j. Condition (4.2) shows the correctness of the definition (4.25) and the validity of condition (4.14). Condition (b) follows from (4.20). Condition (c) follows from (4.19). This proves Proposition 4.3. □

Proposition 4.5. *The mapping F given by formula* (4.13) *induces a mapping F^* of the manifold (M, ρ_M) into the complex Lagrangian Grassmannian $G_{2n}(I_n)$, which is an s-analytic mapping of the s-analytic manifold (M, ρ_M) into the analytic manifold $\Lambda_{\mathbb{C}}(n)$.*

To prove this it is sufficient to notice that all the matrix functions in the proof of Proposition 4.3 are s-analytic functions.

Definition 4.2. A Lagrangian s-analytic manifold (M, ρ_M) in the phase space $\Phi_{\mathbb{C}}$ is called *positive* if the following conditions hold:

a) There exists an s-analytic function $S \in {}^s\mathcal{O}'(M, \rho_M)$ such that

$$dS - f^*(\overline{\omega}) = 0 \in {}^s\Lambda^1(M, \rho_M) \ , \tag{4.26}$$

where

$$\overline{\omega} = \sum_{j=1}^{n} \zeta_j dz^j \ .$$

b) In each canonical chart U_I the function

$$S_I = S - \zeta_{\overline{I}} z^{\overline{I}} \tag{4.27}$$

satisfies the inequality

$$C_1 \rho_M^2(\alpha) \geq \operatorname{Im} S_I(\alpha) \geq C \rho_M^2(\alpha) \ , \tag{4.28}$$

if $\alpha \in U_I$ is a point such that

$$\begin{aligned} \operatorname{Im} z^I(\alpha) &= 0 \ , \\ \operatorname{Im} \zeta_{\overline{I}}(\alpha) &= 0 \ . \end{aligned} \tag{4.29}$$

Proposition 4.6. *If (M, ρ_M) is a positive Lagrangian s-analytic manifold of the phase space $\Phi_{\mathbb{C}}$, and $x \in \Omega(M, \rho_M)$, then the tangent space T_x to the manifold M is a positive Lagrangian plane.*

Proof. By Lemma 1.1 there exists a collection of indices I such that the matrix

$$X = \begin{bmatrix} \partial z / \partial \alpha \\ \partial \zeta / \partial \alpha \end{bmatrix}$$

in the decomposition of phase space $\Phi_{\mathbb{C}} = \mathbb{C}^I \oplus \mathbb{C}^{\overline{I}} \oplus \mathbb{C}_I \oplus \mathbb{C}_{\overline{I}}$ has the form

$$X = \begin{bmatrix} A_1 & A_2 \\ A_3 & A_4 \\ B_1 & B_2 \\ B_3 & B_4 \end{bmatrix}$$

where the matrix $\begin{bmatrix} A_1 & A_2 \\ B_3 & B_4 \end{bmatrix}$ is invertible. It is necessary to show that the matrix

$$\operatorname{Im} \begin{bmatrix} B_1 & B_2 \\ -A_3 & -A_4 \end{bmatrix} \begin{bmatrix} A_1 & A_2 \\ B_3 & B_4 \end{bmatrix}^{-1} \tag{4.30}$$

is nonnegative definite. The function $z^k(\alpha)$, $k \in I$; $\zeta_j(\alpha)$, $j \in \overline{I}$ form a local coordinate system in a neighborhood of the point $x \in \Omega(M, \rho_M)$, that is there exist s-analytic functions

$$\alpha^k = g^k\left(z^I, \zeta_{\overline{I}}\right)$$

such that

$$\begin{aligned} z^k &= z^k\left(g^1\left(z^I, \zeta_{\overline{I}}\right), \ldots, g^n\left(z^I, \zeta_{\overline{I}}\right)\right) , \qquad k \in I , \\ \zeta_j &= \zeta_j\left(g^1\left(z^I, \zeta_{\overline{I}}\right), \ldots, g^n\left(z^I, \zeta_{\overline{I}}\right)\right) , \qquad j \in \overline{I} . \end{aligned}$$

We consider the function S_I as a function of the coordinates $\left(z^I, \zeta_{\overline{I}}\right)$:

$$\begin{aligned} S_I\left(g\left(z^I, \zeta_{\overline{I}}\right)\right) &= \mathcal{F}\left(z^I, \zeta_{\overline{I}}\right) , \\ \mathcal{F}\left(z^I, \zeta_{\overline{I}}\right) &= \mathcal{F}_1 + i\mathcal{F}_2 . \end{aligned}$$

If $z^k = x^k + iy^k$, $\zeta_k = p_k + i\eta_k$, then condition (4.28) means in particular that

$$\mathcal{F}_2\left(x^I, p_{\overline{I}}\right) \geq 0 . \tag{4.31}$$

Then it follows from (4.31) that the matrix of second partial derivatives

$$H = \begin{bmatrix} \dfrac{\partial^2 \mathcal{F}_2}{\partial x^I \partial x^I} & \dfrac{\partial^2 \mathcal{F}_2}{\partial x^I \partial p_{\overline{I}}} \\ \dfrac{\partial^2 \mathcal{F}_2}{\partial p_{\overline{I}} \partial x^I} & \dfrac{\partial^2 \mathcal{F}_2}{\partial p_{\overline{I}} \partial p_{\overline{I}}} \end{bmatrix}$$

is nonnegative definite. By (4.27) the matrix H can be rewritten in the following form:

$$H = \operatorname{Im} \begin{bmatrix} \dfrac{\partial \zeta_I}{\partial z^I} & -\dfrac{\partial z^{\overline{I}}}{\partial z^I} \\ \dfrac{\partial \zeta_I}{\partial \zeta_{\overline{I}}} & -\dfrac{\partial z^{\overline{I}}}{\partial \zeta_{\overline{I}}} \end{bmatrix} \tag{4.32}$$

Since the right-hand side of (4.32) coincides with (4.30), the proposition is proven. □

Corollary 4.7. *The mapping*

$$F^*(M, \rho_M) \to G_{2n}(I_n)$$

from Proposition 4.5 maps a neighborhood $V \supset \Omega(M, \rho_M)$ *in the manifold* (M, ρ_M) *to a neighborhood of the set* $G_{2n}^+(I_n) \subset G_{2n}(I_n)$ *of positive Lagrangian planes.*

Definition 4.3. Let $\sigma : G_{2n}^+(I_n) \to \eta \in (\Lambda^n(\xi_n))^*$ be the section constructed by Theorem 1.4. Since σ is an analytic section, it can be uniquely continued to some neighborhood of the set $G_{2n}^+(I_n)$.

We denote by σ_M the s-analytic form on the manifold (M, ρ_M) induced as the pullback of the section σ:

$$\sigma_M = F^*(\sigma) \in {}^s\Lambda^n(M, \rho_M) \tag{4.33}$$

Suppose (M, ρ_M) is an s-analytic Lagrangian manifold lying in the complex phase space $\Phi_{\mathbb{C}}$, and μ is some nondegenerate s-analytic measure on the manifold M, i.e. an n-dimensional form such that in each local s-analytic coordinate system $(\alpha^1, \ldots, \alpha^n)$ the form μ can be written as

$$\mu = f(\alpha^1, \ldots, \alpha^n)\, d\alpha^1 \wedge \ldots \wedge d\alpha^n \in {}^s\Lambda^n(M, \rho_M) \ ,$$

where $f(\alpha^1, \ldots, \alpha^n) \neq 0$. The function $f(\alpha^1, \ldots, \alpha^n)$ will be called the *derivative of the measure* μ *with respect to the measure* $d\alpha^1 \wedge \ldots \wedge d\alpha^n$ and will be denoted in the following way:

$$f(\alpha^1, \ldots, \alpha^n) = \frac{\partial \mu}{\partial(\alpha^1, \ldots, \alpha^n)} \ .$$

We now formulate a certain condition on the measure μ. Suppose $\{U_I\}$ is an atlas of canonical charts on a positive s-analytic Lagrangian manifold (M, ρ_M) with local coordinate system $(z^I, \zeta_{\overline{I}})$. We consider the function

$$\mu_I = \frac{\partial \mu}{\partial(z^I, \zeta_{\overline{I}})} \ . \tag{4.34}$$

Condition 4.1. *It is possible to choose the argument* $\arg \mu_I$ *of the function* μ_I *in such a way that in the intersections of charts* $U_I \cap U_J$ *the following relation holds:*

$$\arg \mu_I - \arg \mu_J = \sum_k \arg \lambda_{k,IJ} + |I_2| \cdot \pi \tag{4.35}$$

where the eigenvalues $\lambda_{k,IJ}$ *and their arguments are defined in (2.1).*

Theorem 4.8. *Suppose* (M, ρ_M) *is a positive s-analytic Lagrangian manifold in the phase space* $\Phi_{\mathbb{C}}$ *and suppose* μ *is a nondegenerate s-analytic measure on the manifold* (M, ρ_M). *Then Condition* 4.1 *holds if and only if* $\partial\mu/\partial\sigma_M$ *defines a one-dimensional cocycle on the manifold* (M, ρ_M) *which is cohomologous to zero.*

Proof. First suppose $\mu = \sigma_M$. Then the measure μ is induced with the help of the canonical mapping $M \to G^+_{2n}(I_n)$ by the section σ of the bundle η. Consequently, Condition 4.1 for the manifold M follows from the fulfillment of the same condition on $G^+_{2n}(I_n)$.

If $\mu = f \cdot \sigma_M$, $f \neq 0$ and f defines the null class of one-dimensional cohomology on the manifold M, then the function f is homotopic to a constant function in the class of nowhere vanishing functions. Therefore $\arg f$ separates into single-valued branches on the entire manifold M. Consequently, Condition 4.1 holds if one chooses the arguments of the functions μ_I according to the formula

$$\arg \mu_I = \arg \sigma_{M,I} + \arg f \ .$$

Conversely, suppose that Condition 4.1 holds for the measure μ. We will show that the function $\arg f$ separates into single-valued branches. To do this we set

$$\arg_I f = \arg \sigma_{M,I} - \arg \mu_I \ .$$

Using Condition 4.1 for the measures σ_M and μ, we obtain

$$\arg_I f = \arg_K f$$

on the intersections of charts $U_I \cap U_K$. This means that the functions $\arg_I f$ give a single-valued branch on the entire manifold M. This proves Theorem 4.8. □

Corollary 4.9. *Condition 4.1 is fulfilled if and only if the form*

$$i\, d \ln \left[\frac{\left(\dfrac{\partial \mu}{\partial(z - i\zeta)} \right)}{\left| \dfrac{\partial \mu}{\partial(z - i\zeta)} \right|} \right] \tag{4.36}$$

is cohomologous to zero.

Proof. Let

$$f = \frac{\partial\mu}{\partial(z - i\zeta)} \,.$$

Since $f(x) \neq 0$, by Theorem 4.8, Condition 4.1 is equivalent to the statement that the mapping

$$f : M \to \mathbb{C}^* = \mathbb{C} \setminus \{0\}$$

is homotopic to a point mapping. Consequently Condition 4.1 is equivalent to the statement that the function

$$g = \frac{\arg f}{|f|}$$

separates into single-valued branches, i.e. the form (4.36) is cohomologous to zero. □

As is evident from Theorem 4.8, *the Maslov condition of quantization of an appropriate Lagrangian manifold depends in an essential way on the choice of the measure μ on the manifold M.* For example, if $\mu = \sigma_M$, then the quantization condition holds for trivial reasons.

Part II
Maslov's Canonical Operator on a Real Lagrangian Manifold

In this part we present the construction of Maslov's canonical operator. As was already explained in the Introduction, we have adopted a topological method of exposition. We fix a Lagrangian manifold L in phase space. In each canonical chart of the manifold L a local canonical operator, which sends functions on the manifold L with support in the canonical chart into functions of configuration space, can be defined. If the local canonical operators coincide on the intersections of charts, then an operator on the entire Lagrangian manifold L is obtained. However, in general, the local canonical operators do not coincide on the intersections of charts and a comparison of two such operators leads to a series of cocycles, which can be interpreted as obstructions to constructing a global operator.

In Chap. 4 the first two obstructions are considered in the real case. When the conditions of quantization hold these obstructing cocycles can be chosen to be trivial. Correspondingly, we obtain a global canonical operator, defined up to the first order in the parameter h. In this chapter we also derive formulas for the commutation of the canonical operator with pseudodifferential operators up to order h^2.

In order to obtain asymptotics of solutions accurate up to any power of the parameter h, one has to study the entire class of cocycles which obstruct the construction of the canonical operator on a Lagrangian manifold. Towards this end, we derive in Chap. 5 formulas for the expansion of integrals of rapidly oscillating functions in a power series in the parameter h. In Chap. 6 the canonical operator is constructed and formulas are derived for the commutation of the canonical operator with a pseudodifferential operator with an accuracy up to any power of h.

In Chap. 6 we construct a more general complex theory of the canonical operator. For simplicity of exposition, in the first two sections of this chapter we construct the theory of the canonical operator only up to first order in the parameter h and establish the formulas for the commutation of the canonical operator with a pseudodifferential operator up to order h^2 in the complex case. These two paragraphs represent a generalization to the complex case of the results of Chap. 4; however, we have tried to make the presentation autonomous. In the final two sections of this chapter the canonical operator is constructed and the corresponding commutation formulas, accurate up to an arbitrary power of the parameter h, are derived. Here we present the complex case immediately, without initially considering the (simpler) real case, since we hope that the preceding material will prepare the reader adequately to understand the material presented in these sections.

We will show that there are no obstructing cocycles other than the quantization conditions on a Lagrangian manifold considered in the first two sections considered in the first two sections of Chap. 4, i.e. all the higher-order (in powers of the parameter h) obstructions are inconsequential.

Chapter 4
Maslov's Canonical Operator (Real Case)

4.1 The Construction of Maslov's Elementary Canonical Operator

4.1.1 Canonical Cochains. We denote, as before, by $\Phi_{\mathbb{R}}$ the $2n$-dimensional real space with basis $(e_1, \ldots, e_n, f^1, \ldots, f^n)$. We call this space real phase space. Any vector $a \in \Phi_{\mathbb{R}}$ can be represented uniquely as a linear combination of the basis vectors (e, f) with real coefficients

$$a = x^k e_k + p_k f^k \ .$$

The numbers x^k, $k = 1, \ldots, n$, will be called the *x-coordinates*, and the numbers p_k, $k = 1, \ldots, n$, the *p-coordinates* of the vector a. We provide the space $\Phi_{\mathbb{R}}$ with the structure of a real Hamiltonian (symplectic) space, introducing on it a nondegenerate differential form

$$\omega = dp \wedge dx \ . \tag{1.1}$$

Now let L be a smooth n-dimensional manifold.

Definition 1.1.[1] The imbedding

$$i : L \hookrightarrow \Phi_{\mathbb{R}}$$

of the manifold L into phase space $\Phi_{\mathbb{R}}$ is called *Lagrangian* if the restriction $i^*\omega$ of the form ω is equal to zero, i.e. if

$$i^*\omega = 0 \ .$$

In the future we will also call Lagrangian imbeddings Lagrangian manifolds.

[1] We repeat here certain definitions of Chap. 2 in order to make the presentation more autonomous.

Example 1. Suppose $\Phi_{\mathbb{R}}$ is a two-dimensional vector space. Then any smoothly imbedded curve

$$\begin{aligned} x &= x(t) \ , \quad t \in [a,b] \subset R^1 \ , \\ p &= p(t) \ , \end{aligned} \tag{1.2}$$

defines a one-dimensional Lagrangian manifold. Indeed, since any two tangent vectors to an arbitrary point of the curve (1.2) are collinear, the skew-symmetric form (1.1) annihilates them.

Example 2. We consider the manifold in the space $\Phi_{\mathbb{R}}$ given by the system of equations

$$\left(x^k\right)^2 + \left(p_k\right)^2 = 1 \ , \quad k = 1, \dots, n \ . \tag{1.3}$$

For $n = 1$ this manifold determines, obviously, a circle with center at the origin. For arbitrary n the given manifold is a direct product of n circles, i.e. an n-dimensional torus.

We prove that manifold (1.3) is Lagrangian. Indeed, since for each k there is either a smooth function $p_k = p_k\left(x^k\right)$ or a smooth function $x^k = x^k\left(p_k\right)$, all the terms of expression (1.1) cancel out and, consequently, the entire expression (1.1) does as well.

We denote by $[n]$ the set of natural numbers from 1 to n:

$$[n] = \{1, 2, \dots, n\} \ .$$

Let $I \subset [n]$ be an arbitrary subset of the set $[n]$, $\overline{I}$ be the complement of I in the set $[n]$: $\overline{I} = [n] \setminus I$.

Definition 1.2. A *canonical chart* U_I of a Lagrangian imbedding i: $L \hookrightarrow \Phi_{\mathbb{R}}$ is a pair of an open, simply-connected domain (the support of the chart) $U \subset L$ and a projection

$$\pi_I : U \to R^I \oplus R_{\overline{I}} \ ,$$

which is a diffeomorphism onto the open set $\mathbb{R}^I \oplus R_{\overline{I}}$, where $\mathbb{R}^I$, $\mathbb{R}_{\overline{I}}$ are the spaces of coordinates $\left(x^{i_1}, \dots, x^{i_k}\right)$ and $\left(p_{i_{k+1}}, \dots, p_{i_n}\right)$ respectively. Here $\{i_1, \dots, i_k\} = I$, $\{i_{k+1}, \dots, i_n\} = \overline{I}$.

Thus the functions $\left(x^I, p_{\overline{I}}\right)$ serve as coordinates in a canonical chart of a Lagrangian imbedding $i : L \hookrightarrow \Phi_{\mathbb{R}}$. The domains of the chart U_I can be written in the form of equations

$$\begin{aligned} x^I &= x^{\overline{I}}\left(x^I, p_{\overline{I}}\right) \\ p_I &= p_I\left(x^I, p_{\overline{I}}\right) \ . \end{aligned}$$

Definition 1.3. A *canonical atlas* of a Lagrangian manifold L is any atlas consisting of canonical charts.

Lemma 1.1. *On any Lagrangian manifold there exists a canonical atlas.*

Notes. a) The assertion of the lemma can be rephrased in the following way. There exists an atlas of charts such that the support of each chart can be written in the form of equations

$$\begin{aligned} x^{\bar{I}} &= x^{\bar{I}}(x^I, p_{\bar{I}}) \ , \\ p_I &= p_I(x^I, p_{\bar{I}}) \ . \end{aligned} \tag{1.4}$$

b) For a smoothly imbedded curve

$$x = x(t) \ , \quad p = p(t) \ , \quad t \in [a, b] \ ,$$

the assertion of the lemma is obvious. Indeed, since

$$\operatorname{rank}(\dot{x}, \dot{p}) = 1 \ ,$$

where a dot indicates differentiation by a parameter t, then for each value of the parameter t either $\dot{x} \neq 0$ or $\dot{p} \neq 0$. By the implicit function theorem there exists either a smooth function $t_1 = t_1(x)$ or a smooth function $t_2 = t_2(p)$.

Proof of Lemma 1.1. We consider an arbitrary point $l \in L$. It is clear that if the tangent plane to the manifold L at the point l has canonical coordinates of the form $(x^I, p_{\bar{I}})$, then by the implicit function theorem some neighborhood of the point l in the manifold L can be written in the form (1.4).

Thus the analytic problem of the existence of a canonical atlas on the Lagrangian manifold L in fact becomes the algebraic problem of choosing canonical coordinates on Lagrangian hyperplanes (subspaces). This problem is solved in Lemma 1.1 of 2.1 of Chap. 2. □

Suppose μ is some smooth, nonsingular measure on the Lagrangian imbedding L, in other words an element $\mu \in \Lambda^n(M)$ such that in any coordinate system (U, α) on the manifold L, $\mu = \mu(\alpha) d\alpha^1 \wedge \ldots \wedge d\alpha^n$, $\mu \neq 0$. The latter condition, obviously, does not depend on the choice of coordinates. We define two zero-dimensional cochains of a canonical covering (i.e. a covering consisting of canonical charts) of the manifold L. However, unlike th cochains which we met with in the first part of this book, the cochains defined below will take values in the set of smooth functions. Thus, to give such a zero-dimensional cochain f, we assign to each element of the covering U_i a certain smooth function f_i. The boundary δf of such a cochain is defined as a collection of smooth functions on the intersections of chart $U_i \cap U_j$, setting

$$\delta(f)_{ij} = f_i - f_j \quad \text{in} \quad U_i \cap U_j \; .$$

This language, which is used widely in mathematics, will turn out to be useful in our case as well.

And so we choose some canonical atlas of charts on the manifold L and define on each canonical chart U_I, given by the equations

$$\begin{aligned} x^{\overline{I}} &= x^{\overline{I}}(x^I, p_{\overline{I}}) \; , \\ p_I &= p_I(x^I, p_{\overline{I}}) \; . \end{aligned}$$

a function $S_I(x^I, p_{\overline{I}})$, called the action. The function $S_I(x^I, p_{\overline{I}})$ is defined as an arbitrary solution of the equation

$$dS_I(x^I, p_{\overline{I}}) = p_I(x^I, p_{\overline{I}})dx^I - x^{\overline{I}}(x^I, p_{\overline{I}})dp_{\overline{I}} \; . \tag{1.5}$$

Equation (1.5) is solvable because, since the imbedding of the manifold L is Lagrangian, the equation

$$d\,dS = i^* \left(dp_I \wedge dx^I - dx^{\overline{I}} \wedge dp_{\overline{I}}\right) = i^*\left(dp \wedge dx\right) = 0$$

holds, and the support of the chart U_I is simply-connected. We note that the solution of Eq. (1.5) is defined up to a real constant.

Choosing in each chart a certain solution of Eq. (1.5), we thus assign to each element of the covering U_I a smooth function $S_I(x^I, p_{\overline{I}})$. This is our first zero-dimensional cochain.

We denote by μ_I the density of the measure μ relative to the coordinates $(x^I, p_{\overline{I}})$ of the chart U_I, so that in the system of local coordinates of the chart U_I the measure μ has the form

$$\begin{aligned} \mu &= \mu_I(x^I, p_{\overline{I}})dx^I \wedge dp_{\overline{I}} \\ &= \mu_I(x^I, p_{\overline{I}})dx^{i_1} \wedge \ldots \wedge dx^{i_k} \wedge dp_{i_{k+1}} \wedge \ldots \wedge dp_{i_n} \; , \end{aligned}$$

$$I = \{i_1, \ldots, i_k\} \; , \quad \overline{I} = (i_{k+1}, \ldots, i_n\} \; .$$

Here and below we will very often encounter expressions involving the groups of indices I, $\overline{I}$ for certain subsets $I \subset \{1, \ldots, n\}$. To shorten our notation we will adopt the following conventions.

1) An expression of the form $dx^I \wedge dp_{\overline{I}}$ will denote the n-form

$$dx^I \wedge dp_{\overline{I}} = dx^{i_1} \wedge \ldots \wedge dx^{i_k} \wedge dp_{i_{k+1}} \wedge \ldots \wedge dp_{i_n} \; ,$$

if $I = \{i_1, \dots, i_k\}$, $\overline{I} = \{i_{k+1}, \dots, i_n\}$. The same abbreviation will be used in the notation of derivatives and expressions under an integral sign as well.

2) Repeated indices will always be summed. For example,

$$p_I(x^I, p_{\overline{I}})dx^I = \sum_{i \in I} p_i(x^I, p_{\overline{I}})dx^i .$$

As one can see from this example, the second convention does not apply when one of the indices stands inside a function sign and the other is outside. This convention also applies to indices which indicate that an object belongs to a chart U_I. For example, S_I is the action in the chart U_I, μ_I is the density of the measure μ in the chart U_I, k_I is the local canonical operator in the chart U_I, and so on.

3) Sums of the form

$$a_{II}\xi^I \eta^I = \sum_{i,j \in I} a_{ij}\xi^i \eta^j , \quad a_{II}\xi^I = \left\{ \sum_{i \in I} a_{ki}\xi^I , \quad k \in I \right\}$$

will be used only when the matrices $\|a_{ij}\|$, $i \in I$, $j \in I$, are symmetric. In these cases no confusion can arise as to the meaning of the corresponding formulas.

The above conventions apply not only to the sets I, $\overline{I}$ but to any subsets of the set $\{1, \dots, n\}$.

Keeping in mind that we will have to take the square root $\left[\mu_I(x^I, p_{\overline{I}})\right]^{1/2}$, we assign to the density $\mu_I(x^I, p_{\overline{I}})$ a certain value $\arg \mu_I(x^I, p_{\overline{I}})$ of the argument, $0 \le \arg \mu_I \le 4\pi$, and we will consider that $\arg \left[\mu_I(x^I, p_{\overline{I}})\right]^{1/2} = \arg \mu_I(x^I, p_{\overline{I}})/2$.

Hence we assign to each element of the covering U_I the smooth function $\left[\mu_I(x^I, p_{\overline{I}})\right]^{1/2}$. This is our second zero-dimensional cochain.

4.1.2 Maslov's Elementary Canonical Operator. We now define Maslov's elementary canonical operator k_I, setting for each $\varphi \in C_0^\infty(U_I)$ (a function with compact support in the chart U_I)

$$k_I \varphi = \overline{F}^{1/h}_{p_{\overline{I}} \to x^{\overline{I}}} \exp\left\{ \frac{i}{h} S_I(x^I, p_{\overline{I}}) \right\} \sqrt{\mu_I(x^I, p_{\overline{I}})}\, \varphi(x^I, p_{\overline{I}}) . \tag{1.6}$$

Here $\overline{F}^{1/h}$ is the $1/h$-Fourier transform:

$$F^{1/h}_{x \to p}[f](p) = \left(\frac{-i}{2\pi h}\right)^{n/2} \int_{\mathbb{R}^n} \exp\left\{ \frac{-i\langle x, p\rangle}{h} \right\} f(x)dx .$$

$$\overline{F}^{1/h}_{p \to x}[f](x) = \left(\frac{i}{2\pi h}\right)^{n/2} \int_{\mathbb{R}_n} \exp\left\{ \frac{i\langle x, p\rangle}{h} \right\} f(p)dp .$$

The $1/h$-Fourier transform which we apply to our theory has several properties which are analogous to the ordinary Fourier transform. The facts about $1/h$-Fourier transforms that we will need are presented in Appendix I to Chap. 6.

Proposition 1.2. *Suppose U_I and U_J are two canonical charts with a nonempty intersection, and suppose $\varphi \in C_0^\infty(U_I \cap U_J)$. Then there exist constants $c_{IJ}^{(1)} \in \mathbb{R}$ and $c_{IJ}^{(2)} \in \mathbb{C}$ such that the congruence*

$$k_I\varphi \equiv \exp\left(\frac{i}{h}c_{IJ}^{(1)} + i\pi c_{IJ}^{(2)}\right\} k_J\varphi \quad (\mathrm{mod} h) \ . \tag{1.7}$$

holds.

The congruence

$$\psi(x,h) \equiv 0 \quad (\mathrm{mod} h^N)$$

should be interpreted as a system of inequalities

$$\|\widehat{p}^\alpha \psi(x,h)\|_{L_2} \le ch^N \tag{1.8}$$

for all multi-indices $\alpha = (a_1, \ldots, a_n)$, $|\alpha| \le N$. Here

$$\widehat{p}^\alpha = (-ih)^{|\alpha|} \frac{\partial^{|\alpha|}}{(\partial x^1)^{\alpha_1} \ldots (\partial x^n)^{\alpha_n}} ,$$

$|\alpha| = \alpha_1 + \ldots + \alpha_n$, and $c > 0$ is a certain constant.

Proof. We need to compare two functions:

$$k_I\varphi = F_{p_{\overline{I}} \to x^{\overline{I}}} \exp\left\{\frac{i}{h}S_I\right\} \sqrt{\mu_I}\,\varphi ,$$

$$k_J\varphi = F_{p_{\overline{J}} \to x^{\overline{J}}} \exp\left\{\frac{i}{h}S_J\right\} \sqrt{\mu_J}\,\varphi ,$$

where $\mathrm{supp}\varphi \in U_I \cap U_J$. We write $I_2 = I \backslash J$, $I_3 = J \backslash I$, $I_4 = [n] \backslash I_1 \cup I_2 \cup I_3$.

Since $\overline{I} \cap \overline{J} = I_4$, by applying the $1/h$-Fourier transform $F^{1/h}_{x^{I_4} \to p_{\overline{I}_4}}$ to the function k_I and k_J and noting that under this application functions for which $f \equiv 0$ (mod$h$) are transformed into similarfunctions, we find that to prove (1.7) it is enough to compare the expressions

$$\overline{F}^{1/h}_{p_{I_2} \to x^{I_2}} \exp\left\{\frac{i}{h}S_I\right\} \sqrt{\mu_I}\,\varphi \quad \text{and} \quad \overline{F}^{1/h}_{p_{I_2} \to x^{I_2}} \exp\left\{\frac{i}{h}S_J\right\} \sqrt{\mu_J}\,\varphi .$$

Taking the $1/h$-Fourier transform $F^{1/h}_{x^{I_2}\to P_{I_2}}$ of both of these expressions, we obtain that the problem reduces to a comparison of the two expressions:

$$\begin{gathered} F^{1/h}_{\substack{p_{I_3}\to x^{I_3}\\ x^{I_2}\to p_{I_2}}} \exp\left\{\frac{i}{h}S_I\right\}\sqrt{\mu_I}\,\varphi\ , \\ \exp\left\{\frac{i}{h}S_J\right\}\sqrt{\mu_J}\,\varphi\ . \end{gathered} \tag{1.9}$$

Written out in greater detail, expression (1.9) has the form

$$\begin{aligned} (-1)^{|I_2|/2}\left(\frac{i}{2\pi h}\right)^{\frac{I_2+I_3}{2}} \iint \exp\left\{\frac{i}{h}\left(\langle p_{I_3}, x^{I_3}\rangle - \langle x^{I_2}, p_{I_2}\rangle\right)\right\} \\ \times \exp\left(\frac{i}{h}S_I\right)\sqrt{\mu_I}\,\varphi\, dp_{I_3}dx^{I_2}\ . \end{aligned} \tag{1.10}$$

We apply the formula for the asymptotic expansion of a rapidly oscillating function to integral (1.10). This formula asserts that the following asymptotic expansion holds:

$$\begin{aligned} \left(\frac{i}{2\pi h}\right)^{n/2} \int\limits_{\mathbb{R}^n} \exp\left\{\frac{i}{h}\Phi(x,p)\right\}\varphi(x,p)dp \\ \equiv \frac{\exp\left\{\frac{i}{h}\Phi\left(x,p(x)\right)\right\}\varphi\left(x,p(x)\right)}{\sqrt{\det \operatorname{Hess}\left(-\Phi\left(x,p(x)\right)\right)}}\ (\operatorname{mod} h)\ . \end{aligned} \tag{1.11}$$

In this formula $p = p(x)$ denotes the unique stationary point of the function $\Phi(x,p)$ on the support of the function $\varphi(x,p)$, i.e. a point at which the gradient of the function $\Phi(x,p)$ in the variables p is equal to zero. $\operatorname{Hess}\left(-\Phi\left(x,p(x)\right)\right)$ denotes the matrix of second partial derivatives of the funtion $-\Phi(x,p)$ in the variables p. To determine the root $\left[\det \operatorname{Hess}\left(-\Phi\left(x,p(x)\right)\right)\right]^{1/2}$ the argument of the expression under the square-root sign is chosen by the formula

$$\arg \det \operatorname{Hess}\left(-\Phi\left(x,p(x)\right)\right) = \sum_{k=1}^{n} \arg \lambda_k\ ,$$

where λ_k are the eigenvalues of the matrix $\operatorname{Hess}\left(-\Phi\left(x,p(x)\right)\right)$, whose arguments are chosen in the interval $-3\pi/2 < \arg\lambda_k \le \pi/2$. Formula (1.11) will be proven in Chap. 5.

For the integral (1.10) the stationary point can be found from the equations

$$d_{p_{I_3},x^{I_2}}\left(\langle p_{I_3}, x^{I_3}\rangle - \langle p_{I_2}, x^{I_2}\rangle + S_I\right) = 0\ .$$

Since (see (1.5))

$$dS_I = p_I\left(x^I, p_{\overline{I}}\right) dx^I - x^{\overline{I}}\left(x^I, p_{\overline{I}}\right) dp_{\overline{I}} ,$$

the equations for a stationary point take the form

$$\begin{aligned} x^{I_3} + x^{I_3}\left(x^I, p_{\overline{I}}\right) &= 0 , \\ -p_{I_2} + p_{I_2}\left(x^I, p_{\overline{I}}\right) &= 0 . \end{aligned} \tag{1.12}$$

Adding to the system (1.12) the identities

$$x^{I_1} = x^{I_1} , \quad p_{I_4} = p_{I_4} . \tag{1.13}$$

we see that Eqs. (1.12), (1.13) determine the equations of transformation from the coordinates $\left(x^I, p_{\overline{I}}\right)$ to the coordinates $\left(x^J, p_{\overline{J}}\right)$ on the intersection $U_I \cap U_j$. Because of this the system (1.12) is uniquely solvable in terms of the coordinates x^{I_2}, p_{I_3}:

$$\begin{aligned} x^{I_2} &= x^{I_2}\left(x^J, p_{\overline{J}}\right) , \\ p_{I_3} &= p_{I_3}\left(x^J, p_{\overline{J}}\right) . \end{aligned} \tag{1.14}$$

Furthermore, since

$$\operatorname{Hess}_{x^{I_2}, p_{I_3}}(-S_I) = \frac{\partial\left(-p_{I_3}, x^{I_2}\right)}{\partial\left(x^{I_2}, p_{I_3}\right)} , \tag{1.15}$$

at points (1.14) the Hessian (1.15) is defined and different from zero.

By Formula (1.11) we have

$$\begin{aligned} &(-1)^{|I_2|/2}\left(\frac{i}{2\pi h}\right)^{|I_2+I_3|/2} \iint \exp\left\{\frac{i}{h}\left(\langle p_{I_3}, x^{I_3}\rangle - \langle x^{I_2}, p_{I_2}\rangle + S_I(x^I, p_{\overline{I}})\right)\right\} \\ &\quad \times \sqrt{\mu_I(x^I, p_{\overline{I}})}\, \varphi(x^J, p_{\overline{J}})\, dp_{I_3} dx^{I_2} \equiv (-1)^{|I_2|/2} \\ &\quad \times \exp\left\{\frac{i}{h}\left(S_I(x^I, _{\overline{I}})\right)\right\} \exp\left\{\frac{i}{h}\langle p_{I_3}, x^{I_3}\rangle - \langle x^{I_2}, p_{I_2}\rangle\right\} \\ &\quad \times \frac{\sqrt{\mu_I(x^J, p_{\overline{J}})}\, \varphi(x^J, p_{\overline{J}})}{\sqrt{\det \operatorname{Hess}_{x^{I_2}, p_{I_3}}\left(-S_I(x^J, p_{\overline{J}})\right)}} \pmod{h} , \end{aligned}$$

with

$$\arg \frac{\sqrt{\mu_I}}{\sqrt{\det \operatorname{Hess}(-S_I)}} = \frac{\arg \mu_I}{2} - \frac{1}{2} \sum \arg \lambda_{k,IJ} ,$$

where $\lambda_{k,IJ}$ are the eigenvalues of the matrix

$$\mathrm{Hess}_{x^{I_2},p_{I_3}}\left(-S_I(x^J,p_{\overline{J}})\right) = \frac{\partial\left(-p_{I_2},x^{I_3}\right)}{\partial\left(x^{I_2},p_{\overline{I}_3}\right)} .$$

Consequently,

$$k_I\varphi \equiv (-1)^{|I_2|/2}\exp\left\{\frac{i}{h}\left[S_I + \langle p_{I_3},x^{I_3}\rangle - \langle x^{I_2},p_{I_2}\rangle - S_J\right]\right\}$$
$$\times\exp\left\{\frac{i}{2}\left[\arg\mu_I - \arg\mu_J - \sum\arg\lambda_{k,IJ}\right]\right\} k_J\varphi\,(\mathrm{mod}\,h) .$$

Thus the expressions $c_{IJ}^{(1)}$ and $c_{IJ}^{(2)}$ mentioned in the formulation of Proposition 1.2 are respectively equal to

$$c_{IJ}^{(1)} = S_I - S_J + \langle P_{I_3},x^{I_3}\rangle - \langle x^{I_2},p_{I_2}\rangle ,$$
$$c_{IJ}^{(2)} = (2\pi)^{-1}\left[\arg\mu_I - \arg\mu_J - \sum\arg\lambda_{k,IJ} - |I_2|\,\pi\right] .$$

To complete the proof of Proposition 1.2 it is necessary to establish that

a) $c_{IJ}^{(1)}$ is a real number; (in other words the functions $c_{IJ}^{(1)}(x)$ are constant)

b) $c_{IJ}^{(2)}$ is an integer.

We show that

$$dc_{IJ}^{(1)} = 0 . \tag{1.16}$$

By (1.5) we have on the intersection $U_I\cap U_J$

$$\begin{aligned} dc_{IJ}^{(1)} &= d\left(S_I - S_J + \langle p_{I_3},x^{I_3}\rangle - \langle p_{I_2},x^{I_2}\rangle\right)\\ &= p_I dx^I - x^{\overline{I}}dp_{\overline{I}} - p_J dx^J + x^{\overline{J}}dp_{\overline{J}}\\ &\quad + x^{I_3}dp_{I_3} + p_{I_3}dx^{I_3} - p_{I_3}dx^{I_3} - x^{I_2}dp_{I_2} . \end{aligned} \tag{1.17}$$

From the relations

$$\begin{aligned} I &= I_1\cup I_2 , \quad \overline{I} = I_3\cup I_4 ,\\ J &= I_1\cup I_3 , \quad \overline{J} = I_2\cup I_4 \end{aligned} \tag{1.18}$$

it follows that all the terms in (1.17) cancel each other out and Eq. (1.16) holds.

Furthermore, since

$$\mu_I = \mu_J \det\frac{\partial\left(-p_{I_2},x^{I_3}\right)}{\partial\left(x^{I_2},p_{I_3}\right)}(-1)^{|I_2|} = \mu_j\prod_{k=1}^{|I_2+I_3|}\lambda_{k,IJ}(-1)^{|I_2|} , \tag{1.19}$$

it follows that, up to a multiple of 2π, the arguments of the left and right-hand sides of expression (1.19) are equal:

$$\arg \mu_I \equiv \arg \mu_J + \arg \lambda_{k,IJ} + |I(2)|\,\pi\,(\mathrm{mod} 2\pi) \ .$$

This completes the proof of Proposition 1.2. □

4.1.3 Quantized Lagrangian Manifolds and Maslov's Global Canonical Operator. We have defined two one-dimensional cochains of the canonical covering or, for short, canonical cochains.

1) The cochain $c^{(1)} = \left\{c_{IJ}^{(1)}\right\}$:

$$c_{IJ}^{(1)} = \langle p_{I_3} x^{I_3}\rangle - \langle p_{I_2}, x^{I_2}\rangle + S_I - S_J \ . \tag{1.20}$$

2) The cochain $c^{(2)} = \left\{c_{IJ}^{(2)}\right\}$:

$$c_{IJ}^{(2)} = \frac{1}{2\pi}\left[\arg \mu_I - \arg \mu_J - \sum_k \arg \lambda_{k,IJ} - |I_2|\,\pi\right] \ . \tag{1.21}$$

In this passage we will show that cochains (1.20) and (1.21) are cocycles.

Proposition 1.3. *The cochain $c^{(1)}$ is a cocycle.*

Proof. From (1.18) it follows that

$$c_{IJ}^{(1)} = \left(S_I + \langle x^{\bar{I}}, p_{\bar{I}}\rangle\right) - \left(S_J + \langle x^{\bar{J}}, p_{\bar{J}}\rangle\right) \ . \tag{1.22}$$

Thus if one defines a zero-dimensional cochain $f = \{f_I\}$ with values in the space of smooth functions, setting

$$f_I = S_I + \langle x^{\bar{I}}(x^I, p_{\bar{I}}), p_{\bar{I}}\rangle$$

in the chart U_I, then (1.22) is nothing but the coboundary of this cochain:

$$c^{(1)} = \partial f \ .$$

By virtue of the well-known property of the coboundary operator $\left(\partial^2 = 0\right)$, it follows from this that the coboundary $\partial c^{(1)}$ of the cochain is equal to zero. This proves Proposition 1.3. □

Proposition 1.4. *The cochain $c^{(2)}$ is a cocycle.*

The cochain $\left\{c_{IJ}^{(2)}\right\}$ decomposes into a sum of two cochains, $c_{IJ}^{(2)} = \alpha_{IJ} + \beta_{IJ}$, where

$$\alpha_{IJ} = \frac{\arg \mu_I - \arg \mu_J}{2\pi} ,$$
$$\beta_{IJ} = \frac{-\left(\sum_k \arg \lambda_{k,IJ} + |I(2)|\,\pi\right)}{2\pi} .$$

By Theorem 2.2 of Chap. 3

$$\beta_{IJ} = \arg \sigma_J - \arg \sigma_I .$$

Thus both cochains $\{\alpha_{IJ}\}$ and $\{\beta_{IJ}\}$ are coboundaries, and, hence, cocycles. This proves the proposition. □

Definition 1.4. A canonical atlas $\{U_I\}$ is called *quantized* if the canonical cochains $c^{(1)}$ and $c^{(2)}$ are cohomologous to zero.

We note that the condition that the cochain $c^{(1)}$ be cohomologous to zero in the class of cohomology with constant coefficients means that for each canonical chart U_I there exists a constant c_I such that for each pair of charts U_I, U_J such that $U_I \cap U_J \neq \emptyset$ the equation

$$c^{(1)}_{IJ} = c_I - c_J$$

holds. Now we change the zero-dimensional cochain $\{S_I\}$ by replacing the function S_I in each chart U_I with the function $S'_I = S_I - c_I$. We note that the functions S'_I, like S_I, satisfy condition (1.5). For the cochain S'_I we have the equations

$$\begin{aligned} S'_J - S'_I &= (S_J - c_J) - (S_I - c_I) = (c_I - c_J) \\ &\quad - \left(S_I - S_J + \langle p_{I_3}, x^{I_3}\rangle - \langle x^{I_2}, p_{I_2}\rangle\right) + \langle p_{I_3}, x^{I_3}\rangle - \langle x^{I_2}, p_{I_2}\rangle \\ &= (c_I - c_J) - c^{(1)}_{IJ} + \langle p_{I_3}, x^{I_3}\rangle - \langle x^{I_2}, p_{I_2}\rangle = \langle p_{I_3}, x^{I_3}\rangle - \langle x^{I_2}, p_{I_2}\rangle . \end{aligned}$$

Next, the condition that the cochain $c^{(2)}$ be cohomologous to zero in the integral cohomology means that there exist whole numbers n_I for each chart U_I such that

$$c^{(2)}_{IJ} = n_I - n_J .$$

Choosing $\arg' \mu_I = \arg \mu_I - 2\pi n_I$ in each chart U_I, as for the first cochain we obtain

$$\arg' \mu_I - \arg' \mu_J = \sum_k \arg \lambda_{k,IJ} + |I_2|\,\pi$$

holds, where $\{\lambda_{k,IJ}\}$ are as defined before.

Although we defined the quantization of a canonical covering above, in fact this condition is a characteristic of the Lagrangian manifold itself. We notice, first of all, that the cohomology class of the cocyle $c^{(1)}$ is not changed if one changes each function S_I by a constant. Therefore the indicated cohomology class is independent of the choice of S_I. Next we will show that the cohomology class of $c^{(2)}$ is independent of the choice of measure μ. Suppose μ' is another nondegenerate measure. Then the ratio $\mu'/\mu = f$ is a real-valued function which is nonzero on the entire manifold L. Therefore one may choose the argument $\arg f$ of this function (equal to 0 or π depending on its sign) globally on the entire manifold. Moreover,

$$\arg \mu' = \arg \mu + \arg f$$

and formula (1.21) shows that the cochain $c_{IJ}^{(2)}$ is unchanged. It is also clear that the cohomology class $c^{(2)}$ is also independent of the arbitrariness in the choice of argument of the density of the measure μ_I in the chart U_I. We note here that we have used the fact that the measure μ is real in an essential way. As is shown in Sect. 3.2 of Chap. 3, the second quantization condition in the complex case does depend on the measure μ.

Finally, the independence of the classes $c^{(1)}$ and $c^{(2)}$ from the choice of canonical covering follows from Leray's theorem, according to which the cohomology classes of a manifold computed by means of a covering (Čech cohomology) do not depend on the covering if all the elements of the covering and all their intersections are contractible. One can assume that the canonical covering satisfies the stated condition. Consequently, the condition of quantization, i.e. the triviality of certain cohomology classes, is in fact a characteristic of the Lagrangian manifold itself, and we are justified in making the following.

Definition 1.5. The manifold L is called *quantized* if its cohomolgoy classes $c^{(1)}$ and $c^{(2)}$ are trivial.

Now we can formulate the first main theorem of this chapter on the existence of a global canonical operator on a quantized Lagrangian manifold.

Indeed, let $H_r^{1/h}$ be the completion of the set of smooth finite functions $f(x,h)$ with respect to the norm

$$\|f\|_r = \sup_{0<h\leq 1} \left\| \left(1 + |x|^2 + \widehat{p}^2\right)^{r/2} f \right\|_{L_2} ,$$

where $\widehat{p}^2$ is the $1/h$-Laplacian operator:

$$\widehat{p}^2 = + \sum \widehat{p}_i^2, p_i = -ih\frac{\partial}{\partial x} .$$

Furthermore, let $H^{1/h}$ be the intersection of all the spaces $H_r^{1/h}$:

$$H^{1/h} = \bigcap_r H_r^{1/h} .$$

We introduce on the space $H^{1/h}$ an (unbounded) operator A^k of multiplication by h^{-k}.

$$A^k : H^{1/h} \to H^{1/h} , \qquad k = 0, 1, 2, \dots .$$

We denote by $D(A^k)$ the domain of definition of the operator A^k, and we denote the (algebraic) factor space $H^{1/h}/D(A^k)$ by ${}^kH^{1/h} = {}^kH^{1/h}(\mathbb{R}^n)$.

We observe that the set $D(A^k)$ consists of those functions f for which inequality (1.8) holds with $N = k$. This allows one to simplify the notation of certain formulas; for example, the comparison (1.7) can, for quantized manifolds, be rewritten as the equation $k_I\varphi = k_J\varphi$ in the space ${}^1H^{1/h}(R^n)$.

It is in these terms that we will formulate our first theorem on the existence and uniqueness of Maslov's global canonical operator on quantized Lagrangian manifolds.

Theorem 1.5. *For a quantized Lagrangian manifold L there exists a unique operator*

$$k : C_0^\infty(L) \to {}^1H^{1/h}(\mathbb{R}^n) ,$$

which coincides on each chart U_I of the canonical atlas with the operator (1.6), i.e. for any finite function $\varphi \in C_0^\infty(U_I)$ the equation

$$k\varphi = k_I\varphi$$

holds in the space ${}^1H^{1/h}(\mathbb{R}^n)$.

This theorem is implied by the following obvious lemma:

Lemma 1.6. *Suppose L is a smooth manifold. H is a vector space. We assume that an open covering $\{U_I\}$ is given on L, along with a family of linear operators $k_i : C_0^\infty(U_i) \to H$, for which $k_if = k_jf$ for functions $f \in C_0^\infty(U_i \cap U_j)$. Then there exists a unique operator*

$$k : C_0^\infty(L) \to H$$

such that $kf = k_if$ for any smooth function $f \in C_0^\infty(U_i)$.

As such an operator k we may take the operator defined by the formula

$$k\varphi = \sum k_I e_I \varphi , \tag{1.23}$$

where $\{e_I\}$ is a partition of unity subordinate to the canonical covering $\{U_I\}$.

4.2 Commutation of Maslov's Canonical Operator and the Hamiltonian Operator

4.2.1 Formulation of the Main Theorem.

Definition 2.1. A *Hamilton function* is any smooth function

$$H : \Phi_{\mathbb{R}} \to \mathbb{R}$$

on the phase space $\Phi_{\mathbb{R}}$ which satisfies the condition

$$\left|D_x^\alpha D_p^\beta H(x,p)\right| \le C_{\alpha,\beta}\,(1+|x|+|p|)^m \tag{2.1}$$

for some constant $C_{\alpha,\beta} = C_{\alpha,\beta}(H) > 0$ and some natural number m.
Here

$$D_x^\alpha = \frac{\partial^{\alpha_1+\ldots+\alpha_n}}{(\partial x^1)^{\alpha_1}\ldots(\partial x^n)^{\alpha_n}}\,, \qquad D_p^\beta = \frac{\partial^{\beta_1+\ldots+\beta_n}}{\partial p_1^{\beta_1}\ldots\partial p_n^{\beta_n}}$$

and $\alpha_1,\ldots,\alpha_n$; $\beta_1,\ldots,\beta_n$ are arbitrary natural numbers.

Definition 2.2. The vector field $V(H)$ associated with H, given by

$$V(H) = \frac{\partial H}{\partial p_i}\frac{\partial}{\partial x^i} - \frac{\partial H}{\partial x^i}\frac{\partial}{\partial p_i}\,. \tag{2.2}$$

is called the *Hamiltonian vector field.*

Here, as usual, repeated indices are to be summed, $i = 1,\ldots n$.
We will often leave out the indices in expression (2.2) and simply write

$$V(H) = \frac{\partial H}{\partial p}\frac{\partial}{\partial x} - \frac{\partial H}{\partial x}\frac{\partial}{\partial p}\,. \tag{2.3}$$

We consider a smooth imbedding

$$i : L \hookrightarrow \Phi_{\mathbb{R}}$$

of a manifold L into the phase space $\Phi_{\mathbb{R}}$.

Definition 2.3. The manifold $L \subset \Phi_{\mathbb{R}}$ is said to be *invariant with respect to the vector field* (2.3) if this field is tangent to the manifold L:

$$V(H)|_L \subset T(L)\,.$$

Suppose H is a Hamilton function and

$$i : L \hookrightarrow \Phi_{\mathbb{R}} \tag{2.4}$$

is a Lagrangian manifold, i.e. for any vector fields $X, Y \in T(L)$ the equation

$$\langle i^* dx \wedge dp, (X, Y) \rangle = 0 \tag{2.5}$$

holds.

Proposition 2.1. *The manifold* (2.4) *is invariant with respect to a Hamiltonian vector field* $V(H)$ *if and only if it lies on some level surface of the Hamilton function, i.e.*

$$\langle i^* dH, X \rangle = 0 \tag{2.6}$$

for any smooth field $x \in T(L)$.

Proof. We show first of all that for any vector $X \in T(L)$ the following formula holds:

$$\langle dH, X \rangle = \langle dx \wedge dp, (V(H), X) \rangle \ . \tag{2.7}$$

Indeed, suppose $X \in T(L)$ is an arbitrary vector field. We calculate the right-hand side of Eq. (2.7). We have

$$\langle dx \wedge dp, (V(H), X) \rangle = \langle dx^i, V(H) \rangle \langle dp_i, X \rangle - \langle dx^i, X \rangle \langle dp_i, V(H) \rangle \ .$$

Since

$$\langle dx, \partial/\partial x \rangle = E \ , \quad \langle dx, \partial/\partial p \rangle = 0 \ , \quad \langle dp, \partial/\partial x \rangle = 0 \ , \quad \langle dp, \partial/\partial p \rangle = E \ .$$

where E is the identity matrix, by Definition 2.2 the right-hand side of (2.7) transforms into

$$\langle \frac{\partial H}{\partial p} dp + \frac{\partial H}{\partial x} dx, X \rangle = \langle dH, X \rangle \ .$$

This proves formula (2.7).

Now we prove Proposition 2.1. First, suppose the Lagrangian manifold (2.4) is invariant with respect to the field $V(H)$. Then

$$V(H)|_L \in T(L) \ , \tag{2.8}$$

and if

$$X \in T(L) \tag{2.9}$$

then, using formula (2.7) and taking into consideration (2.8) and (2.9), as well as the fact that the imbedding i is Lagrangian, we obtain

$$\langle i^* dH, X \rangle = \langle i^* dx \wedge dp, (V(H), X) \rangle = 0 \ .$$

Conversely, suppose $a \in L$ is an arbitrary point of the manifold L and that for any vector $X_a \in T(L)_a$ Eq. (2.6) holds. Again using (2.7), we obtain (letting i_a^* be the restriction to the point a)

$$\langle i_a^* dH, X_a \rangle = \langle (dx \wedge dp)_a, (V(H)_a, X_a) \rangle = 0 \ .$$

We suppose the contrary, i.e. that the vector $V(H)_a$ does not belong to the tangent space $T_a(L)$ to the manifold L. Since in the previous identity the vector X_a can be chosen arbitrarily in the tangent space $T_a(L)$, the structure form $(dx \wedge dp)_a$ is identically equal to zero on a subspace of dimension $(n+1)$, generated by the set of all vectors of $T_a(L)$ and the additional vector $(V(H))_a$. Indeed, if $X' = X + \lambda V(H)$, $Y' = Y + \mu V(H)$, where $X \in T_a(L)$, $Y \in T_a(L)$, then

$$\begin{aligned}\langle (dx \wedge dp)_a, (X', Y') \rangle &= \langle (dx \wedge dp)_a, (X, Y) \rangle + \lambda \langle (dx \wedge dp)_a, (V(H), Y) \rangle \\ &+ \mu \langle (dx \wedge dp)_a, (X, V(H)) \rangle + \lambda\mu \langle (dx \wedge dp)_a, (V(H), V(H)) \rangle = 0\end{aligned}$$

by virtue of the preceding identity, the fact that the manifold is Lagrangian and the skew-symmetry of the form $dx \wedge dp$. On the other hand, the dimension of a vector space on which the form $(dx \wedge dp)_a$ is identically equal to zero cannot exceed n. Indeed, we fix a basis $(\partial/\partial x^1, \ldots, \partial/\partial x^n, \partial/\partial p_1, \ldots, \partial/\partial p_n)$ on the tangent space to the entire phase space $\Phi_{\mathbb{R}}$. We will consider it an orthonormal basis under a certain direct product, which we will denote by brackets $\langle\, , \rangle$. Then the form $(dx \wedge dp)_a$ on the basis $(\partial/\partial x, \partial/\partial p)$ is given by the skew-symmetric matrix

$$A = \begin{pmatrix} 0 & E \\ -E & 0 \end{pmatrix} . \tag{2.10}$$

Let W be a subspace of the tangent space on which the form $(dx \wedge dp)_a$ vanishes. Then for any vectors $X, Y \in W$ we obtain the equation

$$\langle AX, Y \rangle = 0 \ .$$

Consequently the spaces W and AW are orthogonal with respect to the direct product $\langle\, , \rangle$ and, hence, $W \cap AW = 0$. Since the matrix (2.10) is nondegenerate, $\dim W = \dim AW$. Therefore $2 \dim W \leq \dim \Phi_{\mathbb{R}} = 2n$, i.e. $\dim W \leq n$. This completes the proof of Proposition 2.1. □

Proposition 2.2. *Suppose the Lagrangian manifold $i : M \hookrightarrow \Phi_{\mathbb{R}}$ lies on the null level set of the Hamilton function H:*

$$i^*(H) = 0 \tag{2.11}$$

and suppose U_I is some canonical chart with coordinates $(x^I, p_{\overline{I}})$. Then, if $i_I : U_I \to \Phi_{\mathbb{R}}$ is the restriction of the imbedding i to the chart U_I, then for any function $\psi \in C^\infty(\Phi)$ the formula

$$i_I^*[V(H)\psi] = V(H_I)\, i_I^*\psi$$

holds, where

$$V(H_I) = i_I^*\left(\frac{\partial H}{\partial p_I}\right)\frac{\partial}{\partial x^I} - i_I^*\left(\frac{\partial H}{\partial x^{\overline{I}}}\right)\frac{\partial}{\partial p_{\overline{I}}} \, .$$

Proof. By definition we have

$$\begin{aligned} i_I^*[V(H)\psi] &= i_I^*\left[\left(\frac{\partial H}{\partial p}\frac{\partial}{\partial x} - \frac{\partial H}{\partial x}\frac{\partial}{\partial p}\right)\psi\right] \\ &= i_I^*\left\{\frac{\partial H}{\partial p_I}\frac{\partial \psi}{\partial x^I} + \frac{\partial H}{\partial p_{\overline{I}}}\frac{\partial \psi}{\partial x^{\overline{I}}} - \frac{\partial H}{\partial x^I}\frac{\partial \psi}{\partial p_I} - \frac{\partial H}{\partial x^{\overline{I}}}\frac{\partial \psi}{\partial p_{\overline{I}}}\right\} \, . \end{aligned} \tag{2.12}$$

It follows from condition (2.11) that

$$i_I^* H = H\left(x^I, x^{\overline{I}}(x^I, p_{\overline{I}}), p_I(x^I, p_{\overline{I}}), p_{\overline{I}}\right) = 0 \, .$$

Differentiating this equation in the variables $(x^I, p_{\overline{I}})$, we obtain the following equations for the derivatives $\partial H/\partial x^I$, $\partial H/\partial p_{\overline{I}}$ on L:

$$\begin{aligned} \frac{\partial H}{\partial x^I} &= -\left[\frac{\partial H}{\partial x^{\overline{I}}}\frac{\partial x^{\overline{I}}}{\partial x^I} + \frac{\partial H}{\partial p_I}\frac{\partial p_I}{\partial x^I}\right] , \\ \frac{\partial H}{\partial p_{\overline{I}}} &= -\left[\frac{\partial H}{\partial x^{\overline{I}}}\frac{\partial x^{\overline{I}}}{\partial p_{\overline{I}}} + \frac{\partial H}{\partial p_I}\frac{\partial p_I}{\partial p_{\overline{I}}}\right] . \end{aligned} \tag{2.13}$$

Substituting the values in (2.13) into (2.12), we obtain

$$\begin{aligned} i_I^*[V(H)\psi] = i_I^*\Bigg\{ &\frac{\partial H}{\partial p_I}\frac{\partial \psi_I}{\partial x^I} - \left[\frac{\partial H}{\partial x^{\overline{I}}}\frac{\partial x^{\overline{I}}}{\partial p_{\overline{I}}} + \frac{\partial H}{\partial p_I}\frac{\partial p_I}{\partial p_{\overline{I}}}\right]\frac{\partial \psi}{\partial x^{\overline{I}}} \\ &- \frac{\partial H}{\partial x^{\overline{I}}}\frac{\partial \psi}{\partial p_{\overline{I}}} + \left[\frac{\partial H}{\partial x^{\overline{I}}}\frac{\partial x^{\overline{I}}}{\partial x^I} + \frac{\partial H}{\partial p_I}\frac{\partial p_I}{\partial x^I}\right]\frac{\partial \psi}{\partial p_{\overline{I}}}\Bigg\} \, . \end{aligned} \tag{2.14}$$

Since L is a Lagrangian manifold, $i^*(dx \wedge dp) = 0$. In local coordinates $(x^I, p_{\overline{I}})$ we have

$$i_I^*(dx\wedge dp) = dx^I\wedge dp_I(x^I,p_{\overline{I}}) + dx^{\overline{I}}(x^I,p_{\overline{I}})\wedge dp_{\overline{I}}$$
$$= \frac{\partial p_I}{\partial x^I}dx^I\wedge dx^I + \left[\frac{\partial p_I}{\partial p_{\overline{I}}} + \frac{\partial x^{\overline{I}}}{\partial x^I}\right]dx^I\wedge dp_{\overline{I}} + \frac{\partial x^{\overline{I}}}{\partial p_{\overline{I}}}dp_{\overline{I}}\wedge dp_{\overline{I}} = 0\ ,$$

from which we obtain the symmetry of the matrices $\frac{\partial p_I}{\partial x^I}$, $\frac{\partial x^{\overline{I}}}{\partial p_{\overline{I}}}$ and the relation $\frac{\partial p_I}{\partial p_{\overline{I}}} = -\frac{\partial x^{\overline{I}}}{\partial x^I}$. Therefore in local coordinates $(x^I,p_{\overline{I}})$ we obtain the equation $\partial p_I/\partial p_{\overline{I}} = -\partial x^{\overline{I}}/\partial x^I$, and therefore

$$i_I^*[V(H)\psi] = i_I^*\left\{\frac{\partial H}{\partial p_I}\left[\frac{\partial}{\partial x^I} + \frac{\partial x^{\overline{I}}}{\partial x^I}\frac{\partial}{\partial x^{\overline{I}}} + \frac{\partial p_I}{\partial x^I}\frac{\partial}{\partial p_I}\right]\psi\right.$$
$$\left. - \frac{\partial H}{\partial x^{\overline{I}}}\left[\frac{\partial}{\partial p_{\overline{I}}} + \frac{\partial x^{\overline{I}}}{\partial p_{\overline{I}}}\frac{\partial}{\partial x^{\overline{I}}} + \frac{\partial p_I}{\partial p_{\overline{I}}}\frac{\partial}{\partial p_I}\right]\psi\right\} = V(H_I)i\psi_2\ ,$$

since
$$\frac{\partial}{\partial x^I}(i_I^*\psi) = \frac{\partial}{\partial x^I}\left[\psi\left(x^I,x^{\overline{I}}(x^I,p_{\overline{I}}),p_I(x^I,p_{\overline{I}}),p_{\overline{I}}\right)\right]$$
$$= i_I^*\left[\frac{\partial\psi}{\partial x^I} + \frac{\partial x^{\overline{I}}}{\partial x^I}\frac{\partial\psi}{\partial x^{\overline{I}}} + \frac{\partial p_I}{\partial x^I}\frac{\partial\psi}{\partial p_I}\right]\ ;$$
$$\frac{\partial}{\partial p_{\overline{I}}}(i_I^*\psi) = i_I^*\left[\frac{\partial\psi}{\partial p_{\overline{I}}} + \frac{\partial x^{\overline{I}}}{\partial p_{\overline{I}}}\frac{\partial\psi}{\partial x^{\overline{I}}} + \frac{\partial p_I}{\partial p_{\overline{I}}}\frac{\partial\psi}{\partial p_I}\right]\ .$$ □

On a Lagrangian manifold we will sometimes write $\partial H/\partial p_I$, $\partial H/\partial x^I$ and so on instead of $i_I^*(\partial H/\partial p_I)$, $i_I^*(\partial H/\partial x^I)$, etc. Proposition 2.2 asserts, in particular, that the expression of the field $V(H)$, restricted to the invariant Lagrangian manifold L, coincides with $V(H)$ in the local coordinates $(x^I,p_{\overline{I}})$.

Definition 2.4. A *Hamilton operator* $\widehat{H} = H(x,\widehat{p})$ is an operator which acts on any function $\varphi\in C_0^\infty(\mathbb{R}^n)$ by the formula

$$H(x,\widehat{p})\varphi = \left\{\overline{F}^{1/h}_{p\to y}H(x,p)F^{1/h}_{x\to p}\varphi\right\}_{x=y}\ .$$

Using the properties of the $1/h$-Fourier transform (cf. Appendix I to Chap. 6), it is not hard to show that if the Hamilton function $H(x,p)$ satisfies inequality (2.1) then the operator (2.15) can be extended to a continuous operator

$$\widehat{H} = H(x,\widehat{p}) : H_r^{1/h}(\mathbb{R}^n)\to H_{r-m}^{1/h}(\mathbb{R}^n) \tag{2.15}$$

for any real number r.

Finally we will have to examine the operator $H\left(x^I, \widehat{x}^{\overline{I}}, \widehat{p}_I, p_{\overline{I}}\right)$ on the chart U_I. For each function $\varphi(x^I, p_{\overline{I}}) \in C_0^\infty(U_I)$ this operator is defined by the formula

$$H\left(x^I, \widehat{x}^{\overline{I}}, \widehat{p}_I, p_{\overline{I}}\right)\varphi(x^I, p_{\overline{I}}) = F^{1/h}_{x^{\overline{I}} \to p_{\overline{I}}}$$
$$\cdot \left\{ \overline{F}^{1/h}_{\widetilde{p}_{\overline{I}} \to x^{\overline{I}}} \left[\overline{F}^{1/h}_{p_I \to x^I} H\left(x^I, x^{\overline{I}}, p_I, \widetilde{p}_{\overline{I}}\right) F_{\widetilde{x}^I \to p_I} \varphi\left(\widetilde{x}^I, \widetilde{p}_{\overline{I}}\right)\right]\right\} .$$

We give here for completeness a presentation of the definition of the Lie derivative $\mathcal{L}_X$ of a differential form ω on a manifold M along a vector field X.

Suppose g_I is a local one-parameter group of motions along a vector field X. Then if m is an arbitrary point on the manifold M, by definition

$$(\mathcal{L}_X(\omega))_m = \frac{d}{dt}\left[g_t^*\left(\omega_{g_t(m)}\right)\right]\Big|_{t=0} ,$$

where by $\omega_{g_t(m)}$ we have denoted the restriction of the form ω to the point $g_t(m)$. Here we also mention the useful formula for calculating the Lie derivative of a differential form ω of degree r:

$$\mathcal{L}_X(\omega) = i_X(d\omega) + d(i_X(\omega)) ,$$

where $i_X(\omega)(Y_1, \ldots, Y_{r-1}) = \omega(X, Y_1, \ldots, Y_{r-1})$. We will also denote $i_X(\omega)$ by the symbol $\lrcorner$ of interior multiplication:

$$i_X(\omega) = X \lrcorner\, \omega .$$

Now suppose $i : L \hookrightarrow \Phi_{\mathbb{R}}$ is a Lagrangian manifold and μ is some nondegenerate measure on it.

Definition 2.5. We say that the Hamilton function H and the pair (L, μ) are *associated* if

$$i^*H = 0 , \quad \mathcal{L}_{V(H)}\mu = 0 . \tag{2.16}$$

Let us comment on formula (2.16). The first condition of (2.16) means that the Lagrangian manifold must lie on the null level surface of the Hamilton function:

$$H|_L = 0 .$$

From this condition, as shown in Proposition 2.1, it follows that the manifold L is invariant with respect to the Hamiltonian vector field $V(H)$ and, consequently, it makes sense to differentiate the measure μ on the manifold

L along the field $V(H)$. Under this definition the second condition in (2.16) means that the measure is constant along the field $V(H)$.

If we introduce on the manifold L local coordinates $\alpha_1, \ldots, \alpha_{n-1}, t$ in which $V(H) = \frac{\partial}{\partial t}$ (we will assume that the vector field $V(H)$ is nowhere vanishing), then any measure invariant with respect to $V(H)$ has the form

$$\mu = \mu(\alpha_1, \ldots, \alpha_{n-1})\, dt \wedge d\alpha_1 \wedge \ldots \wedge d\alpha_{n-1} .$$

Indeed, if $\mu = \mu(\alpha_1, \ldots, \alpha_{n-1}, t)\, dt \wedge d\alpha_1 \wedge \ldots \wedge d\alpha_{n-1}$, then

$$\begin{aligned}\mathcal{L}_{\partial/\partial t}\mu &= d\left(\frac{\partial}{\partial t} \lrcorner\, \mu\right) = d(\mu(\alpha_1 \ldots, \alpha_{n-1}, t)\, d\alpha_1 \wedge \ldots \wedge d\alpha_{n-1}) \\ &= \frac{\partial \mu}{\partial t}(\alpha_1, \ldots, \alpha_{n-1}, t)\, dt \wedge d\alpha_1 \wedge \ldots \wedge d\alpha_{n-1} = 0 ,\end{aligned}$$

and therefore $\mu(\alpha_1, \ldots, \alpha_{n-1}, t)$ does not depend on t. Next, if v is some region lying entirely in the given chart, and $g_{t_0}(v)$ is the translation of this region along the trajectories of $V(H)$ for a time t_0, then

$$\int_v \mu = \int_{g_{t_0}(v)} \mu ,$$

since in the given coordinates the density of the measure μ does not depend on t, and g_{t_0} is a translation along the t axis by t_0 units. The condition of invariance of the measure, therefore, is equivalent to the equation

$$\int_v \mu = \int_{g_t(v)} \mu$$

for any region v in the manifold L.

Now we proceed to the formulation of the main theorem. We consider a Lagrangian manifold L with a nondegenerate measure μ on it and a Hamiltonian H associated with the pair (L, μ), i.e. a function H such that $i^*H = 0$, $\mathcal{L}_{V(H)}\mu = 0$. Let k be the canonical operator on the Lagrangian manifold L with measure μ.

Theorem 2.3 (Fundamental theorem). *Suppose the pair (L, μ) is associated with the Hamiltonian function H. Then, if the vector field $V(H)$ does not vanish anywhere, for any function $\varphi \in C_0^\infty(L)$ the congruence*

$$H(x, \widehat{p})k(\varphi) \equiv -ihk\left(V(H) - \frac{1}{2}\frac{\partial^2 H}{\partial x^i \partial p_i}\right)\varphi \pmod{h^2}$$

holds.[2]

[2] Here by k we mean the operator defined by formula (1.23).

The proof of the fundamental theorem will be given below.

4.2.2 The Formula for the Commutation of a $1/h$-Pseudodifferential Operator and a Rapidly-Oscillating Exponent. We consider a Hamilton function $H(x,p)$ and fix some subset I of indices $I \subset [n] = \{1,2,\ldots,n\}$.

Proposition 2.4. *For any functions* $\varphi(x^I, p_{\overline{I}}) \in C_0^\infty(\mathbb{R}^I \oplus \mathbb{R}_{\overline{I}})$, $S(x^I, p_{\overline{I}}) \in C_0^\infty(\mathbb{R}^I \oplus \mathbb{R}_{\overline{I}})$ *the congruence*

$$
\begin{aligned}
&H\left(x^I, \widehat{x}^{\overline{I}}, \widehat{p}_I, p_{\overline{I}}\right) \exp\left\{\frac{i}{h} S(x^I, p_{\overline{I}})\right\} \varphi(x^I, p_{\overline{I}}) \equiv \exp\left\{\frac{i}{h} S(x^I, p_{\overline{I}})\right\} \\
&\times \left\{ H\left(x^I, -\frac{\partial S}{\partial p_{\overline{I}}}, \frac{\partial S}{\partial x^I}, p_{\overline{I}}\right) + ih \left[\frac{\partial H}{\partial x^{\overline{I}}}\frac{\partial}{\partial p_{\overline{I}}} - \frac{\partial H}{\partial p_I}\frac{\partial}{\partial x^I} \right.\right. \\
&- \frac{1}{2}\left(\frac{\partial^2 S}{\partial x^I \partial x^I}\frac{\partial^2 H}{\partial p_I \partial p_I} + \frac{\partial^2 S}{\partial p_{\overline{I}} \partial p_{\overline{I}}}\frac{\partial^2 H}{\partial x^{\overline{I}} \partial x^{\overline{I}}} - 2\frac{\partial^2 S}{\partial x^I \partial p_{\overline{I}}}\frac{\partial^2 H}{\partial x^{\overline{I}} \partial p_I}\right) \\
&\left.\left. + \frac{\partial^2 H}{\partial x^{\overline{I}} \partial p_{\overline{I}}}\right]\right\} \varphi(x^I, p_{\overline{I}}) \; (\mathrm{mod}\, h^2)
\end{aligned}
$$

holds.

A proof of a more general proposition will be given in Chap. 6, Proposition 3.3.

Now suppose k_I is the elementary canonical operator defined by formula 1.6. Then it follows from Proposition 2.4 that the congruence

$$
\begin{aligned}
&H(x\widehat{p})\, k_I \varphi \equiv k_I \left\{ \mu_I^{-1/2} \left(H\left(x^I, -\frac{\partial S_I}{\partial p_{\overline{I}}}, \frac{\partial S_I}{\partial x^I}, p_{\overline{I}}\right) \mu_I^{1/2} \right.\right. \\
&+ ih\mu_I^{-1/2} \left[\frac{\partial H}{\partial x^{\overline{I}}}\frac{\partial}{\partial p_{\overline{I}}} - \frac{\partial H}{\partial p_I}\frac{\partial}{\partial x^I} - \frac{1}{2}\left(\frac{\partial^2 S_I}{\partial x^I \partial x^I}\frac{\partial^2 H}{\partial p_I \partial p_I} \right.\right. \\
&\left.\left.\left. + \frac{\partial^2 S_I}{\partial p_{\overline{I}} \partial p_{\overline{I}}}\frac{\partial^2 H}{\partial x^{\overline{I}} \partial x^{\overline{I}}} - 2\frac{\partial^2 S_I}{\partial x^I \partial p_{\overline{I}}}\frac{\partial^2 H}{\partial x^{\overline{I}} \partial p_I}\right) + \frac{\partial^2 H}{\partial x^{\overline{I}} \partial p_{\overline{I}}}\right] \mu_I^{1/2} \right\} \varphi \; (\mathrm{mod}\, h^2)
\end{aligned}
$$

holds.

Definition 2.6. The operator of multiplication by the function

$$H\left(x^I, -\partial S_I/\partial p_{\overline{I}}, \partial S_I/\partial x^I, p_{\overline{I}}\right)$$

is called the *local Hamilton-Jacobi operator* in the chart U_I.

Definition 2.7. The first-order differential operator

$$\mathcal{P}_I^1 = -\mu_I^{1/2}\left(\frac{\partial H}{\partial x^{\overline{I}}}\frac{\partial}{\partial p_{\overline{I}}} - \frac{\partial H}{\partial p_I}\frac{\partial}{\partial x^I} - \frac{1}{2}\left(\frac{\partial^2 S_I}{\partial x^I \partial x^I}\frac{\partial^2 H}{\partial p_I \partial p_I}\right.\right.$$

$$\left.\left. + \frac{\partial^2 S_I}{\partial p_{\overline{I}}\partial p_{\overline{I}}}\frac{\partial^2 H}{\partial x^{\overline{I}}\partial x^{\overline{I}}} - 2\frac{\partial^2 S_I}{\partial x^I \partial p_{\overline{I}}}\frac{\partial^2 H}{\partial x^{\overline{I}}\partial p_I}\right) + \frac{\partial^2 H}{\partial x^{\overline{I}}\partial p_{\overline{I}}}\right\}\mu_I^{1/2}$$

is called the *local transport operator* in the chart U_I.

4.2.3 Proof of Theorem 2.3. In this section we will consider the hypotheses of Theorem 2.3 to be fulfilled, without any special comment.

Proposition 2.5. *The local Hamilton-Jacobi operator is equal to zero.*

Proof. By definition the function $S_I(x^I, p_{\overline{I}})$ is a solution of the equation

$$dS_I = i^*\left(p_I dx^I - x^{\overline{I}} dp_{\overline{I}}\right) .$$

Hence it follows that

$$\partial S/\partial x^I = i^* p_I , \quad \partial S/\partial p_{\overline{I}} = -i^* x^{\overline{I}} .$$

Therefore,

$$H\left(x^I, -\partial S_I/\partial p_{\overline{I}}, \partial S_I/\partial x^I, p_{\overline{I}}\right) = i^* H\left(x^I, x^{\overline{I}}, p_I, p_{\overline{I}}\right) = i^* H = 0 .$$

This proves Proposition 2.5. □

Proposition 2.6. *The local transport operator $\mathcal{P}_I^1$ has the form*

$$\mathcal{P}_I^1 \varphi = \left[V(H) - \frac{1}{2}\partial^2 H/\partial x \partial p\right]\varphi , \tag{2.17}$$

where

$$\partial^2 H/\partial x \partial p = \partial^2 H/\partial x^i \partial p_i$$

and repeated indices indicate summation.

Proof. By Definition 2.7 the operator $\mathcal{P}_I^1$ equals

$$\mathcal{P}_I^1 = -\mu_I^{-1/2}\left\{\frac{\partial H}{\partial x^{\bar I}}\frac{\partial}{\partial p_{\bar I}} - \frac{\partial H}{\partial p_I}\frac{\partial}{\partial x^I} - \frac{1}{2}\left(\frac{\partial^2 S_I}{\partial x^I \partial x^I}\frac{\partial^2 H}{\partial p_I \partial p_I} + \frac{\partial^2 S_I}{\partial p_{\bar I}\partial p_{\bar I}}\frac{\partial^2 H}{\partial x^{\bar I}\partial x^{\bar I}} - 2\frac{\partial^2 S_I}{\partial x^I \partial p_{\bar I}}\frac{\partial^2 H}{\partial x^{\bar I}\partial p_I}\right) + \frac{\partial^2 H}{\partial x^{\bar I}\partial p_{\bar I}}\right\}\mu_I^{1/2} . \tag{2.18}$$

From Formula (2.18) it follows that $\mathcal{P}_I^1$ can be represented in the form

$$\begin{aligned}\mathcal{P}_I^1(\varphi) = &-\left[V(H_I) - \frac{1}{2}V(H_I)\ln\mu_I\right]\varphi \\ &-\frac{1}{2}\left(\frac{\partial^2 S_I}{\partial x^I\partial x^I}\frac{\partial^2 H}{\partial p_I\partial p_I} + \frac{\partial^2 S_I}{\partial p_{\bar I}\partial p_{\bar I}}\frac{\partial^2 H}{\partial x^{\bar I}\partial x^{\bar I}}\right. \\ &\left.-2\frac{\partial^2 S_I}{\partial x^I\partial p_{\bar I}}\frac{\partial^2 H}{\partial x^{\bar I}\partial p_I}\right)\varphi + \frac{\partial^2 H}{\partial x^{\bar I}\partial p_{\bar I}}\varphi ,\end{aligned} \tag{2.19}$$

where

$$V(H_I) = \frac{\partial H}{\partial p_I}\frac{\partial}{\partial x^I} - \frac{\partial H}{\partial p_{\bar I}}\frac{\partial}{\partial x^{\bar I}}$$

coincides, by Proposition 2.2, with the restriction of the vector field $V(H)$ to the Lagrangian manifold L.

We transform the second term in the expression (2.19) by appealing to the following lemma due to S.L. Sobolev.

Lemma 2.7 (Sobolev). *Suppose that a vector field X and a measure μ satisfying the condition*

$$\mathcal{L}_X\mu = 0$$

are given in $\mathbb{R}^n$ with coordinates $\alpha = (\alpha^1, \dots \alpha^n)$.

Then (with respect to the coordinates α) the following equation holds for the density σ of the measure μ:

$$-X\sigma = \sigma\,\mathrm{div}X , \tag{2.20}$$

where $\mathrm{div}X$ is the divergence of the field X in the local coordinates α, that is

$$\mathrm{div}X = \frac{\partial X^i}{\partial \alpha^i} \tag{2.21}$$

if $X^1, \dots, X^n$ are the components of the field X in the coordinate system $(\alpha^1, \dots, \alpha^n)$.

Proof. Differentiating the measure μ along the field X, we obtain that

$$\mathcal{L}_X = X \lrcorner\, d\mu + d(X \lrcorner\, \mu) .$$

Since $d\mu = 0$, it follows from this formula that $d(X \lrcorner \mu) = 0$. Then using the explicit representation of the form

$$X \lrcorner \mu = X^1 \sigma d\alpha^2 \wedge \ldots \wedge d\alpha^n - X^2 \sigma d\alpha^1 \wedge d\alpha^3 \wedge \ldots \wedge d\alpha^n + \ldots ,$$

we obtain

$$\begin{aligned} d(X \lrcorner \mu) = &\left(X^1 \frac{\partial \sigma}{\partial \alpha^1} + \ldots + X^n \frac{\partial \sigma}{\partial \alpha^n} \right) d\alpha^1 \wedge \ldots \wedge d\alpha^n \\ &+ \sigma \left(\frac{\partial X^1}{\partial \alpha^1} + \ldots + \frac{\partial X^n}{\partial \alpha^n} \right) d\alpha^1 \wedge \ldots \wedge d\alpha^n = 0 . \end{aligned} \tag{2.22}$$

From (2.22) it follows that

$$X\sigma = -\sigma \operatorname{div} X .$$

This proves Sobolev's lemma. □

Formula (2.20) can take the following form:

$$X(\ln \sigma) = -\operatorname{div} X .$$

Now applying Sobolev's lemma to the vector field

$$V(H_I) = \frac{\partial H}{\partial p_I} \frac{\partial}{\partial x^I} - \frac{\partial H}{\partial x^{\bar{I}}} \frac{\partial}{\partial p_{\bar{I}}}$$

and the density μ_I, we obtain

$$\begin{aligned} -V(H_I) \ln \mu_I = \operatorname{div} \left\{ \frac{\partial H}{\partial p_I}, -\frac{\partial H}{\partial x^{\bar{I}}} \right\} = \frac{\partial^2 H}{\partial x^I \partial p_I} + \frac{\partial^2 H}{\partial x^{\bar{I}} \partial p_{\bar{I}}} \frac{\partial x^{\bar{I}}}{\partial x^I} \\ + \frac{\partial^2 H}{\partial p_I \partial p_I} \frac{\partial p_I}{\partial x^I} - \frac{\partial^2 H}{\partial p_I \partial x^I} - \frac{\partial^2 H}{\partial x^{\bar{I}} \partial x^{\bar{I}}} \frac{\partial x^{\bar{I}}}{\partial p_{\bar{I}}} - \frac{\partial^2 H}{\partial x^I \partial p_I} \frac{\partial p_I}{\partial p_{\bar{I}}} . \end{aligned}$$

We can replace the derivatives of the coordinate functions of the manifold U_I with respect to the local coordinates x^I, $p_{\bar{I}}$ by elements of the Hessian matrix of the function S_I, in agreement with the formula

$$dS_I = i^* \left(p_I dx^I - x^{\bar{I}} dp_{\bar{I}} \right) .$$

We obtain the result

$$\begin{aligned} -V(H_I) \ln \mu_I = \frac{\partial^2 H}{\partial p_I \partial x^I} - \frac{\partial^2 H}{\partial x^{\bar{I}} \partial p_I} \frac{\partial^2 S_I}{\partial x^I \partial p_{\bar{I}}} + \frac{\partial^2 H}{\partial p_I \partial p_I} \frac{\partial^2 S_I}{\partial x^I \partial x^I} \\ - \frac{\partial^2 H}{\partial x^{\bar{I}} \partial p_{\bar{I}}} - \frac{\partial^2 H}{\partial x^{\bar{I}} \partial x^{\bar{I}}} \frac{\partial^2 S_I}{\partial p_{\bar{I}} \partial p_{\bar{I}}} - \frac{\partial^2 H}{\partial x^{\bar{I}} \partial p_I} \frac{\partial^2 S_I}{\partial x^I \partial p_{\bar{I}}} . \end{aligned} \tag{2.23}$$

Now, substituting expression (2.23) into (2.19) and applying Proposition 2.2, we obtain Formula (2.17). This proves Proposition 2.6. □

Therefore the following assertion holds.

Proposition 2.8. *Suppose the pair (L, μ) and the Hamiltonian H are associated. Then for any function $\varphi \in C_0^\infty(U_I)$ the congruence*

$$\widehat{H} k_I \varphi \equiv -ihk_I \left[V(H) - \frac{1}{2} \frac{\partial^2 H}{\partial x \partial p} \right] \pmod{h^2} . \tag{2.24}$$

holds.

Now let's finish off the proof of the fundamental theorem (Theorem 2.3). Let $\varphi \in C_0^\infty(L)$ and let $\sum e_I \equiv 1$ be a partition of unity subordinate to the covering $\{U_I\}$. Then by virtue of the linearity of the operator k we have

$$\widehat{H} k \varphi = \widehat{H} \sum k_I e_I \varphi = \sum \widehat{H} k e_I \varphi .$$

We note that because of the compactness of suppφ the above sum has only a finite number of nonzero terms. Furthermore, by Proposition 2.8

$$\sum \widehat{H} k e_I \varphi \equiv -ih \sum k_I i^* \left\{ V(H) - \frac{1}{2} \frac{\partial^2 H}{\partial x \partial p} \right\} e_I \varphi \pmod{h^2} .$$

Now calculating the right-hand side of expression (2.24), we obtain

$$\begin{aligned} -ih \sum k_I \left\{ V(H) - \frac{1}{2} \frac{\partial^2 H}{\partial x \partial p} \right\} e_I \varphi &= -ih \sum k_I e_I \\ \cdot \left\{ V(H) - \frac{1}{2} \frac{\partial^2 H}{\partial x \partial p} \right\} \varphi &+ ih \sum k_I \varphi \left\{ V(H) e_I \right\} . \end{aligned} \tag{2.25}$$

Furthermore, by Theorem 1.5, we have

$$\begin{aligned} ih \sum k_I \varphi \{V(H) e_I\} &\equiv ih \sum k \varphi \{V(H) e_I\} \\ &= ihk\varphi \sum V(H) e_I = ihk\varphi V(H) \sum e_I = 0 \pmod{h^2} . \end{aligned}$$

Therefore we have obtained that

$$\begin{aligned} \sum \widehat{H} k \varphi &\underset{\bmod h^2}{\equiv} -ih \sum k_I e_I \left\{ V(H) - \frac{1}{2} \frac{\partial^2 H}{\partial x \partial p} \right\} \varphi \\ &= -ihk \left\{ V(H) - \frac{1}{2} \frac{\partial^2 H}{\partial x \partial p} \right\} \varphi . \end{aligned}$$

This completes the proof of Theorem 2.3. □

Chapter 5
The Asymptotics of Integrals of Rapidly Oscillating Functions with a Complex Phase

Here we will obtain the formula for the asymptotic expansion with respect to a parameter h of the integral of a rapidly oscillating function.

It has been traditional to derive the asymptotic expansion of the integral of an oscillating function by the stationary phase method, the idea of which is that the main contribution to the integral comes from a neighborhood of a stationary point of the expression under the integral.

We note here that in the proof of certain assertions (Proposition 2.3 and 3.3 in 6.3, Lemma 4.1 in 6.4) we use the method of undetermined coefficients. The idea of this method is that, knowing the general form of the dependence of the coefficients of the asymptotic expansion of the integral of a rapidly-oscillating function on the phase and the amplitude, we can calculate the concrete form of this dependence not in the general case, but only with certain concrete "model" functions in the role of the phase and amplitude of the given integral. However, to carry out this program we must establish the form of dependence of the coefficients of the expansion on the functions under the integral sign. We prove that the coefficients are linear functions in the derivatives of the amplitude with coefficients which depend on the derivatives of the phase as square roots of rational functions. Moreover, the particular roots of rational functions which appear in the expansion are universal, that is they do not depend on the amplitude and the phase of the integral under consideration. This last circumstance is not entirely trivial in the complex case, since the coefficients of the asymptotic expansion for a complete phase function are not uniquely defined.

The contents of this chapter are in two sections. In the first we prove the formula for the asymptotic expansion, leaving the proof of certain technical assertions for the second section of the chapter (Proposition 1.2). The second section is devoted to the proof of this proposition.

5.1 The Formula for Asymptotic Expansion of the Integral of a Rapidly-Oscillating Function

We consider smooth complex-valued functions $\Phi(x,p)$ and $a(x,p)$ in the variables $x=\left(x^1,\ldots,x^m\right)$ and $p=(p_1,\ldots,p_n)$, where the function $a(x,p)$ has compact support $\operatorname{supp} a(x,p)$. We require the function $\Phi(x,p)$ to have nonnegative imaginary part on the support of the function $a(x,p)$, that is

$$\operatorname{Im}\Phi(x,p)\geq 0 \qquad \text{for} \ \ (x,p)\in \operatorname{supp} a(x,p) \ .$$

We denote by Ω the set of points $x\in \mathbb{R}^m$ such that there exists a point $p\in\mathbb{R}_n$ for which the equations

$$\operatorname{Im}\Phi(x,p)=0 \ , \tag{1.1}$$

$$\operatorname{grad}_p \Phi(x,p)=0 \ , \tag{1.2}$$

hold. A point $p\in\mathbb{R}_n$ satisfying conditions (1.1) and (1.2) is called a *real stationary point* of the function $\Phi(x,p)$ for the fixed value $x\in\Omega$.

We suppose that the equation (1.2) has no more than one solution $p=p(x)$ on $\operatorname{supp} a(x,p)$ and that the Hessian of the function $\Phi(x,p)$ at the point $(x,p(x))$ is nondegenerate:

$$\det \operatorname{Hess}_p (x,p(x))\neq 0 \ .$$

We denote by ${}^s\Phi(x,\zeta)$ the s-analytic continuation of the function $\Phi(x,p)$, $\zeta=p+i\eta$. Then from the implicit function theorem it follows that in some neighborhood of the set Ω there exists a unique solution $\zeta=\zeta(x)$ of the system of equations

$${}^s\Phi_\zeta(x,\zeta)=0 \ , \tag{1.3}$$

where

$${}^s\Phi_\zeta=\left\{\frac{\partial^s\Phi(x,\zeta)}{\partial\zeta_1},\ldots,\frac{\partial^s\Phi(x,\zeta)}{\partial\zeta_n}\right\} \ .$$

We will denote by $\mathcal{T}_r(\mathbb{C}_n)$ the finite-dimensional complex vector space whose coordinates (t_α) are denumerated by the multi-indices $\alpha=(\alpha_1,\ldots,\alpha_n)$, where $|\alpha|=\alpha_1+\ldots+\alpha_n\leq r$. We will denote by D the determinant of the matrix whose (ij)-th entry is the variable t_α, where $\alpha=\Big(0,\ldots,0,\underset{i}{1},0,\ldots,0,\underset{j}{1},0,\ldots,0\Big)$ when $i\neq j$ and $\alpha=\Big(0,\ldots,0,\underset{i}{2},0,\ldots,0\Big)$ when $i=j$. It is obvious that $D=D(t_\alpha)$ is a polynomial on the space $\mathcal{T}_r(\mathbb{C}_n)$ for $r\geq 2$.

Theorem 1.1. *There exist functions*

$$W_k(t_\alpha, u_\beta) = \sum_{|\beta|\le 2k} W_k^{(\beta)}(t_\alpha) u_\beta \ ,$$

where $W_k^{(\beta)}(t_\alpha)$ *is a root of a rational function on the space* $\mathcal{T}_{2k+2-|\beta|}(\mathbb{C}_n)$ *with denominator* $[D(t_\alpha)]^{N_\beta}$*, such that for any functions* $\Phi(x,p), a(x,p)$ *satisfying the above conditions, the estimate below holds.*

$$\begin{aligned} I_n(x,h) &= \left(\frac{i}{2\pi h}\right)^{1/2} \int_{\mathbb{R}_n} e^{(i/h)\Phi(x,p)} a(x,p)dp \\ &\equiv \frac{{}^s\Phi(x,\zeta(x)) \exp\left[\left(\frac{i}{h}{}^s\Phi(x,\zeta(x))\right]\right.}{\sqrt{\det \operatorname{Hess}_\zeta\left(-{}^s\Phi(x,\zeta)\right)|_{\zeta=\zeta(x)}}} \sum_{0\le k<(s-1)/2} h^k \qquad (1.4) \\ &\quad \times W_k\left(\frac{\partial^{|\alpha|}\, {}^s\Phi}{\partial\zeta^\alpha}(x,\zeta(x)), \frac{\partial^{|\beta|}\, {}^s a}{\partial\zeta^\beta}(x,\zeta(x))\right)\left(\operatorname{mod} h^{(s-1)/2}\right) . \end{aligned}$$

In the future we will sometimes, for the sake of simplicity, denote the functions $W_k\left[\frac{\partial^{|\alpha|}{}^s\Phi}{\partial\zeta^\alpha}(x,\zeta(x)), \frac{\partial^{|\beta|}{}^s a}{\partial\zeta^\beta}(x,\zeta(x))\right]$ by $W_k[\Phi, \alpha]$.

In the formulation of the theorem the argument of the expression under the radical in formula (1.4) equals[1]

$$\arg\det \operatorname{Hess}_\zeta\left(-{}^s\Phi(x,\zeta)\right)|_{\zeta=\zeta(x)} = \sum_{j=1}^n \arg\lambda_j(x) \ . \qquad (1.5)$$

λ_j are the eigenvalues of the matrix $\operatorname{Hess}_\zeta\left(-{}^s\Phi(x,\zeta)\right)|_{\zeta=\zeta(x)}$, the arguments $\arg\lambda_i(x)$ of which are chosen within the bounds

$$-\frac{3\pi}{2} < \arg\lambda_j(x) < \frac{\pi}{2} \ . \qquad (1.6)$$

In formula (1.4) the estimate

$$f(x,h) \equiv 0 \,(\operatorname{mod} h^r)$$

means that for any $\alpha = (\alpha_1, \dots \alpha_n)$ the inequality

$$\|\widehat{p}^\alpha f(x,h)\|_{L_2} \le c_\alpha h^r \ , \quad \widehat{p} = ih\frac{\partial}{\partial x}$$

holds, with constants c_a independent of h.

Proof of Theorem 1.1. We note, first of all, that the left-hand side of the estimate (1.4) is a linear operator applied to the function of compact

[1] The fact that the argument is well-defined follows from the inequality $\operatorname{Im}\Phi(x,p(x)) \ge 0$.

support a. Consequently, if necessary, we will assume that the support of the function a is as small as desired.

We denote by $A_r(x,h)$ the function

$$A_r(x,h) = \frac{\exp\left[\frac{i}{h}{}^r\Phi(x,\zeta(x))\right]}{\left[\det \operatorname{Hess}_\zeta\left(-{}^r\Phi(x,\zeta)\right)\big|_{\zeta=\zeta^r(x)}\right]^{1/2}} \times \sum_{0\le k<(s-1)/2} W_k\left[{}^r\Phi, {}^r a\right],$$

$r \geq s$, $\zeta^r(x)$ is the solution of the equation ${}^r\Phi_\zeta(x,\zeta)=0$.

The proof of the theorem relies upon the following proposition, whose proof will be given in 5.2.

Proposition 1.2. *Under the hypotheses of the theorem, for any multi-index a there exists a number $N(\alpha)$ such that the inequality*

$$\left\|\widehat{p}^\alpha\left(I_n(x,h) - A_{N(\alpha)}(x,h)\right)\right\|_{L_2} \le C h^{(s-1)/2}$$

holds.

Now we prove Theorem 5.1. We must estimate the expression

$$\|\widehat{p}^\alpha\left(I_n(x,h) - A_s(x,h)\right)\|_{L_2}$$

for any multi-index α. We fix a certain multi-index. By virtue of Proposition 1.2 there exists a number $N = N(\alpha)$ such that

$$\|\widehat{p}^\alpha\left(I_n(x,h) - A_N(x,h)\right)\|_{L_2} \le C h^{(s-1)/2} .$$

On the other hand, we have

$$\begin{aligned}\|\widehat{p}^\alpha\left(I_n(x,h) - A_s(x,h)\right)\|_{L_2} \le\ & \|\widehat{p}^\alpha\left(I_n(x,h) - A_N(x,h)\right)\|_{L_2} \\ & + \|\widehat{p}^\alpha\left(A_n(x,h) - A_s(x,h)\right)\|_{L_2} .\end{aligned}$$

To finish the proof of Theorem 1 it remains to show that for any $N \geq s$ the inequality

$$\|\widehat{p}^\alpha\left(A_N(x,h) - A_s(x,h)\right)\|_{L_2} \le C h^{(s-1)/2} . \tag{1.7}$$

holds. First of all we estimate the difference

$$\zeta^s(x) - \zeta^N(x) .$$

We have, obviously,

$$\left|D^\alpha\left({}^N\Phi_\zeta\left(x,\zeta^s(x)\right)\right)\right| \le C\left|\operatorname{Im}\zeta^s(x)\right|^{s+1-|\alpha|}, \quad {}^N\Phi_\zeta\left(x,\zeta^N(x)\right) = 0 .$$

Therefore

$$\left|D^\alpha\left({}^N\Phi_\zeta\left(x,\zeta^s(x)\right) - {}^N\Phi_\zeta\left(x,\zeta^N(x)\right)\right)\right| \le C\left|\operatorname{Im}\zeta^s(x)\right|^{s+1-|\alpha|} .$$

Using the Lagrange formula for finite increments and taking into consideration the inequality ${}^N\Phi_{\zeta\zeta} \neq 0$, we have (by induction on $|\alpha|$)

$$\left|D_x^\alpha\left(\zeta^s(x)-\zeta^N(x)\right)\right| \le C\left|\mathrm{Im}\zeta^s(x)\right|^{s+1-|\alpha|} .$$

From the last inequality it is clear that if we replace the function $\zeta^N(x)$ by $\zeta^s(x)$ in the expression $A_N(x,h)$, then the function Φ and a change by an element of the ideal ${}^{s+1}I(\mathbb{C}^m,\rho)$, $\rho = |\mathrm{Im}\zeta^s(x)|$, Therefore to prove formula (1.7) it is now sufficient to use the inequalities

$$\left|\exp\left[(-i/h)^N\Phi\left(x,\zeta^s(x)\right)\right]\left|\mathrm{Im}\zeta^s(x)\right|^r\right| \le Ch^{r/2} \tag{1.8}$$

$$\begin{aligned}\left|\exp\left[i/h\, O\left(\left|\mathrm{Im}\zeta^s(x)\right|^{r+1}\right)\right]-1\right| \le \frac{C}{h}\exp\left[c/h\left|\mathrm{Im}\zeta^s(x)\right|^{r+1}\right]\\ \times\left|\mathrm{Im}\zeta^s(x)\right|^{r+1} ,\end{aligned} \tag{1.9}$$

(where C, c are constants).

Inequality (1.9) is obvious. To prove inequality (1.8) we establish two lemmas. Inequality (1.8) is a corollary of the second of these.

Lemma 1.3. *For any s there exist functions $V_{ij}(x,\rho)$ and $W(x,\rho)$ such that the representation*

$$\begin{aligned}\Phi(x,p) = {}^s\Phi\left(x,\zeta(x)\right) + \sum_{i,j}\left(p_i-\zeta_i(x)\right)\left(p_j-\zeta_j(x)\right)\\ \times V_{ij}(x,p) + W(x,p) ,\end{aligned} \tag{1.10}$$

holds, where

$$\left|D_p^\alpha D_x^\beta W(x,p)\right| \le C_{\alpha\beta}\left|\mathrm{Im}\zeta(x)\right|^{s-|\alpha|-|\beta|} . \tag{1.11}$$

Proof. We represent the function ${}^s\Phi(x,\zeta)$ as the integral of a derivative:

$$\begin{aligned}{}^s\Phi(x,\zeta) &= {}^s\Phi\left(x,\zeta(x)\right) + \int_0^1 \frac{d}{dt}{}^s\Phi\left(x,t\zeta+(1-t)\zeta(x)\right)dt\\ &= {}^s\Phi\left(x,\zeta(x)\right) + \sum_k\left(\zeta_k-\zeta_k(x)\right)\\ &\times\int_0^1\frac{\partial^2\Phi}{\partial\zeta_k}\left(x,t\zeta+(1-t)\zeta(x)\right)dt + W(x,\zeta) ,\end{aligned} \tag{1.12}$$

where

$$W(x,\zeta) = \sum_k\left(\overline{\zeta}_k-\overline{\zeta_k(x)}\right)\int_0^1\frac{\partial^s\Phi}{\partial\overline{\zeta}_k}\left(x,t\zeta+(1-t)\zeta(x)\right)dt . \tag{1.13}$$

Then

$$\begin{aligned}|W(x,\zeta)| &\le C\int_0^1 (t\,|\mathrm{Im}\zeta| + (1-t)\,|\mathrm{Im}\zeta(x)|)^s\,dt \\ &\le C\left(|\mathrm{Im}\zeta|^s + |\mathrm{Im}\zeta(x)|^s\right) .\end{aligned} \tag{1.14}$$

Setting $\zeta = \rho$, we obtain (1.11) for $\alpha = 0$, $\beta = 0$. An inequality of the form (1.11) with arbitrary α and β is proven analogously. Applying formula (1.12) to each term in the summation, we obtain condition (1.10) with the estimate (1.11). □

Lemma 1.4. *The inequality*

$$\mathrm{Im}^s\Phi\,(x,\zeta(x)) \ge c\sum_{k=1}^{n} |\mathrm{Im}\zeta_k(x)|^2 \ , \quad c > 0 \ . \tag{1.15}$$

holds.

Proof. We suppose, to start with, that there exist functions

$$Q_{ij}(x,p)\ , \quad i \ge j\ , \quad \mathrm{Re}Q_{ij}(x,p) \ne 0\ ,$$

such that for the functions

$$\xi_k(x,p) = \sum_{l=k}^{n} (p_l - \zeta_l(x))\, Q_{lk}(x,p) \tag{1.16}$$

the equation

$$\Phi(x,p) = {}^s\Phi\,(x,\zeta(x)) + i\left(\sum_{k=1}^{n} \xi_k^2(x,p)\right) + W(x,p)\ , \tag{1.17}$$

holds, where the estimate (1.11) holds for the function $W(x,p)$. In this case we will find real-valued functions

$$p = \widetilde{p}(x)\ , \tag{1.18}$$

satisfying the system of equations

$$\mathrm{Re}\xi_k\,(x,\widetilde{p}(x)) = 0\ . \tag{1.19}$$

Indeed, if $x \in \Omega$, then it is possible to take $p(x) = \zeta(x)|\Omega$ in (1.18), since $\xi(x)|\Omega$ is real and $\xi_k\,(x,p(x)) = 0$ on Ω.

Furthermore, at the points $(x,p(x))$, by virtue of the inequality $\mathrm{Re}Q_{ii} \ne 0$, the Jacobian matrix of the system (1.19) is nonzero, which means that a solution of this system exists and is unique in a neighborhood of Ω. Substituting the value $p = \widetilde{p}(x)$ in (1.17) we obtain

$$\Phi\left(x,\widetilde{p}(x)\right) = {}^{s}\Phi\left(x,\zeta(x)\right) + i\sum_{k=1}^{n}\left(i\,\mathrm{Im}\,\xi_k\left(x,\widetilde{p}(x)\right)\right)^2 + W\left(x,\widetilde{p}(x)\right) \ ,$$

that is

$$\begin{aligned}\mathrm{Im}\,{}^{s}\Phi\left(x,\zeta(x)\right) &= \mathrm{Im}\Phi\left(x,\widetilde{p}(x)\right)\\ &\quad+\sum_{k=1}^{n}\left(\mathrm{Im}\,\xi_k\left(x,\widetilde{p}(x)\right)\right)^2 - \mathrm{Im}W\left(x,\widetilde{p}(x)\right) \ .\end{aligned} \tag{1.20}$$

Since $\det\left(\mathrm{Re}Q_{kl}\right) \neq 0$, by expressing the function $\widetilde{p}(x) - \mathrm{Re}\zeta(x)$ fron Eq. (1.19) and substituting it in the expression for $\mathrm{Im}\xi$, we have

$$\begin{aligned}\mathrm{Im}\,\xi\left(x,\widetilde{p}(x)\right) &= -\mathrm{Im}\,\xi(x)\mathrm{Re}Q\left(x,\widetilde{p}(x)\right)\\ &\quad - \mathrm{Im}\,\zeta(x)\cdot\mathrm{Im}\,Q\left(x,\widetilde{p}(x)\right)\left(\mathrm{Re}Q\left(x,\widetilde{p}(x)\right)\right)^{-1}\cdot\mathrm{Im}\,Q\left(x,\widetilde{p}(x)\right)\\ &= -\mathrm{Im}\,\zeta(x)\left[\mathrm{Re}Q + \mathrm{Im}\,Q\left(\mathrm{Re}Q\right)^{-1}\mathrm{Im}\,Q\right]\left(x,\widetilde{p}(x)\right) \ .\end{aligned}$$

Since the matrix Q is triangular,

$$\det\left(\mathrm{Re}Q + \mathrm{Im}\,Q\left(\mathrm{Re}Q\right)^{-1}\mathrm{Im}\,Q\right) \neq 0 \ .$$

Consequently there exists a constant $C > 0$ such that

$$\sum\left|\mathrm{Im}\,\xi_k\left(x,\widetilde{p}(x)\right)\right|^2 \geq c\sum\left|\mathrm{Im}\,\zeta_k(x)\right|^2 \ . \tag{1.21}$$

Substituting into (1.20) the condition $\mathrm{Im}\,\Phi(x,\rho) \geq 0$, (1.11), and (1.21), for $s \geq 3$ we obtain the assertion of Lemma 1.4.

Thus it remains only to find the functions $Q_{ij}(x,p)$. We note that it is sufficient to define the values of the functions $Q_{lk}(x,p)$ in a neighborhood of the set of points of the form

$$x \in \Omega\,, \quad p = p(x) \ .$$

Then the matrix $(V_{ij}(x,p))$ in Lemma 1.3 is nondegenerate at the indicated points. Therefore, applying the standard procedure of reducing a quadratic form to its principal axes, we construct the matrix Q. Now we observe that Q is a nondegenerate triangular matrix. Furthermore, from (1.17) we have for $x \in \Omega$:

$$\mathrm{Im}\,\frac{\partial^2\Phi}{\partial p_l\partial p_m}(x,p) = 2\,\mathrm{Re}\sum_k\frac{\partial\xi_k}{\partial p_l}(x,p)\frac{\partial\xi_k}{\partial p_m}(x,p) \ . \tag{1.22}$$

Suppose there is a vector $\lambda = (\lambda_k) \neq 0$ with real coordinates such that

$$\sum_k\lambda_k\frac{\partial\,\mathrm{Re}\xi_l}{\partial p_k}(x,p) = 0 \ . \tag{1.23}$$

Since the matrix (Q_{kl}) is nondegenerate,

$$\sum_k \lambda_k \frac{\partial\,\mathrm{Im}\xi_l}{\partial p_k}(x,p) \neq 0 \ . \tag{1.24}$$

Then we obtain from (1.22) and (1.23) that

$$\sum_{k,l} \lambda_k \lambda_l \frac{\partial^2 \mathrm{Im}\Phi}{\partial p_k \partial p_l} = -2 \sum_{k,l,m} \lambda_k \frac{\partial\,\mathrm{Im}\xi_m}{\partial p_k} \lambda_l \frac{\partial\,\mathrm{Im}\xi_m}{\partial p_k} < 0 \ ,$$

which contradicts the condition $\mathrm{Im}\Phi(x,p) \geq 0$. Consequently the matrix $\left(\frac{\partial \mathrm{Re}\xi_l}{\partial p_k}(x,p)\right)$ is nondegenerate, which is equivalent to the condition

$$\det\left(\mathrm{Re}Q_{kl}\right) \neq 0 \ ,$$

or

$$\mathrm{Re}Q_{ii}(x,p) \neq 0 \ .$$

This proves the lemma. □

Lemma 1.4 proves the inequality (1.8), which finishes the proof of Theorem 1.1. □

5.2 Proof of Proposition 1.2

We will carry out the proof of Proposition 1.2 by induction on the dimension of the space.

5.2.1 Basis of the Induction. Suppose $n = 1$. We consider the integral

$$I_1(x,h) = \left(\frac{i}{2\pi h}\right)^{1/2} \int_{-\infty}^{+\infty} e^{(i/h)\Phi(x,p)} a(x,p)dp \ . \tag{2.1}$$

We choose $N = N(\alpha) = 2s + |\alpha| + 7$. The equation for a stationary point has the form

$${}^N\Phi_\zeta(x,\zeta) = 0 \ . \tag{2.2}$$

Equation (2.2) is uniquely solvable in some neighborhood of any point $(x_0, p(x_0))$ satisfying condition (1.2). This assertion follows from the implicit function theorem, since

$${}^N\Phi_\zeta\left(x_0,\zeta\right)|_{\zeta=p(x_0)} = \Phi_p\left(x_0, p(x_0)\right) = 0$$

and

$$\Phi_{pp}(x_0, p(x_0)) \neq 0 ,$$

and, consequently, in some neighborhood of the set $(x_0, p(x_0))$

$$^N\Phi_{\zeta\zeta}(x, \zeta) \neq 0 .$$

Moreover, the solution $\zeta(x)$ of Eq. (2.2) satisfies the condition

$$\zeta(x) = p(x) \qquad \text{for } x \in \Omega .$$

Now we consider the function $^N\Phi(x, \zeta)$. It follows from Lemma 1.3 that the decomposition

$$\Phi(x, p) = {}^N\Phi(x, \zeta(x)) + (p - \zeta(x))^2 F^\Phi(x, p) + G^\Phi(x, p) , \tag{2.3}$$

holds, where

$$\left|D_p^\alpha D_x^\beta(x, p)\right| \leq C_{\alpha\beta} \left|\mathrm{Im}\zeta(x)\right|^{N-|\alpha|-|\beta|} . \tag{2.4}$$

We denote

$$\xi(x, p) = (p - \zeta(x)) \sqrt{F^\Phi(x, p) + \frac{G^\Phi(x, p)}{(p - \zeta(x))^2}} . \tag{2.5}$$

We note that the function under the radical on the right-hand side of (2.5) can have discontinuities only at the points of the set Ω when $p = p(x)$. However, in view of the inequality

$$\left|D_p^\alpha D_x^\beta \frac{G^\Phi(x, p)}{(p - \zeta(x))^2}\right| \leq C \left|\mathrm{Im}\zeta(x)\right|^{N-|\alpha|-|\beta|-2} \tag{2.6}$$

all the derivatives of the function $G^\Phi(x, p)(p - \zeta(x))^{-2}$ up to order $N - 3$ approach zero as $x \to x_0 \in \Omega$. Extending the definition of this function to be zero on Ω, we obtain a function belonging to the class $C^{(N-3)}(\mathbb{R}^m \oplus \mathbb{R}_1)$, which is equal to zero on the set Ω. The function $\xi(x, p)$ defined by formula (2.5) also has $N - 3$ continuous derivatives.

Since $F^\Phi(x, p) \neq 0$ in a neighborhood of the point $(x_0, p(x_0))$, the expression (2.5) allows the choice of a single-valued branch[2], and we fix one such choice. We note that in the neighborhood under consideration

$$\frac{d\xi(x, p)}{dp} \neq 0 .$$

In view of formula (2.5), expression (2.3) takes on the following form:

$$\Phi(x, p) = {}^N\Phi(x, \zeta(x)) + [\xi(x, p)]^2 . \tag{2.7}$$

[2] In fact the expression (2.5) allows a choice of a single-valued branch only locally. However, we will always assume that the support $\mathrm{supp}\, a(x, p)$ is sufficiently small.

Now we denote by $b(x,p)$ the function

$$b(x,p)=a(x,p)\left[\frac{d\xi(x,p)}{dp}\right]^{-1} . \tag{2.8}$$

This function belongs to the class $C^{(N-4)}(\mathbb{R}^m \oplus \mathbb{R}_1)$. Therefore a t-analytic continuation ${}^t b(x,\zeta)$ of this function is possible for $t \leq N-4$ and has $N-t-4$ continuous derivatives. In the following computations we will not keep track of the precise orders of differentiability, assuming that N is however great we need for our arguments to be valid. We note, however, that we will never use infinite differentiability of our functions.

We use Eq. (1.12) for the function $b(x,p)$:

$$b(x,p)={}^t b\,(x,\zeta(x))+(p-\zeta(x))^{\,2}F^b(x,p)+G_0^b(x,p) . \tag{2.9}$$

As is evident from the proof of Lemma 1.3, the decomposition (2.9) is applicable to functions of a finite degree of differentiability, but under these circumstances the degree of differentiability of the functions F^b and G_0^b decreases by one in comparison with the degree of differentiability of the function ${}^t b(x,\zeta)$. The function $G_0^b(x,p)$ satisfies the inequalities

$$\left|D_x^\alpha D_p^\beta G_0^b(x,p)\right| \leq C\left|\mathrm{Im}\zeta(x)\right|^{t-|\alpha|-|\beta|} \tag{2.10}$$

for $|\alpha|+|\beta| \leq t$ (we recall that N is supposed to be large enough that all the derivatives in Eq. (2.10) exist). We rewrite Eq. (2.9) in the form

$$b(x,p)={}^t b\,(x,\zeta(x))+\xi(x,p)\frac{F^b(x,p)}{\left[F^\Phi(x,p)+\frac{G^\Phi(x,p)}{(p-\zeta(x))^2}\right]^{1/2}}+G_0^b(x,p)$$

and we repeat the described procedure with the function

$$\frac{F^b(x,p)}{\left[F^\Phi(x,p)+\frac{G^\Phi(x,p)}{(p-\zeta(x))^2}\right]^{1/2}}$$

in place of $b(x,p)$. Repeated application of decompositions of the form (2.9) lead us to the formula

$$\begin{aligned}b(x,p)=&\sum_{j=0}^{k} b_j(x)\,[\xi(x,p)]^j+[\xi(x,p)]^{k+1}\\&\times F_{k+1}^b(x,p)+G_{k+1}^b(x,p) .\end{aligned} \tag{2.11}$$

Here all the functions in the decomposition (2.11) have an arbitrary finite degree of differentiability for sufficiently large N, and at the same time the inequality

$$\left|D_x^\alpha D_p^\beta G_{k+1}^b(x,p)\right| \leq C\,|\mathrm{Im}\zeta(x)|^{t-|\alpha|-|\beta|} \tag{2.12}$$

holds for $|\alpha|+|\beta|\leq t$. Substituting the expansions (2.7) and (2.11) into the formula (2.1) and using Eq. (2.8), we obtain

$$I_1(x,h)=\sum_{j=0}^{k} b_j(x)I_{1j}^{(1)}(x,h)+I_1^{(2)}(x,h)+I_1^{(3)}(x,h)\ , \tag{2.13}$$

where ($\chi(x,p)\equiv 1$ on $\mathrm{supp}a(x,p)$)

$$\begin{aligned} I_{1j}^{(1)}(x,h) = \left(\frac{i}{2\pi h}\right)^{1/2} & e^{(i/h)^N\Phi(x,\zeta(x))}\int_{-\infty}^{\infty} e^{(i/h)[\xi(x,p)]^2} \\ & \times [\xi(x,p)]^j\,\chi(x,p)d\xi(x,p)\ ; \end{aligned} \tag{2.14}$$

$$\begin{aligned} I_1^{(2)}(x,h) = \left(\frac{i}{2\pi h}\right)^{1/2} & e^{(i/h)^N\Phi(x,\zeta(x))}\int_{-\infty}^{\infty} e^{(i/h)[\xi(x,p)]^2} \\ & \times [\xi(x,p)]^{k+1}\,F_{k+1}^b(x,p)\chi(x,p)d\xi(x,p)\ ; \end{aligned} \tag{2.15}$$

$$\begin{aligned} I_3^{(1)}(x,h) = \left(\frac{i}{2\pi h}\right)^{1/2} & e^{(i/h)^N\Phi(x,\zeta(x))}\int_{-\infty}^{\infty} e^{(i/h)[\xi(x,p)]^2} \\ & \times G_{k+1}^b(x,p)\chi(x,p)d\xi(x,p)\ . \end{aligned} \tag{2.16}$$

Below we will calculate the coefficients $b_j(x)$ of the expansion (2.11) (Lemma 2.4); we will calculate the integrals (2.14) (Lemma 2.3) and we will estimate the integrals (2.15) and (2.16) (Lemmas 2.1 and 2.2).

We choose and fix an arbitrary multi-index α (appearing in the formulation of Proposition (1.2). The following assertions hold.

Lemma 2.1 *There exists a value of k such that the inequality*

$$\left\|\widehat{p}^\alpha I_1^{(2)}(x,h)\right\|_{L_2} \leq Ch^{s/2}\ . \tag{2.17}$$

holds.

Lemma 2.2. *There exists a value of t such that the inequality*

$$\left\|\widehat{p}^\alpha I_1^{(3)}(x,h)\right\|_{L_2} \leq Ch^{s/2}\ . \tag{2.18}$$

holds.

Lemma 2.3 *The congruence*

$$I_{ij}^{(1)}(x,h)\equiv C_j e^{(i/h)^N\Phi(x,\zeta(x))}h^{j/2}\ (\mathrm{mod}h^\infty)\ , \tag{2.19}$$

holds, where C_j *are constants, with* $C_j = 0$ *for odd values of* j.

Lemma 2.4. *There exist polynomials* $S_j^k(t_2, \dots, t_{k-j+2})$ *such that for any functions* $\Phi(x,p)$, $a(x,p)$ *the congruence*

$$\begin{aligned} b_{2\nu}(x) \equiv & \frac{1}{\sqrt{-t\Phi''_\zeta(x,\zeta(x))}} \sum_{j=0}^{2\nu} \frac{1}{\left[{}^t\Phi''_\zeta(x,\zeta(x))\right]^{3\nu-j+(2\nu-j)r_{2\nu-j+1}}} \\ & \times S_j^{2\nu}\left({}^t\Phi''_\zeta(x,\zeta(x)), \dots, {}^t\Phi_\zeta^{(2\nu-j+2)}(x,\zeta(x))\right) \\ & \times \frac{d^j\, {}^ta}{d\zeta^j}(x,\zeta(x)) \left(\mathrm{mod}^{s+1} I(\mathbb{R}^m, |\mathrm{Im}\zeta(x)|)\right) . \end{aligned} \tag{2.20}$$

holds.

We observe that all the assertions of Proposition 1.2 follow from Lemmas 2.1–2.4, except for the equation $W_0^{(1)} = 1$, since, by inequalities (1.8), (1.15), we can replace the t-analytic continuation in formula (2.20) by N-analytic ones for $t \geq s$, changing the decomposition by a term of order $O\left(h^{s/2}\right)$. We will establish the equality $W_0^{(1)} = 1$ simultaneously for all n at the end of the proof.

Now we proceed to the proof of Lemmas 2.1–2.4.

Proof of Lemma 2.1. We observe that the function $\widehat{p}^\alpha I_1^{(2)}(x,h)$ is given by a sum of integrals of the form (2.15) with coefficients which are uniformly bounded in h, while the degree k can decrease by no more than $|\alpha|$. Therefore it is enough to prove the assertion of the lemma for $\alpha = 0$. We transform the integral (2.15) by means of integration by parts:

$$\begin{aligned} I_1^{(2)}(x,h) = & \left(\frac{i}{2\pi h}\right)^{1/2} e^{(i/h)^N\Phi(x,\zeta(x))}(ih) \int_{-\infty}^{\infty} e^{(i/h)[\xi(x,p)]^2} \\ & \times \left[\frac{d\xi(x,p)}{dp}\right]^{-1} \frac{d}{dp}\left\{[\xi(x,p)]^k F_{k+1}^b(x,p)\chi(x,p)\right\} d\xi(x,p) . \end{aligned}$$

We note that the integrals obtained as a result of this transformation are bounded by virtue of the estimate

$$\mathrm{Im}\left\{{}^N\Phi(x,\zeta(x)) + [\xi(x,p)]^2\right\} = \mathrm{Im}\Phi(x,p) \geq 0 \tag{2.21}$$

A repeated application of this formula for sufficiently large k will give estimate (2.17). This proves the lemma. □

Proof of Lemma 2.2. We transform the integral (2.16) by integrating by parts. We have

$$I_1^{(3)}(x,h) = \left(\frac{i}{2\pi h}\right)^{1/2} e^{(i/h)^N\Phi(x,\zeta(x))}(ih)^r \int_{-\infty}^{\infty} e^{(i/h)[\xi(x,p)]^2}$$
$$\times \left\{ \left[\frac{d\xi(x,p)}{dp}\right]^{-1} \frac{d}{dp}\frac{1}{\xi(x,p)} \right\}^r \left[G_{k+1}^b(x,p)\chi(x,p)\right] d\xi(x,p) \, .$$

It is not hard to see that this integral decomposes into a sum of integrals of the form

$$\left(-\frac{1}{2\pi}\right)^{1/2} (ih)^{r-1/2} e^{(i/h)\ ^N\Phi\,(x,\zeta(x))} \int_{-\infty}^{\infty} e^{(1/h)[\xi(x,p)]^2} \frac{\widetilde{G}^\nu(x,p)}{[\xi(x,p)]^\nu} d\xi(x,p) \, , \tag{2.22}$$

where $r \leq \nu \leq 2r$, $G^\nu(x,p)$ satisfies the estimate

$$\left| D_x^\beta D_p^\gamma \widetilde{G}^\nu(x,p) \right| \leq C \left|\mathrm{Im}\zeta(x)\right|^{t+(\nu-2r)-|\beta|-|\gamma|} \, . \tag{2.23}$$

Since $|\xi(x,p)| \geq C\,|\mathrm{Im}\zeta(x)|$, by virtue of (2.23) we have the estimate

$$\left| \frac{G^\nu(x,p)}{[\xi(x,p)]^\nu} \right| \leq C \left|\mathrm{Im}\zeta(x)\right|^{t-2r} \, .$$

Together with inequality (2.21), the latter estimate shows that for $t \geq 2r$ the expression (2.22) does not exceed $Ch^{r-1/2}$. Choosing $r = (s+1)/2$, we obtain the assertion of the lemma. □

Proof of Lemma 2.3. We consider the expression

$$\left(\frac{i}{2\pi h}\right)^{1/2} e^{(i/h)^N\Phi(x,\zeta(x))} \int_{-\infty}^{\infty} e^{(i/h)[\xi(x,p)]^2} \left[\xi(x,p)\right]^j \chi(x,p) d\xi(x,p) \, . \tag{2.24}$$

This integral may be viewed as an integral along a contour $\gamma(x)$ in the complex plane $\mathbb{C}$ (with coordinate ξ), whose equation is

$$\xi = \xi(x,p) \, .$$

We recall that

$$\Phi(x,p) = \left[\xi(x,p)\right]^2 + {}^N\Phi\,(x,\zeta(x)) \, ,$$

and therefore on $\mathrm{supp}\chi(x,p)$ the contour $\gamma(x)$ lies in a region where the inequality

$$\mathrm{Im}\Phi(x,p) \geq 0$$

holds. We will show that it is possible to continue the contour $\gamma(x)$ in such a way that

1) outside a sufficiently small neighborhood of the point $\xi = 0$ it coincides with the real line;

2) on this contour

$$\operatorname{Im}\left[\xi^2 + {}^N\Phi\left(x, \zeta(x)\right)\right] \geq 0 \; ; \tag{2.25}$$

3) the contour $\gamma(x)$ depends continuously on x.

We begin with condition (2). We have

$$\operatorname{Im}\left[\xi^2 + {}^N\Phi\left(x, \zeta(x)\right)\right] = 2\xi_1\xi_2 + \operatorname{Im}{}^N\Phi\left(x, \zeta(x)\right) \; .$$

From Lemma 1.4 it follows that there exists a constant $C > 0$ such that the inequality

$$\operatorname{Im}{}^t\Phi\left(x, \zeta(x)\right) \geq c\left[\operatorname{Im}\zeta(x)\right]^2 \; . \tag{2.26}$$

holds. Therefore the region in the complex plane (ξ_1, ξ_2) in which inequality (2.25) holds is bounded by the hyperbola

$$\xi_1 \cdot \xi_2 = -\frac{\operatorname{Im}{}^N\Phi\left(x, \zeta(x)\right)}{2} \leq 0 \; . \tag{2.27}$$

We denote by U_x the region, bounded by the hyperbola (2.27), where inequality (2.25) holds. If $x_0 \in \Omega$, then U_{x_0} is the region consisting of the first and third quadrants. Obviously, the contour $\gamma(x)$ lies in the domain U_x. We compute $\frac{d\xi(x,p)}{dp}$ for $x \in \Omega$. As will be shown below,

$$\left.\frac{d\xi}{d\zeta}\right|_{\zeta=\zeta(x)} = \left[{}^N\Phi''_\zeta\left(x, \zeta(x)\right)/2\right]^{1/2}$$

In particular, on the set Ω the function $\zeta(x) = p(x)$ and

$$\left.\frac{d\xi}{dp}\right|_{p=p(x)} = \left[\Phi''_p\left(x, p(x)\right)/2\right]^{1/2} \; .$$

Below we will require the contour $\gamma(x)$ to pass from the third quadrant into the first. Therefore the argument of the expression under the square root in the last formula must be taken in the interval from 0 to π. For the following it will be convenient to choose the argument not of the function $\Phi''_p\left(x, p(x)\right)$, but of the function $-\Phi''_p\left(x, p(x)\right)$, so that

$$-\pi < \arg\left[-\Phi''_p\left(x, p(x)\right)\right] < 0 \; .$$

Here

$$\left.\frac{d\xi}{dp}\right|_{p=p(x)} = i\left[-\Phi''_p\left(x, p(x)\right)/2\right]^{1/2}$$

It is not hard to see that on such a choice of the argument the contour passes from the third quadrant into the first.

For $x_0 \in \Omega$ this contour passes through the region U_{x_0} and therefore can be continued in this region uniquely up to homotopy in such a way that condition (1) holds. Obviously the choice of the contours $\gamma(x)$ for

$x \in \Omega$ can be carried out in such a way that the function $\gamma(x)$ depends continuously on x. We note further that $\chi(x,p) = 1$ on some neighborhood of the stationary point $\xi(x,p) = 0$. Thus the expression under the integral is an analytic function in this region. Therefore we may assume that in some smaller neighborhood of the point $\xi(x,p) = 0$ the contour $\gamma(x)$ coincides with the real axis. With an accuracy to order $O(h^\infty)$ it is possible to replace the function $\chi(x,p)$ by any other function of compact support which is identically equal to one in a neighborhood of the point $\xi(x,p) = 0$. We replace $\chi(x,p)$ by the function

$$\chi'(x,p) = \widetilde{\chi}\left(\xi(x,p)\right) ,$$

where suppχ' lies in the region where the contour coincides with the real axis and $\widetilde{\chi}$ is an even function. Up to terms of order $O(h^\infty)$ the integral (2.24) equals

$$e^{(i/h)^N \Phi(x,\zeta(x))} \left[\frac{i}{2\pi h}\right]^{1/2} \int_{-\infty}^{\infty} e^{(i/h)\xi^2} \xi^k \widetilde{\chi}(\xi) d\xi .$$

For odd k the latter integral is equal to zero. For even k this integral reduces by integration by parts to an analytic integral with $k = 0$, multiplied by $h^{k/2}$. In this integral it is possible to replace $\widetilde{\chi}$ by unity with accuracy up to order $0(h^\infty)$, after which it is possible to calculate the integral directly:

$$\left[\frac{i}{2\pi h}\right]^{1/2} \int_{-\infty}^{\infty} e^{(i/h)\xi^2} d\xi = \frac{1}{\sqrt{2}} .$$

This proves the lemma. □

We now turn to the calculation of the functions $b_j(x)$ in expression (2.11). We note that it is enough for us to compute the functions $b_j(x)$ only for even numbers $j = 2\nu$.

Proof of Lemma 2.4. From formulas (2.3) and (2.5) it follows that the equation

$$\Phi(x,p) = \Phi\left(x,\zeta(x)\right) + \left[\xi(x,p)\right]^2$$

holds (we omit the index N). Choosing a t-analytic continuation of the latter formula in p and using the fact that the operator of t-analytic continuation is a ring homomorphism modulo the ideal ${}^{t+1}I(\rho)$, where $\rho = |\mathrm{Im}\zeta|$, we obtain the equation

$$^t\Phi(x,\zeta) = \Phi\left(x,\zeta(x)\right) + \left[\xi(x,\zeta)\right]^2 + F(x,\zeta) , \tag{2.28}$$

where

$$\xi(x,\zeta) = (\zeta - \zeta(x))^{t} \left\{ \left[F_{\Phi}(x,p) + \frac{G^{\Phi}(x,p)}{(\rho - \xi(x))^{2}} \right]^{1/2} \right\} (x,\zeta) ,$$

so that $\xi(x,\zeta(x)) = 0$. Besides this, the function $F(x,\zeta)$ admits an estimate

$$\left| D_x^{\alpha} D_p^{\beta} D_{\eta}^{\gamma} F(x,\zeta) \right| \leq |\mathrm{Im}\zeta|^{t+1-|\alpha|-|\beta|-|\gamma|} .$$

Since t is sufficiently large, and in the final results we will always set $\zeta = \zeta(x)$, we can, with accuracy up to the terms specified in formula (2.20), omit $F(x,\zeta)$ in formula (2.28) and rewrite this equation in the form

$$\Phi(x,\zeta) = \Phi(x,\zeta(x)) + [\xi(x,\zeta)]^{2} . \tag{2.29}$$

Analogously, we rewrite formula (2.11), taking

$$a(x,\zeta) \left[\frac{d\xi(x,\zeta)}{d\zeta} \right]^{-1} = \sum_{k=0}^{s+1} [\xi(x,\zeta)]^{k} b_k(x) + \xi^{s+2} F^{b}_{s+2}(x,\zeta) . \tag{2.30}$$

From the decomposition formula (2.30) it follows that

$$\begin{aligned} b_k(x) &= \frac{1}{k!} \left(\frac{d}{d\xi} \right)^{k} \left\{ a(x,\zeta) \left(\frac{d\xi(x,\zeta)}{d\zeta} \right)^{-1} \right\} \Bigg|_{\zeta=\zeta(x)} \\ &\overset{\mathrm{def}}{=} \frac{1}{k!} \left\{ \left(\frac{d\xi(x,\zeta)}{d\zeta} \right)^{-1} \frac{d}{d\zeta} \right\}^{k} \left\{ a(x,\zeta) \left(\frac{d\xi(x,\zeta)}{d\zeta} \right)^{-1} \right\} \Bigg|_{\zeta=\zeta(x)} . \end{aligned} \tag{2.31}$$

For brevity we will denote these functions without the substitution $\zeta = \zeta(x)$ by $b_k(x,\zeta)$.

To compute the right-hand side of formula (2.31) we must compute the functions

$$\beta_j(x) = \frac{d^j \xi}{d\zeta^j} (x,\zeta(x)) . \tag{2.32}$$

To find these functions we differentiate (2.29) with respect to the variable ζ. The second derivative gives

$$2 \left(\frac{d\xi}{d\zeta} \right)^{2} + 2\xi \frac{d^2\xi}{d\zeta^2} = \frac{d^2 \Phi(x,\zeta)}{d\zeta^2} . \tag{2.33}$$

Further differentiations lead to the formula

$$(2+2j) \frac{d\xi}{d\zeta} \frac{d^j\xi}{d\zeta^j} + P_2^{(j)} \left(\frac{d^2\xi}{d\zeta^2}, \ldots, \frac{d^{j-1}\xi}{d\zeta^{j-1}} \right) + 2\xi \frac{d^{j+1}\xi}{d\zeta^{j+1}} = \frac{d^{j+1}\Phi(x,\zeta)}{d\zeta^{j+1}} . \tag{2.34}$$

where the polynomials $P_2^{(j)}(\alpha_2,\dots,a_{j-1})$ are homogenous second-degree polynomials, defined by the recursive system of relations

$$P_2^{(2)}=0\ , \tag{2.35}$$

$$P_2^{(j+1)}(\alpha_2,\dots,\alpha_j)=\alpha_2\alpha_j+\sum_{k=2}^{j-1}\frac{\partial P_2^{(j)}}{\partial\alpha^k}(\alpha_2,\dots,\alpha_{j-1})\cdot\alpha_{k+1}\ .$$

The proof of formulas (2.34), (2.35) proceeds by induction on j. The expression for the functions (2.32) has the form

$$\beta_1(x)=\left.\frac{d\xi(x,\zeta)}{d\zeta}\right|_{\zeta=\zeta(x)}=\left[\Phi''_\zeta(x,\zeta(x))/2\right]^{1/2}$$

(this formula follows from (2.33)), and

$$\beta_j(x)=\frac{1}{\sqrt{\Phi''_\zeta(x,\zeta(x))}}\frac{Q_j\left(\Phi''_\zeta(x,\zeta(x)),\dots,\Phi_j^{(j+1)}(x,\zeta(x))\right)}{\left[\Phi''_\zeta(x,\zeta(x))\right]^{r_j}} \tag{2.36}$$

where $r_j=2^{j-2}-1$, and the polynomials Q_j are defined by the recursive sequence of relations

$$\begin{gathered}Q_2(\gamma_2,\gamma_3)=\frac{\sqrt{2}}{6}\gamma_2\gamma_3\\ Q_j(\gamma_2,\dots,\gamma_{j+1})=\gamma_2^{r_j}\gamma_{j+1}-P_2^{(j)}\Big(\gamma_2^{r_{j-1}-r_2}Q_2(\gamma_2,\gamma_3)\ ,\\ \gamma_2^{r_{j-1}-r_3}Q_3(\gamma_2,\gamma_3,\gamma_4),\dots,Q_{j-1}(\gamma_2,\dots,\gamma_j)\Big)\ .\end{gathered} \tag{2.37}$$

Formulas (2.36) and (2.37) follow from formula (2.34) in view of the homogeneity of the polynomials $P_2^{(j)}$.

Furthermore, we have the formula

$$\widetilde{b}_k(x,\zeta)=\sum_{j=0}^{k}\frac{d^ja(x,\zeta)}{\partial\zeta^j}\left[\frac{d\xi}{d\zeta}\right]^{-(2k+1-j)}R_{k-j}^k\left(\frac{d\xi}{d\zeta},\frac{d^2\xi}{d\zeta^2},\dots,\frac{d^{k+j-1}\xi}{d\zeta^{k+j-1}}\right)\ . \tag{2.38}$$

where $R_{k-j}^k(\delta_1,\delta_2,\dots,\delta_{k+j-1})$ are homogeneous polynomials of degree $k-j$, defined by the following relations, which are recursive in the upper index of the polynomial (the number of the function b_k):

$$R_0^{(0)}(\delta_1)=1\ , \tag{2.39}$$

$$\begin{aligned}R_{k+1}^{k+1}(\delta_1,\dots,\delta_{k+2})=&-\frac{2k+1}{k+1}\delta_2R_k^k(\delta_1,\dots,\delta_{k+1})\\ &+\frac{1}{k+1}\delta_1\sum_{l=1}^{k+1}\frac{\partial R_k^k}{\partial\delta_l}(\delta_1,\dots,\delta_{k+1})\,\delta_{l+1}\end{aligned} \tag{2.40}$$

$$R^{k+1}_{k+1-j}(\delta_1,\dots,\delta_{k+2-j}) = \frac{1}{k+1}R^k_{k+1-j}(\delta_1,\dots,\delta_{k+2-j})$$
$$-\frac{2k+1-j}{k+1}\delta_2 R^k_{k-j}(\delta_1,\dots,\delta_{k+1-j}) + \frac{1}{k+1}\delta_1 \tag{2.41}$$
$$\times \sum_{l=1}^{k-j+1} \frac{\partial R^k_{k-j}}{\partial \delta_l}(\delta_1,\dots,\delta_{k+1-j})\,\delta_{l+1}\ , \qquad 1 \le j \le k\ .$$

$$R^{k+1}_0(\delta_1) = \frac{1}{k+1}R^k_0(\delta_1)\ . \tag{2.42}$$

The proof of formulas (2.38)–(2.42) proceeds by induction on the number k of the function $b_k(x,\zeta)$, taking into account the equation

$$b_{k+1}(x,\zeta) = \frac{1}{k+1}\left[\frac{d\xi}{d\zeta}\right]^{-1}\frac{d}{d\zeta}b_k(x,\zeta)\ .$$

Substituting $\zeta = \zeta(x)$ in formula (2.38) and taking into account formula (2.36) and the homogeneity of the polynomials R^k_{k-j}, we obtain (2.20) with

$$S^k_j(t_2,\dots,t_{k+2-j}) = iR^k_{k-j}\Big[2^{-1/2}t_2^{r_{k-j+1}}, t_2^{r_{k-j+1}-r_2}Q_2(t_2,t_3)\ ,$$
$$t_2^{r_{k-j+1}-r_3}Q_3(t_2,t_3,t_4),\dots,Q_{k-j+1}(t_2,\dots,t_{k+2+j})\Big]\,2^{(2k+1-j)/2}$$

for $k = 2\nu$. This proves Lemma 2.4. □

Therefore we have established the formula

$$I_1(x,h) = \left(\frac{i}{2\pi h}\right)^{1/2}\int_{-\infty}^{\infty} e^{(1/h)\Phi(x,p)}a(x,p)dp = \frac{e^{(1/h)^N\Phi(x,\zeta)}}{\sqrt{-\Phi''_\xi(x,\zeta)}} \tag{2.43}$$
$$\times \sum_{0\le k<(s-1)/2} h^k W^1_k[\Phi,a]\big|_{\zeta=\zeta(x)}\left(\operatorname{mod} h^{(s-1)/2}\right)\ .$$

5.2.2 Inductive Step. Now suppose formula (1.4) holds for $(n-1)$ dimensions. We will prove it for n dimensions.

We will examine the integral $I_n(x,h)$ in a neighborhood of a point $x_0 \in \Omega$.

$$I_n(x,h) = \left(\frac{i}{2\pi h}\right)^{n/2}\int_{\mathbb{R}_n} e^{(1/h)\Phi(x,p)}a(x,p)dp\ . \tag{2.44}$$

By a nondegenerate real linear transformation of the space $\mathbb{R}_n$ it is possible to make one of the diagonal elements of the matrix $\operatorname{Hess}_p\Phi(x_0,p(x_0))$ nonzero. Without loss of generality we will assume that

$$\frac{\partial^2\Phi}{\partial p_1^2}(x_0,p(x_0)) \ne 0\ . \tag{2.45}$$

We rewrite the integral (2.44) in the following form (again denoting the variables of integration by p):

$$\begin{aligned} I_n(x,h) = &\left(\frac{i}{2\pi h}\right)^{(n-1)/2} \int_{\mathbb{R}_{n-1}} dp' \left(\frac{i}{2\pi h}\right)^{1/2} \\ &\times \int_{-\infty}^{\infty} e^{(i/h)\Phi(x,p_1,p')} a(x,p_1,p')\, dp_1 \ . \end{aligned} \tag{2.46}$$

Here $p' = (p_2, \ldots, p_n)$, so that $p = (p_1, p')$.

Let us verify the hypotheses of Theorem 1.1 for the inside integral

$$\left(\frac{i}{2\pi h}\right)^{1/2} \int_{-\infty}^{\infty} e^{(1/h)\Phi(x,p_1,p')} a(x,p_1,p')\, dp_1 \ . \tag{2.47}$$

Here the parametric variables are (x, p') and the set $\widetilde{\Omega} = \{(x,p) | p = p'(x), x \in \Omega\}$. The conditions on the function $a(x,p_1,p') = a(x,p)$ and $\Phi(x,p_1,p') = \Phi(x,p)$ are obviously satisfied. Furthermore, since Eq. (1.2) has no more than one solution $p = p(x)$ on the support of the function $a(x,p)$ the equation

$$\Phi'_{p_1}(x,p_1,p') = 0 \ , \quad x \in \Omega$$

has no more than one solution

$$p_1 = p_1(x, p'(x)) \ ,$$

and condition (2.45) represents the nondegeneracy of the Hessian of the function $\Phi(x,p_1,p')$ with respect to the variable p_1. We denote by

$$\zeta_1 = \zeta_1(x,p') \tag{2.48}$$

the solution of the equation

$${}^N\Phi_\zeta(x,p',\zeta_1) = 0 \ . \tag{2.49}$$

We apply formula (2.43) to the interior integral in (2.46):

$$\begin{aligned} &\left(\frac{i}{2\pi h}\right)^{1/2} \int_{-\infty}^{\infty} e^{(i/h)\Phi(x,p_1,p')} a(x,p_1,p')\, dp_1 \\ &\equiv \frac{e^{(i/h)^N\Phi(x,\zeta_1(x,p'),p')}}{\sqrt{-\Phi''_{\zeta_1}(x,\zeta_1(x,p'),p')}} \sum_{0 \le k < (s-1)/2} h^k W_k^{(1)}[\Phi, a] \ \left(\operatorname{mod} h^{s-1/2}\right) \end{aligned} \tag{2.50}$$

(The superscript 1 in the function $W_k^{(1)}$ represents the dimension of the space.) We denote

$$\Phi_1(x,p') = {}^N\Phi(x,\zeta_1(x,p'),p') \ . \tag{2.51}$$

The integral (2.46) can now be written in the form

$$I_n(x,h) = \sum_{0\le k<(s-1)/2} h^k \left(\frac{i}{2\pi h}\right)^{(n-1)/2} \int_{\mathbb{R}_{n-1}} \frac{e^{(i/h)\Phi_1(x,p')}}{\sqrt{-\Phi''_{\zeta_1}\left(x,\zeta_1(x,p'),p'\right)}}$$
$$\times W_k^{(1)}[\Phi,a]dp' \left(\operatorname{mod} h^{(s-1)/2}\right) . \tag{2.52}$$

Now we check the assumptions of Theorem 1.1 for the integral (2.52). Here the parametric variables are x, the set Ω coincides with the set Ω introduced in Theorem 1.1, and the condition $\operatorname{Im}\Phi_1(x,p') \ge 0$ is guaranteed to hold by Lemma 1.4. It remains to prove that the system of equations

$$\frac{\partial\Phi_1}{\partial p_1}(x,p') = 0 , \qquad i = 2,\dots,n \tag{2.53}$$

has a unique solution for $x \in \Omega$. Indeed, if $x \in \Omega$, then

$$\zeta_1\left(x,p'(x)\right) = p_1(x) . \tag{2.54}$$

Furthermore

$$\left.\frac{\partial\Phi_1}{\partial p_i}\right|_{p'=p'(x)} = \frac{\partial^s\Phi}{\partial p_i}\left(x,p(x)\right) + \frac{\partial^s\Phi}{\partial\zeta_1}\left(x,p'(x)\right)\frac{\partial\zeta_1}{\partial p_i} = 0 \tag{2.55}$$

for all $i = 2,\dots,n$, since the second summand in (2.55) equals zero by virtue of (2.49), and the first equals zero by the hypothesis of Theorem 1.1.

Now we verify that the inequality

$$\det \operatorname{Hess}_{p'} \Phi_1\left(x_0,p'\left(x_0\right)\right) \ne 0 , \quad x_0 \in \Omega . \tag{2.56}$$

holds. To do this we consider the determinant[3]

$$\det \operatorname{Hess}_p \Phi = \begin{vmatrix} \frac{\partial^2\Phi}{\partial p_1^2} & \frac{\partial^2\Phi}{\partial p_1\partial p_2} & \cdots & \frac{\partial^2\Phi}{\partial p_1\partial p_n} \\ \frac{\partial^2\Phi}{\partial p_2\partial p_1} & \frac{\partial^2\Phi}{\partial p_2^2} & \cdots & \frac{\partial^2\Phi}{\partial p_2\partial p_n} \\ \cdots & \cdots & \cdots & \cdots \\ \frac{\partial^2\Phi}{\partial p_n\partial p_1} & \frac{\partial^2\Phi}{\partial p_n\partial p_2} & \cdots & \frac{\partial^2\Phi}{\partial p_n^2} \end{vmatrix} \tag{2.57}$$

In the determinant (2.57) we apply the following transformations, which obviously do not affect its value: we multiply the first column of (2.57) in turn by $\partial\zeta_1/\partial p_1$, $i = 2,\dots,n$, and add the result to the i-th column $(i = 2,\dots,n)$.

Using Eqs. (2.49) and (2.51) we obtain as a result the following formula:

---

[3] All the functions in the determinant below are evaluated at the point $(x_0, p(x_0))$.

$$\det \operatorname{Hess}_p \Phi(x_0, p(x_0)) = \frac{\partial^2 \Phi}{\partial p_1}(x_0, p(x_0)) \det \operatorname{Hess}_{p'} \Phi_1(x_0, p'(x_0)) \ . \quad (2.58)$$

Now from the hypothesis $\det \operatorname{Hess}_p \Phi(x_0, p(x_0)) \neq 0$ of Theorem 1.1 it immediately follows that

$$\det \operatorname{Hess}_{p'} \Phi_1(x_0, p'(x_0)) \neq 0 \ . \quad (2.59)$$

Thus all the hypotheses of Theorem 1.1 are satisfied, and we have the right to apply the inductive assumption. We have

$$\begin{aligned} &\left(\frac{i}{2\pi h}\right)^{n-1/2} \int_{\mathbb{R}_{n-1}} e^{(i/h)\Phi_1(x,p')} \frac{W_k^{(1)}[\Phi, a]dp'}{\sqrt{-\Phi''_{\zeta_1}(x, p', \zeta_1(x,p'))}} \\ &\equiv \sum_{0 \le k' < \frac{s-1}{2} - k} h^{k'} \frac{1}{\sqrt{\det \operatorname{Hess}_{\zeta'}(-\Phi_1(x, \zeta'(x)))}} \\ &\times W_k^{(n-1)}\left[\Phi_1, \frac{W_k^{(1)}[\Phi, a]dp'}{\sqrt{-\Phi''_{\zeta_1}(x, \zeta', \zeta_1(x,\zeta'))}}\right] \left(\operatorname{mod} h^{\frac{s-1}{2}-k}\right) , \end{aligned} \quad (2.60)$$

where $\zeta'(x)$ is the solution of the system of equations

$$\frac{\partial}{\partial \zeta_i} {}^N\Phi(x, \zeta') = 0 \ , \qquad i = 2, \dots, n \ .$$

On the other hand,

$${}^N\Phi_1(x, \zeta') = {}^N\Phi(x, \zeta_1(x, \zeta'), \zeta') \ .$$

Let $\zeta = \widetilde{\zeta}(x)$ be the solution of the system (2.2). Then

$$\frac{\partial}{\partial \zeta_i} {}^N\Phi(x, {}^N\zeta_1(x, \zeta'(x)), \zeta'(x)) = 0 \ , \qquad i = 2, \dots, n \ ,$$

$$\frac{\partial}{\partial \zeta_1} {}^N\Phi(x, {}^N\zeta_1(x, \zeta'(x)), \zeta'(x)) = 0$$

and

$$\frac{\partial}{\partial \zeta_i} {}^N\Phi\left(x, \widetilde{\zeta}_1(x), \widetilde{\zeta}'(x)\right) = 0 \ , \qquad i = 1, \dots, n \ .$$

From the uniqueness of the solution to the system (2.2) it follows that

$$\widetilde{\zeta}'(x) = \zeta'(x), \widetilde{\zeta}_1(x) = {}^N\zeta_1(x, \zeta'(x))$$

and therefore

$${}^N\Phi_1(x, \zeta')\big|_{\zeta'=\zeta'(x)} = {}^N\Phi(x, \zeta'(x), {}^N\zeta_1(x, \zeta'(x))) = {}^N\Phi\left(x, \widetilde{\zeta}(x)\right) \ .$$

By formulas (2.60), (2.50) and (2.46) we have

$$
\begin{aligned}
I_n(x,h) &\equiv e^{(i/h)^N \Phi\left(x,\widetilde{\zeta}(x)\right)} \frac{1}{\sqrt{\det \operatorname{Hess}_\zeta \left(-\Phi\left(x,\widetilde{\zeta}(x)\right)\right)}} \\
&\times \frac{\sqrt{\det \operatorname{Hess}_\zeta \left(-\Phi\left(x,\widetilde{\zeta}(x)\right)\right)}}{\sqrt{-\Phi''_{\zeta_1}\left(x,\zeta'(x),\zeta_1\left(x,\zeta'(x)\right)\right)}\sqrt{\det \operatorname{Hess}_{\zeta'}\left(-\Phi_1\left(x,\zeta'(x)\right)\right)}} \\
&\times \sum_{0\le l<(s-1)/2} h^l \sum_{k'+k=l} \sqrt{-\Phi''_{\zeta_1}\left(x,\zeta'(x),\zeta_1\left(x,\zeta'(x)\right)\right)} \\
&\times W_{k'}^{(n-1)}\left[\Phi_1, \frac{W_k^{(1)}[\Phi,a]}{\sqrt{-\Phi''_{\zeta_1}\left(x,\zeta'(x),\zeta_1\left(x,\zeta'(x)\right)\right)}}\right] \left(\operatorname{mod} h^{(s-1)/2}\right). \qquad (2.61)
\end{aligned}
$$

Now we note that the expression

$$
\frac{\sqrt{\det \operatorname{Hess}_\zeta \left(-\Phi\left(x,\widetilde{\zeta}(x)\right)\right)}}{\sqrt{-\Phi''_{\zeta_1}\left(x,\zeta_1\left(x,\zeta'(x)\right),\zeta'(x)\right)}\sqrt{\det \operatorname{Hess}_{\zeta'}\left(-\Phi_1\left(x,\zeta'(x)\right)\right)}} = \sigma \qquad (2.62)
$$

is, by formula (2.58), equal to ± 1. Moreover, the expression (2.62) is an analytic function in the values of the second derivatives of the function ${}^N\Phi(x,\zeta)$ at the point $\zeta = \widetilde{\zeta}(x)$ where $\Phi''_{\zeta_1}\left(x,\widetilde{\zeta}(x)\right) \neq 0$ (that is, in the variables t_α, $|\alpha| = 2$). Hence it follows that the function σ is constant on this set and does not depend on the function Φ. We now denote

$$
\begin{aligned}
W_l^{(n)}[\Phi,a] = \sigma \sum_{k+k'=l} &\sqrt{-\Phi''_{\zeta_1}\left(x,\zeta_1\left(x,\zeta'(x)\right),\zeta'(x)\right)} \\
&\times W_{k'}^{(n-1)}\left[\Phi_1, \frac{W_k^{(1)}[\Phi,a]}{\sqrt{-\Phi''_{\zeta_1}\left(x,\zeta_1\left(x,\zeta'(x)\right),\zeta'(x)\right)}}\right]. \qquad (2.63)
\end{aligned}
$$

The latter expression, by the first step of the induction and the inductive assumption concerning the functions $W_k^{(n-1)}[\Phi,a]$, is a linear form in the derivatives of the function a up to order $2l$ at the point $\zeta = \widetilde{\zeta}(x)$. The coefficients of this form are roots of rational functions of the derivatives of the function $\Phi(x,\zeta)$ of order not less than second and not greater than $2l-j+2$ (j is the order of the derivative of the function a). The denominators of these functions are

$$\left[\Phi''_{\zeta_1}\left(x,\widetilde{\zeta}(x)\right)\right]^{\gamma}\left[\det \operatorname{Hess}_{\zeta'}\Phi_1\left(x,\zeta'(x)\right)\right]^{\delta} \tag{2.64}$$

where γ, δ are some natural numbers.

If one uses formula (2.58):

$$\det \operatorname{Hess}_{\zeta'}\Phi_1\left(x,\zeta'(x)\right) = \frac{\det \operatorname{Hess}_{\zeta}\Phi\left(x,\widetilde{\zeta}(x)\right)}{\Phi''_{\zeta_1}\left(x,\widetilde{\zeta}(x)\right)},$$

then the denominator (2.64) can be replaced by

$$\left[\Phi''_{\zeta_1}\left(x,\widetilde{\zeta}(x)\right)\right]^{\gamma'}\left[\det \operatorname{Hess}_{\zeta}\Phi\left(x,\widetilde{\zeta}(x)\right)\right]^{\delta} \tag{2.65}$$

(γ' is some natural number).

The corresponding numerators in these expressions are polynomials which do not depend on the function Φ on the set $t_{(2,0,\dots,0)} \neq 0$, since we assume that $\Phi''_{p_1}(x_0, p(x_0)) \neq 0$.

In a manner analogous to the above one can construct roots of rational functions on the sets $\left\{t_{(0,2,0,\dots,0)} \neq 0\right\}, \dots, \left\{t_{(0,0,\dots,0,2)} \neq 0\right\}$, as well as on the sets $\{t_{(2,0,\dots,0)} = t_{(0,2,0,\dots,0)} = \dots = t_{(0,0,\dots,0,2)} = 0,\ t_{(0,\dots,0,\underset{i}{1},0,\dots,0,\underset{j}{1},0,\dots,0)} \neq 0\}$ for some $i \neq j$. We assert that all these roots of rational functions are identically equal. Indeed, it is sufficient to convince ourselves of this for real values t_α only. If we consider formula (2.61) with a real phase function Φ, then the fact that the asymptotic expansions coincide implies that all their coefficients coincide. On the other hand, the derivatives of all possible real functions take on all possible real values.

From the fact that the roots of rational functions coincide on general domains of definition it follows, in particular, that the exponent γ' in formula (2.65) may be taken to be zero, in other words if $\gamma' \neq 0$ then the corresponding polynomial in the numerator can be divided by $\left[\Phi''_{\zeta_1}\left(x,\widetilde{\zeta}(x)\right)\right]^{\gamma_1}$. To complete the proof of Proposition 1.2 it remains to show that

$$W_0^{(n)}[\Phi, a] = a\left(x,\widetilde{\zeta}(x)\right).$$

Since the coefficient of the function $a\left(x,\widetilde{\zeta}(x)\right)$ depends only on the second derivatives of the function Φ, it is sufficient to prove the assertion for the integral

$$\left(\frac{i}{2\pi h}\right)^{n/2}\int e^{\frac{i}{h}\sum_{|\alpha|=2} t_\alpha p^\alpha}\,dp = \frac{1}{\left[D\left(t_\alpha\right)\right]^{1/2}}$$

(for the definition of $D(t_\alpha)$ see p. 209). The argument of the expression under the square root in the last formula is chosen in correspondence with Proposition 1.2. The last integral can be calculated directly. This completes the proof of Proposition 1.2. □

Chapter 6
Maslov's Canonical Operator (Complex Case)

6.1 Maslov's Elementary Operator on a Complex Lagrangian Manifold

We consider a complex phase space $\Phi_{\mathbb{C}}$. As before (see Sect. 2.2) the space $\Phi_{\mathbb{C}}$ is a (complex) $2n$-dimensional vector space with a fixed basis $(e_1, \dots, e_n, f^1, \dots, f^n)$. By means of this basis the space $\Phi_{\mathbb{C}}$ can be identified with the space $\mathbb{C}^n \oplus \mathbb{C}_n$, whose coordinates we will denote by $(z, \zeta) = (z^1, \dots, z^n, \zeta_1, \dots, \zeta_n)$, so that any vector in $\Phi_{\mathbb{C}}$ can be represented in the form

$$z^1 e_1 + \ldots + z^n e_n + \zeta_1 f^1 + \ldots + \zeta_n f^n = z^i e_i + \zeta_i f^i .$$

(As usual, we mean repeated indices to be summed.)

The space $\Phi_{\mathbb{C}}$ is s-analytic with weight function

$$\rho_\Phi(x, \zeta) = \sqrt{|\mathrm{Im}\, z|^2 + |\mathrm{Im}\, \zeta|^2} = \sqrt{\sum_{j=1}^{n} (\mathrm{Im}\, z^j)^2 + \sum_{j=1}^{n} (\mathrm{Im}\, \zeta_j)^2} \tag{1.1}$$

(see 3.4 of Chap. 3). The real coordinates in $\Phi_{\mathbb{C}}$ are x^j, y^j, p_j, η_j, $j = 1, \dots, n$; they are associated with the complex coordinates by means of the formulas

$$z^j = x^j + iy^j , \quad \zeta_j = p_j + i\eta_j ,$$

so that

$$\rho_\Phi(z, \zeta) = \rho_\Phi(x, y, p, \eta) = \left[|y|^2 + |\eta|^2\right]^{1/2} .$$

The form

$$\omega = d\zeta \wedge dz = d\zeta_j \wedge dz^j \tag{1.2}$$

converts the space $\Phi_{\mathbb{C}}$ into a symplectic s-analytic space.

Now suppose s is some sufficiently large number. (We will not make precise here the minimal possible value of s; in each case this will not be difficult to do.) Let (M, ρ_M) be an $(s+1)$-analytic manifold, and

$$i : (M, \rho_M) \to (\Phi_{\mathbb{C}}, \rho_\Phi) \tag{1.3}$$

an $(s+1)$-analytic imbedding. We recall that the imbedding (1.3) is called *Lagrangian* if the condition

$$i^*\omega = 0 \in {}^s\Lambda^2 (M, \rho_M) \ . \tag{1.4}$$

holds, where ω is defined by formula (1.2). Sometimes in this situation we will simply refer to a Lagrangian manifold. We will also assume that there exists on the manifold M a function

$$S \in {}^{s+1}\mathcal{O}(M, \rho_M) \ , \tag{1.5}$$

which satisfies the condition

$$dS = i^*(\zeta dz) = i^* \left(\zeta_j dz^j\right) \in {}^s\Lambda^1 (M, \rho_M) \ . \tag{1.6}$$

We will call the function (1.5) a nonsingular (or global) action on the manifold M.

Remark 1.1. We note that in the s-analytic case the relation (1.4) *does not follow* from relation (1.6); only the weaker relation

$$i^*\omega = i^* \left(d(\zeta dz)\right) = di^* \left(\zeta dz\right) = d(dS) = 0 \in {}^{s-1}\Lambda^2 (M, \rho_M)$$

follows, and hence we have formulated the requirement (1.4) separately.

We consider an s-analytic canonical atlas $\{U_I\}$ on the manifold (M, ρ_M) (see 3.4). Here, as before, $I \subset \{1, \ldots, n\}$ is some set, $\overline{I}$ is its complement, and the coordinates in the chart U_I are the functions $\left(z^I, \zeta_{\overline{I}}\right)$ (it would be more precise to write $i^*z^I, i^*\zeta_{\overline{I}}$, however we will sometimes omit the operator i^* in order not to over-complicate the exposition). In each chart U_I we introduce the function

$$S_I = S - z^{\overline{I}} \zeta_{\overline{I}} \ , \tag{1.7}$$

which in the future we will call the *action in the chart* U_I. We will assume that the Lagrangian $(s+1)$-analytic manifold (M, ρ_M) is positive (see 3.4), that is in each chart U_I there are some constants c, C such that the inequalities

$$c\rho_M^2 \le \mathrm{Im} S_I \le C\rho_M^2 \quad \text{on} \quad U_I^0 \ , \tag{1.8}$$

hold, where the function S_I is defined by formula (1.7).

Remark 1.2. In the real case (see formula (1.5) in Sect. 4.1 of Chap. 4) the action S_I in the chart U_I was defined as a solution of the equation

$$dS_I = p_I dx^I - x^{\overline{I}} dp_{\overline{I}} \,. \tag{1.9}$$

Then in Sect. 4.1 a one-dimensional co-chain of the canonical covering was defined (cf. Proposition 1.2 of Chap. 4):

$$c_{IJ}^{(1)} = S_I - S_J - x^{\overline{I}} p_{\overline{I}} + x^{\overline{J}} p_{\overline{J}} \,. \tag{1.10}$$

This co-chain $c_{IJ}^{(1)}$ turned out to be a real-valued cocycle, and one of the quantization conditions was that this cocycle be cohomologous to zero. It is not hard to see that when this condition is satisfied there exists a global function S on the manifold L such that $dS = p\,dx$ and the functions $S'_I = S - x^{\overline{I}} p_{\overline{I}}$ differ from the functions S_I by certain constants. In the s-analytic case this construction is not applicable, although the functions S_I defined by formula (1.7) do indeed satisfy the equation

$$dS_I = \zeta_I dz^I - z^{\overline{I}} d\zeta_{\overline{I}} \in {}^s\Lambda^1(M, \rho_M) \tag{1.11}$$

analogous to Eq. (1.10). The reason for this is that without supplementary conditions on the set $\Omega(M, \rho_M)$ the equation $df = 0 \in {}^t\Lambda(M, \rho_M)$ does not, in general, imply the equation $f = \text{const}$ in the ring ${}^{t+1}\mathcal{O}(M, \rho_M)$. Therefore the co-chain defined by a formula analogous to formula (1.10) in the real case would be a cochain not with real but with s-analytic coefficients. For this reason we have adopted a different approach, requiring the existence of a global action. Under these circumstances formula (1.6) is the complete analog of the first quantization condition in the real case. We note, however, that under certain regularity conditions on the function ρ_M the construction of the real case can be applied.

Suppose a certain nondegenerate s-analytic measure $\mu \in {}^s\Lambda^n(M, \rho_M)$ is chosen and fixed on the manifold M. We will denote the density of this measure with respect to the $(s+1)$-analytic coordinates $(z^I, \zeta_{\overline{I}})$ in the chart U_I by μ_I, so that

$$\mu|_{U_I} = \mu_I(z^I, \zeta_{\overline{I}}) dz^I \wedge d\zeta_{\overline{I}} \,. \tag{1.12}$$

Here and below we will assume that any chart U_I is simply connected. Because of this, and also by the nondegeneracy of the measure μ, the complex function $\mu_I(z^I, \zeta_{\overline{I}})$ admits a single-valued choice of a branch of the argument on each chart U_I. We fix in each chart an arbitrary branch

$$\operatorname{Arg}\mu_I(z^I, \zeta_{\overline{I}}) \tag{1.13}$$

of the argument of $\mu_I(z^I, \zeta_{\overline{I}})$. We will define the function $[\mu_I(z^I, \zeta_{\overline{I}})]^{1/2}$ by assuming that $\operatorname{Arg}[\mu_I(z^I, \zeta_{\overline{I}})]^{1/2} = \frac{1}{2}\operatorname{Arg}\mu_I(z^I, \zeta_{\overline{I}})$.

Remark 1.3. Obviously it is sufficient to consider the function (1.13) as being defined modulo 4π. Nevertheless, we will consider a certain concrete value to be fixed.

Definition 1.1. We will call the operator[1]

$$k_I : {}^s\mathcal{O}_0\left(U_I, \rho_M | U_I\right) \to {}^{1/2}H^{1/h}\left(\mathbb{R}^n\right) ,$$

acting by the formula

$$k_I\varphi = \overline{F}^{1/h}_{p_{\overline{I}}\to x^{\overline{I}}} R_I^0 \left\{ e^{(i/h)S_I(z^I,\zeta_{\overline{I}})} \sqrt{\mu_I(z^I,\zeta_{\overline{I}})}\, \varphi(z^I,\zeta_{\overline{I}}) \right\} \tag{1.14}$$

Maslov's elementary canonical operator in the chart U_I of the manifold (M, ρ_M).

We will prove that this object is well-defined. To do this we must show that a change of the functions S_I by an element of the ideal ${}^{s+2}I\left(M, \rho_M\right)$, or a change of the functions $\mu_I(z^I,\zeta_{\overline{I}})$, $\varphi(z^I,\zeta_{\overline{I}})$ by an element of the ideal ${}^{s+1}I\left(M, \rho_M\right)$, leads to a change of expression (1.14) by a function lying in the space $\mathcal{D}\left(A^{s/2}\right)$ (see Sect. 4.1).

We introduce the notation

$$a(z^I,\zeta_{\overline{I}}) = \left[\mu_I(z^I,\zeta_{\overline{I}})\right]^{1/2} \varphi(z^I,\zeta_{\overline{I}}) .$$

To prove that k_I is well-defined it suffices to estimate the expressions

$$\left\| \frac{1}{h^{s/2}} R_I^0 \left\{ e^{(i/h)S_I(z^I,\zeta_{\overline{I}})} \tilde{a}(z^I,\zeta_{\overline{I}}) \right\} \right\|_r , \tilde{a}(z^I,\zeta_{\overline{I}}) \in {}^{s+1}I_0)\left(U_I, \rho_M|_{U_I}\right) ; \tag{1.15}$$

$$\left\| \frac{1}{h^{s/h}} R_I^0 \left\{ e^{(i/h)S_I(z^I,\zeta_{\overline{I}})} \left[1 - e^{(i/h)f(z^I,\zeta_{\overline{I}})}\right] a(z^I,\zeta_{\overline{I}}) \right\} \right\|_r ,$$

$$f(z^I,\zeta_{\overline{I}}) \in {}^{s+2}I\left(U_I, \rho_M|_{U_I}\right) \tag{1.16}$$

for any $r > 0$ and use the unitary property of the $1/h$-Fourier transformation. For the estimate we introduce a positive function $\psi(x^I, p_{\overline{I}})$, which is identically 1 on $\operatorname{supp} a(x^I, p_{\overline{I}})$ and has compact support. Bearing in mind that $R_I^0 f(z^I,\zeta_{\overline{I}}) = f(x^I, p_{\overline{I}})$ for any function f, we have

$$\begin{aligned} \left| e^{(i/h)S_I(x^I,p_{\overline{I}})} \tilde{a}(x^I,p_{\overline{I}}) \right| &\le C_1 \psi(x^I,p_{\overline{I}}) e^{-1/h \operatorname{Im} S_I(x^I,p_{\overline{I}})} \\ \times \rho_M^{s+1} &\le C_2 \psi(x^I,p_{\overline{I}}) e^{-c/h \rho_M^2} \rho_M^{s+1} \le C_3 h^{(s+1)/2} \psi(x^I,p_{\overline{I}}) . \end{aligned} \tag{1.17}$$

---

[1] For the definition of the space ${}^sH^{1/h}\left(\mathbb{R}^n\right)$ see the end of Sect. 4.1; ${}^s\mathcal{O}_0$ is the space of s-analytic functions of compact support.

The derivatives[2] of this expression are estimated in an analogous fashion. Taking the L_2-norm of expression (1.17) and the analogous expressions for the derivatives, we obtain the boundedness of expression (1.15).

The boundedness of expression (1.16) can be proved analogously, by using the estimate

$$\left|1 - e^{(i/h)f}\right| \le \frac{c_1}{h} |f| e^{|f|/h} \le \frac{c_2}{h} e^{c_3(\rho_M)^{s+2}/h} \rho_M^{s+2} ,$$

which is valid in a small enough neighborhood of the set $\Omega(M, \rho_M)$ for $s \ge 1$.

Proposition 1.1. *Let U_I and U_J be two canonical charts having nonempty intersection, and let $\varphi \in {}^s\mathcal{O}_0(U_I \cap U_J, \rho_m | U_I \cap U_J)$. Then there exists a constant $c_{IJ} \in \mathbb{Z}$ such that the congruence*

$$k_I\varphi \equiv e^{i\pi c_{IJ}} k_J \varphi \left(\operatorname{mod} \mathcal{D}\left(A^{s/2}\right)\right) . \tag{1.18}$$

holds. Moreover, the equality

$$c_{IJ} = \frac{1}{2\pi}\left[\operatorname{Arg}\mu_I - \operatorname{Arg}\mu_J - \sum_j \arg \lambda_{j,IJ} - |I_2| \cdot \pi\right]\Bigg|_{\Omega(M,\rho_M)} , \tag{1.19}$$

holds, where $\lambda_{j,IJ}$ are the eigenvalues of the matrix

$$\operatorname{Hess}_{z^{I_2},\zeta_{I_3}}(-S_I) = \frac{\partial\left(-\zeta_{I_2}, z^{I_3}\right)}{\partial\left(z^{I_2}, \zeta_{I_3}\right)} , \tag{1.20}$$

and $\arg \lambda_{j,IJ}$ *is chosen in the interval from* $-3\pi/2$ *to* $\pi/2$.

Here, as before, $I_1 = I \cap J$, $I_2 = I \setminus J$, $I_3 = J \setminus I$, $I_4 = \overline{I} \cap \overline{J} = \{1, \ldots, n\} \setminus (I \cup J)$.

Proof. We need to compare the two functions

$$k_I\varphi = \overline{F}^{1/h}_{p_{\overline{I}} \to x^{\overline{I}}} R_I^0 \left\{ e^{(i/h)S_I(z^I, \zeta_{\overline{I}})} \left[\mu_I(z^I, \zeta_{\overline{I}})\right]^{1/2} \varphi(z^I, \zeta_{\overline{I}}) \right\} ;$$

$$k_J\varphi = \overline{F}^{1/h}_{p_{\overline{J}} \to x^{\overline{J}}} R_J^0 \left\{ e^{(i/h)S_J(z^J, \zeta_{\overline{J}})} \left[\mu_J(z^J, \zeta_{\overline{J}})\right]^{1/2} \varphi(z^J, \zeta_{\overline{J}}) \right\}$$

for a function φ which has compact support in the intersection of charts $U_I \cap U_J$. In these two expressions we take the $1/h$-Fourier transform with respect to the variables $x^{\overline{J}}$; we obtain that the congruence (1.18) is equivalent to the congruence

[2] We have in mind the "quantized" derivatives $\widehat{p}_j = -ih\frac{\partial}{\partial x^j}$.

$$e^{i\pi c_{IJ}} e^{(i/h)S_J(x^J,p_{\overline{J}})} \sqrt{\mu_J(x^J,p_{\overline{J}})}\, \varphi(x^J,p_{\overline{J}})$$
$$\equiv F^{1/h}_{x^{\overline{J}}\to p_{\overline{J}}} \circ \overline{F}^{1/h}_{p_{\overline{I}}\to x^{\overline{I}}} e^{(i/h)S_I(x^I,p_{\overline{I}})} \sqrt{\mu_I(x^I,p_{\overline{I}})}\, \varphi(x^I,p_{\overline{I}}) \;\left(\bmod h^{s/2}\right) . \tag{1.21}$$

Remembering that $\overline{I} \cap \overline{J} = I_4$, $\overline{I} \setminus I_4 = I_3$, $\overline{J} \setminus I_4 = I_2$ and cancelling the mutually inverse Fourier transforms on the right-hand side of Formula (1.21), we obtain that the right-hand side is equal to

$$(-1)^{|I_2|/2} \left(\frac{i}{2\pi h}\right)^{1/2(|I_2|+|I_3|)} \int \exp\left\{\frac{i}{h}\left(\langle p_{I_3}, x^{I_3}\rangle \right.\right.$$
$$\left.\left. - \langle x^{I_2}, p_{I_2}\rangle + S_I(x^I,p_{\overline{I}})\right)\right\} \sqrt{\mu_I(x^I,p_{\overline{I}})}\, \varphi(x^I,p_{\overline{I}}) dp_{I_3} dx^{I_2} . \tag{1.22}$$

We will calculate the integral (1.22) by using the formula for the asymptotic expansion of the integral of a rapidly-oscillating function (see Chap. 5), using only the main term of the asymptotic series.

The amplitude of the integral (1.22), which is equal to $\left[\mu_I(x^I,p_{\overline{I}})\right]^{1/2} \times \varphi(x^I,p_{\overline{I}})$, is not a function of compact support in the variables x^{I_3}, p_{I_2}. However, it is simple to overcome this difficulty by multiplying the integral (1.22) by a cutoff function $\chi\left(x^{I_3},p_{I_2}\right)$, of compact support and identically equal to 1 on the image of the mapping (1.14), where $(z^I,\zeta_{\overline{I}})$ vary in the chart U_I in such a way that $(x^J,p_{\overline{J}})$ are real. The remainder, that is the integral of the form (1.22) with amplitude $\left(1-\chi\left(x^{I_3},p_{I_2}\right)\right)\left[\mu_I(x^I,p_{\overline{I}})\right]^{1/2} \times \varphi(x^I,p_{\overline{I}})$, is easy to estimate by integrating by parts (since in this integral there are no stationary points on the support of the amplitude), using the properties of the Fourier transform.

The phase of the integral (1.22) is

$$\langle p_{I_3}, x^{I_3}\rangle - \langle x^{I_2}, p_{I_2}\rangle + S_I(x^I,p_{\overline{I}}) = \Phi . \tag{1.23}$$

The stationary points of the function (1.23) are found from the formulas

$$x^{I_3} = -\frac{\partial\, {}^{s+1}S_I\left(x^{I_1}, z^{I_2}, \zeta_{I_3}, p_{I_4}\right)}{\partial \zeta_{I_3}} ,$$
$$p_{I_2} = \frac{\partial\, {}^{s+1}S_I\left(x^{I_1}, z^{I_2}, \zeta_{I_3}, p_{I_4}\right)}{\partial z^{I_2}} \tag{1.24}$$

where ${}^{s+1}S_I\left(x^{I_1}, z^{I_2}, \zeta_{I_3}, p_{I_4}\right)$ is an $(s+1)$-analytic continuation of the function $S_I(x^I,p_{\overline{I}})$ with respect to the variables x^{I_2}, p_{I_3}. Up to elements of the ideal ${}^{s+2}I\left(U_I, \rho_M | U_I\right)$ we can replace an $(s+1)$-analytic continuation of the functions $S_I(x^I,p_{\overline{I}})$ by a function $S_I(z^I,\zeta_{\overline{I}})$. But by virtue of Eq. (1.11), we have the equations

$$-\frac{\partial S_I(z^I,\zeta_{\overline{I}})}{\partial \zeta_{I_3}} = z^{I_3}(z^I,\zeta_{\overline{I}}), \frac{\partial S_I(z^I,\zeta_{\overline{I}})}{\partial z^{I_2}} = \zeta_{I_2}(z^I,\zeta_{\overline{I}}) \tag{1.25}$$

Under a change of coordinates in the intersection $U_I \cap U_J$, we have

$$\begin{aligned} -\frac{\partial S_I}{\partial \zeta_{I_3}} \left(z^{I_1}, z^{I_2}\left(z^{I_1}, z^{I_3}, \zeta_{I_2}, \zeta_{I_4}\right), \zeta_{I_3}\left(z^{I_1}, z^{I_3}, \zeta_{I_2}, \zeta_{I_4}\right), \zeta_{I_4}\right) = z^{I_3} , \\ \frac{\partial S_I}{\partial z^{I_2}} \left(z^{I_1}, z^{I_2}\left(z^{I_1}, z^{I_3}, \zeta_{I_2}, \zeta_{I_4}\right), \zeta_{I_3}\left(z^{I_1}, z^{I_3}, \zeta_{I_2}, \zeta_{I_4}\right), \zeta_{I_4}\right) = \zeta_{I_2} . \end{aligned} \tag{1.26}$$

Comparing formulas (1.24) with formulas (1.26) for real-valued $(z^{I_1}, z^{I_3}, \zeta_{I_2}, \zeta_{I_4}) = (z^J, \zeta_{\overline{J}}) = (x^J, p_{\overline{J}})$, we see that the functions

$$X^{I_2} = z^{I_2}(x^J, p_{\overline{J}}) , \quad p_{I_3} = \zeta_{I_3}(x^J, p_{\overline{J}}) .$$

are solutions of Eq. (1.24). Furthermore, the value of the phase (1.23) at the stationary point is equal to

$$\begin{aligned} &\langle \zeta_{I_3}(x^J,p_{\overline{J}}), x^{I_3}\rangle - \langle z^{I_2}(x^J,p_{\overline{J}}), p_{I_2}\rangle + S_I\left(x^{I_1}, z^{I_2}(x^J,p_{\overline{J}}), \zeta_{I_3}(x^J,p_{\overline{J}}), p_{I_4}\right) \\ &= R_J^0\left(\zeta_{\overline{I}} z^{\overline{I}} - z^{\overline{J}}\zeta_{\overline{J}} + S_I\right) = R_J^0 S_J , \end{aligned}$$

where we have used Definition (1.7) of the function S_J. Furthermore, bearing in mind that it follows from (1.25) that

$$\operatorname{Hess}_{z^{I_2},\zeta_{I_3}}\left(-{}^{s+1}\Phi\right) = \operatorname{Hess}_{z^{I_2},\zeta_{I_3}}\left(-S_I\right) = \frac{\partial\left(-\zeta_{I_2}, z^{I_3}\right)}{\partial\left(z^{I_2}, \zeta_{I_3}\right)} ,$$

we obtain, by the formula for the asymptotic expansion (Chap. 5) that the integral (1.22) is equal modulo $h^{s/2}$ to the expression

$$(-1)^{|I_2|/2} e^{(i/h)S_J(x^J,p_{\overline{J}})} \left[\mu_I(x^J,p_{\overline{J}})\right]^{1/2} \left[\det \frac{\partial\left(-\zeta_{I_2}, z^{I_3}\right)}{\partial\left(z^{I_2}, \zeta_{I_3}\right)}\right]^{-1/2} \varphi(x^J,p_{\overline{J}}) . \tag{1.27}$$

Taking into account the equation

$$\mu_I = (-1)^{|I_2|} \frac{\partial\left(-\zeta_{I_2}, z^{I_3}\right)}{\partial\left(z^{I_2}, \zeta_{I_3}\right)} \mu_J , \tag{1.28}$$

we reduce the expression (1.27) to the form

$$\begin{aligned} &e^{(i/h)S_J(x^J,p_{\overline{J}})} \left[\mu_J(x^J,p_{\overline{J}})\right]^{1/2} \varphi(x^J,p_{\overline{J}}) \\ &\times e^{(i/h)\left[\arg\mu_I - \arg\mu_J - \sum_j \lambda_{j,IJ} + |I_2|\pi\right]} . \end{aligned}$$

A comparison of the latter expression with the left-hand side of (1.21), along with the definition (1.19) of the constant c_{IJ}, proves the congruence

(1.18). To finish the proof we must show that the numbers c_{IJ} are integers. However, this follows from Eq. (1.28), since the arguments of equal complex numbers differ from one another by an integer multiple of 2π. This proves the proposition. □

Proposition 1.1 allows us to define an integral cochain $c = \{c_{IJ}\}$ of the canonical covering, where the constants c_{IJ} are defined by formula (1.19).

Proposition 1.2. *The cochain c is a cocyle.*

The proof of Proposition 1.2 is carried out completely analogously to the proof of Proposition 1.4 of Chap. 4. □

The cocycle c defines a one-dimensional cohomology class of the manifold (M, ρ_M) (more precisely, of some neighborhood of the set Ω in the manifold M). This cohomology class is studied in Sect. 3.4. There we give conditions for the triviality of this cohomology class and prove that if the class $[c]$ is zero, then it is possible to choose the arguments $\operatorname{Arg}\mu_I$ in such a way that the constants $c_{IJ} = 0$ on any intersection $U_I \cap U_J$. We note that, in contrast to the real case, the class $[c]$ depends on the measure μ.

Definition 1.2. An $(s+1)$-analytic positive Lagrangian imbedding (1.3) with measure μ is called *quantized* if $[c] = 0$.

Corollary 1.3. *If the imbedding* (1.3) *is a positive* $(s+1)$*-analytic quantized Lagrangian imbedding, then there exists a choice of arguments* $\operatorname{Arg}\mu_I$ *of the densities* μ_I *of the measure* μ *in each canonical chart* U_I *such that in each intersection* $U_I \cap U_J \neq \emptyset$ *the congruence*

$$k_I \varphi \equiv k_J \varphi \,(\operatorname{mod} h) \tag{1.29}$$

holds for any s-analytic function $\varphi \in {}^s\mathcal{O}_0\,(U_I \cap U_J)$.

Now the following theorem, which is the fundamental theorem of this section, can be proven in completely analogous fashion to Theorem 1.5 of Chap. 4.

Theorem 1.4. *Suppose* $i : (M, \rho_M) \to (\Phi_{\mathbb{C}}, \rho_\Phi)$ *is a positive quantized* $(s+1)$*-analytic Lagrangian imbedding. Then there exists a unique operator*

$$k : {}^s\mathcal{O}_0\,(M, \rho_M) \to {}^1H^{1/h}\,(\mathbb{R}^n) \ , \tag{1.30}$$

which coincides with the operator (1.14) *in each canonical chart* U_I *of the manifold* M*, that is for any function* $\varphi \in {}^s\mathcal{O}_0\left(U_I, \rho_M|_{U_I}\right)$ *the congruence*

$$k\varphi \equiv k_I \varphi \,(\operatorname{mod} h) \tag{1.31}$$

holds.

Remark 1.4. In fact there exists an operator

$$k : {}^s\mathcal{O}_0\,(M, \rho_M) \to H^{1/h}\,(\mathbb{R}^n) \tag{1.32}$$

(we will denote it by the same letter k; this will never cause any confusion) satisfying condition (1.31). However, such an operator is not unique, but defined modulo $O(h)$. For example, as the operator (1.32) one may choose an operator of the form

$$k\varphi \sum_I k_I\,(e_I \varphi)\ , \tag{1.33}$$

where $\{e_I\}$ is an s-analytic partition of unity subordinate to the covering $\{U_I\}$. Of course, formula (1.33) defines the operator (1.30) as well, if one considers k_I as an operator acting on the space ${}^1H^{1/h}\,(\mathbb{R}^n)$ (for $s \geq 2$).

We will now dwell in greater detail on the comparison between Definition 1.1 (the definition of the elementary canonical operator) and Proposition 1.1. It is immediately obvious from formula (1.14) that, in order to compute the elementary canonical operator on a chart U_I one must know the values of all the functions with appear in formula (1.14) (the action S_I, the density of the measure μ_I, the amplitude φ) only on the manifold U_I^0, which is defined by the equations $\operatorname{Im} z^I = 0$, $\operatorname{Im} \zeta_{\overline{I}} = 0$. Unfortunately, it is impossible to define the operator (1.30) using only the sets U_I^0. The reason for this is that in the intersections of charts $U_I \cap U_J$ the sets

$$U_I^0 = \left\{\operatorname{Im} z^I = 0\,, \operatorname{Im} \zeta_{\overline{I}} = 0\right\}\ ,$$
$$U_J^0 = \left\{\operatorname{Im} z^J = 0\,, \operatorname{Im} \zeta_{\overline{J}} = 0\right\}$$

do not, in general, coincide. Therefore the comparison of operators of the form (1.14) in distinct charts is carried out according to the following scheme: one takes an s-analytic extension of functions from the manifold U_I^0 to the intersection $U_I \cap U_J$ by using the operator ${}^sA_I^0$ defined by formula (3.10) of Chap. 3, and then restricts the result to the manifold U_J^0. Thus the $(s+1)$-analytic manifold (of real dimension $2n$) is needed in such a derivation only for the comparison of the local elements (1.14), while for the definition of these local elements one only needs the manifolds U_I^0 (of real dimension n). One would like to define the canonical operator on an n-dimensional (in the real sense) manifold in such a way that one would not have to extend the values of the functions to an $(s+1)$-analytic manifold in order to carry out these comparisons. This problem is solved (partially, at least) by the following assertion.

Proposition 1.5. *Suppose* $g_I \in G(U_I)$ *is a nonsingular germ in the chart* U_I *(cf. Definition* 3.11 *of Chap.* 3*). Then the operator defined by Formula* (1.14) *coincides in the sense of the space* ${}^{s/2}H^{1/h}$ *with the operator*

$$k_I^g \varphi = \overline{F}^{1/h}_{p_{\overline{I}} \to x^{\overline{I}}} \left\{ {}^{s+1}T_I^g \circ R_I^g e^{(i/h)S_I(z^I, \zeta_{\overline{I}})} \sqrt{\mu_I(z^I, \zeta_{\overline{I}})}\, \varphi(z^I, \zeta_{\overline{I}}) \right\} \tag{1.34}$$

(for a definition of the operators ${}^tT_I^g, R_I^g$ see Sect. 3.3 of Chap. 3).

The *Proof* follows immediately from the commutativity of diagram (3.70) in Chap. 3, that is from the equation

$${}^tT_I^g \circ R_I^g = R_I^0 \ .$$

To compute the operator (1.34) it is necessary to know the values of the functions entering into it on the (real) n-dimensional manifold U_I^g, which is given by the formulas

$$\begin{cases} z^I = x^I + ig^I(x^I, p_{\overline{I}}) \ , \\ \zeta_{\overline{I}} = p_{\overline{I}} + ig_{\overline{I}}(x^I, p_{\overline{I}}) \ , \end{cases} \tag{1.35}$$

where the functions $(g^I(x^I, p_{\overline{I}}), g_{\overline{I}}(x^I, p_{\overline{I}}))$ define the germ $g \in G(U_I)$. However, on transforming to other canonical coordinates the manifold (1.35) is again defined by some nonsingular germ $g' \in G(U_J)$ (cf. Proposition 3.36 of Chap. 3), and, therefore, a comparison of the expressions (1.34) in different canonical charts is possible without referring to the entire $(s+1)$-analytic manifold.

In certain cases there exists a submanifold $L \subset M$ of real dimension n which is defined by a nonsingular germ in each $(s+1)$-analytic chart U_I. In this case formulas (1.34) allow one to construct a global canonical operator analogous to Theorem 1.4 using only the manifold L. Unfortunately, the question of the existence of such a submanifold L in the general case remains open for the time being.

6.2 Commutation of the Canonical Operator and the Hamiltonian (Elementary Theory)

Suppose $H(x,p) = H_1(x,p) + iH_2(x,p)$ is a complex-valued Hamilton function. This means that the functions $H_1(x,p), H_2(x,p)$ are Hamilton functions in the sense of Definition 2.1 of Chap. 4. Let

$$\widehat{H} = H(x, \widehat{p}) = H_1(x, \widehat{p}) + iH_2(x, \widehat{p})$$

be the corresponding Hamiltonian (see Definition 2.4 of Chap. 4). We will also use the name Hamilton function for an $(s+1)$-analytic continuation of the function $H(x,p)$ to the complex phase space $\Phi_{\mathbb{C}}$:

$$H(z,\zeta) = {}^{s+1}\overset{0}{A}\, H(x,p) = \sum_{k=1}^{s+1} \frac{1}{k!}\left(iy\frac{\partial}{\partial x} + i\eta\frac{\partial}{\partial p}\right)^k H(x,p) \,. \tag{2.1}$$

(It would be more precise to denote the function (2.1) by ${}^{s+1}H(z,\zeta)$, explicitly indicating the order of the $(s+1)$-analytic continuation; however, since the number s will be considered to be fixed, the index $s+1$ of the function ${}^{s+1}H(z,\zeta)$ will be dropped, in order to avoid unwieldy notation.)

Definition 2.1. We will call the s-analytic vector field

$$V(H) = \frac{\partial H}{\partial \zeta_i}\frac{\partial}{\partial z^i} - \frac{\partial H}{\partial z^i}\frac{\partial}{\partial \zeta_i} \tag{2.2}$$

on the complex phase space $\Phi_{\mathbb{C}}$ the *Hamiltonian vector field associated with the function* $H(x,p)$.

In formula (2.2), as usual, repeated indices are to be summed; as in Chap. 4 the index of summation in formula (2.2) will often be omitted.

Now suppose

$$i : (M, \rho_M) \to (\Phi_{\mathbb{C}}, \rho_\Phi) \tag{2.3}$$

is an $(s+1)$-analytic Lagrangian imbedding.

Proposition 2.1. *Suppose the condition*

$$i^* H(z,\zeta) = 0 \in {}^{s+1}\mathcal{O}(M, \rho_M) \,. \tag{2.4}$$

holds for the imbedding (2.3). *Then the imbedding i is invariant with respect to the field* (2.2).

We recall that, according to Definition 3.9 of Chap. 3, the invariance of the imbedding (2.3) with respect to the vector field (2.2) means that there exists an s-analytic vector field $Y \in {}^sT(M,\rho_M)$ such that for any function $\varphi \in {}^{s+1}\mathcal{O}(\Phi_{\mathbb{C}}, \rho_\Phi)$ the relation

$$i^*\left(V(H)\varphi\right) = Y\left(i^*\varphi\right) \in {}^s\mathcal{O}(M, \rho_M) \,. \tag{2.5}$$

is satisfied.

Remark 2.1. A comparison of Proposition 2.1 with Proposition 2.1 of Chap. 4 shows that in the s-analytic case this proposition is weaker. From the invariance of the imbedding (2.3) with respect to the vector field (2.2) it follows, as in the real case, that $i^*(dH) = 0$; however, it is impossible to deduce from this that $H = \text{const}$ (compare Remark 1.2 in 6.1).

Proof of Proposition 2.1. The desired assertion is obviously local; hence it is sufficient to carry out the proof in some canonical chart U_I with local $(s+1)$-analytic coordinates $(z^I, \zeta_{\overline{I}})$. The equation of the submanifold[3] M in the chart U_I is

$$z^{\overline{I}} = z^{\overline{I}}(z^I, \zeta_{\overline{I}}) \ , \quad \zeta_I = \zeta_I(z^I, \zeta_{\overline{I}}) \ . \tag{2.6}$$

where by hypothesis the functions in (2.6) are $(s+1)$-analytic.

The inclusion (2.4) means that

$$i^*H = H\left(z^I, z^{\overline{I}}(z^I, \zeta_{\overline{I}}), \zeta_I(z^I, \zeta_{\overline{I}}), \zeta_{\overline{I}}\right) \in {}^{s+2}I(M, \rho_M) \ . \tag{2.7}$$

Now we compute the left-hand side of the relation (2.5):

$$\begin{aligned} i^*\left(V(H)\varphi\right) &= i^*\left(\frac{\partial H}{\partial \zeta}\frac{\partial \varphi}{\partial z} - \frac{\partial H}{\partial z}\frac{\partial \varphi}{\partial \zeta}\right) \\ &= i^*\left(\frac{\partial H}{\partial \zeta_I}\right) i^*\left(\frac{\partial \varphi}{\partial z^I}\right) + i^*\left(\frac{\partial H}{\partial \zeta_{\overline{I}}}\right) i^*\left(\frac{\partial \varphi}{\partial z^{\overline{I}}}\right) \\ &\quad - i^*\left(\frac{\partial H}{\partial z^I}\right) i^*\left(\frac{\partial \varphi}{\partial \zeta_I}\right) - i^*\left(\frac{\partial H}{\partial z^{\overline{I}}}\right) i^*\left(\frac{\partial \varphi}{\partial \zeta_{\overline{I}}}\right) \ . \end{aligned} \tag{2.8}$$

Differentiating the relation (2.7) with respect to the variables $(z^I, \zeta_{\overline{I}})$, we obtain

$$\begin{aligned} i^*\left(\frac{\partial H}{\partial z^I}\right) &= -\left[i^*\left(\frac{\partial H}{\partial z^{\overline{I}}}\right)\frac{\partial z^{\overline{I}}(z^I, \zeta_{\overline{I}})}{\partial z^I} + i^*\left(\frac{\partial H}{\partial \zeta_I}\right)\frac{\partial \zeta_I(z^I, \zeta_{\overline{I}})}{\partial z^I}\right] \ , \\ i^*\left(\frac{\partial H}{\partial \zeta_{\overline{I}}}\right) &= -\left[i^*\left(\frac{\partial H}{\partial z^{\overline{I}}}\right)\frac{\partial z^{\overline{I}}(z^I, \zeta_{\overline{I}})}{\partial \zeta_{\overline{I}}} + i^*\left(\frac{\partial H}{\partial \zeta_I}\right)\frac{\partial \zeta_I(z^I, \zeta_{\overline{I}})}{\partial \zeta_{\overline{I}}}\right] \ , \end{aligned} \tag{2.9}$$

in the ring ${}^s\mathcal{O}(M, \rho_M)$. Furthermore the condition (1.4) that the imbedding is Lagrangian gives the equation

$$\frac{\partial \zeta_I(z^I, \zeta_{\overline{I}})}{\partial \zeta_{\overline{I}}} = -\frac{\partial z^{\overline{I}}(z^I, \zeta_{\overline{I}})}{\partial z^I} \ . \tag{2.10}$$

Substituting (2.9) into (2.8), taking into account (2.10), we obtain

[3] Allowing a certain looseness of terminology, we will speak of M as a submanifold of $\Phi_{\mathbb{C}}$.

$$i^*\left(V(H)\varphi\right) = i^*\left(\frac{\partial H}{\partial \zeta_I}\right)\left[i^*\left(\frac{\partial \varphi}{\partial z^I}\right) + \frac{\partial z^{\overline{I}}(z^I,\zeta_{\overline{I}})}{\partial z^I}\, i^*\left(\frac{\partial \varphi}{\partial z^{\overline{I}}}\right)\right.$$
$$\left. + \frac{\partial \zeta_I(z^I,\zeta_{\overline{I}})}{\partial z^I}\, i^*\left(\frac{\partial \varphi}{\partial \zeta_I}\right)\right] - i^*\left(\frac{\partial \varphi}{\partial z^{\overline{I}}}\right)\left[i^*\left(\frac{\partial \varphi}{\partial \zeta_{\overline{I}}}\right) + \frac{\partial z^{\overline{I}}(z^I,\zeta_{\overline{I}})}{\partial \zeta_{\overline{I}}}\right. \tag{2.11}$$
$$\left. \times\, i^*\left(\frac{\partial \varphi}{\partial z^{\overline{I}}}\right) + \frac{\partial \zeta_I(z^I,\zeta_{\overline{I}})}{\partial \zeta_{\overline{I}}}\, i^*\left(\frac{\partial \varphi}{\partial \zeta_I}\right)\right] .$$

in the ring ${}^s\mathcal{O}\,(M, \rho_M)$.

However, the expressions in square brackets on the right-hand side of formula (2.11) are, by the chain rule, equal to

$$\frac{\partial}{\partial z^I}\,(i^*\varphi)\ ,\quad \frac{\partial}{\partial \zeta_{\overline{I}}}\,(i^*\varphi)\ ,$$

respectively. Therefore formula (2.11) can be rewritten in the form

$$i^*\left(V(H)\varphi\right) = \left[i^*\left(\frac{\partial H}{\partial \zeta_I}\right)\frac{\partial}{\partial z^I} - i^*\left(\frac{\partial H}{\partial z^{\overline{I}}}\right)\frac{\partial}{\partial \zeta_{\overline{I}}}\right](i^*\varphi) \in {}^s\mathcal{O}\,(M, \rho_M)\ . \tag{2.12}$$

Substituting

$$Y = i^*\left(\frac{\partial H}{\partial \zeta_I}\right)\frac{\partial}{\partial z^I} - i^*\left(\frac{\partial H}{\partial z^{\overline{I}}}\right)\frac{\partial}{\partial \zeta_{\overline{I}}}\ . \tag{2.13}$$

we see that formula (2.12) coincides with formula (2.5). This completes the proof of the proposition. □

Remark 2.2. We will denote the field Y defined by formula (2.13) by $V\,(H_I)$, and sometimes simply by $V(H)$.

We will say that a measure $\mu \in {}^s\Lambda^n\,(M, \rho_M)$ is invariant with respect to the vector field $V(H)$ if

$$\mathcal{L}_{V(H)}\mu = d(V(H)\lrcorner\mu) = 0 \in {}^{s-1}\Lambda^n\,(M, \rho_M)\ . \tag{2.14}$$

Definition 2.2. The pair (M, μ) are said to be *associated* with the Hamilton function H if conditions (2.4), (2.14) hold.

We now formulate the main theorem of this section.

Theorem 2.2. *Suppose the pair (M, μ) is associated with the Hamilton function H and suppose k is the operator defined by formulas* (1.33) *and* (1.14). *If the vector field $V(H)$ is nonvanishing, then for any function $\varphi \in {}^{s+1}\mathcal{O}_0\,(M, \rho_M)$ the congruence*

$$H\left(x,\widehat{p}\right)k\varphi \equiv -ihk\left(V(H)-\frac{1}{2}\frac{\partial^2 H}{\partial z^i\partial\zeta_i}\right)\varphi\,(\operatorname{mod} h^2) \tag{2.15}$$

holds.

Remark 2.3. We wish to emphasize the fact that congruence (2.15) holds not for any operator k defined as in Theorem 1.4, but for the particular operator defined by formula (1.33) for some choice of a partition of unity.

Proposition 2.3. *For any function* $a(z^I,\zeta_{\overline{I}})\in {}^s\mathcal{O}(U_I,\rho_M|U_I)$ *the congruence*

$$\begin{aligned} &H\left(x^I,\widehat{x}^{\overline{I}},\widehat{p}_I,p_{\overline{I}}\right)R_I^0\left\{e^{(i/h)S_I(z^I,\zeta_{\overline{I}})}a(z^I,\zeta_{\overline{I}})\right\}\\ &\equiv R_I^0 e^{(i/h)S_I(z^I,\zeta_{\overline{I}})}\left\{H\left(z^I,-\frac{\partial S_I}{\partial\zeta_{\overline{I}}},\frac{\partial S_I}{\partial z^I},\zeta_{\overline{I}}\right)-ih\left[\frac{\partial H}{\partial\zeta_1}\frac{\partial}{\partial z^I}\right.\right.\\ &\quad-\frac{\partial H}{\partial z^{\overline{I}}}\frac{\partial}{\partial\zeta_{\overline{I}}}+\frac{1}{2}\left(\frac{\partial^2 S_I}{\partial z^I\partial z^I}\frac{\partial^2 H}{\partial\zeta_I\partial\zeta_I}+\frac{\partial^2 S_I}{\partial\zeta_{\overline{I}}\partial\zeta_{\overline{I}}}\frac{\partial^2 H}{\partial z^{\overline{I}}\partial z^{\overline{I}}}\right.\\ &\quad\left.\left.\left.-2\frac{\partial^2 S_I}{\partial z^I\partial\zeta_{\overline{I}}}\frac{\partial^2 H}{\partial z^{\overline{I}}\partial\zeta_I}\right)-\frac{\partial^2 H}{\partial z^{\overline{I}}\partial\zeta_{\overline{I}}}\right]\right\}a(z^I,\zeta_{\overline{I}})\ (\operatorname{mod} h^2)\ , \end{aligned} \tag{2.16}$$

holds, where the derivatives of the function H are evaluated at the point $\left(z^I,-\frac{\partial S_I}{\partial\zeta_{\overline{I}}},\frac{\partial S_I}{\partial z^I},\zeta_{\overline{I}}\right)$ (for the definition of the operator $H\left(x^I,\widehat{x}^{\overline{I}},\widehat{p}_I,p_{\overline{I}}\right)$ see Sect. 4.2 of Chap. 4). As in the real case, we will give some definitions.

Definition 2.3. The operator of multiplication by the function

$$H\left(z^I,-\frac{\partial S_I}{\partial\zeta_{\overline{I}}},\frac{\partial S_I}{\partial z^I},\zeta_{\overline{I}}\right)$$

is called the *local Hamilton-Jacobi operator* in the chart U_I.

Definition 2.4. The first-order differential operator

$$\begin{aligned} \mathcal{P}_I=\mu_I^{-1/2}&\left\{\frac{\partial H}{\partial\zeta_I}\frac{\partial}{\partial z^I}-\frac{\partial H}{\partial z^{\overline{I}}}\frac{\partial}{\partial\zeta_{\overline{I}}}+\frac{1}{2}\left(\frac{\partial^2 S_I}{\partial z^I\partial z^I}\frac{\partial^2 H}{\partial\zeta_I\partial\zeta_I}\right.\right.\\ &\left.\left.+\frac{\partial^2 S_I}{\partial\zeta_{\overline{I}}\partial\zeta_{\overline{I}}}\frac{\partial^2 H}{\partial z^{\overline{I}}\partial z^{\overline{I}}}-2\frac{\partial^2 S_I}{\partial z^I\partial\zeta_{\overline{I}}}\frac{\partial^2 H}{\partial\zeta_I\partial z^{\overline{I}}}\right)-\frac{\partial^2 H}{\partial z^{\overline{I}}\partial\zeta_{\overline{I}}}\right\}\mu_I^{1/2} \end{aligned}$$

is called the *local transport operator* on the chart U_I.

From Proposition 2.3 and Definitions 2.3 and 2.4 we obtain the following assertion.

Corollary 2.4. *The following commutation formula holds for the canonical operator* (1.14):

$$H(x,\widehat{p})\,k_I\varphi \equiv k_I\left\{\left[H\left(z^I,-\frac{\partial S_I}{\partial \zeta_{\overline{I}}},\frac{\partial S_I}{\partial z^I},\zeta_{\overline{I}}\right)-ih\mathcal{P}_I\right]\varphi\right\}\ \left(\operatorname{mod} h^2\right)\ .$$

The proof of the following two propositions can be translated with obvious changes from the real case (cf. Propositions 2.5 and 2.6 of Sect. 4.2 of Chap. 4).

Proposition 2.5. *The local Hamilton-Jacobi operator is equal to zero.*

Proposition 2.6. *The local transport operator $\mathcal{P}_I$ has the form*

$$\mathcal{P}_I\varphi=\left[V(H)-\frac{1}{2}\frac{\partial^2 H}{\partial z\partial\zeta}\right]\varphi\ . \tag{2.17}$$

Corollary 2.4, together with Proposition 2.5 and 2.6, gives the formula

$$H(x,\widehat{p})\,k_I\varphi \equiv -ihk_I\left[V(H)-\frac{1}{2}\frac{\partial^2 H}{\partial z\partial\zeta}\right]\varphi\ \left(\operatorname{mod} h^2\right)\ . \tag{2.18}$$

We note that by formula (2.17) the local transport operator does not in fact depend on the chart U_I.

The proof of Theorem 2.2 now proceeds just as in the real case. □

6.3 Commutation of Maslov's Canonical Operator and the Hamiltonian (General Theory)

Maslov's canonical operator k, described for the real case in Chap. 4 and for the complex case in the first two sections of this chapter, is designed for the calculation of the first term of the asymptotic expansion in terms of h of the solutions of homogeneous $1/h$-pseudodifferential equations of the form

$$H(x,\widehat{p})\,u=0\ . \tag{3.1}$$

In order not to be concerned with questions of estimating the inverse operator to $H(x,\widehat{p})$, we will talk about the asymptotics "on the right-hand side", i.e. about finding a function $\widetilde{u}(x,h)$ which, upon substitution into Eq. (3.1) gives a sufficiently small error:

$$H(x,\widehat{p})\,\widetilde{u}(x,h)=O\left(h^N\right)\ . \tag{3.2}$$

Up until now we have restricted our attention to the case $N = 2$. In this case the function $\widetilde{u}$ is given up to terms of order $O(h)$ by the expression $\widetilde{u}(x, h) = k\varphi$, where φ satisfies the transport equation

$$\left[V(H) - \frac{1}{2}\frac{\partial^2 H}{\partial z \partial \zeta}\right]\varphi = 0\,.$$

Strictly speaking (see Remark 2.3), the function $\widetilde{u}(x, h) = k\varphi$, where k is the operator whose existence is asserted by Theorem 1.4, does not, in general, satisfy Eq. (3.2) even when $N = 2$. However, if we understand k to be the operator defined by formula (1.33), Eq. (3.2) will already be satisfied for any choice of the partition of unity entering into formula (1.33). By Theorem 1.4 itself, the functions $\widetilde{u}(x, h) = k\varphi$ defined under different partitions of unity differ from one another by a quantity of order $O(h)$.

In this section we will undertake the solution of Eq. (3.2) for an arbitrary N. Here the amplitude φ appearing under the sign of the canonical operator will already depend regularly on the parameter h:

$$\varphi = \varphi_0 + h\varphi_1 + \ldots + h^m \varphi_m\ . \tag{3.3}$$

However, it does not seem possible to generalize the operator k described in Theorem 1.4 to a function of the form (3.3), since the operator k is defined by Theorem 1.4 only up to an accuracy $O(h)$. It is possible to remedy this situation in various ways. First, one can fix a concrete operator k (for example, the one given by formula (1.33) for some fixed choice of a partition of unity) – this is the approach we will take in this section. Secondly, one could consider the operator k not to be acting upon functions, but on certain objects of a more complicated nature; under this assumption the elementary canonical operators (1.14) agree up to terms of order $O\left(h^N\right)$ in distinct charts. Thirdly, one may alter the definition (1.14) of the operators k_I in such a way that in different charts these operators coincide on functions with an accuracy up to $O\left(h^N\right)$; the latter two methods will be analyzed in 6.4.

Let us proceed to the first method. We denote by ${}^t\mathcal{O}'[h]\,(M, \rho_M)$ the space of polynomials of the form

$$f_0 + hf_1 + \ldots + h^m f_m\ , \tag{3.4}$$

where

$$f_j \in {}^{t-2j}\mathcal{O}'[h]\,(M, \rho_M)$$

(the latter inclusion is not required if $t - 2j < 0$). The space of functions of the form (3.4) form a *ring*, since

$$\left(f_0 + hf_1 + h^2 f_2 + \ldots + h^m f_m\right)\left(g_0 + hg_1 + h^2 g_2 + \ldots + h^m g_m\right)$$
$$= \sum_k h^k \sum_{i+j=k} f_i g_j$$

(we consider f_j, g_j to be zero for $j > m$), and the sum of $f_i g_j$ lies in the ring ${}^{t-2k}\mathcal{O}(M, \rho_M)$ since $i \le k$, $j \le k$ and

$${}^{t_1}\mathcal{O}(M, \rho_M) \subset {}^{t_2}\mathcal{O}(M, \rho_M) , \qquad t_2 \le t_1 .$$

We denote the ideal in the ring ${}^t\mathcal{O}'[h](M, \rho_M)$ consisting of functions of the form (3.4) for which the inclusion

$$f_j \in {}^{t+1-2j}I(M, \rho_M) , \qquad t+1-2j > 0$$

holds, by ${}^{t+1}I[h](M, \rho_M)$. We denote the factor ring of the ring ${}^t\mathcal{O}'[h](M, \rho_M)$ modulo the ideal ${}^{t+1}I[h](M, \rho_M)$ by ${}^t\mathcal{O}[h](M, \rho_M)$:

$${}^t\mathcal{O}[h](M, \rho_M) = \frac{{}^t\mathcal{O}'[h](M, \rho_M)}{{}^{t+1}I[h](M, \rho_M)} . \tag{3.5}$$

Now we extend the operator (1.33) by linearity to an operator[4]

$$k : {}^s\mathcal{O}_0[h](M, \rho_M) \to {}^{s/2}H^{1/h}(\mathbb{R}^n) . \tag{3.6}$$

Suppose (M, μ) is a positive quantized $(s+1)$-analytic Lagrangian manifold. We will assume that the arguments $\operatorname{Arg}_{\mu_I}$ of the density of the measure μ in the canonical coordinates are chosen in such a way that the constants c_{IJ} defined by formula (1.19) are equal to zero in any nonempty intersection $U_I \cap U_J$.

Theorem 3.1. *Suppose the pair (M, μ) is associated with the Hamilton function H. Then there exists a differential operator*

$$\mathcal{P} = \mathcal{P}_0 + ih\mathcal{P}_1 + (ih)^2\mathcal{P}_2 + \ldots , \tag{3.7}$$

such that the congruence

$$H(x, \widehat{p})\, k\varphi \equiv -ihk\mathcal{P}\varphi \left(\operatorname{mod} h^{s/2}\right) . \tag{3.8}$$

holds. Here

$$\mathcal{P}_0 = V(H) - \frac{1}{2} \left.\frac{\partial^2 H}{\partial z \partial \zeta}\right|_M . \tag{3.9}$$

To prove Theorem 3.1 we will need two auxiliary assertions.

[4] The subscript "0" on a space denotes, as before, the compactness of the support.

Proposition 3.2. *In any intersection $U_I \cap U_J$ there exist differential operators V_{IJ}^r of order $2r$,*

$$V_{IJ} = V_{IJ}^0 + ihV_{IJ}^1 + (ih)^2 V_{IJ}^2 + \dots , \tag{3.10}$$

$$V_{IJ} : {}^s\mathcal{O}[h]\,(U_I \cap U_J, \rho_M\,|U_I \cap U_J) \to {}^s\mathcal{O}[h]\,(U_I \cap U_J, \rho_M|\,U_I \cap U_J)$$

such that for any function $\varphi \in {}^s\mathcal{O}_0[h]\,(U_I \cap U_J,\, \rho_M|\,U_I \cap U_J)$ the congruence

$$k_I\varphi \equiv k_J V_{IJ}\varphi \, \left(\operatorname{mod} h^{s/2}\right) . \tag{3.11}$$

holds. Here V_{IJ}^0 is the identity operator.

Proof. Proceeding as in the proof of Proposition 1.1, we obtain that the congruence (3.11) is equivalent to the congruence

$$\begin{aligned} \mathcal{I} = (-1)^{|I_2|/2} \left(\frac{i}{2\pi h}\right)^{1/2(|I_2|+|I_3|)} & \int \exp\left\{\frac{i}{h}\left(\langle p_{I_3}, x^{I_3}\rangle\right.\right. \\ & - \left.\left. \langle x^{I_2}, p_{I_2}\rangle + S_I(x^I, p_{\overline{I}})\right)\right\} \sqrt{\mu_I(x^I, p_{\overline{I}})}\, \varphi(x^I, p_{\overline{I}}) dp_{I_3} dx^{I_2} \\ & \equiv e^{(i/h)S_J(x^J, p_{\overline{J}})} \sqrt{\mu_J(x^J, p_{\overline{J}})}\, V_{IJ}\varphi(x^J, p_{\overline{J}}) \, \left(\operatorname{mod} h^{s/2}\right) . \end{aligned} \tag{3.12}$$

We compute the integral $\mathcal{I}$ on the left-hand side of formula (3.12) by using the formula for the asymptotic expansion (Chap. 5). We have (see the remark after formula (1.22))

$$\begin{aligned} \mathcal{I} = (-1)^{|I_2|/2} \exp & \left[\frac{i}{h}\left(\langle \zeta_{I_3}(x^J, p_{\overline{J}}), x^{I_3}\rangle - \langle z^{I_2}(x^J, p_{\overline{J}}), p_{I_2}\rangle\right.\right. \\ & + \left.\left. S_I\left(x^{I_1}, z^{I_2}(x^J, p_{\overline{J}}), \zeta_{I_3}(x^J, p_{\overline{J}}), p_{I_4}\right)\right)\right] \times \Big[\det \operatorname{Hess}_{z^{I_2}, \zeta_{I_3}} \\ & \times \left(-S_I\left(x^{I_1}, z^{I_2}(x^J, p_{\overline{J}}), \zeta_{I_3}(x^J, p_{\overline{J}}), p_{I_4}\right)\right)\Big]^{-1/2} \sum_{0 \le r < s/2} h^r \widehat{W}_r \\ & \times \left(\sqrt{\mu_I} \cdot \varphi\right)\left(x^{I_1}, z^{I_2}(x^J, p_{\overline{J}}), \zeta_{I_3}(x^J, p_{\overline{J}}), p_{I_4}\right) \, \left(\operatorname{mod} h^{s/2}\right) . \end{aligned} \tag{3.13}$$

Here $z^{I_2}(x^J, p_{\overline{J}}), \zeta_{I_3}(x^J, p_{\overline{J}})$ is the stationary point of the phase integral $\mathcal{I}, \widehat{W}_r$ are differential operators of order $2r$ with coefficients in ${}^{s-2r}\mathcal{O}\,(M, \rho_M)$, with $\widehat{W}_0 = I$, $\widehat{W}_r(a) = W_r[\Phi, a]$, where $W_r[\Phi, a]$ are the operators defined in the formulation of Theorem 1.1 of Chap. 5, and $\Phi = \langle p_{I_3}, x^{I_3}\rangle - \langle x^{I_2}, p_{I_2}\rangle + S_I(x^I, p_{\overline{I}})$ is the phase of the integral appearing in (3.12). In Proposition 1.1 the first term of the expansion (3.13) was computed. It was shown there that the equations for the stationary point of the phase of the integral $\mathcal{I}$

coincide with the formulas for transforming from the coordinates $(z^J, \zeta_{\overline{J}})$ to coordinates $(z^I, \zeta_{\overline{I}})$, and the equations

$$\langle \zeta_{I_3}(x^J, p_{\overline{J}}) x^{I_3}\rangle - \langle z^{I_2}(x^J, p_{\overline{J}}), p_{I_2}\rangle + S_I\left(x^{I_1}, z^{I_2}(x^J, p_{\overline{J}}), \zeta_{I_3}(x^J, p_{\overline{J}}), p_{I_4}\right) \\ = S_J(x^J, p_{\overline{J}}) \; ; \tag{3.14}$$

$$\frac{\sqrt{\mu_I\left(x^{I_1}, z^{I_2}(x^J, p_{\overline{J}}), \zeta_{I_3}(x^J, p_{\overline{J}}), p_{I_4}\right)}}{\sqrt{\det \operatorname{Hess}_{z^{I_2}, \zeta_{I_3}}\left(-S_I\left(x^{I_1}, z^{I_2}(x^J, p_{\overline{J}}), \zeta_{I_3}(x^J, p_{\overline{J}}), p_{I_4}\right)\right)}} \\ \times (-1)^{|I_2|/2} = \sqrt{\mu_J(x^J, p_{\overline{J}})} \; . \tag{3.15}$$

hold as well. Multiplying and dividing the sum on the right-hand side of relation (3.13) by $\left[\mu_I\left(x^{I_1}, z^{I_2}(x^J, p_{\overline{J}}), \zeta_{I_3}(x^J, p_{\overline{J}}), p_{I_4}\right)\right]^{1/2}$ and taking into account relations (3.14) and (3.15), we obtain

$$\mathcal{I} \equiv e^{(i/h)S_J(x^J, p_{\overline{J}})} \sqrt{\mu_J(x^J, p_{\overline{J}})} \\ \times \left\{ \mu_I^{-1/2} \sum_{0 \le r < s/2} h^r \widehat{W}_r \mu_I^{1/2} \right\} \varphi(x^I, p_{\overline{I}}) \pmod{h^{s/2}} \; . \tag{3.16}$$

Comparing formulas (3.16) and (3.12) we see that formula (3.12) is valid if we set the operators V_{IJ} equal to

$$V_{IJ} = \mu_I^{-1/2}(x^J, p_{\overline{J}}) \sum_{0 \le r < s/2} h^r \widehat{W}_r \mu_I^{1/2}(x^J, p_{\overline{J}}) = \sum_{0 \le r < s/2} (ih)^r V_{IJ}^r \; ,$$

where

$$V_{IJ}^r = (-1)^r \mu_I^{-1/2}(x^J, p_{\overline{J}}) \widehat{W}_r \mu_I^{1/2}(x^J, p_{\overline{J}}) \; .$$

Since $\widehat{W}_0 = 1$, V_{IJ}^0 is the identity operator. This completes the proof of the proposition. □

Proposition 3.3. *For any function* $a(z^I, \zeta_{\overline{I}}) \in {}^s\mathcal{O}_0\left(U_I, \rho_M | U_I\right)$ *the congruence*

$$H\left(x^I, \widehat{x}^{\overline{I}}, \widehat{p}_I, p_{\overline{I}}\right) R_I^0 \left\{ e^{(i/h)S_I(z^I, \zeta_{\overline{I}})} a(z^I, \zeta_{\overline{I}}) \right\} \\ \equiv R_I^0 e^{(i/h)S_I(z^I, \zeta_{\overline{I}})} P_I(z^I, \zeta_{\overline{I}}) \pmod{h^{s/2}} \; . \tag{3.17}$$

holds. Here

$$P_I = \sum_{0 \le r < s/2} (ih)^r P_I^r \; , \tag{3.18}$$

P_I^r *are differential operators of order* r, *where*

$$P_I^0 = H\left(z^I, -\frac{\partial S_I}{\partial \zeta_{\overline{I}}} \frac{\partial S_I}{\partial z^I} \zeta_{\overline{I}}\right) . \tag{3.19}$$

$$\begin{aligned} P_I^1 = & -\left[\frac{\partial H}{\partial \zeta_I}\frac{\partial}{\partial z^I} - \frac{\partial H}{\partial z^{\overline{I}}}\frac{\partial}{\partial \zeta_{\overline{I}}} + \frac{1}{2}\left(\frac{\partial^2 S_I}{\partial z^I \partial z^I}\frac{\partial^2 H}{\partial \zeta_I \partial \zeta_I}\right.\right. \\ & \left.\left. + \frac{\partial^2 S_I}{\partial \zeta_{\overline{I}} \partial \zeta_{\overline{I}}}\frac{\partial^2 H}{\partial z^{\overline{I}} \partial z^{\overline{I}}} - 2\frac{\partial^2 S_I}{\partial z^I \partial \zeta_{\overline{I}}}\frac{\partial^2 H}{\partial \zeta_I \partial z^{\overline{I}}}\right) - \frac{\partial^2 H}{\partial z^{\overline{I}} \partial \zeta_{\overline{I}}}\right] . \end{aligned} \tag{3.20}$$

and the derivatives of the function H appearing on the right-hand side of formula (3.20) *are evaluated at*

$$(z^I, -\partial S_I/\partial \zeta_{\overline{I}}, \partial S_I/\partial z^I, \zeta_{\overline{I}}) .$$

Proof. We compute the left-hand side of congruence (3.17):

$$\begin{aligned} & H\left(x^I, \widehat{x}^{\overline{I}}, \widehat{p}_I, p_{\overline{I}}\right) e^{(i/h)S_I(x^I, p_{\overline{I}})} a(x^I, p_{\overline{I}}) = F^{1/h}_{x^{\overline{I}} \to p_{\overline{I}}} \circ \overline{F}^{1/h}_{p_I \to x^I} \\ & \times \left\{ \overline{F}^{1/h}_{\widetilde{p}_{\overline{I}} \to x^{\overline{I}}} H\left(x^I, x^{\overline{I}}, p_I, \widetilde{p}_{\overline{I}}\right) F^{1/h}_{\widetilde{x}^I \to p_I} e^{(i/h)S_I(\widetilde{x}^I, \widetilde{p}_{\overline{I}})} a\left(\widetilde{x}^I, \widetilde{p}_{\overline{I}}\right) \right\} = I(h) . \end{aligned}$$

Writing out the latter expression in integral form, we obtain

$$\begin{aligned} I(h) = & \left(\frac{1}{2\pi h}\right)^n \int_{\mathbb{R}^{\overline{I}} \times \mathbb{R}_I} e^{(i/h)\left\{\langle x^I, p_I\rangle - \langle x^{\overline{I}}, p_{\overline{I}}\rangle\right\}} dx^{\overline{I}} dp_I \int_{\mathbb{R}_{\overline{I}}} e^{(i/h)\langle x^{\overline{I}}, \widetilde{p}_{\overline{I}}\rangle} \\ & \times H\left(x^I, x^{\overline{I}}, p_I, \widetilde{p}_{\overline{I}}\right) d\widetilde{p}_{\overline{I}} \int_{\mathbb{R}^I} e^{(i/h)\left\{-\langle \widetilde{x}^I, p_I\rangle + S_I\left(\widetilde{x}^{\overline{I}}, \widetilde{p}_{\overline{I}}\right)\right\}} a\left(\widetilde{x}^I, \widetilde{p}_{\overline{I}}\right) d\widetilde{x}^I . \end{aligned} \tag{3.21}$$

The integral (3.21) should be understood as an iterated integral whose corresponding multiple integral diverges. In order to consider (3.21) as a multiple integral, we introduce an open covering of the space $\mathbb{R}^{\overline{I}} \times \mathbb{R}_I$ with coordinates $\left(x^{\overline{I}}, p_I\right)$ by two neighborhoods U_1 and U_2, where

$$U_1 = \left\{ \left(x^{\overline{I}}, p_I\right) \middle| \left[\left|x^{\overline{I}}\right|^2 + |p_I|^2\right]^{1/2} < M + 1/2 \right\} ,$$

$$U_2 = \left\{ \left(x^{\overline{I}}, p_I\right) \middle| \left[\left|x^{\overline{I}}\right|^2 + |p_I|^2\right]^{1/2} > M \right\}$$

and $M = \max\left\{\left|\operatorname{grad}_{\widetilde{x}^I, \widetilde{p}_{\overline{I}}} S_I\left(\widetilde{x}^I, \widetilde{p}_{\overline{I}}\right)\right|\right\} + 1$ (the maximum is taken over the support of the function $a\left(\widetilde{x}^I, \widetilde{p}_{\overline{I}}\right)$. Let

$$e_1 + e_2 = 1$$

be a partition of unity subordinate to the covering $\{U_1, U_2\}$. Writing

$$H_1 = H\left(x^I, x^{\overline{I}}, p_I, \widetilde{p_{\overline{I}}}\right) e_1\left(x^{\overline{I}}, p_I\right) ,$$
$$H_2 = H\left(x^I, x^{\overline{I}}, p_I, \widetilde{p_{\overline{I}}}\right) e_2\left(x^{\overline{I}}, p_I\right) ,$$

we obtain that the integral $I(h)$ defined by formula (3.21) can be divided into a sum of two integrals $I_1(h)$ and $I_2(h)$. Moreover, there are no stationary points of the phase on the support of the amplitude in the integral $I_2(h)$; therefore by integrating by parts we obtain an estimate for the integral $I_2(h)$:

$$I_2(h) \equiv 0 \pmod{h^\infty} . \tag{3.22}$$

Analogously to the remark after formula (1.22), we can consider the amplitude in the integral $I_1(h)$ to be bounded in all variables. Now we have

$$I_1(h) = \left(\frac{1}{2\pi h}\right)^n \int \exp\left\{\frac{i}{h}\left[\langle x^{\overline{I}}, \widetilde{p_{\overline{I}}} - p_{\overline{I}}\rangle + \langle p_I, x^I - \widetilde{x}^I\rangle + S_I\left(\widetilde{x}^I, \widetilde{p_{\overline{I}}}\right)\right]\right\}$$
$$\times H_1\left(x^I, x^{\overline{I}}, p_I, \widetilde{p_{\overline{I}}}\right) a\left(\widetilde{x}^I, \widetilde{p_{\overline{I}}}\right) dx^{\overline{I}} dp_I d\widetilde{x}^I d\widetilde{p_{\overline{I}}} . \tag{3.23}$$

We apply to the integral (3.23) the formula for the asymptotic decomposition of the integral (Chap. 5). The equations for the stationary point have the form

$$\begin{aligned}
\frac{\partial \Phi}{\partial \widetilde{z}^I} &= -\zeta_I + \frac{\partial S_I\left(\widetilde{z}^I, \widetilde{\zeta_{\overline{I}}}\right)}{\partial \widetilde{z}^I} = 0 , \\
\frac{\partial \Phi}{\partial \widetilde{\zeta_{\overline{I}}}} &= z^{\overline{I}} + \frac{\partial S_I\left(\widetilde{z}^I, \widetilde{\zeta_{\overline{I}}}\right)}{\partial \widetilde{\zeta_{\overline{I}}}} = 0 , \\
\frac{\partial \Phi}{\partial z^{\overline{I}}} &= \widetilde{\zeta_{\overline{I}}} - p_{\overline{I}} = 0 , \\
\frac{\partial \Phi}{\partial \zeta_I} &= x^I - \widetilde{z}^I = 0 .
\end{aligned} \tag{3.24}$$

where by Φ and S_I we denote the $(s+1)$-analytic continuations of the corresponding functions.

The system of Eq. (3.24) has, obviously, a unique solution with respect to $z^{\overline{I}}, \zeta_I, \widetilde{z}^I, \widetilde{\zeta_{\overline{I}}}$, namely

$$\widetilde{z}^I = x^I , \quad \widetilde{\zeta_{\overline{I}}} = p_{\overline{I}} ,$$

$$\zeta_I = \frac{\partial S_I(x^I, p_{\overline{I}})}{\partial x^I} , \quad z^{\overline{I}} = \frac{\partial S_I(x^I, p_{\overline{I}})}{\partial p_{\overline{I}}} . \tag{3.25}$$

We now demonstrate that the stationary point (3.25) is non-degenerate. To do this we compute the determinant of the Hessian of the phase of the integral (3.23) at the stationary point (3.25). The Hessian matrix is equal to

$$\begin{bmatrix} \frac{\partial^2 S_I(x^I,p_{\overline{I}})}{\partial x^I \partial x^I} & \frac{\partial^2 S_I(x^I,p_{\overline{I}})}{\partial x^I \partial p_{\overline{I}}} & 0 & -1_I \\ \frac{\partial^2 S_I(x^I,p_{\overline{I}})}{\partial p_{\overline{I}} \partial x^I} & \frac{\partial^2 S_I(x^I,p_{\overline{I}})}{\partial p_{\overline{I}} \partial p_{\overline{I}}} & 1_{\overline{I}} & 0 \\ 0 & 1_{\overline{I}} & 0 & 0 \\ -1_I & 0 & 0 & 0 \end{bmatrix} \tag{3.26}$$

In formula (3.26) the symbols 1_I and $1_{\overline{I}}$ denote the identity matrices of dimension $|I|$ and $|\overline{I}|$ respectively. It is not hard to see that the determinant of the matrix (3.26) is equal to one. Finally, the imaginary part of the phase of the integral (3.23) equals the imaginary part of the function $S_I(x^I, p_{\overline{I}})$ and is therefore nonnegative. We have verified all the hypotheses of Theorem 1.1 of Chap. 5. Applying this theorem and using congruence (3.22), we have

$$I(h) = H\left(x^I, \hat{x}^{\overline{I}}, \hat{p}_I, p_{\overline{I}}\right) R_I^0 \left\{e^{(i/h)S_I(z^I,\zeta_{\overline{I}})} a(z^I, \zeta_{\overline{I}})\right\} \equiv e^{(i/h)S_I(x^I,p_{\overline{I}})} \tag{3.27}$$

$$\times \sum_{0\le j<s/2} h^j W_j \left[S_I(x^I,p_{\overline{I}}), H\left(x^I, x^{\overline{I}}, p_I, p_{\overline{I}}\right) a(x^I,p_{\overline{I}})\right] \left(\bmod h^{s/2}\right) .$$

Here we have taken into account that (where Φ is the phase of the integral (3.23)):

1) the determinant of the Hessian of the function $(-\Phi)$ is equal to one;

2) the second derivatives of Φ are equal to either one or the second derivatives of the function S_I.

From Theorem 1.1 of Chap. 5 it follows that W_j is a bilinear form in the derivatives of the function $a(x^I, p_{\overline{I}})$ and the derivatives of the function H up to order $2j$ evaluated at the point

$$\left(x^I, -\partial S_I(x^I,p_{\overline{I}})/\partial p_{\overline{I}}, \partial S_I(x^I,p_{\overline{I}})/\partial x^I, p_{\overline{I}}\right)$$

and that the coefficients of this form are universal polynomials in the function S_I and its derivatives. This proves formula (3.18), if at the same time one shows that the expression

$$W_j\left[S_I(x^I,p_{\overline{I}}), H\left(x^I, x^{\overline{I}}, p_I, p_{\overline{I}}\right) a(x^I,p_{\overline{I}})\right]$$

does not contain derivatives of the function $a(x^I, p_{\overline{I}})$ of order higher than j. However, the coefficients of such derivatives will be linear forms in the derivatives of the Hamiltonian function which depend polynomially on the derivatives of the function S_I. Such forms are uniquely determined by their values on real values of the argument. Therefore it is sufficient to check that the coefficients of derivatives of the function a of order higher than j are zero only in the case of real-valued functions S_I. Furthermore, it suffices to consider only Hamiltonian functions which are polynomial in the variables $(x^I, p_{\overline{I}})$, since their derivatives take on all possible values. However, the expressions

$$\left(\widehat{x^{\overline{I}}}\right)^{\alpha} (\widehat{p}_I)^{\beta} e^{(i/h)S_I(x^I, p_{\overline{I}})} a(x^I, p_{\overline{I}}) = \left(ih\frac{\partial}{\partial p_{\overline{I}}}\right)^{\alpha} \times \left(-ih\frac{\partial}{\partial x^I}\right)^{\beta} e^{(i/h)S_I(x^I, p_{\overline{I}})} a(x^I, p_{\overline{I}}) \,. \tag{3.28}$$

(α, β are multi-indices) can be calculated directly, and the coefficients of h^j in these expressions do not contain derivatives of the function a of order higher than j. On the other hand, we can calculate expression (3.28) in the manner described above; the two expressions we obtain in this way must agree up to order $O\left(h^{s/2}\right)$. It remains only to note that for real S_I the coefficients of the asymptotic expansion are uniquely defined.

From the same considerations it is sufficient to verify conditions (3.19) and (3.20) for real-valued functions S_I and for polynomial Hamiltonian functions of at most second degree.

We have from the general theorem of the asymptotic expansion

$$P_I^0 = A \cdot H\left(x^I, -\frac{\partial S_I(x^I, p_{\overline{I}})}{\partial p_{\overline{I}}}, \frac{\partial S_I(x^I, p_{\overline{I}})}{\partial x^I}, p_{\overline{I}}\right) \,. \tag{3.29}$$

$$\begin{aligned} P_I^1 = a_0 &+ \frac{\partial H}{\partial z^{\overline{I}}} a^{\overline{I}} + \frac{\partial H}{\partial \zeta_I} a_I + \frac{\partial H}{\partial \zeta_{\overline{I}}} a_{\overline{I}} \\ &+ \frac{\partial^2 H}{\partial \zeta_I \partial \zeta_I} b_{II} + \frac{\partial^2 H}{\partial z^{\overline{I}} \partial z^{\overline{I}}} b^{\overline{I}\overline{I}} + \frac{\partial^2 H}{\partial \zeta_{\overline{I}} \partial \zeta_{\overline{I}}} b_{\overline{I}\overline{I}} \\ &+ \frac{\partial^2 H}{\partial \zeta_{\overline{I}} \partial z^{\overline{I}}} b_{\overline{I}}^{\overline{I}} + \frac{\partial^2 H}{\partial \zeta_I \partial \zeta_{\overline{I}}} b_{I\overline{I}} + \frac{\partial^2 H}{\partial z^{\overline{I}} d\zeta_I} b_I^{\overline{I}} \,. \end{aligned} \tag{3.30}$$

In formula (3.29) A is a polynomial in the derivatives of the function S_I. In formula (3.30) the derivatives of the function H are evaluated at the same points as function H in formula (3.29), and a and b are first-order differential operators whose coefficients are polynomial functions in the derivatives of the function S_I. Consequently we will choose polynomial Hamiltonians such that we can compute the coefficients in the formulas (3.29), (3.30). The computations can be carried out by direct differentiation. We will not carry them out here, but will present the results in the following table.

Hamilton function	Coefficient
1	$A = 1, \quad a_0 = 0$
$p_{\overline{I}}$	$a_{\overline{I}} = 0$
$x^{\overline{I}}$	$a^{\overline{I}} = \partial/\partial\zeta_{\overline{I}}$
p_I	$a_I = -\partial/\partial z^I$
$p_{\overline{I}}p_{\overline{I}}$	$b_{\overline{I}\,\overline{I}} = 0$
$p_I p_{\overline{I}}$	$b_{I\overline{I}} = 0$
$p_I p_I$	$b_{II} = -\frac{1}{2}\partial^2 S_I/\partial z^I \partial z^I$
$x^{\overline{I}}x^{\overline{I}}$	$b^{\overline{I}\,\overline{I}} = -\frac{1}{2}\partial^2 S_I/\partial\zeta_{\overline{I}}\partial\zeta_{\overline{I}}$
$x^{\overline{I}}p_I$	$b_I^{\overline{I}} = \partial^2 S_I/\partial z^I \partial\zeta_{\overline{I}}$
$x^{\overline{I}}p_{\overline{I}}$	$b_{\overline{I}}^{\overline{I}} = \delta_{\overline{I}}^{\overline{I}}$

In the last row of the table, $\delta_{\overline{I}}^{\overline{I}}$ is the Kronecker delta symbol. This completes the proof of the proposition. □

Proof of Theorem 3.1. We consider the expression on the left-hand side of formula (3.8). By the definition (3.6) of the operator k and formula (1.33), we have

$$H(x,\widehat{p})\,k\varphi = H(x,\widehat{p})\sum_I k_I(e_I\varphi)\ . \tag{3.31}$$

Since $\operatorname{supp}\varphi$ is compact, the sum contains only a finite number of terms, and we can place the operator $H(x,\widehat{p})$ under the summation sign. Then using Proposition 3.3 and writing

$$\mathcal{P}_I = \sum_{0\le j<s/2}(ih)^j\mathcal{P}_I^j\ , \quad \mathcal{P}_I^j = \mu_I^{-1/2}p_I^j\mu_I^{1/2}\ . \tag{3.32}$$

we obtain the congruence

$$
\begin{aligned}
H(x,\widehat{p})\,k\varphi &\equiv \sum_I H(x,\widehat{p})\,k_I(e_I\varphi) = \sum_I \overline{F}^{1/h}_{p_{\overline{I}}\to x^{\overline{I}}} H\left(x^I,\widehat{x}^{\overline{I}},\widehat{p}_I,p_{\overline{I}}\right)\\
&\times R_I^0\left\{e^{(i/h)S_I(z^I,\zeta_{\overline{I}})}\sqrt{\mu_I(z^I,\zeta_{\overline{I}})}\,\varphi(z^I,\zeta_{\overline{I}})\right\}\\
&= \sum_I \overline{F}^{1/h}_{p_{\overline{I}}\to x^{\overline{I}}} R_I^0\left\{e^{(i/h)S_I(z^I,\zeta_{\overline{I}})}P_I\sqrt{\mu_I(z^I,\zeta_{\overline{I}})}\right.\\
&\left.\times\,\varphi(z^I,\zeta_{\overline{I}})\right\} = \sum_I k_I\{\mathcal{P}_I e_I\varphi\}\ \left(\operatorname{mod} h^{s/2}\right)\,.
\end{aligned}
\tag{3.33}
$$

To transform the right-hand side of formula (3.33) we introduce the notation

$$V_{IJ} = 1 + ih\widetilde{V}_{IJ}\ ,$$

where V_{IJ} is the operator defined in Proposition 3.2 (we bear in mind that V_{IJ}^0 is the identity operator). We have

$$
\begin{aligned}
H(x,\widehat{p})\,k\varphi &= \sum_I k_I\{\mathcal{P}_I e_I\varphi\} = \sum_{I,J} k_I\{e_J\mathcal{P}_I e_I\varphi\}\\
&\equiv \sum_{I,J} k_J\left\{\left(1+ih\widetilde{V}_{IJ}\right)e_J\mathcal{P}_I e_I\varphi\right\} = \sum_J k_J\left\{e_J\sum_I \mathcal{P}_J e_I\varphi\right\}\\
&+ ih\sum_{I,J} k_J\left\{\widetilde{V}_{IJ}e_J\mathcal{P}_I e_I\varphi\right\}\ \left(\operatorname{mod} h^{s/2}\right)\,.
\end{aligned}
\tag{3.34}
$$

The second summand in formula (3.34) can be transformed analogously. Indeed, using the partition of unity $\{e_K\}$ and Proposition 3.2 again, we have

$$
\begin{aligned}
&\sum_{I,J} k_J\left\{\widetilde{V}_{IJ}e_J\mathcal{P}_I e_I\varphi\right\} = \sum_{I,J,K} k_J\left\{e_K\widetilde{V}_{IJ}e_J\mathcal{P}_I e_I\varphi\right\}\\
&\equiv \sum_{I,J,K} k_K\left\{\left(1+ih\widetilde{V}_{JK}\right)e_K\widetilde{V}_{IJ}e_J\mathcal{P}_I e_I\varphi\right\} = \sum_K k_K\left\{e_K\sum_{I,J}\widetilde{V}_{IJ}e_J\mathcal{P}_I e_I\varphi\right\}\\
&+ ih\sum_{I,J,K} k_K\left\{\widetilde{V}_{JK}e_K\widetilde{V}_{IJ}e_J\mathcal{P}_I e_I\varphi\right\}\ \left(\operatorname{mod} h^{(s/2)-1}\right)\,.
\end{aligned}
\tag{3.35}
$$

Now we observe that the first summands in formulas (3.34), (3.35) result from applying the operator (3.6) to the functions

$$\sum_I \mathcal{P}_I e_I\varphi\ ,\quad \sum_{I,J}\widetilde{V}_{IJ}e_J\mathcal{P}_I e_I\varphi$$

respectively. Repeating the procedure described as many times as necessary, we arrive at the relation

$$H(x,\widehat{p})\,k\varphi \equiv k\left\{\sum_I \mathcal{P}_I e_I + \sum_{I,J} \widetilde{V}_{IJ} e_J \mathcal{P}_I e_I \right.$$
$$\left. + (ih)^2 \sum_{I,J,K} \widetilde{V}_{JK} e_K \widetilde{V}_{IJ} e_J \mathcal{P}_I e_I + \dots \right\} \varphi \ \left(\operatorname{mod} h^{s/2}\right) . \tag{3.36}$$

By Propositions 2.5 and 3.3 (formula (3.19)) the operator $\mathcal{P}_I^{(0)}$ equals zero, since the pair (M,μ) is associated with the Hamiltonian H. Therefore

$$\mathcal{P}_I = -ih\widetilde{\mathcal{P}}_I , \tag{3.37}$$

where the operator $\widetilde{\mathcal{P}}_I$ is defined by the last formula and formula (3.32). If we now denote

$$\mathcal{P} = \sum_I \widetilde{\mathcal{P}}_I e_I + ih \sum_{I,J} \widetilde{V}_{IJ} e_J \widetilde{\mathcal{P}}_I e_I + (ih)^2 \sum_{I,J,K} \widetilde{V}_{JK} e_K \widetilde{V}_{IJ} e_J \widetilde{\mathcal{P}}_I e_I + \dots \tag{3.38}$$

then formula (3.36), along with (3.37), can be rewritten in the form (3.8). Taking into account the definition (3.37), (3.32) of the operators $\widetilde{\mathcal{P}}_I$, we see that the operator $\mathcal{P}$ defined by formula (3.38) can be represented in the form (3.7), and the operator $\mathcal{P}_0$ appearing in formula (3.7) can be obtained from the first summand in (3.38), if one restricts oneself only to the coefficient of $(ih)^0$ in the operator $\widetilde{\mathcal{P}}_I$. Hence we obtain

$$\mathcal{P}_0 = \sum_I -\mathcal{P}_I^{(1)} e_I = -\sum_I \mu_I^{-1/2} P_I^{(1)} \mu_I^{1/2} e_I .$$

However, the operator $\mu_I^{-1/2} P_I^{(1)} \mu_I^{1/2}$, where $P_I^{(1)}$ is given by formula (3.20), was calculated in Proposition 2.6 and is equal to $V(H) - \frac{1}{2}\frac{\partial^2 H}{\partial z \partial \zeta}\Big|_M$. Therefore the last formula yields

$$\mathcal{P}_0 = \sum_I \left(V(H) - \frac{1}{2}\frac{\partial^2 H}{\partial z \partial \zeta}\bigg|_M \right) e_I = \left(V(H) - \frac{1}{2}\frac{\partial^2 H}{\partial z \partial \zeta}\bigg|_M \right) \sum_I e_I$$
$$= V(H) - \frac{1}{2}\frac{\partial^2 H}{\partial z \partial \zeta}\bigg|_M$$

which agrees with formula (3.9). This proves the theorem. □

6.4 Other Approaches

In this section we will need more precise information about the operators V_{IJ} than that which is contained in Proposition 3.2.

Lemma 4.1. *The congruences*

$$V_{II} \equiv 1 \ , \quad V_{JK} \circ V_{IJ} \equiv V_{IK} \ \left(\operatorname{mod} h^{s/2}\right) \tag{4.1}$$

are valid.

Proof. From the proof of Proposition 3.2 it is evident that the operators V_{IJ} are differential operators whose coefficients are universal polynomials in the derivatives of the function S_I and the derivatives of the function μ_I up to a certain order. We would like to show that the coefficients of $(ih)^j$, $j < s/2$, in the expression $V_{II} - 1$ are identically equal to zero, and also that the coefficients of the same powers of ih in the expressions $V_{JK} \circ V_{IJ}$, V_{IK} coincide identically. To do this it suffices to verify that the corresponding universal polynomials are equal. However, as in the proof of Proposition 3.3, we will use the fact that two polynomials are identically equal if they are identically equal on real arguments. Therefore it suffices to verify equation (4.1) for real functions $S_I(x^I, p_{\overline{I}})$. Proposition 3.2 gives us the congruences

$$k_I \varphi \equiv k_I V_{II} \varphi \left(\operatorname{mod} h^{s/2}\right) \ , \tag{4.2}$$

$$k_K V_{IK} \varphi \equiv k_I \varphi \equiv k_J V_{IJ} \varphi \equiv k_K V_{JK} \circ V_{IJ} \varphi \left(\operatorname{mod} h^{s/2}\right) \ , \tag{4.3}$$

for $\varphi \in {}^s\mathcal{O}_0[h](U_I, \rho_M | U_I)$, $\varphi \in {}^s\mathcal{O}_0[h](U_I \cap U_J \cap U_K, \rho_M | U_I \cap U_J \cap U_K)$ respectively. Applying the Fourier transform $F^{1/h}_{x^{\overline{I}} \to p_{\overline{I}}}$ to the left- and right-hand sides of formula (4.2) and the transform $F^{1/h}_{x^{\overline{K}} \to p_{\overline{K}}}$ to formula (4.3) and using the unitary property of the Fourier transform, we obtain the congruence

$$e^{(i/h)S_I(x^I, p_{\overline{I}})} \varphi(x^I, p_{\overline{I}}) \equiv e^{(i/h)S_I(x^I, p_{\overline{I}})} V_{II} \varphi(x^I, p_{\overline{I}}) \ , \tag{4.4}$$

$$e^{(i/h)S_K(x^K, p_{\overline{K}})} V_{IK} \varphi(x^K, p_{\overline{K}}) \equiv e^{(i/h)S_K(x^K, p_{\overline{K}})} V_{JK} \circ V_{IJ} \varphi(x^K, p_{\overline{K}}) \tag{4.5}$$

modulo $h^{s/2}$. However, for real-valued functions S_I, S_K the congruences (4.4), (4.5) imply that

$$\varphi(x^I, p_{\overline{I}}) \equiv V_{II} \varphi(x^I, p_{\overline{I}}) \ \left(\operatorname{mod} h^{s/2}\right) \ ,$$

$$V_{IK} \varphi(x^K, p_{\overline{K}}) \equiv V_{JK} \circ V_{IJ} \varphi(x^K, p_{\overline{K}}) \ \left(\operatorname{mod} h^{s/2}\right) \ ,$$

from which the assertion of the lemma follows. □

It is possible to give an invariant definition of the canonical operator as an operator on the space of sections of a certain algebraic sheaf.

We remind the reader that an algebraic sheaf over a topological space X is a contravariant functor F, which assigns to each open subset $W \subset X$ a group $\Gamma(F, W)$, called the group of sections of the sheaf F over the set W, and assigns to each imbedding $i : W_1 \subset W_2$ of one open set W_1 into another W_2 a group homomorphism

$$\Gamma(F, i) : \Gamma(F, W_2) \to \Gamma(F, W_1) ,$$

called the restriction of the sections of the sheaf F to the subset W_1. The image of a section s under the homomorphism $\Gamma(F, i)$ will be denoted by $s|W_1$. Moreover, the following conditions must hold:

If the set W is represented as a union

$$W = \bigcup_{\alpha} W_\alpha ,$$

and $s_\alpha \in \Gamma(F, W_\alpha)$ are sections such that

$$s_\alpha |W_\alpha \cap W_\beta = s_\beta| W_\alpha \cap W_\beta \in \Gamma(F, W_\alpha \cap W_\beta) ,$$

then there exists a unique section $s \in \Gamma(F, W)$ such that

$$s|W_\alpha = s_\alpha .$$

An example of an algebraic sheaf is the sheaf of germs of sections of a vector bundle ξ over a topological space X. Indeed, if ξ is a vector bundle over a space X, then we define the algebraic sheaf $\underset{\sim}{\xi}$ by setting

$$\Gamma(\underset{\sim}{\xi}, W) = \Gamma(\xi, W) ,$$

and for the imbedding $i : W_1 \subset W_2$ the homomorphism $\Gamma(\underset{\sim}{\xi}, i)$ denotes the usual restriction of sections in the bundle ξ.

For example, if ξ is the one-dimensional trivial bundle, then $\Gamma(\underset{\sim}{\xi}, W)$ coincides with the space of continuous functions on the set W.

If X is a smooth manifold and ξ is a vector bundle, the we may consider the bundle $\underset{\sim}{\xi}^{\infty}$ of germs of smooth sections of the bundle ξ:

$$\Gamma(\underset{\sim}{\xi}^{\infty}, W) = C^{\infty}(W, \xi) .$$

A mapping f of an algebraic sheaf F_1 into an algebraic sheaf F_2 is a system of homomorphisms

$$\Gamma(f, W) : \Gamma(F_1, W) \to \Gamma(F_2, W)$$

which commute with the restriction homomorphisms, in other words for the imbedding $i : W_1 \to W_2$ the following diagram commutes.

$$\begin{array}{ccc} \Gamma(F_1,W_1) & \overset{\Gamma(f,W_1)}{\longrightarrow} & \Gamma(F_2,W_1) \\ \Gamma(F_1,i)\downarrow & & \downarrow \Gamma(F_2,i) \\ \Gamma(F_1,W_2) & \overset{\Gamma(f,W_2)}{\longrightarrow} & \Gamma(F_2,W_2) \end{array} .$$

For example, if ξ_1, ξ_2 are two vector bundles over a smooth manifold X, and A is a differential operator mapping smooth sections of the bundle ξ_1 into smooth sections of the bundle ξ_2, then the operator A induces a mapping of the algebraic sheaf $\underset{\sim}{\xi}{}_1^{\infty}$ into the algebraic sheaf $\underset{\sim}{\xi}{}_2^{\infty}$.

If F is an algebraic sheaf over a space X and W is an open subset of X, then we denote by $F|W$ the "restriction" of F to W, by setting

$$\Gamma\left(F|W,W'\right) = \Gamma\left(F,W'\right) , \quad W' \subset W .$$

For any section $s \in \Gamma(F,X)$ the support of the section is defined in the natural way.

It is possible to construct algebraic sheaves by a method similar to the construction of vector bundles, using "gluing functions". Indeed, let the space X be represented in the form of a union of its open subspaces. Suppose $X = \cup_\alpha W_\alpha$, and suppose an algebraic sheaf F_α is given over each subset W_α. Suppose, moreover, that sheaf isomorphisms

$$\varphi_{\alpha\beta} : F_\alpha\,|W_\alpha \cap W_\beta \to F_\beta|\,W_\alpha \cap W_\beta$$

are given, which satisfy the equation

$$\varphi_{\beta\gamma} \circ \varphi_{\alpha\beta} = \varphi_{\alpha\gamma}$$

for the restrictions of the three sheaves F_α, F_β, F_γ to the set $W_\alpha \cap W_\beta \cap W_\gamma$. Then there exists a unique algebraic sheaf F over X and unique isomorphisms

$$\varphi_\alpha : F_\alpha \to F|W_\alpha$$

such that

$$\varphi_{\alpha\beta} = \varphi_\beta^{-1} \circ \varphi_\alpha$$

over the set $W_\alpha \cap W_\beta$.

We denote by $\Gamma(F,W)$ the group whose elements are collections of sections $\{s_\alpha, s_\alpha \in \Gamma(F_\alpha, W \cap W_\alpha)\}$, where

$$\varphi_{\alpha\beta}\left(s_\alpha|W_\alpha \cap W \cap W_\beta\right) = s_\beta|W_\alpha \cap W \cap W_\beta \ .$$

It is not hard to verify that the collection of groups $\{\Gamma(F,W)\}$ and the obvious restriction homomorphisms satisfy the axioms of an algebraic sheaf.

Now let us apply the language of algebraic sheaves to the definition of the canonical operator.

The ring ${}^s\mathcal{O}[h](M,\rho_M)$ induces an algebraic sheaf ${}^s\underset{\sim}{\mathcal{O}}[h](M,\rho_M)$ of germs of s-analytic polynomials in h on the manifold M. We cover the

Lagrangian manifold (M, ρ_M) by an atlas of canonical charts $\{U_I\}$. We give the gluing functions by the formula

$$V_{IJ} : {}^s\mathcal{O}[h]\,(U_I \cap U_J, \rho_M|U_I \cap U_J) \to {}^s\mathcal{O}[h]\,(U_I \cap U_J, \rho_M|U_I \cap U_J)\ .$$

Let F^s be the algebraic sheaf over the manifold (M, ρ_M) corresponding to the indicated gluing functions. Thus, there exist sheaf isomorphisms

$$\psi_I : {}^s\underset{\sim}{Q}\,[h]\left(U_I, \rho_M|_{U_I}\right) \to F^s|_{U^I}\ , \tag{4.6}$$

such that

$$V_{IJ} = \psi_J^{-1} \circ \psi_I\ .$$

It turns out that formulas (1.14) determine a well-defined operator

$$\underset{\sim}{K} : \Gamma\left(F^s, M\right) \to {}^{s/2}H^{1/h}\left(\mathbb{R}^n\right)\ ,$$

such that if $f \in \Gamma(F^s, M)$, supp$f \in U_I$, then

$$k_I\psi_I^{-1}(f) = \underset{\sim}{K}\,(f) \in {}^{s/2}H^{1/h}\left(\mathbb{R}^n\right)\ .$$

Suppose (M, ρ_M) is, as before, a quantized $(s+1)$-analytic positive Lagrangian manifold and suppose the pair (M, μ) is associated with the Hamilton function H.

Theorem 4.1. *There exists a differential operator*

$$\underset{\sim}{P} : \Gamma\left(F^s, M\right) \to \Gamma\left(F^{s-2}, M\right)\ , \tag{4.7}$$

such that the congruence

$$H\left(x, \widehat{p}\right)\underset{\sim}{K}\,f \equiv -ih\underset{\sim}{K}\,\underset{\sim}{P}\,f \ \left(\mathrm{mod}\, h^{s/2}\right)\ , \tag{4.8}$$

holds.

Proof. Let $f \in \Gamma(F^s, M)$, supp$f \subset U_I$. We denote by φ the preimage of f under the isomorphism (4.6). Then

$$H\left(x, \widehat{p}\right)\underset{\sim}{K}\,f = H\left(x, \widehat{p}\right)k_I\varphi$$

where k_I is defined by formula (1.14). As in the proof of Theorem 3.1, we obtain

$$H\left(x, \widehat{p}\right)k_I\varphi \equiv -ihk_I\widetilde{\mathcal{P}}_I\varphi\ \left(\mathrm{mod}\, h^{s/2}\right)\ , \tag{4.9}$$

where $\widetilde{\mathcal{P}}_I$ are defined by formulas (3.37), (3.32).

We now define the operator (4.7) as the collection of operators $\left\{\widetilde{\mathcal{P}}_I\right\}$, given in each chart U_I of the canonical atlas. To prove that this operator is well-defined, we need to establish the congruence

$$V_{IJ}\widetilde{\mathcal{P}}\varphi \equiv \widetilde{\mathcal{P}}_J V_{IJ}\varphi \left(\operatorname{mod} h^{(s-2)/2}\right) \tag{4.10}$$

for $\varphi \in {}^s\mathcal{O}[h]\left(U_I \cap U_J, \rho_M | U_I \cap U_J\right)$. The proof of congruence (4.10) is carried out using the relations

$$H\left(x,\widehat{p}\right) k_I\varphi \equiv -ihk_I\widetilde{\mathcal{P}}_I\varphi \equiv -ihk_J V_{IJ}\widetilde{\mathcal{P}}_I\varphi \left(\operatorname{mod} h^{s/2}\right) ,$$

$$H\left(x,\widehat{p}\right) k_I\varphi \equiv H\left(x,\widehat{p}\right) k_J V_{IJ}\varphi \equiv -ihk_J\widetilde{\mathcal{P}}_J V_{IJ}\varphi \left(\operatorname{mod} h^{s/2}\right) .$$

From these identities it follows that

$$k_J V_{IJ}\widetilde{\mathcal{P}}_I\varphi \equiv k_J\widetilde{\mathcal{P}}_J V_{IJ}\varphi \left(\operatorname{mod} h^{(s-2)/2}\right) .$$

The deduction of relations (4.10) from the latter congruence follows from universality considerations exactly analogous to the proof of Lemma 4.1. This completes the proof of the theorem. □

We now proceed to the third method of constructing the canonical operator, discussed at the start of 6.3.

Definition 4.1. We call the operator

$$K_I : {}^s\mathcal{O}_0[h]\left(U_I, \rho_M|_{U_I}\right) \to {}^{s/2}H^{1/h}\left(\mathbb{R}^n\right) , \tag{4.11}$$

acting by the formula

$$k_I\varphi = \sum_J k_I V_{IJ} e_J\varphi \tag{4.12}$$

the *local canonical operator* in the chart U_I. The following assertion follows easily from Lemma 4.1.

Proposition 4.3. *For elements* $\varphi \in {}^s\mathcal{O}_0[h]\left(U_I \cap U_J, \rho_M | U_I \cap U_J\right)$ *the congruence*

$$K_I\varphi \equiv K_J\varphi \left(\operatorname{mod} h^{s/2}\right) \tag{4.13}$$

holds. (We recall that we are assuming the pair (M, μ) to be quantized, and that the arguments $\operatorname{Arg}\mu_I$ are chosen such that the constants c_{IJ} defined in Proposition 1.1 are zero.)

From Proposition 4.3, in a manner completely analogous to 6.1, follows the theorem

Theorem 4.4. *Suppose $i : (M, \rho_M) \to (\Phi_{\mathbb{C}}, \rho_\Phi)$ is a proper quantized $(s+1)$-analytic Lagrangian imbedding. Then there exists a unique operator*

$$K : {}^s\mathcal{O}_0[h](M, \rho_M) \to {}^{s/2}H^{1/h}(\mathbb{R}^n) \ , \tag{4.14}$$

which coincides with the operator (1.14) in each canonical chart U_I of the manifold M, that is for any function $\varphi \in {}^s\mathcal{O}_0[h](U_I, \rho_M|U_I)$ the congruence

$$K\varphi \equiv K_I\varphi \ \left(\operatorname{mod} h^{s/2}\right) \tag{4.15}$$

holds.

We suppose further that the pair (M, μ) is associated with a Hamilton function $H(x,p)$.

Theorem 4.5. *There exists a differential operator*

$$\begin{aligned} &\mathcal{P}' : {}^s\mathcal{O}[h](M, \rho_M) \to {}^s\mathcal{O}[h](M, \rho_M) \ , \\ &\mathcal{P}' = \mathcal{P}'_0 + ih\mathcal{P}'_1 + (ih)^2\mathcal{P}'_2 + \dots \ , \end{aligned} \tag{4.16}$$

such that for any element ${}^s\mathcal{O}[h](M, \rho_M)$ the congruence

$$H(x, \widehat{p})\, K\varphi \equiv -ihK\mathcal{P}'\varphi \ \left(\operatorname{mod} h^{s/2}\right) \tag{4.17}$$

holds. Here $\mathcal{P}'_0 = V(H) - \frac{1}{2}\frac{\partial^2 H}{\partial z \partial \zeta}\Big|_M$.

Proof. For $\varphi \in {}^s\mathcal{O}_0[h](U_I, \rho_M|U_I)$ we have:

$$\begin{aligned} H(x,\widehat{p})\, K\varphi &\equiv H(x,\widehat{p})\, K_I\varphi = H(x,\widehat{p}) \sum_J k_I V_{IJ} e_J \varphi \\ &\equiv -ih \sum_J k_I \widetilde{\mathcal{P}} V_{IJ} e_J \varphi = -ih \sum_{J,K} k_I e_K \widetilde{\mathcal{P}}_I V_{IJ} e_J \varphi \\ &\equiv -ih \sum_{J,K} k_I V_{IK} e_K \widetilde{\mathcal{P}}_I V_{IJ} e_J \varphi + (ih)^2 \sum_{J,K} k_I \widetilde{V}_{IK} e_K \\ &\quad \times \widetilde{\mathcal{P}}_I V_{IJ} e_J \varphi \ \left(\operatorname{mod} h^{s/2}\right) \ . \end{aligned} \tag{4.18}$$

From here on the proof proceeds in the same way as the proof of Theorem 3.1. Indeed, we apply to the second summand on the right-hand side of formula (4.18) the transformation which was used in the last step of the derivation of this formula and repeat this process. In this way we arrive at the result:

$$H(x,\widehat{p})\, K_I\varphi \equiv -ihK_I\mathcal{P}'_I\varphi \ \left(\operatorname{mod} h^{s/2}\right) \ , \tag{4.19}$$

where

$$\mathcal{P}'_I = \sum_J \widetilde{\mathcal{P}}_I V_{IJ} e_J - (ih) \sum_{J,K} \widetilde{V}_{IK} e_K \widetilde{\mathcal{P}}_I V_{IJ} e_J + \dots \ . \tag{4.20}$$

Furthermore, we can show the relation

$$\mathcal{P}'_I = \mathcal{P}'_J$$

analogously to the proof of Theorem 4.2. Hence it follows that the collection of operators $\{\mathcal{P}'_I\}$ defines a global operator $\mathcal{P}'$. The equation $\mathcal{P}'_0 = V(H) - \frac{1}{2} \frac{\partial^2 H}{\partial z \partial \zeta}\Big|_M$ is proved exactly the same way as Eq. (3.9) of Theorem 3.1. This proves the theorem. □

The relationship between the operators K and $\underset{\sim}{K}$ defined in this section is established in the following theorem.

Theorem 4.6. *Each partition of unity $\{e_I\}$ induces a sheaf isomorphism*

$$Q : {}^s\mathcal{O}[h]\,(M, \rho_M) \to F^s$$

such that

a) *Q is a local differential operator.*

b) $K\varphi \equiv \underset{\sim}{K}\, Q\varphi \,(\operatorname{mod} h^{s/2})$, $\varphi \in {}^s\mathcal{O}_0\,(M, \rho_M)$.

Proof. Let $\{e_I\}$ be a partition of unity subordinate to the covering $\{U_I\}$. In order to define the mapping Q it is sufficient to define the mapping $\Gamma(Q, W)$ of the spaces of sections of the sheaves over an open set W. Suppose $\varphi \in \Gamma\,({}^s\mathcal{O}[h]\,(M, \rho_M)\,, W)$. Then the support of the section $e_I\varphi$ lies in U_I, that is

$$e_I\varphi \in \Gamma\,({}^s\mathcal{O}[h]\,(M, \rho_M)\,, W \cap U_I) \ .$$

Then we set

$$\Gamma\,(Q, W)\,\varphi = \sum_I \Gamma\,(\psi_I, U_I \cap W)\,(e_I\varphi) \ ,$$

interpreting each term on the right-hand side as a section on the set W, continued beyond the support of the section by the zero function. Assertion a) then follows by the equation

$$\Gamma\,(\psi_I, U_I \cap U_J)\,\varphi = \Gamma\,(\psi_J, U_I \cap U_J)\,V_{IJ}\varphi$$

for $\operatorname{supp}\varphi \subset U_I \cap U_J$.

To show that Q is an isomorphism, it suffices to verify that the homomorphisms $\Gamma\,(Q, U_I)$ are isomorphisms. Then

$$\Gamma(Q, U_I)\varphi = \Gamma(\psi_I, U_I)\left(\sum_J V_{IJ} e_J \varphi\right) .$$

The operator V_{IJ}, as a polynomial in h, has an h^0 coefficient of 1:

$$V_{IJ} = 1 + ih\widetilde{V}_{IJ} .$$

Therefore

$$\Gamma(Q, U_I)\varphi = \Gamma(\psi_I, U_I)(\varphi + hR\varphi) ,$$

where R is some differential operator. Since the ring ${}^s\mathcal{O}[h](U_I, \rho_M|U_I)$ is the ring of truncated polynomials in h, we can easily construct an operator inverse to $\Gamma(Q, U_I)$.

Finally, assertion (b) follows directly from the definition. □

Theorem 4.6 signifies that the fact that the canonical operator K is not invariant under changes of the partition of unity is associated with the non-uniqueness of the choice of isomorphism Q between the sheaves ${}^s\mathcal{O}[h](M, \rho_M)$ and F^s. Moreover, if we only consider the initial terms of the sections of these sheaves, then the isomorphism Q is unique, i.e. the canonical operator K is also invariantly defined.

6.5 Appendix. The $1/h$-Fourier Transform

Suppose $h \in (0, 1]$. We will introduce the $1/h$ analogs of the Fourier transform $F^{1/h}$ and the Sobolev spaces $H_k^{1/h}(\mathbb{R}^n)$.

We will denote by $\mathbb{R}^n$ and n-dimensional real vector space and by $\mathbb{R}_n$ its dual, $\mathbb{R}_n = (\mathbb{R}^n)^*$. We define the Laplacian operators by setting

$$\Delta_x = +\sum_{i=1}^n \widehat{p}_i^2 , \quad \Delta_p = \sum_{i=1}^n \left(\widehat{x}^i\right)^2 ,$$

where

$$\widehat{p}_i = -ih\,\partial/\partial x^i , \qquad i = 1, \ldots, n ,$$
$$\widehat{x}^i = ih\,\partial/\partial p_i , \qquad i = 1, \ldots, n .$$

Now let k be some number. We introduce the structure of a Hilbert space on the space of complex valued, smooth functions of compact support $f = f\left(x^1, \ldots, x^n\right)$ in $\mathbb{R}^n$ which depend smoothly on the parameter h, by setting

$$(f, g)_k = \sup\left(\left[1 + |x|^2 + \Delta_x\right]^{k/2} f, \left[1 + |x|^2 + \Delta_x\right]^{k/2} g\right) ,$$

where

$$(f,g)_0 = \int_{\mathbb{R}^n} f(x)\overline{g(x)}dx \ .$$

We denote the corresponding complete normed space by $H_k^{1/h}(\mathbb{R}^n)$.

We note that for any natural number k the functions of the form $\exp\left(\frac{i}{h}S(x)\right)\varphi(x)$ belong to the space $H_k^{1/h}(\mathbb{R}^n)$, where $S(x)$ and $\varphi(x)$ are smooth complex-valued functions with compact support on $\mathbb{R}^n$, $\mathrm{Im}S(x) \geq 0$. We define the $1/h$-Fourier transform $F_{x\to p}^{1/h}$ by setting, for any function $\varphi \in C_0^\infty(\mathbb{R}^n)$,

$$\widehat{\varphi}(p) = \left[F_{x\to p}^{1/h}\varphi(x)\right](p) = \left(\frac{-i}{2\pi h}\right)^{n/2} \int_{\mathbb{R}^n} \exp\left\{-\frac{i}{h}\langle x,p\rangle\right\}\varphi(x)dx \ , \quad (1)$$

where $\langle x,p\rangle = x^i p_i$.

The inversion formula for the Fourier transform is

$$\varphi(x) = \left[\overline{F}_{p\to x}^{1/h}\widehat{\varphi}(p)\right](x) = \left(\frac{i}{2\pi h}\right)^{n/2} \int_{\mathbb{R}_n} \exp\left\{\frac{i}{h}\langle x,p\rangle\right\}\widehat{\varphi}(p)dp \quad (2)$$

Here $\arg i = \pi/2$.

Plancherel's formula

$$(f,g)_0 = \left(F^{1/h}f, F^{1/h}g\right)$$

holds for the $1/h$-Fourier transform, as can be verified by a direct calculation. The following commutation formula for any natural number k follows from (1):

$$F_{x\to p}\left(1+|x|^2+\Delta_x\right)^k = \left(1+\Delta_p+|x|^2\right)^k F_{x\to p} \ .$$

Proposition. *The transform* (1) *can be extended to a bounded operator*

$$F^{1/h} : H_k^{1/h}(\mathbb{R}^n) \to H_k^{1/h}(\mathbb{R}^n) \quad (3)$$

which is an isomorphism.

Proof. Indeed, suppose $f \in H_k^{1/h}(\mathbb{R}^n)$. Then

$$\begin{aligned}\|Ff\|_k^2 &= \sup_h \left(\left[1+\Delta_p+|p|^2\right]^{k/2} Ff, \left[1+\Delta_p+|p|^2\right] Ff\right)_0 \\ &= \sup_h \left(F\left[1+|x|^2+\Delta_x\right]^{k/2} f, F\left[1+|x|^2+\Delta_k\right]^{k/2} f\right)_0 \quad (4) \\ &= \sup_h \left(\left[1+|x|^2+\Delta_x\right]^{k/2} f, \left[1+|x|^2+\Delta_x\right]^{k/2} f\right)_0 \ .\end{aligned}$$

From (4) it follows that the operator (1) is continuous and is, moreover, an isometry. The existence of an inverse mapping follows from the inversion formula (2). This proves the proposition. □

Finally, we define the space $H^{1/h}(\mathbb{R}^n)$ as the intersection

$$H^{1/h}(\mathbb{R}^n) = \bigcap_{k=0}^{\infty} H_k^{1/h}(\mathbb{R}^n)$$

of all the spaces $H_k^{1/h}(\mathbb{R}^n)$ and provide it with the weakest topology in which the mapping

$$H_k^{1/h}(\mathbb{R}^n) \to H_0^{1/h}(\mathbb{R}^n)\ , \quad f \to \Delta^{k/2} f$$

is continuous for all $k = 0, 1, 2, \ldots$.

Chapter 7
Some Applications

The goal of this chapter is to illustrate the solution of some problems of mathematical physics using the method of the canonical operator. In order to make the technical side of matters as simple as possible, and place the emphasis on the more essential points, we will only consider the case of a real Hamilton function. Generalizing the results of this chapter to the complex case does not present any particular difficulty, except for the Cauchy problem discussed in the first section of this chapter: for this we refer the reader to the works of A.S. Mishchenko, B.Yu. Sternin, and V.E. Shatalov [2] and F. Treves [1]. As we have mentioned, the first section of this chapter is devoted to the construction of asymptotic solutions to a Cauchy problem containing a small parameter. The second section concerns the construction of the asymptotics of the spectrum of $1/h$-pseudodifferential operators. Finally, in the third section we study the problem of constructing asymptotic solutions for systems of equations.

7.1 Asymptotic Solutions of the Cauchy Problem

7.1.1 Maslov's Canonical Operator in Extended Phase Space. We consider the Cauchy problem for the equation

$$ih\frac{\partial\psi}{\partial t} = H\left(x, \widehat{p}, t\right)\psi \tag{1.1}$$

with initial conditions of the special type

$$\psi(x,0) = \exp\left(\frac{i}{h}S_0(x)\right)\psi_0(x)\ , \quad x \in \mathbb{R}^n \tag{1.2}$$

and apply the canonical operator method to find an asymptotic solution of this problem. Of course, we will attempt to satisfy a congruence mod $\left(h^N\right)$ rather than the equality (1.1). On the other hand, we can replace the initial conditions (1.2) by more general ones, where the right-hand side of condition (1.2) will be the value of a canonical operator on some Lagrangian manifold L_0, imbedded in the phase space $\Phi = \mathbb{R}^n \oplus \mathbb{R}_n$ with coordinates $(x,p) = (x^1, \ldots, x^n, p_1, \ldots, p_n)$. Thus we seek a function ψ which satisfies the congruence

$$ih\frac{\partial \psi}{\partial t} = H(x, \widehat{p}, t)\, \psi \pmod{h^N}$$

and the initial conditions $\psi|_{t=0} = K_0\varphi_0$, where φ_0 is a function on the Lagrangian manifold L_0.

We will denote $-ih\partial/\partial t = E$. Then problem (1.1)–(1.2) can be rewritten in the form

$$\left[\widehat{E} + H(x, \widehat{p}, t)\right] \psi = 0 \ , \quad \psi|_{t=0} = K_0\varphi_0 \ . \tag{1.3}$$

To solve this problem we define the extended phase space

$$\Phi' = \mathbb{R}^n \oplus \mathbb{R}_n \oplus \mathbb{R}^1 \oplus \mathbb{R}_1$$

with coordinates (x, p, t, E) and we define a symplectic form $dp \wedge dx + dE \wedge dt$. The Hamilton function corresponding to Problem (1.3) is

$$\mathcal{H}(x, p, t, E) = E + H(x, p, t) \ . \tag{1.4}$$

Our next goal is to construct a quasi-classical object (a Lagrangian manifold with measure) in the extended phase space Φ', and use it to construct an asymptotic solution of Problem (1.3).

Let us consider the Hamiltonian vector field corresponding to the Hamilton function (1.4):

$$\begin{aligned} &\frac{\partial \mathcal{H}}{\partial p}\frac{\partial}{\partial x} + \frac{\partial \mathcal{H}}{\partial E}\frac{\partial}{\partial t} - \frac{\partial \mathcal{H}}{\partial x}\frac{\partial}{\partial p} - \frac{\partial \mathcal{H}}{\partial t}\frac{\partial}{\partial E} \\ &\qquad = \frac{\partial}{\partial t} + \frac{\partial \mathcal{H}}{\partial p}\frac{\partial}{\partial x} - \frac{\partial \mathcal{H}}{\partial x}\frac{\partial}{\partial p} - \frac{\partial \mathcal{H}}{\partial t}\frac{\partial}{\partial E} \ . \end{aligned} \tag{1.5}$$

We will sometimes denote this vector field by $\partial/\partial t$. We assume that the system of Hamilton's equation corresponding to the vector field (1.5),

$$\begin{aligned} \dot{x} &= \frac{\partial H}{\partial p}(x, p, t) \ , \qquad \dot{p} = -\frac{\partial H}{\partial x}(x, p, t) \ , \\ \dot{E} &= -\frac{\partial H}{\partial t}(x, p, t) \ , \qquad \dot{t} = 1 \end{aligned} \tag{1.6}$$

(a dot indicating differentiating with respect to τ) has a solution for $0 \le \tau \le T$ for any initial conditions. This means that for $0 \le \tau \le T$ there is a local group of diffeomorphisms $g^\tau(d/d\tau)$ of the extended phase space, corresponding to the vector field $d/d\tau$ in the following way: if

$$x = X(x_0, p_0, E_0, t_0, \tau) \ , \quad p = P(x_0, p_0, E_0, t_0, \tau) \ ,$$

$$E = E(x_0, p_0, E_0, t_0, \tau) \ , \quad t = t_0 + \tau$$

is a solution of system (1.6) with initial data (x_0, p_0, E_0, t_0), then

$$g^\tau (x_0, p_0, E_0, t_0) = (x, p, E, t) \ .$$

Now we will consider an imbedding of the original Lagrangian manifold L_0 in the extended phase space Φ'. We define this as follows. Suppose

$$x = x(\alpha) \ , \quad p = p(\alpha) \ , \quad \alpha \in L_0 \ ,$$

are the functions which define the imbedding $i_0 : L_0 \to \Phi$. Then $i : L_0 \to \Phi'$ is given by the functions

$$x = x(\alpha) \ , \quad p = p(\alpha), \quad t = 0 \ , \quad E = -H(x(\alpha), p(\alpha), 0) \ . \qquad (1.7)$$

We note that the imbedding $i : L_0 \to \Phi'$ depends on the Hamiltonian $H(x, p, t)$, and hence is not generated by any standard imbedding. We let L_τ denote a translation of the manifold $L_0 \subset \Phi'$ along the trajectories of the Hamiltonian system (1.6): $L_\tau = g^\tau(L_0)$, and let L be the manifold swept out by L_0 over the time T: $L = \cup_{0 \le \tau \le T} L_\tau$. We observe that L is a manifold, since the vector field (1.5) is transversal to L_0.

At this point we make an important observation about the choice of parameters. By the last equation in system (1.6) and the initial conditions (1.7) we have $\tau = t$ on the trajectories we are considering. Therefore we will use the notation: $L = \cup_{0 \le t \le T} L_t$ instead of $L = \cup_{0 \le \tau \le T} L_\tau$ and $L_t = g^t L_0$ instead of $L_\tau = g^\tau L_0$ respectively. We also note that the solution of system (1.6) with the initial data of (1.7),

$$x = X(\alpha, t) \ , \quad p = P(\alpha, t) \ , \quad E(\alpha, t) = E \ , \quad t = t \ ,$$

defines a diffeomorphism of the manifold L with the manifold $L_0 \times [0, T]$. Thus if $(\beta_0^1, \dots, \beta_0^n)$ is a system of coordinates on an open set $U \subset L_0$, then $(\beta_0^1, \dots, \beta_0^n, t)$ are coordinates on the phase flow of the chart U, which is itself an open set in the manifold L. We will often use this fact in the future to give coordinate descriptions of objects on the manifold L.

Proposition 1.1. *The manifold L is a Lagrangian manifold in the extended phase space, and $\mathcal{H}|_L = 0$.*

Proof. The function $\mathcal{H}$ is constant along the trajectories of system (1.6), since

$$\frac{d}{d\tau}\mathcal{H}(x,p,t,E) = d\mathcal{H}\left(\frac{d}{d\tau}\right) = \frac{\partial\mathcal{H}}{\partial x}\,dx\left(\frac{d}{d\tau}\right) + \frac{\partial\mathcal{H}}{\partial p}\,dp\left(\frac{d}{d\tau}\right)$$
$$+\frac{\partial\mathcal{H}}{\partial t}\,dt\left(\frac{d}{d\tau}\right) + \frac{\partial\mathcal{H}}{\partial E}\,dE\left(\frac{d}{d\tau}\right) = \frac{\partial H}{\partial x}\frac{\partial H}{\partial p} - \frac{\partial H}{\partial p}\frac{\partial H}{\partial x} + \frac{\partial H}{\partial t} - \frac{\partial H}{\partial t} = 0\ .$$

It follows from Formulas (1.7) that the function $\mathcal{H}$ is identically zero on the submanifold L_0, hence $\mathcal{H}|_L \equiv 0$.

To prove the manifold is Lagrangian it is necessary to show that the restriction of the symplectic form to the manifold L is equal to zero, i.e.

$$dp \wedge dx + dE \wedge dt|_L = 0\ .$$

We show first that for each value of the parameter $t \in [0,T]$ the equation

$$dx \wedge dp + dE \wedge dt|_{L_t} = 0$$

holds. We observe $dx \wedge dp + dE \wedge dt|_{L_t} = dx \wedge dp|_{L_t}$, since on the manifold L_t, t is constant. For $t = 0$ the desired assertion follows from the fact that the manifold L_0 is Lagrangian in the phase space $(\Phi, dx \wedge dp)$ and by Formulas (1.7) giving the imbedding. Furthermore, taking the Lie derivative of the form $dp \wedge dx + dE \wedge dt$ along the trajectories of system (1.6), we obtain[1]

$$\begin{aligned}\mathcal{L}_{d/dt}(dp \wedge dx + dE \wedge dt) &= d\left(\frac{d}{d\tau}\lrcorner\, dp \wedge dx + dE \wedge dt\right)\\ &+ \frac{d}{d\tau}\lrcorner\, d(dp \wedge dx + dE \wedge dt) = -d(d\mathcal{H}) + 0 = 0\ ,\end{aligned} \tag{1.8}$$

since

$$\begin{aligned}\frac{d}{d\tau}\lrcorner\, dp \wedge dx + dE \wedge dt &= dp\left(\frac{d}{d\tau}\right)dx - dx\left(\frac{d}{d\tau}\right)dp\\ &+ dE\left(\frac{d}{d\tau}\right)dt - dt\left(\frac{d}{d\tau}\right)dE = -d\mathcal{H}\ .\end{aligned}$$

Formula (1.8) shows that

$$dx \wedge dp + dt \wedge dE|_{L_t} = 0\ .$$

---

[1] If ω is a differential form and X is a vector field, then $X \lrcorner\, \omega$ denotes a differential form Ω of one less dimension, whose values are calculated using the following formula:

$$\Omega(Y_1,\ldots,Y_s) = \sum_{i=1}^{s+1}(-1)^i\omega(Y_1,\ldots,Y_{i-1},X,Y_i,\ldots,Y_s)\ .$$

In other words, the form $X \lrcorner\, \omega$ is a function with a smaller number of vector fields as its arguments, obtained by fixing one of the arguments of the original form to be equal to the vector field X.

Now let $m \in L$ be an arbitrary point of the manifold L, and let X_m, Y_m be arbitrary tangent vectors to L at the point m. Then the vectors X_m and Y_m can be decomposed into a sum

$$X_m = X'_m + \lambda \frac{d}{d\tau} , \quad Y_m = Y'_m + \mu \frac{d}{d\tau} ,$$

where the vectors X'_m, Y'_m are tangent to the submanifold L_t at m.

Then

$$\begin{aligned} &\langle dp \wedge dx + dE \wedge dt; X_m, Y_m \rangle = \langle dp \wedge dx + dE \wedge dt; X'_m, Y'_m \rangle \\ &+ \lambda \langle \left(\frac{d}{d\tau} \lrcorner\, dp \wedge dx + dE \wedge dt \right), X'_m \rangle - \mu \langle \left(\frac{d}{d\tau} \lrcorner\, dp \wedge dx + dE \wedge dt \right), Y'_m \rangle \\ &= -\lambda d\mathcal{H}(X'_m) + \mu d\mathcal{H}(Y'_m) = 0 , \end{aligned} \tag{1.9}$$

since the first summand in the middle expression in Eq. (1.9) is equal to zero by the above argument, and $dH|_L = 0$. This proves the proposition. □

Note. The manifold L is invariant with respect to the trajectories of the system (1.6), by construction.

Now let μ_0 be some measure on the Lagrangian manifold L_0 and let K_0 be the canonical operator on L_0 associated with the measure μ_0. Obviously, the new measure

$$\mu = \mu_0 \wedge dt = \mu_0(\alpha) \wedge dt , \quad \alpha \in L_0 , \tag{1.10}$$

is a measure on L which is invariant with respect to the vector field $d/d\tau$. Moreover, if $(x_0^I, p_{0\bar{I}})$ are canonical coordinates on the chart U_I of the manifold L_0, then

$$\frac{\mu}{dx_0^I \wedge dp_{0\bar{I}} \wedge dt} = \frac{\mu_0}{dx_0^I \wedge dp_{0\bar{I}}} .$$

We observe that it is possible to choose an covering of L by canonical charts in such a way that the function E is not a coordinate in any chart of that covering (this follows from the last equation in (1.6)). Furthermore, since any cycle in the manifold L is retractible to the manifold L_0, the fact that L is quantized follows from the fact that L_0 is. Therefore some canonical operator K, associated with the measure μ, is defined on the manifold L.

Proposition 1.2. *The formula*

$$K_{t_0}(\varphi | L_{t_0}) = [K\varphi]_{t=t_0} \tag{1.11}$$

correctly defines a canonical operator K_{t_0} on the manifold L_{t_0} for any fixed value t_0 on the interval $[0, T]$.

Proof. Let U_I be a canonical chart on the manifold L with coordinates $(x^I, p_{\overline{I}}, t)$ which has a nonempty intersection with the manifold L_{t_0}. Obviously, the set $U_{t_0} = U \cap L_{t_0}$ is an open set in the manifold L_{t_0}, and the functions $(x^I, p_{\overline{I}})$ form a system of coordinates in the chart U_{t_0}. Suppose $\varphi_{t_0} \in C_0^\infty(U_{t_0})$ is a function on L_{t_0}, and let φ be an arbitrary continuation of φ_{t_0} to a function in $C_0^\infty(U)$. We calculate the right-hand side of (1.11) for the function φ:

$$[K\varphi]_{t=t_0} = \left\{ \overline{F}^{1/h}_{p_{\overline{I}} \to x^{\overline{I}}} \exp\left(\frac{i}{h} S_I(x^I, p_{\overline{I}}, t)\right) \sqrt{\mu_I(x^I, p_{\overline{I}}, t)} \right.$$
$$\left. \times \left(\sum_{U_I \cap U_J \neq \emptyset} V_{IJ}\left(e_J(x^I, p_{\overline{I}}, t)\varphi(x^I, p_{\overline{I}}, t)\right)\right) \right\}_{t=t_0} . \tag{1.12}$$

Here $e_I(x^I, p_{\overline{I}}, t)$ is a partition of unity subordinate to the covering $\{U_I\}$.

We note that the differential operators V_{IJ} which arise in the comparison of the elementary canonical operators on the charts U_I and U_J do not contain any derivative with respect to the variable t. This is due to the special form of the canonical atlas on the manifold L, for which t is a coordinate in every canonical chart. Therefore

$$\left[V_{IJ}\left(e_J(x^I, p_{\overline{I}}, t)\varphi(x^I, p_{\overline{I}}, t)\right)\right]_{t=t_0} = V_{IJ}^{t_0}\left(e_J(x^I, p_{\overline{I}}, t)\varphi_{t_0}(x^I, p_{\overline{I}})\right) , \tag{1.13}$$

where the operator $V_{IJ}^{t_0}$ is obtained from the operator V_{IJ} by restricting all the coefficients of that operator to the submanifold L_{t_0}. Substituting in Eq. (1.13), we can rewrite Eq. (1.12) in the form

$$K\varphi|_{t=t_0} = \overline{F}^{1/h}_{p_{\overline{I}} \to x^{\overline{I}}} \left[\exp\left\{\frac{i}{h} S_I^{t_0}(x^I, p_{\overline{I}})\right\} \sqrt{\mu_I^{t_0}(x^I, p_{\overline{I}})} \right.$$
$$\left. \times \left(\sum_{U_J \cap U_J \neq \emptyset} V_{IJ}^{t_0}\left(e_J^{t_0}(x^I, p_{\overline{I}})\varphi_{t_0}(x^I, p_{\overline{I}})\right)\right) \right] , \tag{1.14}$$

where $S_I^{t_0}(x^I, p_{\overline{I}}) = S_I(x^I, p_{\overline{I}}, t_0)$, $\mu_I^{t_0}(x^I, p_{\overline{I}}) = \mu_I(x^I, p_{\overline{I}}, t_0)$, $e_J^{t_0}(x^I, p_{\overline{I}}) = e_J(x^I, p_{\overline{I}}, t_0)$ are the restrictions of the corresponding functions to the manifold L_{t_0}. Formula (1.14) shows that the right-hand side of Eq. (1.11) is independent of the choice of an extension of φ_{t_0} to a function φ; thus the operator K_{t_0} is well-defined by Formula (1.11). Now we will show that K_{t_0} is the canonical operator associated with the manifold L_{t_0} and the measure μ_{t_0} defined by the relations

$$\mu = \mu_t \wedge dt , \quad \mu_{t_0} = \mu_t|_{t=t_0} . \tag{1.15}$$

It is not hard to see that the family of functions $\{e_I^{t_0}\}$ is a partition of unity subordinate to the canonical covering $\{U_{I,t_0}\}$ of the manifold L_{t_0}.

Furthermore, the cochain $\{S_I^{t_0}\}$ satisfies all the conditions demanded of the action on a Lagrangian manifold L_{t_0}. Indeed,

$$dS_I^{t_0} = [dS_I]_{t=t_0} = \left[p_I dx^I - x^{\overline{I}} dp_{\overline{I}} + E dt\right]_{t=t_0} = p_I dx^I - x^{\overline{I}} dp_{\overline{I}} ,$$
$$S_I^{t_0} - S_J^{t_0} = [S_I - S_J]_{t=t_0} = \left[x^{I_3} p_{I_3} - x^{I_2} p_{I_2}\right]_{t=t_0} = x^{I_3} p_{I_3} - x^{I_2} p_{I_2} ,$$
$$\mu_I^{t_0} = [\mu_I]_{t=t_0} = \left[\frac{\mu}{dx^I \wedge dp_{\overline{I}} \wedge dt}\right]_{t=t_0} = \left[\frac{\mu_t}{dx^I \wedge dp_{\overline{I}}}\right]_{t=t_0} = \frac{\mu_{t_0}}{dx^I \wedge dp_{\overline{I}}} . \tag{1.16}$$

The condition on the choice of arguments can be checked analogously. Since in Formula (1.14) the argument of $\mu_I^{t_0}$ must be chosen equal to the argument of $\mu_I(x^I, p_{\overline{I}}, t_0)$,

$$\mathrm{Arg}\mu_I^{t_0} - \mathrm{Arg}\mu_J^{t_0} = [\mathrm{Arg}\mu_I - \mathrm{Arg}\mu_J]_{t=t_0} = \left[\sum_k \arg \lambda_{k,IJ}\right]_{t=t_0} + |I_2|\pi , \tag{1.17}$$

where $\lambda_{k,IJ}$ are the eigenvalues of the matrix

$$\partial\left(-p_{I_2}, x^{I_3}\right) / \partial\left(x^{I_2}, p_{I_3}\right) . \tag{1.18}$$

Taking into account the fact that the latter expression does not involve any t-derivatives, we obtain

$$\left[\sum_k \arg \lambda_{k,IJ}\right]_{t=t_0} = \sum_k \arg \lambda_{k,IJ}^{t_0} , \qquad -3\pi/2 < \arg \lambda_{k,IJ}^{t_0} \leq \pi/2 ,$$

where $\lambda_{k,IJ}^{t_0}$ are the eigenvalues of the matrix

$$\partial\left(-p_{I_2}, x^{I_3}\right) / \partial\left(x^{I_2}, p_{I_3}\right)$$

on the manifold L_{t_0}.

Thus Formula (1.11) defines a local canonical operator on each chart U_{I,t_0}, and, moreover, conditions (1.16) and (1.17) guarantee that the operator cochain defined in this way is a cocycle. This completes the proof of the proposition. □

Corollary. *Thus one can choose an operator K on the manifold L such that the operator $K_t|_{t=0}$ is equal to the operator K_0 which appears in* (1.3).

Proof. Suppose $\widetilde{K}$ is some canonical operator on the manifold L. Then the operators K_0 and $\widetilde{K}_T|_{t=0}$ are canonical operators on the manifold L_0. Consequently, the operators K_0 and $\widetilde{K}_t|_{t=0}$ differ by a constant factor of the form $\exp\{i(C_1 + C_2/h)\}$, i.e.

$$K_0 = \exp\{i(C_1 + C_2/h)\}\widetilde{K}_t\Big|_{t=0} .$$

The desired operator K is given by the formula

$$K = \exp\{i(C_1 + C_2/h)\}\widetilde{K} .$$

This proves the corollary. □

7.1.2 Construction of the Asymptotic Solution of the Cauchy Problem. Now we will apply the operator K we have constructed to find asymptotic solutions to Problem (1.3). We will look for solutions of the form

$$\psi(x,t,h) = K\varphi . \qquad (1.19)$$

According to Proposition 1.1, the Note after that proposition, and Formula (1.10) for the measure μ, the manifold L with the measure μ is associated with the Hamiltonian H. We apply the commutation formula (Proposition 3.1 of Chap. 6)

$$\mathcal{H}\left(x,\widehat{p},t,\widehat{E}\right)\varphi = \left(\widehat{E} + H(x,\widehat{p},t)\right)\varphi = \mathcal{H}\left(x,\widehat{p},t,\widehat{E}\right)K\varphi = -ihK\mathcal{P}\varphi ,$$

where $\mathcal{P}$ is a polynomial in the variable h whose coefficients are differential operators:

$$\mathcal{P} = \mathcal{P}_0 + h\mathcal{P}_1 + \dots . \qquad (1.20)$$

The principal term of the operator $\mathcal{P}$ has the form

$$\mathcal{P}_0 = \frac{d}{d\tau} - \frac{1}{2}\left[\frac{\partial^2\mathcal{H}}{\partial x\partial p} + \frac{\partial^2\mathcal{H}}{\partial t\partial E}\right] = \frac{d}{d\tau} - \frac{1}{2}\frac{\partial^2 H}{\partial x\partial p} . \qquad (1.21)$$

In order for the function $\psi(x,t,y)$ to satisfy Eq. (1.1) modulo h^N, it is sufficient to require the condition

$$\mathcal{P}\varphi \equiv 0 \pmod{h^N} \qquad (1.22)$$

to hold.

To solve the congruence (1.22) we will seek φ in the form of a polynomial in h:

$$\varphi = \varphi^{(0)} + h\varphi^{(1)} + h^2\varphi^{(2)} + \dots . \qquad (1.23)$$

A substitution of (1.23) into the congruence (1.22) gives, in view of Formulas (1.20) and (1.21), a recursive formula for the functions $\varphi^{(j)}$:

$$\left[\frac{d}{d\tau} - \frac{1}{2}\frac{\partial^2 H}{\partial x \partial p}\right]\varphi^{(0)} = 0 \,, \tag{1.24}$$

$$\left[\frac{d}{d\tau} - \frac{1}{2}\frac{\partial^2 H}{\partial x \partial p}\right]\varphi^{(1)} = -\mathcal{P}_1\varphi^{(0)}, \ldots \,, \tag{1.25}$$

$$\left[\frac{d}{d\tau} - \frac{1}{2}\frac{\partial^2 H}{\partial x \partial p}\right]\varphi^{(j)} = -\sum_{\substack{k+l=j \\ k\geq 1}} \mathcal{P}_k\varphi^{(l)} \,. \tag{1.26}$$

This system can be solved by integration along the trajectories of the vector field $d/d\tau$, since along the trajectories of this field each of the Eqs. (1.24), (1.25), (1.26) is an ordinary differential equation of first order. The initial conditions for Eqs. (1.24), (1.25), (1.26) are obtained from the initial conditions of the problem (1.3). By Proposition 1.2 these conditions can be rewritten in the form

$$\begin{aligned}\psi(x,t,h)|_{t=0} &= K\left[\varphi^{(0)} + h\varphi^{(1)} + h^2\varphi^{(2)} + \ldots\right]_{t=0} \\ &= K_0\left[\varphi^{(0)} + h\varphi^{(1)} + \ldots|_{t=0}\right] = K_0\varphi_0 \,,\end{aligned}$$

from which we obtain

$$\varphi^{(0)}|_{t=0} = \varphi_0 \,, \quad \varphi^{(j)}|_{t=0} = 0 \,, \qquad j \neq 0 \,.$$

Hence we have shown the following theorem.

Theorem 1.3. *An asymptotic solution of Problem* (1.3) *is given by Formula* (1.19) *with a function* φ *of the form* (1.23)*, where the coefficients* $\varphi^{(0)}, \varphi^{(1)}, \ldots$ *for the various powers of* h *are defined by the Eqs.* (1.24)–(1.26) *with initial conditions*

$$\varphi^{(0)}|_{t=0} = \varphi_0 \,, \quad \varphi^{(j)}|_{t=0} = 0 \,, \qquad j \neq 0 \,.$$

Now we will analyse the construction of the operator K (or of the operators K_{t_0} for all $t_0 \in [0,T]$) more closely. We note that the operator K_{t_0} for each t_0 must be associated with the manifold L_{t_0} and the measure μ_{t_0} defined by Formula (1.15). All such operators can be obtained from any particular one of them by multiplication by a constant of the form

$$\exp\left\{i\left(C_1 + \frac{C_2}{h}\right)\right\} \,.$$

Obviously the values of the constants C_1 and C_2 can be found by using the values of the action and the argument of the Jacobian at any point of the manifold L_{t_0}.

First we consider the constant C_2. Suppose $\alpha_0 \in L_0$ is some fixed point on the manifold L_0. We let α_{t_0} denote the image $g^{t_0}\alpha_0$ under the diffeomorphism g^{t_0}. The point α_{t_0} is joined to the point α_0 by the path $\{g^t\alpha_0 : 0 \le t \le t_0\} = \gamma_{t_0}$. To find C_2 we observe that, by conditions (1.16), the cochain S_I is defined by a global function S (a nonsingular action). Indeed, we define the function S in the chart U_I by the formula

$$S = S_I + x^{\overline{I}} p_{\overline{I}} \, . \tag{1.27}$$

Then conditions (1.16) guarantee the independence of S from the choice of charts U_I, and the cochain S_I can be reconstructed from the function S via Formula (1.27). We note also that the function S satisfies the Pfaffian equation

$$dS = p\,dx + E\,dt = p\,dx - H\,dt$$

and therefore the constant C_2 can be found from the condition

$$S^{t_0}(\alpha_{t_0}) = S^0(\alpha_0) + \int_0^{t_0} p\,dx - H\,dt \, ,$$

where the integral is taken along the path γ_{t_0}.

Now let us consider the constant C_1. Here the situation is substantially different from the one considered just above, since condition (1.17) does not allow one to define a global nonsingular argument $\mathrm{Arg}\mu$. This is due to the fact that the matrix (1.18), which is nondegenerate on the intersection of charts $U_I \cap U_J$, can become equal to zero on U_I; therefore the functions $\lambda_{k,IJ}$ can change discontinuously. We will study the constant C_1 under the following simplifying assumption[2]:

The singular points of the projection $\pi : L_{t_0} \to \mathbb{R}^n_x$ along $\mathbb{R}_{np}$ form a submanifold of L_{t_0} of codimension 1.

We will denote this submanifold by Σ_{t_0} and call it the cycle of singularities of the manifold L_{t_0}.

On each connected component of the set $L_{t_0} \setminus \Sigma_{t_0}$ the formula

$$\mathrm{Arg}\mu^{t_0} = -\mathrm{Arg}\mu_I^{t_0} + \sum_k \arg \lambda_{k,I}^{t_0} \, , \qquad -3\pi/2 < \arg \lambda_{k,I}^{t_0} < \pi/2 \, ,$$

where $\lambda_k^{t_0}$ are the eigenvalues of the matrix

$$\left(\frac{\partial x^{\overline{I}}}{\partial p_{\overline{I}}} \right) , \tag{1.28}$$

uniquely defines a number which is a multiple of π. We also observe that

---

[2] Of "general position" type.

$$\sum_k \arg \lambda_k^{t_0} = -\pi \operatorname{index}_{(-)} \left(\frac{\partial x^{\overline{I}}}{\partial p_{\overline{I}}} \right) ,$$

where $\operatorname{index}_{(-)} \left(\partial x^{\overline{I}} / \partial p_{\overline{I}} \right)$ is the negative index of inertia of the matrix (1.28). The function $\operatorname{Arg}\mu^{t_0}$ is constant along any curve which does not intersect the cycle of singularities $\cup_{0\leq t\leq T}\Sigma_t$. Therefore this function has discontinuities on the curve γ_{t_0} only at points where it intersects $\cup_{0\leq t\leq T}\Sigma_t$. Now we study how this function changes when it intersects the cycle of singularities. Suppose the curve γ_{t_0} intersects $\cup_{0\leq t\leq T}\Sigma_t$ in the points $t_1, t_2, \ldots, t_m$. Then $\operatorname{Arg}\mu^t$ changes while crossing the point t_i in accordance with the formula

$$\begin{aligned}
\operatorname{Arg}\mu^{t_0+0} &= \operatorname{Arg}\mu_I^{t_i} + \sum_k \arg \lambda_{k,I}^{t_i+0} = \operatorname{Arg}\mu_I^{t_i} - \pi \operatorname{index}_{(-)} \left(\frac{\partial x^{\overline{I}}}{\partial p_{\overline{I}}} \right) \\
&= \operatorname{Arg}\mu_I^{t_i} - \pi \operatorname{index}_{(-)} \left(\frac{\partial x^{\overline{I}}}{\partial p_{\overline{I}}} \right)_{t_i-0} + \pi \left\{ \operatorname{index}_{(-)} \left(\frac{\partial x^{\overline{I}}}{\partial p_{\overline{I}}} \right)_{t_i-0} \right. \\
&\quad \left. - \operatorname{index}_{(-)} \left(\frac{\partial x^{\overline{I}}}{\partial p_{\overline{I}}} \right)_{t_i+0} \right\} = \operatorname{Arg}\mu^{t_i-0} + \pi\sigma_i .
\end{aligned}$$

Here we have used the notation

$$\sigma_i = \operatorname{index}_{(-)} \left(\frac{\partial x^{\overline{I}}}{\partial p_{\overline{I}}} \right)_{t_i-0} - \operatorname{index}_{(-)} \left(\frac{\partial x^{\overline{I}}}{\partial p_{\overline{I}}} \right)_{t_i+0} . \tag{1.29}$$

It is not hard to check that Formula (1.29) defines an integer for each point t_i which is independent of the choice of the chart U_I.

Now if the value of t_0 is such that $\alpha_{t_0} \in \Sigma_{t_0}$, then it is possible to define the index of the path γ_{t_0} by the formula

$$\operatorname{ind}\gamma_{t_0} = \sum_{t_i<t_0} \sigma_i .$$

Then the formula for $\operatorname{Arg}\mu^{t_0}$ (or, which is the same thing, for C_1) takes the form

$$\operatorname{Arg}\mu^{t_0}(\alpha_{t_0}) = \operatorname{Arg}\mu^0(\alpha_0) + \mu \operatorname{ind}\gamma_{t_0} .$$

In order to construct the asymptotic solution to Eq. (1.3) we have used the definition of the canonical operator given in the second part of Sect. 6.4. We could have equally well used the canonical operator described in Sect. 6.3; then the operators $\mathcal{P}_k$ in formulas (1.24)–(1.26) would have changed. Furthermore, applying the canonical operator introduced in the first part of Sect. 6.4 would have led us to consider the system of Eqs. (1.24)–(1.26) with respect to the sections of the bundle F^s defined at the beginning of 6.4. All these approaches are equivalent and lead to the same asymptotic solution of problem (1.3).

7.2 Asymptotics of the Spectrum of $1/h$-Pseudodifferential Operators

7.2.1 Example. Under this subheading we will illustrate the method for finding the spectrum of $1/h$-pseudodifferential operators in the simplest example. A precise construction of the theory in the general case will be presented in subheadings 2–5 of this section.

Let us consider the problem of finding the eigenvalues of the operator

$$-\frac{h^2}{2}\frac{d^2}{dx^2}+x^2$$

(the energy operator for a one-dimensional quantum-mechanical harmonic oscillator). We denote the eigenvalues for this problem by E, and the eigenfunctions by φ. The problem we are considering, therefore, can be written as follows:

$$-\frac{h^2}{2}\frac{d^2\psi}{dx^2}+x^2\psi=E\psi\ .$$

We will attempt to find the eigenfunctions $\psi(x,h)$ of this problem in the form $k\psi$, where k is the canonical operator on some Lagrangian manifold L. Then, as in Sect. 4.2, we obtain that the manifold L must lie on the null level surface of the Hamilton function

$$H_E(x,p)=H(x,p)-E=\frac{p^2+x^2}{2}-E\ .$$

But the level curves of the function H_E in this case are one-dimensional curves in phase space. Since the dimension of the manifold L in the case $n=1$ must also be equal to one, we obtain that the Lagrangian manifold L must coincide with some level curve of the Hamilton function H_E:

$$L=L_E=\left\{(x,p)|H_E(x,p)=\frac{p^2+x^2}{2}-E=0\right\}\ .$$

Now we will determine whether the manifold L_E is quantized for any values of E. According to 2.3, we must verify the conditions

$$\int_\gamma p\,dx=0\ ,\quad \int_\gamma\frac{1}{2\pi i}d\ln\left[\det\frac{\partial(x+ip)}{\partial\alpha}\bigg/\left|\det\frac{\partial(x+ip)}{\partial\alpha}\right|\right]\equiv 0\pmod 2$$

for any cycle γ in L_E. But the manifold L_E is a circle in the phase space (x, p) with center at the origin and radius $\sqrt{2E}$. Therfore it is necessary to verify the conditions for γ equal to L_E. We have

$$\int_{L_E} p\,dx = \int_0^{2\pi} \sqrt{2E}\sin\theta\, d\left(\sqrt{2E}\cos\theta\right) = -2\pi E \neq 0\ ;$$

$$\int_{L_E} \frac{1}{2\pi i} d\ln\left[\det\frac{\partial(x+ip)}{\partial\alpha} \Big/ \left|\det\frac{\partial(x+ip)}{\partial\alpha}\right|\right]$$

$$= \frac{1}{2\pi i_0}\int_0^{2\pi} d\ln\left[\frac{d}{d\theta}(\cos\theta + i\sin\theta) \Big/ \left|\frac{d}{d\theta}(\cos\theta + i\sin\theta)\right|\right] = 1 \neq 0 \pmod 2$$

These two equations show that the theory of Maslov's canonical operator, in the form in which it was developed in Chap. 4, cannot be applied to the problem of the asymptotics of the spectrum.

Let us analyse the Maslov quantization conditions in more detail. According to Proposition 1.2 of Sect. 4.1, the congruence

$$k_I\varphi \equiv e^{(i/h)c_{IJ}^{(1)} + i\pi c_{IJ}^{(2)}} k_J\varphi \pmod h$$

holds for Maslov's elementary canonical operators k_I, k_J in the charts U_I, U_J. Thus, in order for the elementary canonical operators in the charts of the atlas $\{U_I\}$ to coincide, it is necessary to choose the constant functions S_I and choose the arguments $\operatorname{Arg}\mu_I$ in such a way that the equation

$$(i/h)c_{IJ}^{(1)} + i\pi c_{IJ}^{(2)} \equiv 0 \pmod{2\pi i}$$

holds. In the case considered in Sect. 4.1 the Lagrangian manifold L does not depend on h; therefore it follows from the last equation that $c_{IJ}^{(1)} = 0$ and $c_{IJ}^{(2)} \equiv 0 (\bmod 2)$. However, in our case the manifold L depends on the eigenvalue E which, in general, depends on the parameter h. Therefore in considering spectral problems there is no natural way of dividing the Maslov quantization conditions into a first and second, and the quantization condition for the manifold L_E should be written in the form

$$(1/h)\int_\gamma p\,dx + \frac{1}{2i}\int_\gamma d\ln\left[\det\frac{\partial(x+ip)}{\partial\alpha} \Big/ \left|\det\frac{\partial(x+ip)}{\partial\alpha}\right|\right] = 2\pi j\ ,$$

where j is a whole number, in accordance with the equation $e^{2\pi ij} = 1$. Using the values of the integrals calculated above, we have for the example under consideration (orienting L_E in the negative direction)

$$\frac{2\pi E}{h} - \pi = 2\pi j \quad \text{or} \quad E = h\left(j + \frac{1}{2}\right) = E_j(h)\ .$$

For these values of E there exists a canonical operator k_j on the system of manifolds $L_{E_j(h)}$; the corresponding approximate eigenfunction is equal to $\psi_j = k_j(\varphi)$, where φ satisfies the transport equation. In other words, the congruence

$$-\frac{h^2}{2}\frac{d^2\psi_j(x,h)}{dx^2} + x^2\psi_j(x,h) = E_j(h)\psi_j(x,h) \pmod{h^2}$$

holds. Let us compute the functions $\psi_j(x,h)$ for the given example. To do this we note that a global coordinate on the manifold L_E is the polar angle $\theta(\bmod 2\pi)$, and the Hamiltonian vector field on L_E equals

$$V(H) = p\frac{\partial}{\partial x} - x\frac{\partial}{\partial p} = -\frac{d}{d\theta} .$$

Therefore it is possible to choose the measure $d\theta$ as the measure μ, invariant with respect to the vector field $V(H)$. Let us denote by $U_N^{\pm}$ the charts on L_E where $\pm p > 0$ and with local coordinate x, and by $U_S^{\pm}$ the charts where $\pm x > 0$ with local coordinate p. The densities $\mu_N^{\pm}$ and $\mu_S^{\pm}$ of the nonsingular and singular charts respectively are

$$\mu_N^{\pm} = \frac{d\theta}{dx} = \left[\frac{dx}{d\theta}\right]^{-1} = -\frac{1}{\sqrt{2E}}\sin\theta = \left(2E - x^2\right)^{-1/2} ,$$

$$\mu_S^{\pm} = \frac{d\theta}{dp} = \left[\frac{dp}{d\theta}\right]^{-1} = \frac{1}{\sqrt{2E}}\cos\theta = \pm\left(2E - p^2\right)^{-1/2} ,$$

and it is possible to take the function φ to be unity, since the transport equation is $d\varphi/dt = 0$.

Note. To find the asymptotics of the spectrum in the general case, it is also necessary to modify the transport equation, replacing it by the equation

$$\left[V(H) - \frac{1}{2}\frac{\partial^2 H}{\partial x \partial p}\bigg|_L\right]\varphi = \epsilon\varphi .$$

However, in the example under consideration we may restrict ourselves to the null eigenvalue $\epsilon = 0$ of the transport operator (compare Example 5 of this section).

The actions $S_N^{\pm}$ and $S_S^{\pm}$ in the nonsingular and singular charts are, up to constants, equal to

$$S_N^{\pm} = \pm\left[\frac{x}{2}\left(2E - x^2\right)^{1/2} + E\arcsin\frac{x}{\sqrt{2E}}\right] ,$$

$$S_S^{\pm} = \pm\left[\frac{p}{2}\left(2E - p^2\right)^{1/2} + E\arcsin\frac{p}{\sqrt{2E}}\right] .$$

Therefore the functions $\psi_j(x,h)$ are

$$\begin{aligned}
&\psi_j(x,h)\\
&= A_j\left\{\exp\left[\frac{i}{h}\left[\frac{x}{2}\left(2E-x^2\right)^{1/2}+E\sin^{-1}\frac{x}{\sqrt{2E}}+C_n^+\right]\right]ie_N^+(x)\left(2E-x^2\right)^{-1/4}\right.\\
&+\exp\left[\frac{i}{h}\left[\frac{x}{2}\left(2E-x^2\right)^{1/2}-E\sin^{-1}\frac{x}{\sqrt{2E}}+C_N^-\right]\right]e_N^-(x)\left(2E-x^2\right)^{-1/4}\\
&+F_{p\to x}^{1/h}\left\{\exp\left[\frac{i}{h}\left[-\frac{p}{2}\left(2E-p^2\right)^{1/2}-E\sin^{-1}\frac{p}{\sqrt{2E}}+C_S^+\right]\right]e_S^+(p)\left(2E-p^2\right)^{-1/4}\right.\\
&\left.\left.+\exp\left[\frac{1}{h}\left[\frac{p}{2}\left(2E-p^2\right)^{1/4}+E\sin^{-1}\frac{p}{\sqrt{2E}}+C_S^-\right]\right]ie_S^-(p)\left(2E-p^2\right)^{-1/2}\right\}\right\}\Bigg|_E\\
&= h\left(j+\frac{1}{2}\right)
\end{aligned}$$

(Here A_j is a normalizing constant, chosen so that $\|\psi_j(x,h)\|_{L_2}=1$.) Here the possibility of choosing the constants $C_N^{\pm}$ and $C_S^{\pm}$ (depending on h), to guarantee the independence of the expression $\psi_j(x,h)(\operatorname{mod} h)$ from the choice of the partition of unity $\{e_N^{\pm}, e_S^{\pm}\}$, is a consequence of the quantization conditions when $E = h\left(j+\frac{1}{2}\right)$. We will not compute here the constants $c_N^{\pm}$ and $c_S^{\pm}$, since their values are not essential for the arguments to follow.

We note that when we estimated the remainder on the right-hand side of the equation in Sect. 4.2 we made use in an essential way of the fact that the smooth functions S_I, μ_I and φ do not depend on h, and therefore they and their derivatives in local coordinates are uniformly bounded in h. In the case we are considering, the functions S_I and μ_I depend on h, since they depend on $E = h\left(j+\frac{1}{2}\right)$. Moreover, it is clear that as $E \to 0$ or $E \to \infty$ the functions S_I and μ_I or their derivatives grow without bound (which is evident from the explicit expression for $\psi_j(x,h)$). Consequently, in the present case in order to guarantee the uniformity of the asymptotic expansions it is necessary to consider the values E lying in the bounded segment $0 < E_1 \le E \le E_2$.

The latter restriction, however, requires more attention. Indeed, the inequality

$$0 < E_1 \le h\left(j+\frac{1}{2}\right) \le E_2$$

cannot hold as $h \to 0$, if one considers the number j to be fixed. In other words, our constructions do not allow us to calculate the asymptotics of a *given* eigenvalue (for a fixed index j of this eigenvalue). Our formulas give us the *asymptotic values of the part of the spectrum lying in the segment* $[E_1, E_2]$. In other words we construct not the asymptotics of an eigenvalue of the operator, but the asymptotics of the set of eigenvalues in a certain segment. Here the corresponding set of indices of these eigenvalues depends on the parameter h:

$$E_j(h) = h\left(j + \frac{1}{2}\right) \in [E_1, E_2] \leftrightarrow j \in J(h)$$
$$= \left\{ j \in Z \mid (E_1/h) - \frac{1}{2} \leq j \leq (E_2/h) - \frac{1}{2} \right\} .$$

A precise proof of the fact that the set $\{E_j(h) | j \in J(h)\}$ gives the asymptotics of the spectrum of the operator $-\frac{h^2}{2}\frac{d^2}{dx^2} + x^2$ on the segment $[E_1, E_2]$ will be given in subheadings 2–4 of this section. However, the example we have just considered already enables us to point out certain properties of the theory of Maslov's canonical operator as applied to problems of finding the asymptotics of spectra.

1. Maslov's canonical operator, adapted to spectral problems, must be constructed not on an individual Lagrangian manifold but on a family of Lagrangian manifolds, which depend on a parameter which takes values in a finite set depending on h.

2. The quantization conditions for such a Maslov's canonical operator cannot be separated, as before, into two autonomous quantization conditions, but must be considered as a unified condition which relates the choice of constants in the action S_I and the choice of the argument of the measure μ_I.

3. Since the Lagrangian manifolds in the present situation depend on h, extra conditions need to be imposed on the family of Lagrangian manifolds in order to guarantee the uniformity of the asymptotic expansions (cf. conditions (2.3) and (2.4) below).

7.2.2 Packets of Lagrangian Manifolds. Now we proceed to the precise definitions.

Definition 2.1. A *packet of Lagrangian manifolds* is a triple $\{L, J(h), i_{(h,j)}\}$, where L is a compact manifold; $J(h)$ is a family of finite sets depending on the parameter $h \in (0, 1]$; $i_{(h,j)} : L \to \Phi$, $j \in J(h)$ is a family of imbeddings, for which the equation

$$i^*_{(h,j)} dp \wedge dx = 0 \tag{2.1}$$

holds for each h and $j \in J(h)$.

From condition (2.1) it follows that a family of canonical atlases $\{U_I\}_{(h,j)}$ exists on the manifold. The existence of such an atlas for each h, $j \in J(h)$ is equivalent to the condition

$$\sum_{I\in[n]} \left| \det \frac{\partial\left(i^*_{(h,j)}x^I, i^*_{(h,j)}p_{\bar{I}}\right)}{\partial\xi} \right|^2 \neq 0 \tag{2.2}$$

in any local system of coordinates ξ.

We will require a strengthening of condition (2.2). To wit, we fix a certain finite atlas (V_i, ξ_i), $i = 1, \ldots, m$. We call a packet of Lagrangian manifolds *uniformly* Lagrangian if there exists a constant $\delta > 0$ such that

$$\min_{i=1,2,\ldots,m} \inf_{\xi_i\in V_i'} \sum_{I\subset[n]} \left| \det \frac{\partial\left(i^*_{(h,j)}x^I, i^*_{(h,j)}p_{\bar{I}}\right)}{\partial\xi_i} \right|^2 \geq \delta > 0 , \tag{2.3}$$

$$\max_{i=1,2,\ldots,m} \sup_{\xi_i\in V_i'} \left\{ \left|D^\alpha i^*_{(h,j)}x\right|, \left|D^\alpha i^*_{(h,j)}p\right| \right\} \leq M_\alpha \tag{2.4}$$

for any multi-index α, where $\{V_i'\}$ is an open covering of L such that V_i' is compactly imbedded in V_i for each i: $V_i' \subset\subset V_i$.[3]

Note. In view of the fact that

$$\sum_{I\subset[n]} \left| \det \frac{\partial\left(i^*_{(h,j)}x^I, i^*_{(h,j)}p_{\bar{I}}\right)}{\partial\xi_k'} \right|^2 = \left| \det \frac{\partial\xi_i}{\partial\xi_k'} \right|^2 \sum_{I\subset[n]} \left| \frac{\partial\left(i^*_{(h,j)}x^I, h^*_{(h,j)}p_{\bar{I}}\right)}{\partial\xi_i} \right|^2$$

condition (2.3) is independent of the choice of a finite atlas and the sets V_i'.

The requirement that a packet be uniformly Lagrangian must be imposed in order to guarantee that the asymptotic expansions are uniform, as was discussed at the end of subheading 1 (conclusion 3). In connection with this we should point out that, as a result of this requirement, the sizes of the supports of the elements of the partition of unity corresponding to the canonical atlas do not decrease without bound (cf. Lemma 2.1 below). We note that in the example studied in subheading 1 the requirement of a uniformly Lagrangian packet is satisfied if, for the given packet of Lagrangian manifolds (consisting of circles of radius $\sqrt{2E}$), the inequality $E \geq \delta > 0$ is satisfied. This condition, consequently, forbids contracting the circles to a point in that example.

Now we will define two function spaces which are essential to construct the theory of the canonical operator on Lagrangian packets.

Definition 2.2. We denote by $C_h^\infty(L)$ the space whose elements are families of functions $f_{(h,j)}(\xi)$ parametrized by $h \in (0,1]$ and an index $j \in J(h)$

[3] We note that we have imposed no regularity conditions on the dependence of $i_{(h,j)}$ on h and j.

such that the derivatives of the functions $f_{(h,j)}$ with respect to local coordinates of the manifold L are uniformly bounded in (h,j), $j \in J(h)$.

Now we consider the ring of polynomials in the variable h of the form

$$\sum_{k=0}^{N'} h^k f_{k(h,j)}(\xi) ,$$

where $f_{k(h,j)}(\xi) \in C_h^\infty(L)$. (Here N' depends on the polynomial.) We denote this space by $C_h^\infty(L)[h]$. The set I consisting of those elements of $C_h^\infty(L)[h]$ of the form

$$\sum_{k=N}^{N'} h^k f_{k(h,j)}(\xi)$$

is, as is not hard to see, an ideal in the ring $C_h^\infty(L)[h]$.

Definition 2.3. We denote by $C_h^\infty(L)[h]_N$ the factor ring

$$C_h^\infty(L)[h]_N = C_h^\infty(L)[h]/I .$$

We fix some Riemannian metric on the manifold L. We will denote the distance in this metric by the symbol $r = r(\xi', \xi'')$.

Lemma 2.1. *There exists a family of canonical atlases $\{U_I\}_{(h,j)}$, induced by the imbeddings $i_{(h,j)}$, on the manifold L such that all the charts U are open balls. The radii of these balls are bounded below by a positive constant, which is independent of h and $j \in J(h)$.*

Proof. We use the following form of the implicit function theorem. Suppose $F_k(x_1, \dots, x_n, y_1, \dots, y_m)$, $1 \le k \le m$, is a system of functions such that the estimates

$$|\partial F_k/\partial x_i| \le N , \quad |\partial F_k/\partial y_j| \le N , \quad |\partial^2 F_k/\partial y_i \partial y_l| \le M ,$$
$$|\det \partial F_k/\partial y_j| \ge \delta ,$$

hold uniformly in the variables $(x_1, \dots, x_n, y_1, \dots, y_m)$. Suppose the point $(x_0, y_0) = (x_{10}, \dots, x_{n0}, y_{10}, \dots, y_{m0})$ is a solution of the system $F_k(x_0, y_0) = 0$. Then there exists a constant $C_{m,n}$ depending only on m, n, such that for

$$|x - x_0| \le \frac{C_{mn}\delta^2}{MN^{2m+1}} \tag{2.5}$$

there exists a unique solution

$$y_i = y_i(x) , \qquad i = 1, 2, \dots, m ,$$

of the system of equations

$$F_1(x_1,\dots,x_n,y_1,\dots,y_m) = 0\,,$$
$$\dots$$
$$F_m(x_1,\dots,x_n,y_1,\dots,y_m) = 0\,,$$

which equals y_0 at $x = x_0$.

Now suppose that $a > 0$ is a number small enough that the open ball of radius a with respect to the metric r with center at any point of the manifold L is entirely contained in some neighborhood V_j'. It is obvious that for any number b, $b > 0$, one can choose $a > 0$ such that the open ball of radius a with respect to the metric r about any point of the manifold L lies in the ball of radius b with respect to the metric defined in a neighborhood of its center by the local coordinates ξ. Now we choose

$$b = \frac{C_{nn}\delta}{M^{2n+2}2^n}\,,$$

where $M = \max M_\alpha$. M_α and δ are the constants appearing in the estimates (2.3) and (2.4), and C_{nn} is the constant which appears in expression (2.5) when $m = n$. We assume without loss of generality that $M > 1$. We will show that, for all h and $j \in J(h)$, any ball of radius a (in the metric r) admits functions $i^*_{(h,j)}x^I$ and $i^*_{(h,j)}p_{\overline{I}}$ as coordinate functions for some $I \subset [n]$. Indeed, suppose $K_a(\xi_0)$ is such a ball with center at the point ξ_0. By choice of the number a, this ball lies entirely within the subset V_j' of some chart (V_i, ξ_i). We consider the system of equations

$$\begin{aligned} x^I - i^*_{(h,j)}x^I\left(\xi_i^1,\dots,\xi_i^n\right) &= 0\,,\\ p_{\overline{I}} - i^*_{(h,j)}p_{\overline{I}}\left(\xi_i^1,\dots,\xi_i^n\right) &= 0\,. \end{aligned} \tag{2.6}$$

By condition (2.3) there exists a set $I \subset [n]$ such that

$$\left|\det\frac{\partial\left(i^*_{(h,j)}x^I, i^*_{(h,j)}p_{\overline{I}}\right)}{\partial\xi_i}\right| \geq \sqrt{\frac{\delta}{2^n}} = \frac{\sqrt{\delta}}{2^{n/2}}\,.$$

Moreover, if we let

$$x_0^I = i^*_{(h,j)}x^I\left(\xi_{i_0}^1,\dots,\xi_{i_0}^n\right)\,,$$
$$p_{\overline{I}_0} = i^*_{(h,j)}p_{\overline{I}}\left(\xi_{i_0}^1,\dots,\xi_{i_0}^n\right)\,,$$

where $\left(\xi_{i_0}^1,\dots,\xi_{i_0}^n\right)$ are the coordinates of the point ξ_0 in the system of coordinates (ξ_i), then at the point $\left(x_0^I, p_{\overline{I}_0}\right)$ the left-hand sides of the system (2.6) are equal to zero. Furthermore, the first derivatives of the left-hand sides with respect to the variables $(x^I, p_{\overline{I}})$ are equal to either one or zero, and the first- and second-order derivatives with respect to the variables ξ_i are no greater than M, by virtue of conditions (2.4). Therefore the conditions of the implicit function theorem are met by the system (2.6) with $M = N$ and δ replaced by $\delta^{1/2}/2^{n/2}$. The implicit function theorem guarantees that, within a neighborhood of ξ_0 of radius (with respect to the metric induced by the coordinates ξ_i)

$$\frac{C_{nn}}{M \cdot M^{2n+1}} \left(\frac{\sqrt{\delta}}{2^{n/2}} \right)^2 = \frac{C_{nn}\delta}{M^{2n+2}2^n} = b$$

the system of Eq. (2.6) has a unique solution. But by the choice of a, the ball of radius a lies completely within this neighborhood. This proves the lemma. □

Note. In essence we have shown that the family of canonical atlases $\{U_I\}_{(h,j)}$, $h \in (0,1]$, $j \in J(h)$, can be chosen in such a way that the set of supports of the charts of this family is independent of the parameter h and the index $j \in J(h)$. As such a universal set of supports one can take the system of balls of radius a (in the metric r). Here, of course, the coordinates in each such ball do depend on the choice of h and j.

It is possible, by the compactness of the manifold L, to choose a finite subcovering of the family of charts of canonical atlases. We fix one such family, U_i, $i = 1, 2, \ldots, l$. Obviously, in each set U_i there exist coordinates ξ_i such that for any $h \in (0,1]$ and $j \in J(h)$, if $x^I, p_{\overline{I}}$ are the canonical coordinates in U_i, then

$$\left|\det \partial \left(x^I, p_{\overline{I}}\right) / \, \partial\xi_i\right| \geq \frac{\delta^{1/2}}{2^{n/2}} \,. \tag{2.7}$$

It is not hard to see that none of the constructions which we will carry out below depend on the choice of the finite subcovering $\{U_i\}$, $i = 1, \ldots, 1$.

Corollary. *There exists a family of finite partitions of unity on L, $\{e_i\}_{(h,j)}$, subordinate to the covering U_i, $i = 1, 2, \ldots l$, such that the derivatives with respect to the canonical coordinates for each h, $j \in J(h)$ are uniformly bounded with respect to h, $j \in J(h)$.*

Proof. It is sufficient to consider a particular partition of unity $\{e_i\}$ subordinate to the covering U_i. The boundedness of the corresponding derivatives follows from inequality (2.7). This proves the corollary. □

Now we define the concept of a quantized Lagrangian packet.

Definition 2.4. We denote by ${}^{N}I_h\,(U,\mathbb{Z}_2)$ the set of elements $f\in C_h^\infty(U)$ for which there exists an integer $k\in\mathbb{Z}$ such that

$$f-2\pi k=O\left(h^N\right)$$

uniformly with respect to all variables.

Now suppose a family of measures $\mu_{(h,j)}$ is given on L, with the property that the density of the measure in each local system of coordinates U_i belong to $C_h^\infty(U)$.

Definition 2.5. The packet $\left(L,J(h),i_{(h,j)}\right\}$, together with the family of measures $\mu_{(h,j)}$, is called a *quantized packet with measure* if there exist families of cocycles $\{S_I\}_{(h,j)}$ and $\{\mathrm{Arg}\mu_I\}_{(h,j)}$ of the covering $\{U_i\}$ such that

1.

$$dS_{I(h,j)}=i^*_{(h,j)}\left(p_I dx^I-x^{\overline{I}}dp_{\overline{I}}\right)\;;$$

2. $\mathrm{Arg}\mu_I(h,j)$ is one of the values of the density of the measure μ with respect to the coordinates $\left(x^I,p_{\overline{I}}\right)$ and

$$\left\{h^{-1}\left[S_I-S_J-x^{\overline{J}}p_J+x^{\overline{I}}p_{\overline{I}}\right]-\frac{1}{2}\left[(\mathrm{Arg}\mu)_I-(\mathrm{Arg}\mu)_J\right]\right.$$
$$\left.-\sum_k\arg\lambda_k+|I_2|\,\pi\right\}_{(h,j)}\in{}^{N}I_h\left(U_i\cap U_j,\mathbb{Z}_2\right)\;, \tag{2.8}$$

where λ_k are the eigenvalues of the matrix (1.18),

$$\mathrm{Hess}_{\left(x^{I_2},p_{I_3}\right)}(-S_I)\;,\qquad -3\pi/2<\arg\lambda_k\le\pi/2\;.$$

By the considerations presented at the end of the preceding section (which can be carried over without changes to the case of Lagrangian packets), it is possible to choose coclycles $S_{I(h,j)}$ and $\mathrm{Arg}\mu_{I(h,j)}$ in the indicated way if and only if the following condition holds:

$$h^{-1}\int_\gamma p\,dx+(\pi/2)\mathrm{ind}\gamma\equiv 2\pi k\;\left(\mathrm{mod}h^N\right) \tag{2.9}$$

for any generating cycle γ of the manifold L.

7.2.3 Maslov's Canonical Operator on Lagrangian Packets. Let $\{L,J(h),$ $i_{(h,j)}\}$ be a quantized packet and $\{U_i\}$ be a finite set of supports of canonical charts. As was shown in 7.2.2, this set is independent of h and

$j \in J(h)$. Now we define a local canonical operator in the chart U_i. Let $\varphi_{(h,j)} \in C_h^\infty(U_i)$ have compact support in U_i.

We define the local canonical operator by the formula

$$k_{U_I}\varphi_{(h,j)} = \overline{F}_{p_{\overline{I}} \to x^{\overline{I}}} \exp\left\{\frac{i}{h} S_{I(h,j)}(x^I, p_{\overline{I}})\right\} \times \left[\mu_{I(h,j)}(x^I, p_{\overline{I}})\right]^{-1/2} \varphi_{(h,j)}(x^I, p_{\overline{I}}) \,. \tag{2.10}$$

The value of the elementary canonical operator at an element $\varphi_{(h,j)} \in C_h^\infty(U_i)$ is a family of functions in the variable $x \in \mathbb{R}^n$, parametrized by the parameters $h \in (0,1]$ and $j \in J(h)$. Let $H_r^{1/h}(\mathbb{R}_x^n)$ denote the completion of the space of functions in $C_0^\infty(\mathbb{R}_x^n)$, depending on the parameters h and $j \in J(h)$, with respect to the norm

$$\sup_{\substack{h\in(0,1)\\ j\in J(h)}} \left\| \left(1 + \Delta_x + x^2\right)^{r/2} f_{(h,j)} \right\|_{L_2} = \|f\|_{r,h} \,.$$ [4]

We denote by $H^{1/h}(\mathbb{R}_x^n)$ the projective limit of the spaces $H_r^{1/h}(\mathbb{R}_x^n)$ as $r \to \infty$.

The $1/h$-Fourier transform defines for any r a continuous mapping

$$F_{x\to p}^{1/h} : H_r^{1/h}(\mathbb{R}_x^n) \to H_r^{1/h}(\mathbb{R}_{n,p}) \,. \tag{2.11}$$

Definition 2.6. We denote by ${}^N\! IH^{1/h}(\mathbb{R}_x^n)$ the subspace of $H^{1/h}(\mathbb{R}_x^n)$ such that $f_{(h,j)} \in {}^N\! IH^{1/h}(\mathbb{R}_x^n)$ when and only when for any r,

$$\left\|f_{(h,j)}\right\|_{r,h} \le C_r h^N \,,$$

where the constant C_r does not depend on $h \in (0,1]$ or $j \in J(h)$. We denote by ${}^N\! H^{1/h}(\mathbb{R}_x^n)$ the factor space

$${}^N\! H^{1/h}(\mathbb{R}_x^n) = \frac{H^{1/h}(\mathbb{R}_x^n)}{{}^N\! IH^{1/h}(\mathbb{R}_x^n)} \,.$$

Proposition 2.2. *The elementary Maslov's elementary operator* (2.10) *defines a continuous map of the spaces*

---

[4] Here $\Delta_x = \widehat{p}^{\,2}$ is the "quantum" Laplace operator:

$$\widehat{p}^{\,2} = -\frac{\Delta}{h^2} \,.$$

$$k_{U_i} : C_h^\infty(U_i) \to {}^N H^{1/h}(\mathbb{R}_x^n) .$$

Proof. For even r we have

$$\left(1 + \Delta_{(x^I, p_{\overline{I}})} + \left|x^I\right|^2 + \left|p_{\overline{I}}\right|^2\right)^{r/2} \exp\left\{\frac{i}{h} S_{I(h,j)}\right\} \times \left[\mu_{I(h,j)}\right]^{-1/2} \varphi_{(h,j)}(x^I, p_{\overline{I}}) = \sum_{k=0}^{r} h^k \Phi_k\left(x^I, p_{\overline{I}}, h\right) , \tag{2.12}$$

where the functions $\Phi_k\left(x^I, p_{\overline{I}}, h\right)$ are uniformly bounded in (h, j) by the definition of the spaces $C_h^\infty(U_i)$. Moreover, the functions $\Phi_k\left(x^I, p_{\overline{I}}, h\right)$ can be expressed as a linear combination of functions of the form $x^\alpha \widehat{p}^\beta\left(\varphi_{(h,j)}\right)$, $|\alpha| + |\beta| \leq r$, with coefficients belonging to $C^\infty(U_i)$. Consequently, for any even r the L_2 norm of the right-hand side of Formula (2.12) can be bounded by the norm $\left\|\varphi_{(h,j)}\right\|_{r,h}$, times a constant which is independent of h and $j \in J(h)$. Moreover, it is also obvious that the functions Φ_k have compact support. Therefore the L_2-norm of the right-hand side of (2.12) is bounded by a constant independent of (h, j). Now the continuity of the operator (2.11) shows that

$$k_{U_i} \varphi_{(h,j)} \in H^{1/h}(\mathbb{R}_x^n) ,$$

and the operator k_{U_i} is continuous. This shows the proposition. □

Corollary. *The operator k_{U_i} defines a continuous mapping of the space*

$$k_{U_i} : C_h^\infty(U_i)\,[h]_N \to H^{1/h}(\mathbb{R}_x^n) .$$

Proof. The desired assertion follows from the continuity of the mapping

$$h^k : {}^s H^{1/h}(\mathbb{R}_x^n) \to {}^{s+k} H^{1/h}(\mathbb{R}_x^n) ,$$

which can be checked directly. □

Now we will define for Lagrangian packets certain operators V_{li}, analogous to the operators V_{IJ}. To simplify the proofs we will need a slightly different representation of the operator k_{U_i}, which is completely equivalent so that of (2.10). Namely,

$$k_{U_i} \varphi_{(h,j)} = \overline{F}^{1/h}_{p_{\overline{I}} \to x^{\overline{I}}} \exp\left\{\frac{i}{h} S_{I(h,j)}(x^I, p_{\overline{I}}) - \frac{i}{2} (\operatorname{Arg}\mu)_{I(h,j)}(x^I, p_{\overline{I}})\right\} \left|\mu_{I(h,j)}\left(x^{\overline{I}}, p_{\overline{I}}\right)\right|^{-1/2} \varphi_{(h,j)}(x^I, p_{\overline{I}}) .$$

The operator V_{li}, defined for any two charts U_i, U_l with $U_i \cap U_l \neq \emptyset$,

$$V_{li} : C_h^\infty (U_i \cap U_l) [h]_N \to C_h^\infty (U_i \cap U_l) [h]_N ,$$

is composed of a family of operators $V_{li(h,j)}$. These operators $V_{li(h,j)}$ must satisfy the condition

$$k_{U_i} : V_{li(h,j)}\varphi_{(h,j)} - K_{U_l}\varphi_{(h,j)} = O\left(h^N\right) \tag{2.13}$$

for any function $\varphi_{(h,j)} \in C_h^\infty (U_i \cap U_l) [h]_N$ with support in the intersection $U_i \cap U_l$. Condition (2.13) will be satisfied if the congruence

$$\begin{aligned} &\exp\left\{i\left[\frac{1}{h}S_{I(h,j)}(x^I, p_{\overline{I}}) - \frac{1}{2}(\operatorname{Arg}\mu)_{I(h,j)}(x^I, p_{\overline{I}})\right]\right\} \\ &\quad \times \left|\mu_{I(h,j)}(x^I, p_{\overline{I}})\right|^{-1/2} \left(V_{li(h,j)}\varphi_{(h,j)}(x^I, p_{\overline{I}})\right) \\ &\equiv F^{1/h}_{x^I \to p_{\overline{I}}} \overline{F}^{1/h}_{p_{\overline{J}} \to x^J} \exp\left\{\frac{i}{h}S_{J(h,j)}(x^J, p_{\overline{J}}) - \frac{1}{2}(\operatorname{Arg}\mu)_{J(h,j)}(x^J, p_{\overline{J}})\right\} \\ &\quad \times \left|\mu_{J(h,j)}(x^J, p_{\overline{J}})\right|^{-1/2} \varphi_{(k,j)}(x^J, p_{\overline{J}}) \pmod{h^N} \end{aligned} \tag{2.14}$$

holds. Here U_I is the chart corresponding to the set U_i for the given (h, j); U_J is the chart corresponding to U_l for the same (h, j). As before, we introduce the sets

$$\begin{aligned} &I_1 = I \cap J\ , \quad I_2 = I \setminus I_1\ , \quad I_3 = J \setminus I_1\ , \\ &I_4 = [n] \setminus (I_1 \cup I_2 \cup I_3)\ . \end{aligned}$$

Then the right-hand side of Eq. (2.14) can be rewritten in the form

$$\begin{aligned} &\left(\frac{i}{2\pi h}\right)^{|I_2+I_3|} (-1)^{|I_2|} \iint \exp\left\{\frac{i}{h}\left(-x^{I_3}p_{I_3} + x^{I_2}p_{I_2} + S_{J(h,j)}(x^J, p_{\overline{J}})\right)\right\} \\ &\quad \times \left[\mu_{J(h,j)}(x^J, p_{\overline{J}})\right]^{-1/2} \varphi_{(h,j)}(x^J, p_{\overline{J}}) dx^{I_3} dp_{I_3} = I\ . \end{aligned} \tag{2.15}$$

We apply to the integral in (2.15) the formula for the asymptotic decomposition of an integral of a rapidly oscillating function:

$$\begin{aligned} I = &\exp\left\{\frac{i}{h}\left(-x^{I_3}p_{I_3} + x^{I_2}p_{I_2} + S_{J(h,j)}\left(x^{I_1}, x^{I_3}, p_{I_2}, p_{I_4}\right)\right)\right\} \\ &\times \left[\det \operatorname{Hess}_{x^{I_3}, p_{I_2}}\left(-S_{J(h,j)}\left(x^{I_1}, x^{I_2}, p_{I_2}, p_{I_4}\right)\right)\right]^{-1/2} W_k[\varphi, a] + O\left(h^N\right)\ , \end{aligned}$$

where

$$\Phi = -x^{I_3}p_{I_3} + x^{I_2}p_{I_2} + S_{J(h,j)}\left(x^{I_1}, x^{I_3}, p_{I_2}, p_{I_4}\right) ,$$

$$a = \left[\mu_{J(h,j)}\left(x^{I_1}, x^{I_3}, p_{I_2}, p_{I_4}\right)\right]^{1/2} \varphi(h,j)\left(x^{I_1}, x^{I_3}, p_{I_2}, p_{I_4}\right)$$

noting that $x^{I_3} = x^{I_3}(x^I, p_{\overline{I}})$, $p_{I_2} = p_{I_2}(x^I, p_{\overline{I}})$, where x^{I_3}, p_{I_2} are a solution of the system of equations

$$\begin{aligned} -p_{I_3} + \frac{\partial S_{J(h,j)}}{\partial x^{I_3}}\left(x^{I_1}, x^{I_3}, p_{I_2}, p_{I_4}\right) = 0 , \\ x^{I_2} + \frac{\partial S_{J(h,j)}}{\partial p_{I_2}}\left(x^{I_1}, x^{I_3}, p_{I_2}, p_{I_4}\right) = 0 . \end{aligned} \tag{2.16}$$

As before, it can be checked that the solutions of the system (2.16) are the functions which define the change of coordinates from U_J to U_I. Next, our usual calculations lead to the formula

$$\begin{aligned} I \equiv \exp\Bigg\{ & \frac{i}{h} x^{I_3}p_{I_3} + x^{I_2}p_{I_2} + S_J\left(x^{I_1}, x^{I_3}, p_{I_2}, p_{I_4}\right) \\ & + \frac{i}{2}\Big((\operatorname{Arg}\mu)_{J(h,j)} + \sum_k \arg\lambda_k\Big)\Bigg\} \left[\mu_{I(h,j)}\left(x^{I_1}, x^{I_3}, p_{I_2}, p_{I_4}\right)\right]^{1/2} \\ & \times \left[\mu_{J(h,j)}\left(x^{I_1}, x^{I_3}, p_{I_2}, p_{I_4}\right)\right]^{-1/2} \sum_{k=0}^{N} \{W_k[\varphi, a]\} , \end{aligned} \tag{2.17}$$

where all the congruences are understood to be modulo h^N uniformly in (h,j), $h \in (0,1]$, $j \in J(h)$.

Formula (2.17) along with the congruences (2.9) shows that one can use the operators V_{JI} defined before as the operators $V_{li(h,j)}$, for each fixed $h \in (0,1]$ and $j \in J(h)$.

Now we study certain properties of the operators $V_{li(h,j)}$. We will show the congruences

$$\begin{aligned} V_{li(h,j)}V_{il(h,j)}\varphi \equiv \varphi \pmod{h^N} , \\ V_{li(h,j)}V_{kl(h,j)}\varphi \equiv V_{ik(h,j)}\varphi \pmod{h^N} . \end{aligned} \tag{2.18}$$

To do this we consider the chart U_i as a Lagrangian manifold. We also consider its covering by the sets $V_l = U_l \cap U_i$. As the cochain S we choose the cochain $\widetilde{S}_J = S_I + x^I p_{\overline{I}} - x^J p_{\overline{J}}$. (All of this discussion takes place for fixed (h,j), with U_I being the type of the chart U_i for the given (h,j).) The formula for the decomposition of the integral of a rapidly oscillating function shows that the operators V_{IJ} corresponding to U_i are differential operators whose coefficients depend only on derivatives of the functions $\widetilde{S}_J$. As was shown, the operators V_{IJ} satisfy conditions (2.18). Furthermore, the derivatives of the functions $\widetilde{S}_J$ coincide with the derivatives of the functions S_J up to $O\left(h^N\right)$. This follows from Formula (2.8), if one uses the equation

$$(\text{Arg}\mu)_I - (\text{Arg}\mu)_J - \sum_k \arg \lambda_k + |I_2|\,\pi = \text{const} .$$

The congruence (2.18) follows from this.

We now define a local canonical operator in the chart U_i by the formula

$$K_{U_i}\varphi_{(h,j)} = \sum_l k_{U_i} V_{li(h,j)} e_l \varphi_{(h,j)} , \quad \text{for all } \varphi_{(h,j)} \in C_h^\infty(U_i)\,[h]_N . \quad (2.19)$$

The following assertions hold.

Proposition 2.3. *The restriction of the operators*

$$K_{U_i} : C_h^\infty(U_i)\,[h]_N \to {}^N H^{1/h}(\mathbb{R}_x^n) ,$$
$$K_{U_l} : C_h^\infty(U_l)\,[h]_N \to {}^N H^{1/h}(\mathbb{R}_x^n)$$

to the set of functions in $C_h^\infty(U_i \cap U_l)\,[h]_N$ *with support in the intersection* $U_i \cap U_l$ *coincide.*

Theorem 2.4. *There exists a unique operator*

$$K : C_h^\infty(L)[h]_N \to {}^N H^{1/h}(\mathbb{R}_x^n)$$

whose restriction to functions in $C_h^\infty(U_i)\,[h]_N$ *with support in* U_i *coincide with the operators* (2.19).

The *Proofs* of these assertions are identical with the proofs of the corresponding assertions in Chap. 6. □

We proceed now to formulate the commutation theorem in the case of Lagrangian packets.

Definition 2.7. A quantized packet $\{L, J(h), i_{(h,j)}\}$ with measure $\mu_{(h,j)}$ is said to be *associated with the Hamiltonian* $H(x,p)$ if there exists a system of constants $E_{(h,j)}$ (which we will call energy levels) such that

1) $E_{(h,j)}$ are uniformly bounded with respect to $h \in (0,1]$ and $j \in J(h)$;

2) $i_{(h,j)}H(x,p) = E_{(h,j)}$, and the measure $\mu_{(h,j)}$ is invariant with respect to the vector field $V(H)$, i.e.

$$\mathcal{L}_{V(H)}\mu_{(h,j)} = 0 . \quad (2.20)$$

We note that the requirement that (2.20) hold makes sense because, by Lemma 1.2 of Chap. 2, the manifold L is invariant with respect to $V(H)$.

The condition that the packet $\{L, J(h), i_{(h,j)}\}$ with measure $\mu_{(h,j)}$, associated with the Hamiltonian H, is *quantized* can be viewed as an equation with respect to the energy levels $E_{(h,j)}$. Indeed, by relation (2.9) and the definition of the index $\operatorname{ind}\gamma$ of the cycle γ, quantization is equivalent to the system of congruences

$$\frac{i}{h}\int_{\gamma_j} p\,dx + \int_{\gamma_j}(2i)^{-1}d\ln\frac{\det\frac{\partial(x+ip)}{\partial\alpha}}{\left|\det\frac{\partial(x+ip)}{\partial\alpha}\right|} \equiv 2\pi m_j \pmod{h^N}\ ,$$

where $m_j \in \mathbb{Z}$ are whole numbers and $\{\gamma_j\}$ is a basis for the one-dimensional homology of the manifold L. The latter congruence is an equation in $E_{h,j}$, since the integrals on the left-hand side of this congruence depend on the imbedding $i_{(h,j)}$ and this is associated with $E_{h,j}$ by the relation $i^*_{(h,j)}H(x,p) = E_{h,j}$.

Theorem 2.5. *Suppose the quantized packet $\{L, J(h), i_{(h,j)}\}$ with measure $\mu_{(h,j)}$ is associated with the Hamiltonian $H(x,p)$. Then there exists an operator*

$$\mathcal{P} : C_h^\infty(L)[h]_N \to C_h^\infty(L)[h]_N$$

such that $\mathcal{P} = \{\mathcal{P}_{(h,j)}\}$,

$$H(x,\widehat{p})\,K\varphi_{(h,j)} \equiv -ihK\mathcal{P}_{(h,j)}\varphi_{(h,j)} + E_{h,j}K\varphi_{(h,j)} \pmod{h^N} \qquad (2.21)$$

for functions in $C_h^\infty(L)[h]_N$. Here the operators $\mathcal{P}_{(h,j)}$ are defined locally for each (h,j) as in Sect. 6.2.

The *Proof* is carried out completely analogously to the proof of the corresponding assertion for an individual Lagrangian manifold (Sect. 6.2). □

7.2.4 Asymptotic Spectra. In this article we will prove a theorem on the convergence of the spectrum of a $1/h$-pseudodifferential operator $\widehat{H} = H(x,\widehat{p})$. For this we will impose certain conditions on the operator $\widehat{H}$.

Theorem 2.6. *Suppose $\{L, J(h), H(h,j)\}$ is a Lagrangian packet with measure, associated with the Hamiltonian H; let $E_{(h,j)}$ be the corresponding system of energy levels. Suppose $\epsilon_{(h,j),k} = \sum \epsilon^\sigma_{(h,j)}h^\sigma$, $k \in \mathcal{B}$ is a system of eigenvalues of the operator $\mathcal{P}$ corresponding to eigenfunctions $\varphi_{(h,j),k}$ such that the norms of the functions $K\varphi_{(h,j)}$ in $L_2(\mathbb{R}^n)$ are bounded below by a positive constant independent of $h \in (0,1]$ and $j \in J(h)$. We assume that*

1) The numbers $\epsilon^\sigma_{(h,j)}$ are uniformly bounded with respect to $h \in (0,1]$ and $j \in J(h)$.

2) *The derivatives of the functions* $\varphi_{(h,j),k}$ *with respect to local coordinates on the manifold* L *are uniformly bounded with respect to* $h \in (0,1]$, $j \in J(h)$, $k \in \mathcal{B}$.

3) *There exist positive constants* C *and* M *such that*

$$\|(H - \lambda I)^{-1}\| \leq \frac{C}{d^M} ,$$

where d *is the distance from the point* λ *to the spectrum* $\mathrm{spec}\widehat{H}$ *of the operator* $\widehat{H}$. *Then there exists a constant* $C_2 > 0$ *such that the set*

$$\Lambda_h = \{\lambda : \lambda = E_{(h,j)} - ih\epsilon_{(h,j),k}\}$$

lies in a $C_1 h^{N/M}$*-neighborhood of the set* $\mathrm{spec}\widehat{H}$ *for every* $h \in (0,1]$. We note that for self-dual operators condition (3) is satisfied for $M = 1$.

Proof. We apply the operator $\widehat{H}$ to the functions $\varphi_{(h,j),k}$. By (2.21) we have

$$\begin{aligned} \widehat{H} K\varphi_{(h,j),k} &= H(x,\widehat{p})\, K\varphi_{(h,j),k} = -ihK\mathcal{P}_{(h,j)}\varphi_{(h,j),k} + E_{h,j}K\varphi_{(h,j),k} \\ &+ \varphi_{(h,j),k} = \left(E_{h,j} - ih\epsilon_{(h,j),k}K\varphi_{(h,j),k}\right) + \psi_{(h,j),k} , \end{aligned} \tag{2.22}$$

where

$$\psi_{(h,j),k} \in {}^N I H^{1/h}(\mathbb{R}^n_x) .$$

In particular,

$$\|\psi_{(h,j),k}\|_{L_2(\mathbb{R}^n_x)} \leq \widetilde{C} h^N \tag{2.23}$$

for some constant $\widetilde{C}$ independent of (h,j) and k. We rewrite Eq. (2.22) in the form

$$\left[\widehat{H} - \left(E_{(h,j)} - ih\epsilon_{(h,j),k}\right)\right] K\varphi_{(h,j),k} = \psi_{(h,j),k} . \tag{2.24}$$

Now let us break up the set Λ_h into two parts: Λ_1 and Λ_2, where $\Lambda_1 = \Lambda_h \cap \mathrm{spec}\widehat{H}$, $\Lambda_2 = \Lambda_h - \Lambda_1$. The set Λ_1 lies in any neighborhood of the spectrum $\mathrm{spec}\widehat{H}$. We will show that there exists a constant C_1 such that Λ_2 lies in a $C_1 h^{N/M}$-neighborhood of the set $\mathrm{spec}\widehat{H}$ for any $h \in (0,1]$. Applying the operator $\left(\widehat{H} - \left(E_{(h,j)} - ih\epsilon_{(h,j),k}\right)\right)^{-1}$ to Eq. (2.24) and taking the norm of the left- and right-hand sides, we conclude that

$$\begin{aligned} C_2 &\leq \|K\varphi_{(h,j),k}\|_{L_2} = \left\|\left[H - \left(E_{(h,j)} - ih\epsilon_{(h,j),k}\right)\right]^{-1}\psi_{(h,j),k}\right\|_{L_2} \\ &\leq \frac{C}{\left[d\left(\lambda, \mathrm{spec}\widehat{H}\right)\right]^M}\widetilde{C}h^N , \end{aligned} \tag{2.25}$$

where in the last step we have used the estimate (2.23). From Eq. (2.25) we obtain for any $j \in J(h)$, $k \in B_h$ the following estimate:

$$d\left(\lambda, \mathrm{spec}\widehat{H}\right) \leq \left(C\widetilde{C}/C_2\right)^{1/M} h^{[N]/M}, \tag{2.26}$$

where the constants in (2.26) do not depend on $h \in (0,1]$, $j \in J(h)$, $k \in B_h$. Taking the least upper bound of the function $d\left(\lambda, \mathrm{spec}\widehat{H}\right)$ over all $\lambda \in \Lambda_2$ and observing that, for $\lambda \in \Lambda_1$, $d\left(\lambda, \mathrm{spec}\widehat{H}\right) = 0$, we obtain

$$\sup d\left(\lambda, \mathrm{spec}\widehat{H}\right) \leq \left(C\widetilde{C}/C_2\right)^{1/M} h^{[N]/M}.$$

The latter inequality proves the theorem with $C_1 = \left(C\widetilde{C}/C_2\right)^{1/M}$. □

Corollary. *Suppose the hypotheses of Theorem 2.6 hold and, moreover, all the energy levels $E_{(h,j)}$ lie in a domain E in which the operator $\widehat{H}$ has a purely discrete spectrum for each $h \in (0,1]$. Then the set Λ_h asymptotically approaches the part of the discrete spectrum of the operator $\widehat{H}$ lying in the region E.*

7.2.5 An Observation on Finding the Asymptotics of the Eigenvalues of Differential Operators. As we saw, the construction of the asymptotics of the eigenvalues (the spectrum) of Hamiltonian operators by the method of the canonical operator consists of two stages. First one finds an approximation of zero-th order from the equation of quantization, and then this is sharpened by using the eigenvalues of the transport operator.

In practice, however, *the situation arises where the spectrum of the transport operator occupies some everywhere dense set*, while at the same time the spectrum of the fundamental Hamiltonian operator is discrete. This forces one to find a way to isolate a discrete series of values of the transport operator. This sort of procedure was first carried out by E.M. Vorob'ev for the case of the spectrum of the Dirichlet problem for Laplace's equation in a ring. E.M. Vorob'ev demonstrated a certain condition of the form of (2.30) (see below) which allows one to choose a finite series of eigenvalues of the transport operator. The hypotheses of Theorem 2.6 enable one to make such a choice for problems of a general nature.

The following example shows how the first two requirements lead to the specification of a discrete part of the spectrum of the transport operator. We consider the problem

$$H\psi = E\psi$$

for the eigenvalues for the operator $H(x, \widehat{p})$ with Hamiltonian

$$H(x,p) = \frac{1}{2}\left\{\left(p_1^2 + x_1^2\right) + \lambda\left(p_2^2 + x_2^2\right)\right\} .$$

Here λ is some number. The most interesting case occurs when the number λ is irrational. The following arguments will be carried out under that assumption. The Hamiltonian vector field associated with the Hamiltonian H has the form

$$V(H) = p_1\frac{\partial}{\partial x^1} + \lambda p_2\frac{\partial}{\partial x^2} - x^1\frac{\partial}{\partial p_1} - \lambda x^2\frac{\partial}{\partial p_2} .$$

To construct a system of Lagrangian manifolds which are invariant with respect to the vector vield $V(H)$, we consider the phase flow of the family of manifolds $M^0_{\rho,\sigma}$ along the field $V(H)$:

$$\begin{aligned} x^1 &= \rho\cos\alpha , \quad & x^2 &= \sigma , \\ p_1 &= \rho\sin\alpha , \quad & p_2 &= 0 , \quad \dim M^0_{\rho,\sigma} = 1 . \end{aligned}$$

The solution of the Hamiltonian system

$$\begin{aligned} \dot{x}^1 &= p_1 , \quad & \dot{x}^2 &= \lambda p_2 , \\ \dot{p}_1 &= -x^1 , \quad & \dot{p}_2 &= -\lambda x^2 \end{aligned}$$

with initial data on the manifold $M^0_{\rho,\sigma}$ has the form

$$\begin{aligned} x^1 &= x_0^1\cos t + p_{10}\sin t , \quad & x_0^1 &= \rho\cos\alpha , \\ p_1 &= -x_0^1\sin t + p_{10}\cos t , \quad & p_{10} &= \rho\sin\alpha , \\ x^2 &= x_0^2\cos\lambda t + p_{20}\sin\lambda t , \quad & x_0^2 &= \sigma , \\ p_2 &= -x_0^2\sin\lambda t + p_{20}\cos\lambda t , \quad & p_{20} &= 0 . \end{aligned}$$

We rewrite these equations in the following way:

$$\begin{aligned} x^1 &= \rho(\cos\alpha\cos t + \sin\alpha\sin t) = \rho\cos(t-\alpha) , \\ p_1 &= -\rho(\cos\alpha\sin t - \sin\alpha\cos t) = -\rho\sin(t-\alpha) , \end{aligned} \qquad (2.27)$$

$$x^2 = \sigma\cos\lambda t , \quad p_2 = -\sigma\sin\lambda t .$$

If one considers the parameters α and t to be coordinates, then Eq. (2.27) define a system of two-dimensional Lagrangian manifolds $M_{\rho,\sigma}$ which are invariant with respect to the vector field $V(H)$.

To describe the conditions of quantization and the subsequent solution of the transport equation it is useful to give explicitly the generators of the fundamental group $\pi_1(M_{\rho,\sigma})$. Since $M_{\rho,\sigma}$ is homeomorphic to the two-torus T^2, $\pi_1(M_{\rho,\sigma}) = \mathbb{Z}\oplus\mathbb{Z}$. Consequently the generators of this group are:

1) the cycle γ_1:

$$x^1 = \rho\cos\alpha\ , \quad x^2 = \sigma\ ,$$
$$p_1 = \rho\sin\alpha\ , \quad p_2 = 0\ ,$$

2) the cycle γ_2:

$$x^1 = \rho\ , \quad x^2 = \sigma\cos\lambda t\ ,$$
$$p_1 = 0\ , \quad p_2 = -\sigma\sin\lambda t\ .$$

The quantization conditions give

$$\rho_n^2 = h(2n+1)\ , \quad \sigma_m^2 = h(2m+1)$$

therefore

$$E_{mn} = h\left\{\left(n+\frac{1}{2}\right) + \lambda\left(m+\frac{1}{2}\right)\right\}\ .$$

Now we study the eigenvalues of the transport operator. In the (t,α) coordinates the vector field $V(H) = \partial/\partial t$ (identical to the transport operator $\mathcal{P}$) is $\partial/\partial t$. The equation for the eigenfunctions and eigenvalues of the operator $\partial/\partial t$ has the form

$$\frac{\partial\varphi}{\partial t} = \epsilon\varphi\ .$$

Solving this equation, we obtain that

$$\varphi = C(\alpha)\exp(\epsilon t)\ ,$$

where $C(\alpha)$ is a 2π-periodic function.

We choose the function $C(\alpha)$ in such a way that $\varphi(\alpha,t)$ becomes a function on the torus $M_{\rho,\sigma}$. As is simple to see from Eq. (2.27), if the value of the parameter t changes by $2\pi/\lambda$ while the value of α remains fixed, then the cycle γ_1 is mapped into itself. Under this map any point with coordinate α is mapped to a point with coordinate $\alpha - 2\pi/\lambda$. The condition that the function $\varphi(\alpha,t)$ be well-defined on the torus $M_{\rho,\sigma}$ is

$$\varphi(\alpha,t) = \varphi\left(\alpha - \frac{2\pi}{\lambda}, t + \frac{2\pi}{\lambda}\right)\ .$$

This leads to the following equation for the function $C(\alpha)$:

$$C(\alpha) = C\left(\alpha - \frac{2\pi}{\lambda}\right)\exp\left(\frac{2\pi\epsilon}{\lambda}\right)\ ,$$

which can be rewritten as

$$\exp\left\{i\left(\frac{2\pi i}{\lambda}\frac{\partial}{\partial\alpha}\right)\right\}C(\alpha) = \exp\left(-\frac{2\pi\epsilon}{\lambda}\right)C(\alpha)\ . \tag{2.28}$$

Therefore the function $C(\alpha)$ must be an eigenfunction of the translation operator $\exp\{i\,(2\pi i/\lambda)\,\partial/\partial\alpha\}$.

The eigenfunctions of this operator coincide with the eigenfunctions of the operator $i\partial/\partial\alpha$, and the eigenvalues are obtained by substituting the eigenvalues z of the operator $i\partial/\partial\alpha$ into the function $\exp\{2\pi iz/\lambda\}$. Therefore

$$C_s(\alpha) = \exp(is\alpha)\ , \quad k_s = \exp\left\{-\frac{2\pi is}{\lambda}\right\}\ , \quad s \in \mathbb{Z}\ .$$

On the other hand, by Formula (2.28) the eigenvalues of the translation operator are equal to $\exp\{-2\pi\epsilon/\lambda\}$. Therefore

$$\exp\left\{-\frac{2\pi is}{\lambda}\right\} = \exp\left\{-\frac{2\pi\epsilon}{\lambda}\right\}$$

and the following equation for the eigenvalues of the transport operator results:

$$\epsilon_{sq} = i(s + \lambda q)\ , \quad s \in \mathbb{Z}\ , \ q \in \mathbb{Z}\ . \tag{2.29}$$

The points $\{\epsilon_{sq}\}$ for s, $q \in \mathbb{Z}$ form an everywhere dense set on the axis $\mathbb{R}$. However, for large s the derivatives of the function $C_s(\alpha) = \exp(is\alpha)$ with respect to α are large, and therefore the system of eigenfunctions

$$\varphi_{s,q}(\alpha, t) = C_s(\alpha)\exp(\epsilon_{sq}t) = \exp(is\alpha)\exp\{i(s + \lambda q)t\}$$

does not have derivatives which are uniformly bounded with respect to s, α, if the value of s is unbounded. Therefore it is necessary to require that $|s| \leq A$, where A is a constant. Furthermore, since the numbers $|\epsilon_{sq}|$ must be bounded, it follows necessarily from conditions (2.29) that

$$s = O(1)\ , \quad q = O(1) \tag{2.30}$$

and, consequently, the everywhere dense set disappears. The asymptotic limit of the eigenvalues then has the form

$$\begin{aligned} E_{mn} - ih\epsilon_{sq} &= h\left(n + \frac{1}{2}\right) + \lambda h\left(m + \frac{1}{2}\right) + h(s + \lambda q) \\ &= h\left[(n + s) + \frac{1}{2}\right] + \lambda h\left[(m + q) + \frac{1}{2}\right]\ , \end{aligned} \tag{2.31}$$

where the integers n and m are chosen such that

$$0 < E_1 \leq E_{mn} \leq E_2 < +\infty\ . \tag{2.32}$$

We note that the latter requirement is quite substantial. If one does not introduce such a requirement, then it is possible from Formula (2.31) to obtain negative values for a positive definite operator, which is absurd. For $s = q = O(1)$ this is impossible. Indeed, in this case, for small values of the parameter h the right-hand side in Formula (2.31) is always positive by virtue of the inequalities (2.32) for E_{mn} and the conditions s, $q = O(1)$.

Let us give a resume of the results of the example we have worked out. In certain cases the spectrum of the transport operator can form an everywhere dense set on the axis $\mathbb{R}$. However, the eigenfunctions corresponding to these eigenvalues will not be uniformly bounded over the entire range of the parameters defining the eigenvalues. A consequence of this is that the estimate of the distance of the approximate spectrum to the exact one is non-uniform. This has to do with the fact that the remainder term contains derivatives of the eigenfunctions of the transport operator not only with respect to the parameter of the Hamiltonian trajectories, but also with respect to the transverse coordinates. By specifying a subsystem of eigenvalues, whose eigenfunctions are uniformly bounded, a "discretization" of the spectrum takes place; from the everywhere dense spectrum of the transport operator a subsystem of eigenvalues is chosen which is no longer everywhere dense.

Notwithstanding the fact that these conclusions have been drawn on the basis of the study of a single concrete example, they remain valid for a general class of problems. For example, all the statements made are true if the equation

$$H(x, \widehat{p})\,\psi = E\psi$$

has a discrete spectrum for each value of the parameter h. Here, if the distance between points of the spectrum has order $O(h^m)$, it is necessary to consider the transport operator expanded out to the $(m-1)$-th term in h:

$$\mathcal{P} = \mathcal{P}_0 + h\mathcal{P}_1 + \ldots + h^{m-1}\mathcal{P}_{m-1}\ .$$

This observation follows directly from Theorem 2.6. Indeed, if, for example, the distance between points of the spectrum were proportional to h, and the spectrum of the operator $\mathcal{P}_0$ were everywhere dense on the axis and the eigenfunctions corresponding to the spectrum bounded in their entirety, then the set

$$E_{(h,j)} + ih\epsilon_{(h,j),k} \tag{2.33}$$

would approach the spectrum. Choosing a sufficiently large constant C in the inequality

$$\left|\epsilon_{(h,j),k}\right| < C\ ,$$

we would obtain that the set (2.33) is everywhere dense in the segment $[E_1, E_2]$. On the other hand, by Theorem 2.6 at least one point of the spectrum lies in a Ch^2-neighborhood of any point of this set. This contradicts the assumption that the distance between points of the spectrum is proportional to h.

7.3 Systems of Equations

We will consider the system of equations

$$\widehat{A}u = A(x, \widehat{p})\, u = 0 \;, \tag{3.1}$$

where $A(x,p)$ is an $m \times m$ matrix consisting of $1/h$-differential operators

$$A(x, \widehat{p}) = (a_{ij}(x, \widehat{p})) \;, \qquad i, j = 1, \ldots, m \;.$$

We will assume that the following condition is satisfied.

Condition. *There exists a smooth matrix $\Sigma(x,p)$ and a function $H(x,p)$ (Hamilton function) such that*

$$A(x,p)\Sigma(x,p) = \Sigma(x,p)A(x,p) = H(x,p)E \;, \tag{3.2}$$

where

$$dH(x,p) \neq 0 \;, \quad (x,p) \in \mathrm{char} H = \{(x,p) | H(x,p) = 0\} \tag{3.3}$$

(E is the identity matrix).

We will look for approximate local solutions to Eq. (3.1) in the form

$$u(x,h) = \overline{F}^{1/h}_{p_{\overline{I}} \to x^{\overline{I}}} \left\{ e^{(i/h)S_I(x^I, p_{\overline{I}})} \left[u^{(0)} - ihu^{(1)} \right] (x^I, p_{\overline{I}}) \right\} \;, \tag{3.4}$$

where $u^{(0)}(x^I, p_{\overline{I}})$, $u^{(1)}(x^I, p_{\overline{I}})$ are m-vectors.

Lemma 3.1. *Suppose the functions $S_I(x^I, p_{\overline{I}})$ satisfy the Hamilton-Jacobi equation*

$$H\left(x^I, -\frac{\partial S_I}{\partial p_{\overline{I}}}, \frac{\partial S_I}{\partial x^I}, p_{\overline{I}} \right) = 0 \;, \tag{3.5}$$

and the function $u^{(0)}(x^I, p_{\overline{I}})$ has the form

$$u^{(0)}(x^I, p_{\overline{I}}) = \Sigma\left(x^I, -\frac{\partial S_I}{\partial p_{\overline{I}}}, \frac{\partial S_I}{\partial x^I}, p_{\overline{I}} \right) v^{(0)}(x^I, p_{\overline{I}}) \tag{3.6}$$

where the function $v^{(0)}(x^I, p_{\overline{I}})$ satisfies the congruence

$$\begin{aligned} &\left\{ \left[\frac{\partial H}{\partial p_I} \frac{\partial}{\partial x^I} - \frac{\partial H}{\partial x^{\overline{I}}} \frac{\partial}{\partial p_{\overline{I}}} + \frac{1}{2} \frac{\partial^2 H}{\partial p_I \partial p_I} \frac{\partial^2 S_I}{\partial x^I \partial x^I} + \frac{1}{2} \frac{\partial^2 H}{\partial x^{\overline{I}} \partial x^{\overline{I}}} \frac{\partial^2 S_I}{\partial p_{\overline{I}} \partial p_{\overline{I}}} \right. \right. \\ &\left. \left. - \frac{\partial^2 H}{\partial x^{\overline{I}} \partial p_I} \frac{\partial^2 S_I}{\partial p_{\overline{I}} \partial x^I} - \frac{\partial^2 H}{\partial x^{\overline{I}} \partial p_{\overline{I}}} \right] E + A_p \Sigma_x \right\} v^{(0)}(x^I, p_{\overline{I}}) \equiv 0 \;\; (\mathrm{mod Im} A) \;, \end{aligned} \tag{3.7}$$

where by $\operatorname{Im} A$ *we have denoted the image of the operator* A, *and the functions* H, A, Σ *in formula* (3.7) *are evaluated at the point* $\left(x^I, -\frac{\partial S_I}{\partial p_{\overline{I}}}, \frac{\partial S_I}{\partial x^{\overline{I}}}, p_{\overline{I}}\right)$.

Then there exists a function $u^{(1)}(x^I, p_{\overline{I}})$ *such that formula* (3.4) *gives an asymptotic solution of the system* (3.1) *with accuracy up to* $O(h^2)$.

Proof. Let us denote by $\widehat{A}_I$ the operator

$$\widehat{A}_I = F^{1/h}_{x^{\overline{I}} \to p_{\overline{I}}} \widehat{A} \overline{F}^{1/h}_{p_{\overline{I}} \to x^{\overline{I}}} = A\left(x^I, \widehat{x}^{\overline{I}}, \widehat{p}_I, p_{\overline{I}}\right) .$$

By the unitary property of the Fourier operator, in order to solve the problem we must choose a function $u^{(1)}$ such that the congruence

$$\widehat{A}_I e^{(i/h)S_I} \left\{u^{(0)} - ihu^{(1)}\right\} \equiv 0 \pmod{h^2} . \tag{3.8}$$

Computing the left-hand side of congruence (3.8), we obtain:

$$\begin{aligned}
\widehat{A}_I e^{(i/h)S_I} \left\{u^{(0)} - ihu^{(1)}\right\} &\equiv e^{(i/h)S_I} \left\{A\left(x^I, -\frac{\partial S_I}{\partial p_{\overline{I}}}, \frac{\partial S_I}{\partial x^{\overline{I}}}, p_{\overline{I}}\right)\right. \\
&\times \left[u^{(0)} - ihu^{(1)}\right] - ih\left[\frac{\partial A}{\partial p_I}\frac{\partial}{\partial x^I} - \frac{\partial A}{\partial x^{\overline{I}}}\frac{\partial}{\partial p_{\overline{I}}}\right. \\
&+ \frac{1}{2}\left[\frac{\partial^2 S_I}{\partial x^I \partial x^I}\frac{\partial^2 A}{\partial p_I \partial p_I} + \frac{\partial^2 S_I}{\partial p_{\overline{I}} \partial p_{\overline{I}}}\frac{\partial^2 A}{\partial x^{\overline{I}} \partial x^{\overline{I}}} - 2\frac{\partial^2 S_I}{\partial x^I \partial p_{\overline{I}}}\frac{\partial^2 A}{\partial p_I \partial x^{\overline{I}}}\right] - \left.\frac{\partial^2 A}{\partial x^{\overline{I}} \partial p_{\overline{I}}}\right] \\
&\times \left.\left[u^{(0)} - ihu^{(1)}\right]\right\} \equiv e^{(i/h)S_I}\left(\mathcal{L}_0 - ih\widehat{\mathcal{L}}_1\right)\left(u^{(0)} - ihu^{(1)}\right) \pmod{h^2} .
\end{aligned}$$

Here, as before, the functions A are evaluated at the point $\left(x^I, -\frac{\partial S_I}{\partial p_{\overline{I}}}, \frac{\partial S_I}{\partial x^I}, p_{\overline{I}}\right)$. We will adhere to this convention from now on.

Hence the congruence (3.8) is equivalent to the system

$$\begin{cases} Au^{(0)} = 0 , \\ Au^{(1)} = -\widehat{\mathcal{L}}_1 u^{(0)} . \end{cases} \tag{3.9}$$

The validity of the first of the relations (3.9) follows from (3.2) and (3.5).

The condition for the second of the relations (3.9) to be solvable takes the form

$$\widehat{\mathcal{L}}_1 \Sigma v^{(0)} \equiv 0 \pmod{\operatorname{Im} A} . \tag{3.10}$$

The latter relation is an equation (congruence) in the vector $v^{(0)}$. We transform the left-hand side of this equation. We have

$$\begin{aligned}\widehat{\mathcal{L}}_1 \Sigma v^{(0)} = &\left[\frac{\partial A}{\partial p_I}\frac{\partial}{\partial x^I} - \frac{\partial A}{\partial x^{\overline{I}}}\frac{\partial}{\partial p_{\overline{I}}} + \frac{1}{2}\left(\frac{\partial^2 S_I}{\partial x^I \partial x^I}\right.\right.\\ &\times\frac{\partial^2 A}{\partial p_I \partial p_I} + \frac{\partial^2 S_I}{\partial p_{\overline{I}}\partial p_{\overline{I}}}\frac{\partial^2 A}{\partial x^{\overline{I}}\partial x^{\overline{I}}} - 2\left.\frac{\partial^2 S_I}{\partial x^I \partial p_{\overline{I}}}\frac{\partial^2 A}{\partial p_I \partial x^{\overline{I}}}\right)\\ &\left.-\frac{\partial^2 A}{\partial x^{\overline{I}}\partial p_{\overline{I}}}\right]\Sigma v^{(0)} = \left[\frac{\partial A}{\partial p_I}\Sigma\frac{\partial}{\partial x^I} + \frac{\partial A}{\partial p_I}\frac{\partial \Sigma}{\partial x^I}\right.\\ &- \frac{\partial A}{\partial p_I}\frac{\partial \Sigma}{\partial x^{\overline{I}}}\frac{\partial^2 S_I}{\partial p_{\overline{I}}\partial x^I} + \frac{\partial A}{\partial p_I}\frac{\partial \Sigma}{\partial p_I}\frac{\partial^2 S_I}{\partial x^I \partial x^I}\\ &- \frac{\partial A}{\partial x^{\overline{I}}}\Sigma\frac{\partial}{\partial p_{\overline{I}}} - \frac{\partial A}{\partial x^{\overline{I}}}\frac{\partial \Sigma}{\partial p_{\overline{I}}} + \frac{\partial A}{\partial x^{\overline{I}}}\frac{\partial \Sigma}{\partial x^{\overline{I}}}\frac{\partial^2 S_I}{\partial p_{\overline{I}}\partial p_{\overline{I}}}\\ &- \frac{\partial A}{\partial x^{\overline{I}}}\frac{\partial \Sigma}{\partial p_I}\frac{\partial^2 S_I}{\partial p_{\overline{I}}\partial x^I} + \frac{1}{2}\frac{\partial^2 S_I}{\partial x^I \partial x^I}\frac{\partial^2 A}{\partial p_I \partial p_I}\Sigma\\ &\left.+\frac{1}{2}\frac{\partial^2 S_I}{\partial p_{\overline{I}}\partial p_{\overline{I}}}\frac{\partial^2 A}{\partial x^{\overline{I}}\partial x^{\overline{I}}}\Sigma - \frac{\partial^2 S_I}{\partial x^I \partial p_{\overline{I}}}\frac{\partial^2 A}{\partial p_I \partial x^{\overline{I}}}\Sigma - \frac{\partial^2 A}{\partial x^{\overline{I}}\partial p_{\overline{I}}}\Sigma\right] v^{(0)} .\end{aligned} \tag{3.11}$$

Differentiating Eq. (3.2) with respect to the variables $p_I, x^{\overline{I}}$, we obtain

$$\begin{aligned}\frac{\partial A}{\partial p_I}\Sigma + A\frac{\partial \Sigma}{\partial p_I} &= \frac{\partial H}{\partial p_I}E ,\\ \frac{\partial A}{\partial x^{\overline{I}}}\Sigma + A\frac{\partial \Sigma}{\partial x^{\overline{I}}} &= \frac{\partial H}{\partial x^{\overline{I}}}E .\end{aligned} \tag{3.12}$$

Taking into account Eq. (3.12), formula (3.11) gives, modulo $\operatorname{Im} A$:

$$\begin{aligned}\widehat{\mathcal{L}}_1 \Sigma_v^{(0)} \equiv &\left[\left(\frac{\partial H}{\partial p_I}\frac{\partial}{\partial x^I} - \frac{\partial H}{\partial x^{\overline{I}}}\frac{\partial}{\partial p_{\overline{I}}}\right)E\right.\\ &- \frac{\partial A}{\partial P_I}\frac{\partial \Sigma}{\partial x^{\overline{I}}}\frac{\partial^2 S_I}{\partial x^I \partial p_{\overline{I}}} + \frac{\partial A}{\partial p_I}\frac{\partial \Sigma}{\partial p_I}\frac{\partial^2 S_I}{\partial x^I \partial x^I}\\ &- \frac{\partial A}{\partial x^{\overline{I}}}\frac{\partial \Sigma}{\partial p_{\overline{I}}} + \frac{\partial A}{\partial x^{\overline{I}}}\frac{\partial \Sigma}{\partial x^{\overline{I}}}\frac{\partial^2 S_I}{\partial p_{\overline{I}}\partial p_{\overline{I}}} - \frac{\partial A}{\partial x^{\overline{I}}}\frac{\partial \Sigma}{\partial p_I}\frac{\partial^2 S_I}{\partial p_{\overline{I}}\partial x^I}\\ &+ \frac{1}{2}\frac{\partial^2 S_I}{\partial x^I \partial x^I}\frac{\partial^2 A}{\partial p_I \partial p_I}\Sigma + \frac{1}{2}\frac{\partial^2 S_I}{\partial p_{\overline{I}}\partial p_{\overline{I}}}\frac{\partial^2 A}{\partial x^{\overline{I}}\partial x^{\overline{I}}}\Sigma\\ &\left.-\frac{\partial^2 S_I}{\partial x^I \partial p_{\overline{I}}}\frac{\partial^2 A}{\partial p_I \partial x^{\overline{I}}}\Sigma - \frac{\partial^2 A}{\partial x^{\overline{I}}\partial p_{\overline{I}}}\Sigma\right] v^{(0)} \pmod{\operatorname{Im} A} .\end{aligned} \tag{3.13}$$

Now using the relations

$$A_{p_i p_j}\Sigma + A_{p_i}\Sigma_{p_j} + A_{p_j}\Sigma_{p_i} + A\Sigma_{p_i p_j} = H_{p_i p_j}E , \tag{3.14}$$

$$A_{x^i x^j}\Sigma + A_{x^i}\Sigma_{x^j} + A_{x^j}\Sigma_{x^i} + A\Sigma_{x^i x^j} = H_{x^i x^j}E \ , \tag{3.15}$$

$$A_{p_i x^j}\Sigma + A_{p_i}\Sigma_{x^j} + A_{x^j}\Sigma_{p_i} + A\Sigma_{p_i x^j} = H_{p_i x^j}E \ , \tag{3.16}$$

obtained, as before, by differentiating formula (3.2), and taking into account the symmetry of the matrices $\partial^2 S_I/\partial x^I \partial x^I$, $\partial^2 S_I/\partial p_{\overline{I}}\partial p_{\overline{I}}$, we obtain that the right-hand side of the relation (3.13) is equal to the left-hand side of relation (3.7) modulo $\operatorname{Im} A$. This proves the lemma. □

Theorem 3.1. *Suppose L is a Lagrangian manifold lying in the manifold* $\operatorname{char} H$, *and suppose μ is a nondegenerate measure on L which is invariant with respect to the Hamiltonian vector field $V(H)$. Suppose the function φ satisfies the transport equation*

$$\left[V(H) + \{A,\Sigma\}\Big|_L - \frac{1}{2}A_{xp}\Sigma\Big|_L\right]\varphi = 0 \ . \tag{3.17}$$

where $\{\ ,\}$ represent the Poisson bracket:

$$\{A,\Sigma\} = A_{p_i}\Sigma_{x^i} - A_{x^i}\Sigma_{p_i} \ .$$

Then there exists a vector function $\varphi^{(1)}$ on L such that

$$u(x,h) = K_{(L,\mu)}\left(\Sigma|_L\,\varphi - ih\varphi^{(1)}\right) \tag{3.18}$$

is an asymptotic solution of Eq. (3.1) up to h^2.

Proof. In accordance with the definition of the canonical operator, when we set

$$v^{(0)} = \varphi\sqrt{\mu_I} \ ,$$

where $\mu_i = \mu_I(x^I, p_{\overline{I}})$ is the density of the measure with respect to the coordinates $(x^I, p_{\overline{I}})$, and multiply congruence (3.7) by $(\mu_I)^{-1/2}$, we obtain, carrying out calculations analogous to the scalar case, that the congruence (3.7) is equivalent to the congruence

$$\left\{\left(V(H) - \frac{1}{2}H_{xp}\right)\Big|_L E + A_p\Sigma_x\Big|_L\right\}\varphi \equiv 0 \ (\operatorname{mod} \operatorname{Im} A) \ . \tag{3.19}$$

Now using Eq. (3.16), we obtain that congruence (3.19) holds if the transport Eq. (3.17) holds.

Next, we have

$$A(x,\widehat{p})\,K_{(L,\mu)}\left(\Sigma\Big|_L\varphi\right) \equiv A(x,\widehat{p})\sum_I k_I\left(e_I\Sigma\Big|_L\varphi\right)$$

$$= -ih\sum_I k_I A(x,p)\Big|_L\left(\mathcal{P}_I^{\Sigma}e_I\varphi\right) \ (\operatorname{mod} h^2) \ ,$$

where

$$\mathcal{P}_I^{\Sigma} = (\mu_I)^{-1/2} \left[\Sigma_{p_I}\Big|_L \frac{\partial}{\partial x^I} - \Sigma_{x^{\overline{I}}}\Big|_L \frac{\partial}{\partial p_{\overline{I}}} \right.$$

$$+ \frac{1}{2} \left[\frac{\partial^2 S_I}{\partial x^I \partial x^I} \frac{\partial^2 \Sigma}{\partial p_I \partial p_I}\Big|_L + \frac{\partial^2 S_I}{\partial p_{\overline{I}} \partial p_{\overline{I}}} \frac{\partial^2 \Sigma}{\partial x^{\overline{I}} \partial x^{\overline{I}}}\Big|_L - 2 \frac{\partial^2 S_I}{\partial x^I \partial p_{\overline{I}}} \frac{\partial^2 \Sigma}{\partial x^{\overline{I}} \partial p_I}\Big|_L \right]$$

$$\left. - \frac{1}{2} \frac{\partial^2 \Sigma}{\partial x^{\overline{I}} \partial p_{\overline{I}}}\Big|_L + \frac{1}{2} \frac{\partial^2 \Sigma}{\partial x^I \partial p_I}\Big|_L \right] (\mu_I)^{1/2} , \tag{3.20}$$

and $\{e_I\}$ is the partition of unity which appears in the definition of the canonical operator.

Each summand appearing on the right-hand side of formula (3.20) coincides up to terms to order h^2 with the expression

$$- iHK_{(L,\mu)} A(x,p)\Big|_L \mathcal{P}_I^{\Sigma} (e_I \varphi) ,$$

by the theorem on cocycles (see Chap. 4). Now setting

$$\varphi^{(1)} = \sum_I \mathcal{P}_I^{\Sigma} (e_I \varphi) ,$$

we obtain the assertion of the theorem. □

Remark 3.1. *Equation* (3.17) *may be replaced by a congruence modulo* Im A.

Appendix Fourier-Maslov Integral Operators (The Smooth Theory of Maslov's Canonical Operator)

*V.E. Nazaikinskij, V.G. Oshchmian, B.Yu. Sternin, V.E. Shatalov**

Introduction

The theory of Fourier integral operators, which arose at the end of the 1960's, is presently undergoing rapid growth. Dozens of works which have appeared in the world mathematical literature expound, generalize and utilize the theory of Fourier integral operators. And this is natural. After the fascination with the elliptic theory, which caused the theory of pseudodifferential operators to flourish, there appeared at first a timid, and then stronger and stronger interest in the non-elliptic theory – equations with real characteristics. However, the technique of pseudodifferential operators, which works well in the elliptic theory, turned out to be unsuitable for the solution of this new group of problems.

The essential novelty in the theory of equations with real characteristics lies in the fact that, as opposed to the elliptic case, here the almost inverse operator is not a pseudodifferential operator. At first there were attempts to correct or somehow to add on to the old techniques, in such a way as to make them applicable to the new situation. One of these attempts was the application of (co)boundary operators – in retrospect, the first (nontrivial) Fourier integral operators. Next, several generalizations of pseudodifferential operators appeared in a series of works, primarily by Soviet mathematicians. The linear phase in the Fourier integral was replaced by an arbitrary homogeneous function. This was the first step in

---

* *Translator's Note:* This appendix appeared initially as an article in *Uspekhi Mat. Nauk*, Vol. 36 No. 2 (1981) [translated by D. Mathon in *Russian Math. Surveys*, Vol. 36 No. 2 (1981), 93–161]. Some minor changes have been made by the authors.

the right direction. Gradually experience with this sort of operator accumulated. A whole series of problems repeatedly indicated the existence of some sort of general technical apparatus. The future Fourier integral operators appeared in seemingly very unexpected situations, for example in the study of the transformations of pseudodifferential operators induced by a canonical diffeomorphism of phase space, etc. The general principles of the new technique began to show through more and more clearly. Only one step remained to be taken. At this point, finally, in 1971 appeared the publication of the Swedish mathematician L. Hörmander [3], in which the mathematical apparatus which made it possible to solve the necessary problems, and which he called the method of Fourier integral operators, was presented. The new technique developed rapidly, and soon Fourier integral operators won wide popularity among specialists. A stream of articles on the application and generalization of the method of Fourier integrals sprang forth.

In another, apparently remote field of mathematics – the problem of construction of asymptotic solutions to equations with a small parameter – matters stood more or less the same way. The old method for constructing asymptotic solutions – the WKB method – gave satisfactory results only "in the small" and was completely unsuitable for global considerations. Many attempts were made to generalize the WKB method, adapting it to the study of solutions near singular (focal, caustic) points, and although this was successfuly done on occasion by using a series of artificial and very ingenious methods (see, for example, Babich, V.M., Buldyrev, V.S. [1] and others), the general scheme for constructing the asymptotics in the large was unknown.

In 1965 V.P. Maslov proposed a noteworthy solution to this problem, which not only gave a general method for finding the asymptotic solutions "in the whole", but also made it evident that several classical problems, for example the problem of propagation "in the large" of singularities of the initial data in hyperbolic equations, could be solved using his method. We note that the latter problem includes, in particular, the problem of constructing asymptotic expansions by degrees of smoothness, i.e. the problem for which the method of Fourier integral operators was invented.

Thus it turned out that at the time of the publication by L. Hörmander mentioned above, a method and developed technical appratus adequate to the theory of Fourier integral operators was already on hand. Moreover, it was later clarified that Maslov's canonical operator method, when applied in the situation of Fourier integral operators, precisely coincided with the latter. Moreover, the apparent difference in the final form is purely external: by a series of identity transformations the Fourier integral operator can be reduced to Maslov's canonical operator.

We point out, however, that the methods of construction of the two theories have several significant differences. From this point of view a presentation of the theory of Fourier integral operators by the method of Maslov seems to be of considerable interest, especially since, as it seems to us, the latter is much more transparent, geometric, and allows one to obtain the answer in a more finished form.

Let us illustrate the ideas and the main concepts of the method of Fourier-Maslov integral operators in the example of the simplest problem.

Let us consider the Cauchy problem

$$i\frac{\partial u}{\partial t} = H\left(x, \widehat{p}, t\right) u + f(x,t) \ . \tag{1}$$

Here, $H(x, \widehat{p}, t)$ is a pseudodifferential operator with a homogeneous symbol $H(x,p,t)$ of the first order in p, $x = \left(x^1, \ldots, x^n\right)$, $p = (p_1, \ldots, p_n)$, $\widehat{p} = -i\frac{\partial}{\partial x}$.

We impose the initial conditions

$$u(x,0) = u_0(x) \ . \tag{2}$$

According to the Duhamel principle, it is sufficient to solve the problem (1)–(2) for $f(x,t) \equiv 0$.

We will look for a function $K(x,y,t)$ such that the formula

$$u(x,t) = K(x,y,t)u_0(y)dy \tag{3}$$

gives a solution of the Cauchy problem (1)–(2) for any initial data $u_0(x)$. We obtain the following problem for the kernel $K(x,y,t)$:

$$\begin{cases} i\frac{\partial}{\partial t}K(x,y,t) = H(x,\widehat{p},t)K(x,y,t) \ , \\ K(x,y,0) = \delta(x-y) \ . \end{cases} \tag{4}$$

If a kernel $K(x,y,t)$ satisfying (4) can be found to within sufficiently smooth functions, then the problem (1)–(2) reduces to a Volterra integral equation.

We apply a Fourier transform with respect to the variable y to (4). Writing

$$U(x,q,t) = F_{y\to q}K(x,y,t) \ ,$$

we reduce problem (4) to the form

$$\begin{cases} i\frac{\partial}{\partial t}U(x,q,t) = H(x,\widehat{p},t)U(x,q,t) \ , \\ U(x,q,0) = e^{iqx} \ . \end{cases} \tag{5}$$

The function $K(x,y,t)$ can be expressed in terms of the solutions (5) by the formula

$$K(x,y,t) = F_{q\to y}U(x,q,t) = \left(\frac{1}{2\pi}\right)^{n/2} \int_{\mathbb{R}_n} e^{-iqy}U(x,q,t)dq \ . \tag{6}$$

Below we construct a solution of (5) (asymptotic as $|q| \to \infty$) that is of class C^∞ in all its variables. Therefore, as is clear from (6), the degree of smoothness of $K(x,y,t)$ is governed by the rate of decrease of $U(x,q,t)$ at infinity in q. In other words, the quantity $|q|^{-1}$ ($|q| \to \infty$) constitutes the small parameter of our asymptotic expansion.

Let us now solve (5). Using the WKB method, we look for the function $U(x,q,t)$ in the form

$$U(x,q,t) = e^{iS(x,q,t)}A(x,q,t) \ , \tag{7}$$

where the phase $S(x,q,t)$ is a positively homogeneous function of degree 1 in q; the amplitude $A(x,q,t)$ can be expanded in a formal series by homogeneous functions:

$$A(x,q,t) = \sum_{j=0}^{-\infty} A_j(x,q,t) \ , \tag{8}$$

where the degree of homogeneity of A_j is j.

Substituting (7) in (5) and comparing terms of equal degree of homogeneity, we obtain the following equations for the phase S and the amplitude A:

$$\left\{\begin{aligned}
&\frac{\partial S}{\partial t} + H\left(x, \frac{\partial S}{\partial x}, t\right) = 0 \ , \\
&\frac{\partial A_0}{\partial t} + \sum_{i=1}^{n} H_{p_i}\left(x, \frac{\partial S}{\partial x}, t\right)\frac{\partial A_0}{\partial x^i} \\
&\quad + \frac{1}{2}\sum_{i,j=1}^{n} \frac{\partial^2 S}{\partial x^i \partial x^j} H_{p_i p_j}\left(x, \frac{\partial S}{\partial x}, t\right) A_0 = 0 \ , \\
&\dots\dots\dots\dots\dots\dots\dots\dots\dots\dots \\
&\frac{\partial A_j}{\partial t} + \sum_{i=1}^{n} H_{p_i}\left(x, \frac{\partial S}{\partial x}, t\right)\frac{\partial A_j}{\partial x^i} \\
&\quad + \frac{1}{2}\sum_{i,k=1}^{n} \frac{\partial^2 S}{\partial x^i \partial x^k} H_{p_i p_k}\left(x, \frac{\partial S}{\partial x}, t\right) A_j = F_j \ ,
\end{aligned}\right. \tag{9}$$

where $F_j = F_j[A_0, \dots, A_{j+1}]$ is an expression depending on the functions $A_0, \dots, A_{j+1}$.

The initial conditions of the problem (5) yield the following initial conditions for the system (9):

$$S(x,q,0)=\sum_{i=1}^{n} q_i x^i\ ,\quad A_0(x,q,0)=1\ ,\quad A_j(x,q,0)=0\quad (j=-1,-2,\ldots)\ .$$

We now concentrate on the first equation of (9), the Hamilton-Jacobi equation.

The corresponding Cauchy problem has been described above:

$$\begin{cases} \frac{\partial S}{\partial t}+H\left(x,\frac{\partial S}{\partial x},t\right)=0\ , \\ S(x,q,0)=xq=\sum_{i=1}^{n} x^i q_i\ . \end{cases} \tag{10}$$

To solve the first-order non-linear problem (10) we use the method of trajectories, which associates whith (10) the Hamiltonian system

$$\begin{cases} \dot{x}=H_p(x,p,t)\ , \\ \dot{p}=-H_x(x,p,t)\ . \end{cases} \tag{11}$$

Assuming that $x=x(x_0,t)$ and $p=p(x_0,t)$ is a solution of (11) with the initial conditions

$$x(0)=x_0\ ,\quad p(0)=\left.\frac{\partial S_0(x)}{\partial x}\right|_{x=x_0}=q\ ,$$

then the solution of (10) is given by

$$S(x,t)=S_0(x_0)+\left.\int_0^t [pdx-Hdt]\right|_{x_0=x_0(x,t)}\ , \tag{12}$$

where $x_0(x,t)$ is the solution of the equation

$$x=(x_0,t) \tag{13}$$

with respect to x_0. It is also well known that (10) is, in general, soluble only for small t, since the Jacobian Dx/Dx_0 may vanish at a certain instant t_0, and (13) ceases to have a smooth solution. We now turn to the equations for the functions $A_j(x,q,t)$. The operator on the left-hand sides of this system represents differentiation along a vector field

$$\frac{\partial}{\partial t}+\sum_{i=1}^{n} H_{p_i}\left(x,\frac{\partial S}{\partial x},t\right)\frac{\partial}{\partial x^i}\ ,$$

and the equations for the A_j are thus ordinary differential equations along the trajectories of this vector field. As is not hard to see, the equations of these trajectories are $x=x(x_0,t)$, that is, they are projections of the trajectories of the Hamiltonian system (11) onto the x-space. Thus, we obtain a solution $U(x,q,t)$ of (5) for small t.

At a point where the Jacobian $Dx/Dx_0 = 0$ (in what follows we call such a point *focal*), this solution ceases to exist, as follows from (12) (moreover, of these points the amplitude tends to infinity). Consequently in a neighbourhood of a focal point our expression for the solution cannot be used. This fact is of fundamental significance and indicates that the solution of (5) is not of the form (7) in a neighbourhood of a focal point.

However, it can be shown that in a neighbourhood of such a point the solution has the form[1]

$$U(x,q,t) = \overline{F}_{p_{\overline{I}} \to x^{\overline{I}}} \left\{ e^{iS_I(x^I, p_{\overline{I}}, q, t)} A_I\left(x^I, p_{\overline{I}}, q, t\right) \right\} , \tag{14}$$

where I is a subset of $\{1, \ldots, n\}$ and $\overline{I}$ is its complement. (Later we shall give some motivation for the expression (14).) It is clear from (14) that the parameter "responsible" for the smoothness of the solutions is $\left(\left|p_{\overline{I}}\right|^2 + q^2\right)^{-1/2}$, therefore, the amplitude in (14) can be expanded in terms of the degree of homogeneity in the variables $(p_{\overline{I}}, q)$.

The resulting Hamilton-Jacobi and transport equations are analogous to the system (9).

To obtain initial conditions for the Hamilton-Jacobi and transport equations we must rewrite the asymptotic expansion (7) in the form (14) on the common part of the domains of definition of these local solutions. A calculation by means of the stationary phase method shows that necessary conditions for the equality of (7) and (14) are (modulo some class of functions)

$$\begin{cases} S_I\left(x^I, p_{\overline{I}}, q, t\right) = S(x,q,t) - \sum\limits_{i \in \overline{I}} x^i p_i \Bigg|_{x^{\overline{I}} = x^{\overline{I}}\left(x^I, p_{\overline{I}}, q, t\right)} \\ A_{I0}\left(x^I, p_{\overline{I}}, q, t\right) = \dfrac{i^{|\overline{I}|} A_0(x,q,t)}{\sqrt{\det \operatorname{Hess}_{x^{\overline{I}}}(-S(x,q,t))}} \Bigg|_{x^{\overline{I}} = x^{\overline{I}}\left(x^I, p_{\overline{I}}, q, t\right)} \end{cases} \tag{15}$$

where the $x^{\overline{I}}\left(x^I, p_{\overline{I}}, q, t\right)$ are determined by

$$p_{\overline{I}} = \frac{\partial S(x,q,t)}{\partial x^{\overline{I}}} . \tag{16}$$

[1] The number of elements in the set I is related to the degree of degeneracy of the matrix $(\partial x / \partial x_0)$.

The argument of the square root in (15) is chosen in a special way, and $|\overline{I}|$ denotes the number of elements in the set $\overline{I}$.

It is clear that the formula (16) defines the change of variables $(x,q) \to (x^I, p_{\overline{I}}, q)$ on a portion of the $2n$-dimensional surface

$$p = \frac{\partial S(x,q,t)}{\partial x} \,, \quad y = \frac{\partial S(x,q,t)}{\partial q} \,, \tag{17}$$

embedded in the $4n$-dimensional space with the coordinates (x,p,y,q) (for each fixed value of t). This is part of the surface formed by the trajectories (phase flow) of the Hamiltonian system (11). It can be shown that the whole phase flow L can be covered by neighbourhoods described by equations of the form

$$p_I = \frac{\partial S_I\left(x^I, p_{\overline{I}}, q, t\right)}{\partial x^I} \,, \quad x^{\overline{I}} = -\frac{\partial S_I\left(x^I, p_{\overline{I}}, q, t\right)}{\partial p_{\overline{I}}} \,, \quad y = \frac{\partial S_I\left(x^I, p_{\overline{I}}, q, t\right)}{\partial q} \,.$$

The manifolds with such a property are said to be Lagrangian, and we shall now interpret the results obtained above in terms of the Lagrangian phase flow.

Since

$$\det \operatorname{Hess}_{x^{\overline{I}}} \left(-S(x,q,t)\right) = (-1)^{|\overline{I}|} \det \frac{\partial\left(x^I, p_{\overline{I}}, q\right)}{\partial(x,q)} \,,$$

must hold on the intersection of two neighbourhoods with local coordinates (x,q) and $(x^I, p_{\overline{I}}, q)$, the set of amplitudes $A_0, A_{0I}, \ldots$ specifies a density of order 1/2 on the neighbourhoods of the Lagrangian manifold, that is, an object that is multiplied by the square root of the Jacobian under the change of variables (see, for example, L. Hörmander [3]). It is therefore natural to introduce a certain standard measure μ on L such that locally

$$A_{I0}\left(x^I, p_{\overline{I}}, q, t\right) = \varphi\left(x^I, p_{\overline{I}}, q, t\right) \sqrt{\mu_I\left(x^I, p_{\overline{I}}, q, t\right)} \,, \tag{18}$$

where φ is a function on L, and μ_I is the density of μ with respect to local coordinates. This explains that if μ is invariant with respect to the Hamiltonian vector field (11), then the transport equation for φ simplifies to the form

$$\frac{d\varphi}{dt} - \frac{1}{2}\sum_{j=1}^{n} H_{p_j x^j} \varphi|_L = 0 \,, \tag{19}$$

where d/dt denotes the derivative along the trajectories of (11).

We sum up briefly the outlined procedure. To construct an asymptotic solution of the problem (4) we have introduced the Lagrangian manifold L

formed by the trajectories of the dynamical system (11) and have defined a function $K(x,y,t)$, which can be written locally in the form

$$F_{q\to y}\circ \overline{F}_{p_{\overline{I}}\to x^{\overline{I}}}\left\{e^{iS_I(x^I,p_{\overline{I}},q,t)}\varphi\left(x^I,p_{\overline{I}},q,t\right)\sqrt{\mu_I\left(x^I,p_{\overline{I}},q,t\right)}\right\} . \quad (20)$$

This is called the *canonical distribution* and, as we said above, it determines the kernel of the integral operator

$$\widehat{\Phi}_{(L,\mu,\varphi)}(f)=\int K(x,y,t)f(y)dy , \quad (21)$$

which we call the Fourier integral operator depending on the parameter t. As follows from what we have said, the operator (21) satisfies the asymptotic formula

$$\begin{cases}\left[-i\frac{\partial}{\partial t}+H(x,\widehat{p},t)\right]\circ\widehat{\Phi}_{(L,\mu,\varphi)}\equiv\widehat{\Phi}_{(L,\mu,\mathcal{P}_0\varphi)} ,\\ \mathcal{P}_0\varphi=\left[\frac{d}{dt}-\frac{1}{2}\sum_{j=1}^{n}H_{p_jx^j}|_L\right]\varphi .\end{cases} \quad (22)$$

If now φ satisfies the transport equation $\mathcal{P}_0\varphi=0$, then the right-hand side of (22) vanishes modulo sufficiently smooth functions, thus, the function $\widehat{\Phi}_{(L,\mu,\varphi)}u_0$ provides an asymptotic form for the smoothness of the solution of the Cauchy problem. Naturally, we need the relevant theorems on boundedness of Fourier integral operators in the scale of Sobolev spaces.

The above analysis of the Cauchy problem suggests a method for the development of a general theory of Fourier integral operators. Let $T_0^*M\times T_0^*M\subset T_0^*(M\times M)$ be the cotangent space to $M\times M$ with the zero section deleted. We consider a homogeneous Lagrangian manifold L in this space, that is, a manifold invariant under the action of the multiplicative group $\mathbb{R}_+$ on the fibres. Following the outlined procedure, we construct a canonical distribution on L representing the kernel of a Fourier integral operator.

We mention here that if $K(x,y,t)$ is the kernel of an integral operator defined by (3), then $K(x,y,t)$ is not a function on the manifold M, that is, the law of transformation of $K(x,y,t)$ under a change of coordinates on the manifold differs from the law of transformation of functions. In fact, if Eq. (1) on M is an equation in u, then (3) indicates that the kernel $K(x,y,t)$ is a function in the variables x and t and a density of a measure with respect to y. On the other hand, the equations of quantum mechanics concern densities of order 1/2, that is, objects that are multiplied by the square root of the Jacobian when transformed to new coordinates. The wave function ψ represents such a density, therefore, $K(x,y,t)$ in (3) for the Schrödinger equation is a density of order 1/2 in both the variables x and y. We include in our paper both these cases within a unified approach, by considering densities of order α in (1). Functions are then obtained for $\alpha = 0$ and densities of order 1/2 for $\alpha = 1/2$. The kernel $K(x, y, t)$ is then a density of order α in the first variable and of order $1 - \alpha$ in the second.

Recently the theory of Maslov integral operators has found application to a wide range of questions associated with the study of local and global solvability of linear equations, the study of wavefronts of solutions, the construction of asymptotic expansions, etc. We will illustrate, for example, how the technique of integral operators works in the problem of cosntructing the pseudodifferential operators corresponding to a canonical diffeomorphism of phase space. As is well known, the solution to this problem given by Yu. V. Egorov led to important and deep results in the theory of local solvability of pseudodifferential equations.

§ 1. Densities, Pseudodifferential Operators, and Asymptotic Expansions

The material of this section is preparatory. In it we have gathered fairly well known definitions and theorems on which the subsequent account is based. We also introduce here the notation that will be used in the rest of the paper.

In the definition of pseudodifferential operators on a manifold we follow essentially Hörmander [2], [4], except for some slight formal differences. We consider a slightly narrower class of pseudodifferential operators than Hörmander, to avoid complicating our discussion by technical details.

1°. Notation. We always denote by the letter M (possibly with an index) a real connected smooth orientable manifold of dimension n. It is assumed that an orientation is specified on M and only coordinate systems on M consistent with this orientation are considered (admissible coordinate systems). Thus, if $x = (x^1, \ldots, x^n)$ and $y = (y^1, \ldots, y^n)$ are two coordinate systems in an open set $U \subset M$, then the Jacobian

$$\det \frac{\partial y}{\partial x} = \det \frac{\partial (y^1, \ldots, y^n)}{\partial (x^1, \ldots, x^n)}$$

is positive.

We denote by v the positive volume form on M; in local coordinates $(x^1, \ldots x^n)$ it is given by

$$v = v_x (x^1, \ldots, x^n)\, dx^1 \wedge \ldots \wedge dx^n ,$$

where $v_x (x^1, \ldots, x^n) > 0$ in all admissible coordinate systems.

Let $x = (x^1, \dots, x^n)$ be a local coordinate system on M. We denote by $p = (p_1, \dots, p_n)$ the dual coordinate system in $T_x^* M$. The set of functions

$$(x, p) = (x^1, \dots, x^n, p_1, \dots, p_n)$$

forms a local coordinate system on $T^* M$.

We denote by ω_M^2 the standard Hamiltonian structure (see Arnol'd [4]) in $T^* M$, that is, a closed non-degenerate differential 2-form on $T^* M$, which in local coordinates has the following form:

$$\omega_M^2 = dp \wedge dx = \sum_{i=1}^{n} dp_i \wedge dx^i \ . \tag{1.1}$$

We also denote by ω_M^1 the invariantly defined differential form on $T^* M$ whose expression in local coordinates is

$$\omega_M^1 = \sum_{i=1}^{n} p_i dx^i \ . \tag{1.2}$$

Clearly the forms ω_M^1 and ω_M^2 are connected by the relation $\omega_M^2 = d\omega_M^1$.

Let $I = \{i_1, \dots, i_k\}$ be a subset of $[n] = \{1, 2, \dots, n\}$. We use the following notation: $\overline{I} = [n] \setminus I$ is the complement of I in $[n]$, and $|I| = k$ is the number of elements in I. If q is an n-dimensional vector, then we denote by q^I (or q_I, depending on the context) the $|I|$-dimensional (co)vector consisting of the components of q labelled $i_1, \dots, i_k$. If, for example, $q = (q^1, \dots, q^n)$, then $q^I = \{q^{i_1}, \dots, q^{i_k}\}$. We assume throughout that the indices are ordered in an increasing sequence, for example, if

$$[n] = \{1, 2, 3, 4\} \ , \quad I = \{1, 4\} \ , \quad \overline{I} = \{2, 3\} \ ,$$

then

$$(x^I, p_{\overline{I}}) = (x^1, p_2, p_3, x^4) \ .$$

We adhere to the standard convention about repeated indices: summation is assumed to be carried out over an index repeated in a single term, once as a superscript and once as subscript. If $I \subset [n]$ is such an index, the summation is over all $i \in I$. Thus,

$$xp = x^i p_i = \sum_{i=1}^{n} x^i p_i \ ,$$

$$x^I p_I = \sum_{i \in I} x^i p_i \ .$$

If the same superscript or the same subscript occurs twice, there is no summation over this index; for example, $\partial^2 F / \partial x^I \partial x^I$ denotes the $(|I| \times |I|)$-matrix

$$\begin{pmatrix} \dfrac{\partial^2 F}{\partial x^{i_1}\partial x^{i_1}} & \cdots & \dfrac{\partial^2 F}{\partial x^{i_1}\partial x^{i_k}} \\ \cdots & \cdots & \cdots \\ \dfrac{\partial^2 F}{\partial x^{i_k}\partial x^{i_1}} & \cdots & \dfrac{\partial^2 F}{\partial x^{i_k}\partial x^{i_k}} \end{pmatrix} .$$

A notation such as $a_{II}\xi^I$ and $a_{II}\xi^I\eta^I$ is used only for symmetric matrices a_{II}, and it denotes the sums

$$a_{II}\xi^I = \sum_{i\in I} a_{ij}\xi^i , \quad j \in I ,$$
$$a_{II}\xi^I\eta^I = \sum_{i,j\in I} a_{ij}\xi^i\eta^j .$$

The convention described above does not extend to indices one of which is inside the function symbol and the other is not; for example, there is no summation over $i \in I$ in the expression $p_I(x^I, p_{\overline{I}})$.

The symbol $F_{x^I\to p_I}$ denotes the Fourier transformation with respect to the variables x^I:

$$\left[F_{x^I\to p_I} f\right]\left(p_I, x^{\overline{I}}\right) = (1/2\pi i)^{|I|/2} \int e^{-ix^I p_I} f(x) dx^I \tag{1.3}$$

and $\overline{F}_{p_I\to x^I}$ denotes the inverse Fourier transformation:

$$\left[\overline{F}_{p_I\to x^I}\varphi\right](x) = (i/2\pi)^{|I|/2} \int e^{ix^I p_I} \varphi\left(p_I, x^{\overline{I}}\right) dp_I . \tag{1.4}$$

In (1.3) and (1.4), $\arg i = \pi/2$. The symbol $\lambda(t)$ denotes everywhere a smooth function of $t \in \mathbb{R}^1$ subject to the conditions

$$\begin{cases} \lambda(t) = 0 & \text{for } t \le 1/2 , \\ \lambda(t) = 1 & \text{for } t \ge 1 . \end{cases} \tag{1.5}$$

Finally, $\mathbb{R}_+$ denotes the (multiplicative) group of positive numbers.

2°. Densities on a Manifold M and the Spaces $H^s_{\alpha,\mathrm{loc}}(M)$. Let α be an arbitrary real number. We consider a one-dimensional complex vector bundle such that the transition function from a chart (U, x) to $\left(\widetilde{U}, \widetilde{x}\right)$, is

$$\left(\det \frac{\partial x}{\partial \widetilde{x}}\right)^\alpha = \left(\frac{Dx}{D\widetilde{x}}\right)^\alpha .$$

(The Jacobian $Dx/D\widetilde{x}$ is positive. If M is not oriented, then in the latter expression the modulus of the Jacobian should be taken.) Smooth sections of this bundle are called densities of order α on M.

Thus, a density ρ of order α on M can be regarded as a family of functions $\{f_\rho\}$ defined on coordinate neighbourhoods of M and satisfying the following gluing conditions on their intersections:

$$\widetilde{f}_\rho(\widetilde{x}) = \left(\frac{Dx}{d\widetilde{x}}\right)^\alpha f_\rho(x(\widetilde{x})) \ . \tag{1.6}$$

The space of densities of order α is denoted by $C^\infty_\alpha(M)$ and the space of finite densities of order α by $\mathcal{D}_a(M)$. The function $f_\rho(x)$ is called the *localization* of ρ with respect to the coordinates $\widetilde{x}$. Densities of order 0 are simply functions on M.

Definition 1.1. A *generalized density* of order α on M is a continuous linear functional on the space[2] $\mathcal{D}_{1-\alpha}(M)$ of finite densities of order $1-\alpha$.

There is a natural embedding

$$C^\infty_\alpha(M) \subset \mathcal{D}'_\alpha(M) \ ,$$

where $\mathcal{D}'_\alpha(M)$ is the space of generalized densities of order α on M. In fact, if $\varphi \in C^\infty_\alpha(M)$ and $\rho \in \mathcal{D}_{1-\alpha}(M)$, and if the support of ρ lies entirely in a local coordinate system $x = (x^1, \ldots, x^n)$, then we set

$$\langle \rho, \varphi \rangle = \int f_\rho(x) f_\varphi(x) dx \ . \tag{1.7}$$

By (1.6), the integral (1.7) is independent of the choice of the local coordinate system. By linearity (1.7) extends uniquely to the whole of $\mathcal{D}_{1-\alpha}(M)$.

If $\varphi \in \mathcal{D}'_\alpha(M)$ but $\varphi \notin C^\infty_\alpha(M)$, then $f_\varphi(x)$ is a generalized function. As in the case of a smooth density, we call $f_\varphi(x)$ the *localization* of φ with respect to the local coordinate system $(x^1, \ldots, x^n)$.

When there is no risk of misunderstanding, we do not distinguish between a generalized density and its localization. The concepts of support $\operatorname{supp}\varphi$ and singular support $\operatorname{sing\,supp}\varphi$ are defined for generalized densities in exactly the same way as for ordinary distributions.

Let $K \subset M$ be a compact set, and s a real number. We define the Sobolev space $H^s_\alpha(K)$ as the space of elements $\varphi \in \mathcal{D}'_\alpha(K)$ whose localization in any coordinate system (U, x) belongs to $H^s(U)$. The norm $\|\varphi\|_{s,K}$ is defined as follows. Let $\{e_i\}_{i=1}^m$ be a partition of unity on K subordinate to a covering of K by coordinate neighbourhoods U_i. We set

[2] Convergence in $\mathcal{D}_\alpha(M)$ is defined just as in $\mathcal{D}(\mathbb{R}^n)$ (see, for example, Gel'fand and Shilov [1]).

$$\|\varphi\|_{s,K} = \sum_{i=1}^{m} \|e_i\varphi\|_s \ , \tag{1.8}$$

where $\|e_i\varphi\|$ is the norm of the localization of $e_i\varphi$ in the Sobolev space $H^s(\mathbb{R}^n)$. Up to equivalence, the norm defined by (1.8) is independent of the choice of the partition of unity $\{e_i\}$. We denote by $H^s_{\alpha,\mathrm{loc}}(M)$ the space of densities $\varphi \in \mathcal{D}'_\alpha(M)$ such that $\chi\varphi \in H^s_\alpha(K)$ for any $\chi \in C_0^\infty(M)$. (Here K is a compact set including $\operatorname{supp}\chi$.) A linear operator

$$A : H^s_{\alpha,\mathrm{loc}}(M) \to H^\tau_{\alpha,\mathrm{loc}}(M) \tag{1.9}$$

is said to be bounded if for any $\chi \in C_0^\infty(M)$ there is a function $\chi_1 \in C_0^\infty(M)$ such that

$$\|\chi A\varphi\|_{\tau,K} \le c\,\|\chi_1\varphi\|_{s,K_1} \ , \qquad \varphi \in H^s_{\alpha,\mathrm{loc}}(M) \tag{1.10}$$

with a constant c independent of φ. Here K and K_1 are two compact sets such that $K \supset \operatorname{supp}\chi$ and $K_1 \supset \operatorname{supp}\chi_1$. The space $H^s_{0,\mathrm{loc}}(M)$ is denoted simply by $H^s_{\mathrm{loc}}(M)$.

If M_1 and M_2 are two manifolds we denote by $C^\infty_{\alpha,\beta}(M_1 \times M_2)$ the space of densities of order (α, β); if

$$\left(U_1 \times U_2; x^1, \ldots, x^{n_1}, y^1, \ldots, y^{n_2}\right) \ , \quad \left(U_1 \times U_2; \widetilde{x^1}, \ldots, \widetilde{x^{n_1}}, \widetilde{y^1}, \ldots, \widetilde{y^{n_2}}\right)$$

are two coordinate systems on $M_1 \times M_2$, then the localizations $f_\rho(x,y)$ and $\widetilde{f}_\rho(\widetilde{x},\widetilde{y})$ in these coordinate systems are related by

$$f_\rho(\widetilde{x},\widetilde{y}) = f_\rho(x,y)\,(Dx/D\widetilde{x})^\alpha\,(Dy/D\widetilde{y})^\beta \ . \tag{1.11}$$

The spaces $\mathcal{D}_{\alpha,\beta}(M_1, M_2)$, $\mathcal{D}'_{\alpha,\beta}(M_1, M_2)$, and $H^s_{\alpha,\beta,\mathrm{loc}}(M_1 \times M_2)$ can be introduced along the same lines. With each $k \in C^\infty_{\alpha,\beta}(M_1 \times M_2)$ we can associate an operator

$$\widehat{\Phi}(k) : \mathcal{D}_{1-\beta}(M_2) \to C^\infty_\alpha(M_1)$$

by the formula

$$\widehat{\Phi}(k) = \int k(x,y) f(y)\,dy \ . \tag{1.12}$$

If $k \in \mathcal{D}'_{\alpha,\beta}(M_1, M_2)$, then

$$\langle \widehat{\Phi}(k)f, \varphi\rangle = \langle k, f(y)\varphi(x)\rangle \ , \qquad \varphi \in \mathcal{D}_{1-\alpha}(M_1)$$

defines an operator

$$\widehat{\Phi}(k) : \mathcal{D}_{1-\beta}(M_2) \to \mathcal{D}'_\alpha(M_1) \ .$$

It is not hard to see that if $k \in H^s_{\alpha,\beta,\mathrm{loc}}(M_1 \times M_2)$, then

$$\widehat{\Phi}(k) : H^{-s}_{1-\beta,\mathrm{comp}}(M_2) \to H^s_{\alpha,\mathrm{loc}}(M_1) \ .$$

(Here $H^s_{\alpha,\mathrm{comp}}(M)$ denotes the subspace of $H^s_{\alpha,\mathrm{loc}}(M)$ consisting of the elements with compact support.) If the support of $k \in H^s_{\alpha,\beta,\mathrm{loc}}(M_1 \times M_2)$ is such that the set $\pi^{-1}(A) \cap \operatorname{supp} k$ is compact for any compact set $A \subset M_2$ (where π: $M_1 \times M_2 \to M_2$ is the projection), then

$$\widehat{\Phi}(k) : H^{-s}_{1-\beta,\mathrm{loc}}(M_2) \to H^s_{\alpha,\mathrm{loc}}(M_1) \ .$$

3°. Homogeneous Manifolds and Symbols

Definition 1.2. A smooth manifold L is said to be *homogeneous* if a free infinitely differentiable action[3] of $\mathbb{R}_+$ is defined on L and if the set of orbits $L/\mathbb{R}_+$ admits the structure of a smooth manifold for which the projection $L \to L/\mathbb{R}_+$ is a smooth map.

Examples of homogeneous manifolds:

1) The space $\mathbb{R}^n \times \mathbb{R}_{m^*} = \mathbb{R}^n \times \{\mathbb{R}_m \setminus \{0\}\}$ with the action of $\mathbb{R}_+$ defined by

$$t(x,\xi) = (x, t\xi) \ , \qquad t \in \mathbb{R}_+ \ , \quad x \in \mathbb{R}^n \ , \quad \xi \in \mathbb{R}_{m^*} \ . \tag{1.13}$$

2) $T^*_0 M$, the contangent bundle of M excluding the zero section. In local coordinates (x,p) the action of the group has the form

$$t(x,p) = (x, tp) \ . \tag{1.14}$$

A *homogeneous chart* on L is a diffeomorphism, commuting with the action of $\mathbb{R}_+$, of an R_+-invariant open set $U \subset L$ onto an open subset of $\mathbb{R}^{l'} \times \mathbb{R}_{l-l'^*}$ (where $l = \dim L$ and l', $0 \le l' \le l$, is arbitrary). It is not hard to verify that on an homogeneous manifold L there is always an atlas consisting of homogeneous charts. For it can be obtained as the inverse image of any atlas on the manifold $L/\mathbb{R}_+$ under the projection $L \to L/\mathbb{R}_+$. Such an atlas is said to be homogeneous. Let s be a real number.

Definition 1.3. A function f on a homogeneous manifold is called *homogeneous* of degree s if for all $x \in L$ and $t \in \mathbb{R}_+$

$$f(tx) = t^s f(x) \ . \tag{1.15}$$

[3] We recall that the action of a group G on a set M is said to be free if from $gm = m$, $g \in G$, for some $m \in M$ it follows that $g = 1$.

We denote the space of homogeneous functions of degree s by $\mathcal{O}_s(L)$. If $L = M \times \mathbb{R}_m$, where M is a manifold on which $\mathbb{R}_+$ acts trivially, then we also say that f is homogeneous of degree s in the variables (or parameters) $\xi \in \mathbb{R}_m$.

For a fixed $t \in \mathbb{R}_+$ we denote by

$$t : T \to L$$

the homothetic map defined by the action of $\mathbb{R}_+$. A differential form $\omega \in \Lambda^k(L)$ is said to be homogeneous of degree s if

$$t^*\omega = t^s\omega .$$

A differential operator P on a homogeneous manifold L is called homogeneous of degree k if for any smooth homogeneous function φ on L of degree s the function $P\varphi$ is homogeneous of degree $k + s$.

We say that a set $K \subset L$ is $\mathbb{R}_+$-compact if K is $\mathbb{R}_+$-invariant and $K/\mathbb{R}_+$ is compact.

Using a homogeneous atlas and a partition of unity, we can easily verify that everywhere on L there is a strictly positive homogeneous function χ of degree 1.

If χ_1 and χ_2 are two strictly positive homogeneous functions of degree 1 on L, then on any $\mathbb{R}_+$-compact set

$$c_1\chi_1 \leq \chi_2 \leq c_2\chi_1 . \tag{1.16}$$

We assume that χ is chosen an fixed.

Definition 1.4. A *symbol* of degree s on a homogeneous manifold L is a smooth function

$$f = \sum_{k=0}^{N} f_{s-k} , \tag{1.17}$$

where $f_{s-k} \in \mathcal{O}_{s-k}(L)$. We denote the set of all symbols of degree s by $\mathfrak{S}_s(L)$.

The number N in the expression (1.17) is called the length of f.

It should be noted that $\mathfrak{S}_0(L)$ is closed under addition, multiplication, and multiplication by a number; therefore, we call $\mathfrak{S}_0(L)$ the *algebra of symbols* of degree 0 on L. However, the space $\mathfrak{S}_s(L)$ is not closed under multiplication. The product of two functions determines a bilinear map

$$\mathfrak{S}_s(L) \times \mathfrak{S}_{s'}(L) \to \mathfrak{S}_{s+s'}(L) .$$

The space $\mathfrak{S}_s(L)$ can also be described as follows: $f \in \mathfrak{S}_s(L)$ if and only if

$$f = \chi^s f_0 , \qquad f_0 \in \mathfrak{S}_0(L) .$$

Definition 1.5. An element $f \in \mathfrak{S}_s(L)$ is said to *approximate asymptotically* a function $\varphi \in C^\infty(L)$ if for any homogeneous differential operator P of degree t

$$|P(\varphi - f)| \le c\chi^{s+t-N-1} \quad \text{for } \chi > 1 \tag{1.18}$$

on any $\mathbb{R}_+$-compact set $K \subset L$, where N is the length of f.

Then we say that f is the *asymptotic expansion* of φ up to the order $s - N$, and we write

$$\varphi \sim f = \sum_{k=0}^{N} f_{s-k} .$$

This relation clearly implies that

$$\varphi \sim \sum_{k=0}^{N'} f_{s-k}$$

for any $N' \le N$. It is easy to see that the asymptotic expansion of a given length is uniquely determined by φ.

If $\varphi \in C^\infty(L)$ has an asymptotic expansion of any length, then we say that φ is asymptotically homogeneous. It is not hard to see that in this case we can associate with φ the formal series

$$\varphi \sim \sum_{k=0}^{\infty} f_{s-k}$$

such that the partial sums of any length approximante φ asymptotically.

Let $S \in \mathcal{O}_1(L)$ be a real function. By definition, the relation

$$\varphi \sim e^{iS} f \tag{1.19}$$

for $\varphi \in C^\infty(L)$ and $f \in \mathfrak{S}_s(L)$ means that

$$e^{-iS}\varphi \sim f .$$

It follows from (1.19) that

$$\left| P\left(\varphi - e^{iS}\sum_{k=0}^{N} f_{s-k}\right)\right| \le c\chi^{s+m+t-N-1} , \qquad \chi > 1 , \tag{1.20}$$

for any differential operator P of order m that is homogeneous of degree t. In local coordinates (x,ξ) the formulae (1.18) and (1.20) are equivalent (for $|\xi|>1$) to

$$\left|\frac{\partial^\alpha}{\partial x^\alpha}\frac{\partial^\beta}{\partial \xi^\beta}\left(\varphi(x,\xi)-\sum_{k=0}^{N} f_{s-k}(x,\xi)\right)\right| \le c\,|\xi|^{s-N-|\beta|-1}\ , \quad (1.21)$$

$$\left|\frac{\partial^\alpha}{\partial x^\alpha}\frac{\partial^\beta}{\partial \xi^\beta}\left(\varphi(x,\xi)-e^{iS}\sum_{k=0}^{N} f_{s-k}(x,\xi)\right)\right| \le c\,|\xi|^{s-N+|\alpha|+1}\ . \quad (1.22)$$

respectively. The constants in (1.18) and (1.20)–(1.22) depend on m,t,α and β.

Remark 1.1. In the space $\mathbb{R}^n \times \mathbb{R}_{m^*}$ with the coordinates (x,ξ) the relation $\varphi \sim \sum_{k=0}^{N} f_{s-k}$, $f_{s-k} \in \mathcal{O}_{s-k}(L)$, is equivalent to

$$\varphi(x,t\xi) \equiv \sum_{k=0}^{N} t^{s-k} f_{s-k}(x,\xi) \left(\operatorname{mod} t^{s-N-1}\right)$$

as $t \to +\infty$, where the congruence in the last formula is understood in the sense that

$$\left|D_x^\alpha D_\xi^\beta \left[\varphi(x,t\xi) - \sum t^{s-k} f_{s-k}(x,\xi)\right]\right| \le ct^{s-N-1}$$

uniformly in ξ for $\delta_1 \le |\xi| \le \delta_2$, $\delta_1 > 0$.

Remark 1.2. The definition (1.19) of an asymptotic expansion cannot be used for complex-valued functions S with non-negative imaginary part. In this situation we regard the estimate (1.20) as the definition of the asymptotic expansion (1.19).

4°. Pseudodifferential Operators on Manifolds

Definition 1.6. A *pseudodifferential operator* of order m on M is a continuous linear operator

$$\widehat{H} : \mathcal{D}_\alpha(M) \to C_\alpha^\infty(M)\ ,$$

satisfying the following conditions:

1) For any density $\rho \in \mathcal{D}_\alpha(M)$ and a smooth real function $S(x,\xi)$, $x \in M$, $\xi \in \mathbb{R}_{k^*}$, that is homogeneous of degree 1 in $\xi \in \mathbb{R}_{k^*}$, with $dS \ne 0$ on $\operatorname{supp}\rho$, the density $e^{-iS}\widehat{H}\left(e^{iS}\rho\right)$ is asymptotically homogeneous of degree m in the parameters $\xi \in \mathbb{R}_{k^*}$, that is,

$$e^{-iS}\widehat{H}\left(e^{iS}\rho\right) \sim \sum_{k=0}^{\infty} f_k(\rho, S) \ . \tag{1.23}$$

Here $f_k(\rho, S)$ is a homogeneous density of degree $m - k$ in $\xi \in \mathbb{R}_{k^*}$, and the asymptotic expansion (1.23) is uniform in $S \in K$, where $K \subset C^\infty(M \times \mathbb{R}_{k^*})$ is an arbitrary compact set.

2) Let $K \subset M$ be a compact set. Then there are compact sets K', $K'' \subset M$ such that

$$\begin{cases} \operatorname{supp}\rho \subset K \Rightarrow \operatorname{supp}\widehat{H}\rho \subset K' \ , \\ \operatorname{supp}\rho \bigcap K'' = \emptyset \Rightarrow \operatorname{supp}\widehat{H}\rho \bigcap K = \emptyset \ . \end{cases} \tag{1.24}$$

An operator satisfying 2) is said to be ***properly supported.*** The definition of a properly supported operator can be restated in the following equivalent form. Let $\mathcal{K}(x,y) \in \mathcal{D}'_{\alpha,1-\alpha}(M \times M)$ be the kernel of $\widehat{H}$. For any compact set $K \subset M$ the sets $\pi_i^{-1}(K) \cap \operatorname{supp}\mathcal{K}$, where π_i: $M \times M \to M (i = 1$ or $2)$ is the projection on the i-th factor, are compact.

A pseudodifferential operator of order m is continuous in the spaces

$$\widehat{H} : H^s_{\alpha,\mathrm{loc}}(M) \to H^{s-m}_{\alpha,\mathrm{loc}}(M)$$

for any real s (see Hörmander [2]).

(To avoid complicating our account with unnecessary technical details we consider only the so-called classical pseudodifferential operators (see Hörmander [5]).)

Let $(x^1, \dots, x^n)$ be a local coordinate system on $U \subset M$. Up to an operator with a smooth kernel, in these coordinates the result of the action of $\widehat{H}$ on an element $u \in H^s_{\alpha,\mathrm{loc}}(M)$ whose support lies in U can be written in the form[4]

$$\widehat{H}u(x) = (i/2\pi)^{n/2} \int e^{ixp} H(x,p)\widetilde{u}(p)dp \equiv H\left(\overset{2}{x}, \overset{1}{\widehat{p}}\right) u(x) \ , \tag{1.25}$$

where $\widetilde{u}(p)$ is the Fourier transform of $u(x)$. Here $H(x,p)$ is a function asymptotically homogeneous of degree m in p and is called the total symbol of $\widehat{H}$ in the local coordinates $\left(x^1, \dots, x^n\right)$. The symbol $H(x,p)$ has the following asymptotic expansion:

$$H(x,p) \sim \sum_{k=0}^{\infty} H_k(x,p) \ , \tag{1.26}$$

$$H_k(x,p) = f_k\left(1_x, S_p\right) \ , \tag{1.27}$$

[4] We recall that we use the same notation for the density of a distribution and its localization.

where 1_x is a function identically equal to 1 in a neighbourhood of x such that $\operatorname{supp} 1_x \subset U$, and S_p is of the form $S_p = x^i p_i$, and the f_k are defined by (1.23).

The function $H_0(x,p)$ is called the principal symbol (the Hamiltonian function) of the pseudodifferential operator $\widehat{H}$. It is defined invariantly on T^*M and belongs to $\mathcal{O}_m(T^*M)$. If $v \in \Lambda^n(M)$ is a positive volume form on M, then the function

$$H^v_{\text{sub}}(x,p) = iH_1(x,p) - \frac{1}{2}\left[\frac{\partial^2 H_0(x,p)}{\partial x^k \partial p_k} + (2\alpha - 1)\, V(H_0) \log v_x(x)\right] , \tag{1.28}$$

where $V(H_0)$ is a Hamiltonian vector field corresponding to the Hamiltonian H_0:

$$V(H_0) = \frac{\partial H_0}{\partial p_i}\frac{\partial}{\partial x^i} - \frac{\partial H_0}{\partial x^i}\frac{\partial}{\partial p_i} , \tag{1.29}$$

is called the *subprincipal symbol* of $\widehat{H}$ with respect to the volume form v. The function (1.28) is invariantly defined on T^*M and belongs to $\mathcal{O}_{m-1}(T^*M)$ (see Duistermaat and Hörmander [1]).

Subprincipal symbols calculated with respect to different volume forms v_1 and v_2 differ by the term

$$\frac{2\alpha - 1}{2} V(H_0) \log F ,$$

where $F = v_1/v_2$ is a function defined invariantly on M.

As we showed in § 1.3°, the expansion (1.26) in a local coordinate system defines symbols of $\widehat{H}$ of arbitrary length N:

$$H^{(N)}(x,p) = \sum_{k=0}^{N} H_k(x,p) .$$

It is well known (see, for example, Hörmander [2]) that $\widehat{H}$ can be recovered from its symbol of length N up to operators of order $m - N - 1$. The total symbol determines $\widehat{H}$ modulo integral operators with infinitely differentiable kernel.

5°. Auxiliary Lemmas. Here we derive three lemmas, which we shall require later.

Lemma 1.1. *Let $f(x^I, p_{\bar{I}})$ be a smooth function of compact support in x^I such that for some real s and $\epsilon > 0$*

$$\left|\frac{\partial^{|k|}}{\partial(x^I)^k} f\left(x^I, p_{\overline{I}}\right)\right| \leq c_k \left(1+\left|p_{\overline{I}}\right|\right)^{-s-\frac{|\overline{I}|}{2}-\epsilon+|k|}, \qquad k=1,2,\ldots \tag{1.30}$$

Then

$$\varphi(x) = \left(\overline{F}_{p_{\overline{I}}\to x^{\overline{I}}} f\right)(x)$$

belongs to $H^s(\mathbb{R}^n)$.

Remark 1.3. If instead of the compactness of the support with respect to the variables x^I we require the estimate (1.30) to hold whenever x^I lies in an arbitrary compact set, we obtain that $\varphi(x)$ belongs to the space $H^s_{\text{loc}}(\mathbb{R}^n)$.

The proof is obvious.

Now let $\Phi(x,y,\xi,\eta)$ be a real smooth function on $\mathbb{R}^k \times \mathbb{R}^l \times R_{m^*} \times R_{n^*}$, that is homogeneous of degree 1 (the phase function). Any solution $(y(x,\xi), \eta(x,\xi))$ of the equation

$$d_{y,\eta}\Phi(x,y,\xi,\eta) = 0$$

is called a *stationary point* of the phase Φ. We assume that all the stationary points of Φ are non-degenerate, that is, at the stationary points we have

$$\det \operatorname{Hess}_{y,\eta} \Phi(x,y,\xi,\eta) = \det\left(\frac{\partial^2\Phi}{\partial y\,\partial\eta}\right) \neq 0\,.$$

Let $\varphi(x,y,\xi,\eta)$ be a homogeneous function of degree m in the variables (ξ,η). We assume that the support of φ for fixed (x,ξ) is compact and that for each (x,ξ) there is at most one stationary point (y,η) of Φ such that $(x,y,\xi,\eta) \in \operatorname{supp}\varphi$.

Lemma 1.2. *Under the conditions stated above, we have*

$$\begin{aligned} &(i/2\pi)^{(n+l)/2} \int_{\mathbb{R}^l\times\mathbb{R}_n} e^{i\Phi(x,y,\xi,\eta)} \varphi(x,y,\xi,\eta)\, dy\, d\eta \\ &\sim \sum_{k=0}^{\infty} e^{i\Phi(x,y(x,\xi),\xi,\eta(x,\xi))} \{M_k\varphi\}(x,y(x,\xi),\xi,\eta(x,\xi))\,, \end{aligned} \tag{1.31}$$

where the M_k *are homogeneous differential operators of degree* $-k$ *in the variables* y *and* η *of order at most* $2k$*, with coefficients depending on* Φ *and its derivatives.*

The leading term of the asymptotic expansion has the form

$$\begin{aligned} &[\exp\{i\Phi(x,y,(x,\xi),\xi,\eta(x,\xi))\}]\, M_0\varphi(x,y(x,\xi),\xi,\eta(x,\xi)) \\ &\quad = [\exp\{i\Phi(x,y(x,\xi),\xi,\eta(x,\xi))\}] \\ &\quad\times \frac{\varphi(x,y(x,\xi),\xi,\eta(x,\xi))}{\sqrt{\det \operatorname{Hess}_{y,\eta}(-\Phi(x,y(x,\xi),\xi,\eta(x,\xi)))}}\,. \end{aligned} \tag{1.32}$$

The argument in (1.32) is chosen as follows:

$$\arg\det \mathrm{Hess}_{y,\eta}(-\Phi) = \sum_{k-1}^{n+l} \arg \lambda_k \ , \tag{1.33}$$

where $\{\lambda_k\}$ are the eigenvalues of the matrix $\mathrm{Hess}_{y,\eta}(-\Phi)$ and the arguments of λ_k are chosen in the interval

$$-3\pi/2 < \arg\lambda_k \leq \pi/2 \ .$$

By means of Remark 1.1 the proof of Lemma 1.2 reduces to obtaining the asymptotic expansion as $t \to +\infty$ of the integral

$$\left(\frac{it}{2\pi}\right)^{(n+l)/2} \int_{\mathbb{R}^l \times \mathbb{R}_n} e^{it\Phi(x,y,\xi,\eta)} \varphi(x,y,\xi,\eta)dy\, d\eta \ ,$$

which can be found, for example, in Fedoryuk [1] or Chap. 5 of this book.

We can relax the condition of the compactness of $\operatorname{supp}\varphi$ with respect to (y,η) for fixed (x,ξ) replacing it by

$$\varphi(x,y,\xi,\eta) = 0 \qquad \text{for} \ \ y \notin K$$

for some compact set K. It is then necessary to regularize the integral (1.31) by repeated integration by parts outside some neighbourhood of the stationary point. This procedure shows that only the integral over a certain homogeneous neighbourhood of the stationary point $(y(x,\xi),\eta(x,\xi))$ contributes to the expansion (1.31). Using a homogeneous partition of unity (of degree 0) we can reduce the proof of (1.31) to the case discussed earlier.

Let $\widehat{H}$ be a pseudodifferential operator of order m on $\mathbb{R}^n$. We denote by $\widehat{H}_I$ the operator

$$\widehat{H}_I = F_{x^{\bar{I}} \to p_{\bar{I}}} \circ \widehat{H} \circ \overline{F}_{p_{\bar{I}} \to x^{\bar{I}}} \ , \tag{1.34}$$

acting on functions of the variables $(x^I, p_{\bar{I}})$.

Lemma 1.3 (see Maslov [1]). *Let $S(x^I, p_{\bar{I}})$ be a smooth real homogeneous function of degree 1 in the variables $p_{\bar{I}}$. Then for any homogeneous function $u(x^I, p_{\bar{I}})$ with $\mathbb{R}_+$-compact support we have the following asymptotic expansion:*

$$\widehat{H}_I e^{iS(x^I,p_{\bar{I}})} u(x^I,p_{\bar{I}}) \sim \sum_{k=0}^{\infty} e^{iS(x^I,p_{\bar{I}})} P_k u(x^I,p_{\bar{I}}) \ . \tag{1.35}$$

Here, the P_k are homogeneous differential operators of degree $m-k$ and of order at most k. In particular, the operators P_0 and P_1 have the form

$$P_0 = H_0\left(x^I, -\frac{\partial S}{\partial p_{\overline{I}}}, \frac{\partial S}{\partial x^I}, p_{\overline{I}}\right), \tag{1.36}$$

and

$$\begin{aligned} P_1 = H_1 + i\Bigg\{ & \frac{\partial H_0}{\partial x^{\overline{I}}}\frac{\partial}{\partial p_{\overline{I}}} - \frac{\partial H_0}{\partial p_I}\frac{\partial}{\partial x^I} - \frac{1}{2}\left(\frac{\partial^2 S}{\partial x^I \partial x^I}\frac{\partial^2 H_0}{\partial p_I \partial p_I}\right. \\ & \left. + \frac{\partial^2 S}{\partial p_{\overline{I}}\partial p_{\overline{I}}}\frac{\partial^2 H_0}{\partial x^{\overline{I}}\partial x^{\overline{I}}} - 2\frac{\partial^2 S}{\partial x^I \partial p_{\overline{I}}}\frac{\partial^2 H_0}{\partial x^{\overline{I}}\partial p_I}\right) + \frac{\partial^2 H_0}{\partial x^{\overline{I}}\partial p_{\overline{I}}}\Bigg\}. \end{aligned} \tag{1.37}$$

The values of the functions and of their derivatives in (1.37) *are taken at the point* $(x^I, -\partial S/\partial p_{\overline{I}}, \partial S/\partial x^I, p_{\overline{I}})$.

Proof (see Maslov and Fedoryuk [2]). We consider the expression

$$\begin{aligned} \widehat{H}_I e^{iS(x^I, p_{\overline{I}})} u(x^I, p_{\overline{I}}) = \left(\frac{1}{2\pi}\right)^n \int \exp\Big\{ i \Big[\left(x^I - \widetilde{x}^I\right) p_I \\ + x^{\overline{I}}\left(\widetilde{p}_{\overline{I}} - p_I\right) + S\left(\widetilde{x}^I, \widetilde{p}_{\overline{I}}\right)\Big]\Big\} H\left(x^I, \widetilde{x}^{\overline{I}}, p_I, \widetilde{p}_{\overline{I}}\right) \\ \times u\left(\widetilde{x}^I, \widetilde{p}_{\overline{I}}\right) d\widetilde{p}_{\overline{I}} dx^I d\widetilde{x}^I dp_I . \end{aligned} \tag{1.38}$$

We may assume without loss of generality that $I = \{1, \dots k\}$ and $\overline{I} = \{k+1, \dots, n\}$. We introduce the following notation:

$$\begin{aligned} I^l &= \{l, \dots, k\}, & I_l &= \{1, \dots, l\}, & l &\le k; \\ \overline{I}_j &= \{k+1, \dots, j\}, & \overline{I}^j &= \{j, \dots, n\}, & j &> k. \end{aligned}$$

We rewrite (1.38) in the form

$$\begin{aligned} \widehat{H}_I e^{iS(x^I, p_{\overline{I}})} u(x^I, p_{\overline{I}}) = \left(\frac{1}{2\pi}\right)^{n-1} \int d\widetilde{x}^I dp_I d\widetilde{p}_{\overline{I}_{n-1}} \exp\left\{ i \left[(x^I - \widetilde{x}^I) p_I \right.\right. \\ \left.\left. - x^{\overline{I}_{n-1}}\left(p_{\overline{I}_{n-1}} - \widetilde{p}_{\overline{I}_{n-1}}\right)\right]\right\} \left\{ \frac{1}{2\pi} \int \exp\left\{ i \left[S\left(\widetilde{x}^I, \widetilde{p}_{\overline{I}}\right) \right.\right.\right. \\ \left.\left.\left. - x^n (p_n - \widetilde{p}_n)\right]\right\} H\left(x^I, x^{\overline{I}}, p_I, \widetilde{p}_{\overline{I}}\right) u\left(\widetilde{x}^I, \widetilde{p}_{\overline{I}}\right) d\widetilde{p}_n dx^n \right\}. \end{aligned} \tag{1.39}$$

Generally speaking, the integrals on the right-hand side of (1.38) and (1.39) are divergent, but they converge in the space S'. To prove this it is enough to use any of the standard regularizations of integrals of this kind (see, for example Hörmander [3]).

We now transform the integral in braces in (1.39):

$$
\begin{aligned}
&\frac{1}{2\pi}\int \exp\left\{i\left[S\left(\widetilde{x}^I,\widetilde{p}_{\overline{I}}\right)-x^n\left(p_n-\widetilde{p}_n\right)\right]\right\}\\
&\quad\times\left\{H\left(x^I,x^{\overline{I}_{n-1}},-\frac{\partial S\left(\widetilde{x}^I,\widetilde{p}_{\overline{I}}\right)}{\partial\widetilde{p}_n},p_I,\widetilde{p}_{\overline{I}}\right)\right.\\
&\quad\left.+\left[H\left(x^I,x^{\overline{I}},p_I,\widetilde{p}_{\overline{I}}\right)-H\left(x^I,x^{\overline{I}_{n-1}},-\frac{\partial S\left(\widetilde{x}^I,\widetilde{p}_{\overline{I}}\right)}{\partial\widetilde{p}_n},p_I,\widetilde{p}_{\overline{I}}\right)\right]\right\}\\
&\quad\times u\left(\widetilde{x}^I,\widetilde{p}_{\overline{I}}\right)d\widetilde{p}_n dx^n=e^{iS\left(\widetilde{x}^I,\widetilde{p}_{\overline{I}_{n-1}},p_n\right)}\\
&\quad\times H\left(x^I,x^{\overline{I}_{n-1}},-\frac{\partial S\left(\widetilde{x}^I,\widetilde{p}_{\overline{I}_{n-1}},p_n\right)}{\partial p_n},p_I,\widetilde{p}_{\overline{I}_{n-1}},p_n\right)\\
&\quad\times u\left(\widetilde{x}^I,\widetilde{p}_{\overline{I}_{n-1}},p_n\right)+\frac{1}{2\pi}\int e^{i\left[S\left(\widetilde{x}^I,\widetilde{p}_{\overline{I}}\right)-x^n\left(p_n-\widetilde{p}_n\right)\right]}\\
&\quad\times\left[H\left(x^I,x^{\overline{I}},p_I,\widetilde{p}_{\overline{I}}\right)-H\left(x^I,x^{\overline{I}_{n-1}},-\frac{\partial S\left(\widetilde{x}^I,\widetilde{p}_{\overline{I}}\right)}{\partial\widetilde{p}_n},p_I,\widetilde{p}_{\overline{I}}\right)\right]\\
&\quad\times u\left(\widetilde{x}^I,\widetilde{p}_{\overline{I}}\right)d\widetilde{p}_n dx^n\,.
\end{aligned}
\tag{1.40}
$$

Here we have used the fact that the term is the composite of a direct and an inverse Fourier transform. In the integral over $\widetilde{p}_n$ we integrate the second term by parts and find that this term is equal to

$$
\begin{aligned}
&\frac{i}{2\pi}\int \exp\left\{i\left[S\left(\widetilde{x}^I,\widetilde{p}_{\overline{I}}\right)-x^n\left(p_n-\widetilde{p}_n\right)\right]\right\}\\
&\times\frac{\partial}{\partial\widetilde{p}_n}\left[\frac{H\left(x^I,x^{\overline{I}},p_I,\widetilde{p}_{\overline{I}}\right)-H\left(x^I,x^{\overline{I}_{n-1}},-\frac{\partial S\left(\widetilde{x}^I,\widetilde{p}_{\overline{I}}\right)}{\partial p_n},p_I,p_{\overline{I}_{n-1}},p_n\right)}{x^n+\frac{\partial S\left(\widetilde{x}^I,\widetilde{p}_{\overline{I}}\right)}{\partial p_n}}u\left(\widetilde{x}^I,\widetilde{p}_{\overline{I}}\right)\right]\\
&\cdot d\widetilde{p}_n dx^n\,.
\end{aligned}
$$

Substituting this expression and (1.40) in (1.39) and using the notation

$$
\begin{aligned}
&\Delta_{x^n}\left[H\left(x^I,x^{\overline{I}},p_I,\widetilde{p}_{\overline{I}}\right)\right]\\
&\quad=\frac{H\left(x^I,x^{\overline{I}},p_I,\widetilde{p}_{\overline{I}}\right)-H\left(x^I,x^{\overline{I}_{n-1}},-\frac{\partial S\left(\widetilde{x}^I,\widetilde{p}_{\overline{I}}\right)}{\partial p_n},p_I,\widetilde{p}_{\overline{I}}\right)}{x^n+\frac{\partial S\left(\widetilde{x}^I,\widetilde{p}_{\overline{I}}\right)}{\partial p_n}}\,,
\end{aligned}
$$

we obtain

$$\begin{aligned}
&\widehat{H}_I e^{iS(x^I,p_{\overline{I}})} u(x^I,p_{\overline{I}}) = \left(\frac{1}{2\pi}\right)^{n-1} \int \exp\left\{i\left[\left(x^I-\widetilde{x}^I\right)p_I\right.\right.\\
&\quad \left.\left. -x^{\overline{I}_{n-1}}\left(p_{\overline{I}_{n-1}}-\widetilde{p}_{\overline{I}_{n-1}}\right)+S\left(\widetilde{x}^I,\widetilde{p}_{\overline{I}_{n-1}},p_n\right)\right]\right\}\\
&\quad \times H\left(x^I,x^{\overline{I}_{n-1}},-\frac{\partial S\left(\widetilde{x}^I,\widetilde{p}_{\overline{I}_{n-1}},p_n\right)}{\partial p_n},p_I,\widetilde{p}_{\overline{I}_{n-1}},p_n\right)\\
&\quad \times u\left(\widetilde{x}^I,p_{\overline{I}_{n-1}},p_n\right) d\widetilde{x}^I dp_I d\widetilde{p}_{\overline{I}_{n-1}} dx^{\overline{I}_{n-1}}\\
&\quad +\left(\frac{1}{2\pi}\right)^n \int \exp\left\{i\left[S\left(\widetilde{x}^I,\widetilde{p}_{\overline{I}}\right)+\left(x^I-\widetilde{x}^I\right)p_I-x^{\overline{I}}\left(p_{\overline{I}}-\widetilde{p}_{\overline{I}}\right)\right]\right\}\\
&\quad \times\frac{\partial}{\partial\widetilde{p}_n}\left\{\Delta_{x^n}\left[H\left(x^I,x^{\overline{I}},p_I,\widetilde{p}_{\overline{I}}\right)\right]u\left(\widetilde{x}^I,\widetilde{p}_{\overline{I}}\right)\right\} d\widetilde{x}^I dp_I d\widetilde{p}_{\overline{I}} dx^{\overline{I}}\,.
\end{aligned} \tag{1.41}$$

We repeat this procedure in the first term with respect to the variables $(\widetilde{p}_{n-1},x^{n-1})$, then for $(\widetilde{p}_{n-2},x^{n-2}),\ldots,(\widetilde{p}_{k+1},x^{k+1})$, $(\widetilde{x}^k,p_k),\ldots,(\widetilde{x}^1,p_1)$. To describe the final result we introduce the notation

$$\left\{\begin{aligned}
&\Delta_{x^j}\left[H\left(x^I,x^{\overline{I}},p_I,\widetilde{p}_{\overline{I}}\right)\right]\\
&\quad =\Delta_{x^j}\left[H\left(x^I,x^{\overline{I}_j},-\frac{\partial S\left(\widetilde{x}^I,\widetilde{p}_{\overline{I}_j},p_{\overline{I}^{j+1}}\right)}{\partial p_{\overline{I}^{j+1}}},p_I,\widetilde{p}_{\overline{I}_j},p_{\overline{I}^{j+1}}\right)\right];\\
&\Delta_{p_l}\left[H\left(x^I,x^{\overline{I}},p_I,\widetilde{p}_{\overline{I}}\right)\right]\\
&\quad =\Delta_{p_l}\left[H\left(x^I,-\frac{\partial S\left(\widetilde{x}^{I_l},x^{I^{l+1}},p_{\overline{I}}\right)}{\partial p_{\overline{I}}},p_{I_l},\right.\right.\\
&\quad\quad \left.\left.\frac{\partial S\left(\widetilde{x}^{I_l},x^{I^{l+1}},p_{\overline{I}}\right)}{\partial x^{I^{l+1}}},p_{\overline{I}}\right)\right],
\end{aligned}\right. \tag{1.42}$$

where the symbols Δ_{x^j} and Δ_{p_l} on the right-hand sides of (1.42) denote the standard (difference) derivatives with respect to the variables x^j and p_l at the points

$$x^j=\frac{\partial S}{\partial p_j}\,,\quad p_l=-\frac{\partial S}{\partial x^l}$$

respectively.

In the notation (1.42) we obtain at the first stage

$$
\begin{aligned}
&\widehat{H}_I e^{iS(x^I,p_{\overline{I}})} u(x^I,p_{\overline{I}}) \\
&= e^{iS(x^I,p_{\overline{I}})} H\left(x^I, -\frac{\partial S}{\partial p_{\overline{I}}}, \frac{\partial S}{\partial x^I}, p_{\overline{I}}\right) u(x^I,p_{\overline{I}}) + \sum_{j=k+1}^{n} \left(\frac{1}{2\pi}\right)^j \\
&\times \int \exp\left\{i\left[S\left(\widetilde{x}^I, \widetilde{p}_{\overline{I}_j}, p_{\overline{I}^{j+1}}\right) + \left(x^I - \widetilde{x}^I\right) p_I - x^{\overline{I}_j}\left(p_{\overline{I}_j} - \widetilde{p}_{\overline{I}_j}\right)\right]\right\} \\
&\times \frac{\partial}{\partial \widetilde{p}_j}\left\{\Delta_{x^j}\left[H\left(x^I, x^{\overline{I}}, p_I, \widetilde{p}_{\overline{I}}\right)\right] u\left(\widetilde{x}^I, \widetilde{p}_{\overline{I}_j}, p_{\overline{I}^{j+1}}\right)\right\} d\widetilde{x}^I dp_I d\widetilde{p}_{\overline{I}_j} dx^{\overline{I}_j} \\
&- \sum_{l=1}^{n}\left(\frac{1}{2\pi}\right)^l \int \exp\left\{i\left[S\left(\widetilde{x}^{I_l}, x^{I^{l+1}}, p_{\overline{I}}\right) + \left(x^{I_l} - \widetilde{x}^{I_l}\right) p_{I_l}\right]\right\} \\
&\times \frac{\partial}{\partial \widetilde{x}^l}\left\{\Delta_{p_l}\left[H\left(x^I, x^{\overline{I}}, p_I, \widetilde{p}_{\overline{I}}\right)\right] u\left(\widetilde{x}^{I_l}, x^{I^{l+1}}, p_{\overline{I}}\right)\right\} d\widetilde{x}^{I_l} dp_{I_l} .
\end{aligned}
\tag{1.43}
$$

The first term in (1.43) yields (1.36) and the term H_1 in the expression (1.37). The operations $\frac{\partial}{\partial \widetilde{p}_j}\Delta_{x^j}$ and $\frac{\partial}{\partial \widetilde{x}^l}\Delta_{p_l}$ reduce the degree of homogeneity by one. Applying the described operations once more to the sums on the right-hand side of (1.43), we obtain the next terms of the asymptotic expansion (1.35). Evaluating the integrands in the sums in (1.43) at the point $\left(x^I, -\frac{\partial S}{\partial p_{\overline{I}}}, \frac{\partial S}{\partial x^I}, p_{\overline{I}}\right)$, we come to the expression in braces in (1.37). If we truncate this procedure at some stage, we must estimate the remainder. This estimate follows readily from Lemma 1.1 if we take into account our remark about the degree of homogeneity of the integrands. To be quite rigorous we should introduce a cut-off function in the integrals in question to remove the singularities at the origin.

§ 2. Homogeneous Lagrangian Immersions

Let M_1 and M_2 be two manifolds, $\dim M_i = n_i$. We denote by

$$\pi_i : T_0^* M_1 \times T_0^* M_2 \to T_0^* M_i \qquad (i = 1, 2)$$

the projection of the direct product onto its i-th factor. We define a Hamiltonian structure on $T_0^* M_1 \times T_0^* M_2$ by

$$\omega^2_{M_1,M_2} = \pi_1^* \omega^2_{M_2} - \pi_2^* \omega^2_{M_2} \; ; \quad \omega^1_{M_1,M_2} = \pi_1^* \omega^1_{M_1} - \pi_2^* \omega^1_{M_2} \; . \tag{2.1}$$

Let (x, p) and (y, q) be the coordinate systems in $T_0^* M_1$ and $T_0^* M_2$ induced by the x and y coordinate systems in M_1 and M_2. In these coordinates the forms (2.1) are

$$\omega^1_{M_1,M_2} = p_i dx^i - q_i dy^i \ , \tag{2.2}$$

$$\omega^2_{M_1,M_2} = dp_i \wedge dx^i - dq_i \wedge dy^i \ . \tag{2.3}$$

The symplectic manifold $\left(T_0^* M_1 \times T_0^* M_2, \omega^2_{M_1,M_2}\right)$ is clearly homogeneous. If M_2 consists of a single point, this symplectic manifold is $T_0^* M_1$.

Let L be a homogeneous manifold.

Definition 2.1. A homogeneous immersion (that is, one commuting with the action of $\mathbb{R}_+$)

$$i : L \to T_0^* M_1 \times T_0^* M_2 \tag{2.4}$$

is said to *Lagrangian* if $\dim L = \dim M_1 + \dim M_2$ and

$$i^* \left(\omega^2_{M_1,M_2}\right) = 0 \ . \tag{2.5}$$

Example 2.1. Let $M_1 = \mathbb{R}^{n_1}$ and $M_2 = \mathbb{R}^{n_2}$, let $\left(x^1, \ldots, x^{n_1}\right)$ be the coordinates in M_1, and $\left(y^1, \ldots, y^{n_2}\right)$ in M_2, and let $I \subset [n_1]$ and $J \subset [n_2]$. Let $S\left(x^I, p_{\overline{I}}, y^J, q_{\overline{J}}\right)$ be a homogeneous function of degree 1 in the variables $\left(p_{\overline{I}}, q_{\overline{J}}\right)$. The immersion

$$i_s : \mathbb{R}^{|I|} \times \mathbb{R}_{|\overline{I}|} \times \mathbb{R}^{|J|} \times \mathbb{R}_{|\overline{J}|} \to T_0^* \mathbb{R}^{n_1} \times T_0^* \mathbb{R}^{n_2} \ ,$$

defined by

$$i_S\left(\left(x^I, p_{\overline{I}}, y^J, q_{\overline{J}}\right)\right) = \left(x^I, -\frac{\partial S}{\partial p_I}, \frac{\partial S}{\partial x^I}, p_{\overline{I}}, y^J, \frac{\partial S}{\partial q_{\overline{J}}}, -\frac{\partial S}{\partial y^J}, q_{\overline{J}}\right) \ , \tag{2.6}$$

is Lagrangian. For clearly,

$$dS = i_S^* \left(\omega^1_{M_1,M_2} - d\left(x^{\overline{I}} p_{\overline{I}}\right) + d\left(y^{\overline{J}} q_{\overline{J}}\right)\right) \ ,$$

therefore,

$$i_S^* \omega^2_{M_1,M_2} = i_S^* d\omega^1_{M_1,M_2} = d i_S^* \omega^1_{M_1,M_2} = d(dS) = 0 \ .$$

As we shall see later, Example 2.1 is general in the sense that every Lagrangian immersion (2.4) can be described locally by (2.6).

Example 2.2. Let $n_1 = n_2$ and let

$$g : T_0^* M_2 \to T_0^* M_1 \tag{2.7}$$

be a homogeneous canonical transformation (see Arnol'd [4]), that is, a smooth map commuting with the action of $\mathbb{R}_+$ and preserving the Hamiltonian structure:

$$g^*\omega^2_{M_1} = \omega^2_{M_2} \ . \tag{2.8}$$

The map

$$i_g : T_0^*M_2 \to T_0^*M_1 \times T_0^*M_2 \ , \tag{2.9}$$

defined by $i_g(\alpha) = (g(\alpha), \alpha)$ (the graph of g) is a Lagrangian immersion, since

$$i_g^*\omega^2_{M_1,M_2} = g^*\omega^2_{M_1} - \omega^2_{M_2} = 0$$

by (2.8).

Lemma 2.1 (see Hörmander [3]). *If $i : L \to T_0^*M_1 \times T_0^*M_2$ is a homogeneous Lagrangian immersion, then $i^*\omega^1_{M_1,M_2} = 0$.*

Proof. Let (x, p, y, q) be coordinates in $T_0^*M_1 \times T_0^*M_2$. Since i is $\mathbb{R}_+$-invariant, the vector field $X = p_j \frac{\partial}{\partial p_j} + q^j \frac{\partial}{\partial q^j}$ is tangent to i. But

$$\omega^1_{M_1,M_2}(Y) = \omega^2_{M_1,M_2}(X, Y)$$

for any vector Y tangent to i. Since the right-hand side of this equality vanishes identically in Y, we see that $i^*\omega^1_{M_1,M_2} = 0$. This completes the proof.

Let $i : L \to T_0^*M_1 \times T_0^*M_2$ be a homogeneous Lagrangian immersion. We now define a canonical atlas of L.

Definition 2.2. A *canonical atlas* of L is a homogeneous atlas $\{U_j\}$ of L satisfying the following conditions:

1) $\{U_j\}$ is a locally finite covering of L.

2) The sets U_j and arbitrary intersections of them are simply-connected domains.

3) The coordinate maps have the form

$$\begin{cases} \varphi_j : U_j \to \mathbb{R}^{|I|} \times \mathbb{R}_{|\bar{I}|} \times \mathbb{R}^{|J|} \times \mathbb{R}_{|\bar{J}|} \ , \\ \alpha \to \left(x^I(\alpha), p_{\bar{I}}(\alpha), y^J(\alpha), q_{\bar{J}}(\alpha)\right) \end{cases} \tag{2.10}$$

for some $I = I(j) \subset [n_1]$ and $J = J(j) \subset [n_2]$. The functions $x = x(\alpha)$, $p = p(\alpha)$, $y = y(\alpha)$, and $q = q(\alpha)$ in (2.10) define an immersion i in some local coordinate system (x, p, y, q) on $T_0^*M_1 \times T_0^*M_2$, which depends, in general, on j.

In particular, $i|_{U_j}$ is an embedding. Henceforth we use the notation $U_j = U_{IJ}$, $\varphi_j = \varphi_{IJ}$, $I = I(j)$, and $J = J(j)$. This should not lead to misunderstanding, although one has to bear in mind that the sets $I(j)$ and $J(j)$ may be identical for distinct j. The coordinates $(x^I, p_{\bar{I}}, y^J, q_{\bar{J}})$ in U_{IJ} are said to be *canonical*, and the sets U_{IJ} the *canonical charts*.

We shall meet objects labelled by indices IJ, which indicate that the object belongs to the chart U_{IJ}; for example, the action S_{IJ}, which will be defined later, is such an object. The summation convention described in § 1.1° does not apply to such indices. The role of each index is clear from the context.

Lemma 2.2 (lemma on local coordinates, (see Arnol'd [4] and Maslov [1]). *Let i be a homogenenous Lagrangian immersion. Then there is a canonical atlas on L.*

Proof. Let (x, y) be a coordinate system on $M_1 \times M_2$. It suffices to show that for every $\alpha \in L$ there are sets $I \subset [n_1]$ and $J \subset [n_2]$ such that $(x^I, p_{\overline{I}}, y^J, q_{\overline{J}})$ is a coordinate system in a neighbourhood of α. This assertion is local, therefore, we may regard L as a submanifold of $T_0^*M_1 \times T_0^*M_2$ and assume that $M_1 = \mathbb{R}^{n_1}$ and $M_2 = \mathbb{R}^{n_2}$. Let

$$\pi : T_0^*\mathbb{R}^{n_1} \times T_0^*\mathbb{R}^{n_2} \to \mathbb{R}^{n_1} \times \mathbb{R}^{n_2} , \quad \pi(x,p,y,q) = (x,y) ;$$

$$\pi_{IJ} : T_0^*\mathbb{R}^{n_1} \times T_0^*\mathbb{R}^{n_2} \to \mathbb{R}^{|I|} \times R^{|J|} , \quad \pi_{IJ}(x,p,y,q) = \left(x^I, y^J\right) .$$

We choose I and J so that the system of forms $\{dx^I, dy^J\}$ is a maximum linearly independent subsystem of $\{dx^1, \dots, dx^{n_1}, dy^1, \dots, dy^{n_2}\}$ on T_α^*L. Thus,

$$\begin{cases} dx^{\overline{I}} = \alpha_I^{\overline{I}} dx^I + \beta_J^{\overline{J}} dy^J , \\ dy^{\overline{J}} = \gamma_I^{\overline{J}} dx^I + \delta_J^{\overline{J}} dy^J \end{cases} \tag{2.11}$$

on T_α^*L. Since the immersion is Lagrangian, the restriction of $\omega^2_{M_1,M_2}$ to $T_\alpha L$ is zero. Using (2.11) we obtain

$$\begin{aligned} 0 = \omega^2_{M_1,M_2}\big|_L = dx^I \wedge \left(dp_I + \alpha_I^{\overline{I}} dp_{\overline{I}} + \gamma_I^{\overline{J}} dq_{\overline{J}}\right) \\ + dy^J \wedge \left(dq_J + \beta_J^{\overline{I}} dp_{\overline{I}} + \delta_{J}^{\overline{J}}, dq_{\overline{J}}\right) . \end{aligned} \tag{2.12}$$

The choice of I and J shows that the restriction of π_{IJ_*} to $T_\alpha L$ is an epimorphism, therefore, there are vectors X_i and Y_j, $i \in I$, $j \in J$, tangent to L at α such that

$$\pi_{IJ_*}(X_i) = \frac{\partial}{\partial x^i} , \quad \pi_{IJ_*}(Y_j) = \frac{\partial}{\partial y^j} .$$

Substituting the vectors X_i and Y_j in (2.12) we find that the forms dp_I and dq_J can be expressed linearly in terms of the form $(dx^I, dy^J, dp_{\overline{I}}, dq_{\overline{J}})$. Hence, this system of forms is a generating set on T_α^*L. Since $\dim L = n_1 + n_2$, it is, in fact, a basis in T_α^*L. This completes the proof.

Remark 2.1. Since L is $\mathbb{R}_+$-invariant, we have $\overline{I} \cup \overline{J} \neq \emptyset$, because (x, y) is clearly not a coordinate system on L.

Remark 2.2. When the Lagrangian immersion is determined by a canonical transformation, we can select canonical coordinates so that $J = \emptyset$ for any chart of the canonical atlas. For let (y, q) be a local coordinate system on $T_0^* M_2$ in a neighbourhood of a point (y_0, q^0). Since g is a canonical transformation, the differential dg takes Lagrangian planes into Lagrangian planes. Hence it follows easily that the map $y \to g(y, q)$ is a homogenenous Lagrangian immersion for each fixed q. We can now use Lemma 2.2 to prove the assertion.

Throughout what follows we assume (for a Lagrangian immersion associated with a canonical transformation) that the canonical atlas satisfies the condition in Remark 2.2.

Remark 2.3. We call a subspace $W \subset T_\alpha(T_0^* M_1 \times T_0^* M_2)$ *isotropic* if $\omega_{M_1, M_2}(X, Y) = 0$ for any pair of vectors $X, Y \in W$. It follows from the proof of Lemma 2.2 that $\dim W \leq n_1 + n_2$.

We claim that any Lagrangian immersion in a canonical chart U_{IJ} is defined by (2.6).

Lemma 2.3. *In each canonical chart U_{IJ} there is a unique homogeneous function S_{IJ} of degree* 1 *such that*

$$dS_{IJ} = i^* \left\{ p_I dx^I - x^{\overline{I}} dp_{\overline{I}} - q_J dy^J + y^{\overline{J}} dq_{\overline{J}} \right\} . \tag{2.13}$$

Proof. If S_{IJ} is a solution of (2.13), then

$$d\left(S_{IJ} + i^* \left(x^{\overline{I}} p_{\overline{I}} - y^{\overline{J}} q_{\overline{J}}\right)\right) = i^* \left(p_i dx^i - q_j dy^j\right) = 0$$

by Lemma 2.1. Therefore, $S_{IJ} + i^* \left(x^{\overline{I}} p_{\overline{I}} - y^{\overline{J}} q_{\overline{J}}\right) = c$. Since S_{IJ} is homogeneous of degree 1, we have $c = 0$. Hence,

$$S_{IJ} = i^* \left(y^J q_{\overline{J}} - x^{\overline{I}} p_{\overline{I}}\right)$$

of if $x^{\overline{I}} = x^{\overline{I}}\left(x^I, p_{\overline{I}}, y^J, q_{\overline{J}}\right)$ and $y^{\overline{J}} = y^{\overline{J}}\left(x^I, p_{\overline{I}}, y^J, q_{\overline{J}}\right)$ are the functions that the determine the immersion i in the chart U_{IJ}, then

$$S_{IJ} = -x^{\overline{I}}\left(x^I, p_{\overline{I}}, y^J, q_{\overline{J}}\right) p_{\overline{I}} + y^{\overline{J}}\left(x^I, p_{\overline{I}}, y^J, q_{\overline{J}}\right) q_{\overline{J}} \,. \tag{2.14}$$

Conversely, if S_{IJ} is given by (2.14), then bearing in mind that $i^* \omega^1_{M_1, M_2} = 0$, we find that S_{IJ} satisfies (2.13). This completes the proof.

It is a direct consequence of Lemma 2.3 that in the canonical coordinates on U_{IJ} the immersion i is given by

$$\begin{cases} x^I = x^I \ , \\ x^{\overline{I}} = -\dfrac{\partial S_{IJ}\left(x^I, p_{\overline{I}}, y^J, q_{\overline{J}}\right)}{\partial p_{\overline{I}}} \ , \\ p_I = \dfrac{\partial S_{IJ}\left(x^I, p_{\overline{I}}, y^J, q_{\overline{J}}\right)}{\partial x^I} \ , \\ p_{\overline{I}} = p_{\overline{I}} \ , \end{cases} \qquad \begin{cases} y^J = y^J \ , \\ y^{\overline{J}} = \dfrac{\partial S_{IJ}\left(x^I, p_{\overline{I}}, y^J, q_{\overline{J}}\right)}{\partial q_{\overline{J}}} \ , \\ q_J = -\dfrac{\partial S_{IJ}\left(x^I, p_{\overline{I}}, y^J, q_{\overline{J}}\right)}{\partial y^J} \ , \\ q_{\overline{J}} = q_{\overline{J}} \ . \end{cases} \tag{2.15}$$

Let us consider, in particular, the case of a Lagrangian immersion associated with a canonical transformation g. Let U_I be a canonical chart on $T_0^* M_2$ and $S_I\left(x^I, p_{\overline{I}}, q\right)$ the action on U_I. It then follows from (2.15) and Remark 2.2 that locally g can be defined as a solution of the system of equations

$$\begin{cases} y = \dfrac{\partial S_I}{\partial q}\left(x^I, p_{\overline{I}}, q\right) \ , \\ p_I = \dfrac{\partial S_I}{\partial x^I}\left(x^I, p_{\overline{I}}, q\right) \ , \\ x^{\overline{I}} = -\dfrac{\partial S_I}{\partial p_{\overline{I}}}\left(x^I, p_{\overline{I}}, q\right) \ . \end{cases} \tag{2.16}$$

The function $S_I\left(x^I, p_{\overline{I}}, q\right)$ is called a *generating function* of g (see Arnol'd [4]).

Definition 2.3. A *homogeneous Lagrangian immersion with measure* is a triple (L, i, μ), where $i : L \to T_0^* M_1 \times T_0^* M_2$ is a homogeneous Lagrangian immersion and μ is a differential form (not vanishing at the origin) of degree $n_1 + n_2$ on L, which is homogeneous of some degree m.

Example 2.3. Let i_g be the homogeneous Lagrangian immersion defined in Example 2.2, and

$$\mu = \omega_{M_2}^2 \wedge \ldots \wedge \omega_{M_2}^2 \quad (n \text{ times}) \ .$$

Then the triple $(T_0^* M_2, i_g, \mu)$ is a homogeneous Lagrangian immersion with measure.

Let (L, i, μ) be a homogeneous Lagrangian immersion with measure, and $\{U_{IJ}\}$ a canonical atlas on L. We define in each canonical chart U_{IJ} the density of the measure μ_{IJ}.

Definition 2.4. The *density* of μ in a chart U_{IJ} is a function $\mu_{IJ}\left(x^I, p_{\overline{I}}, y^J, q_{\overline{J}}\right)$, defined by

$$\mu|_{U_{IJ}} = \mu_{IJ} dx^I \wedge dp_{\overline{J}} \wedge dy^J \wedge dq_{\overline{J}} \ . \tag{2.17}$$

It is not hard to see that $\mu_{IJ}\left(x^I, p_{\overline{I}}, y^J, q_{\overline{J}}\right)$ is a homogeneous function of degree $m - |\overline{I}| - |\overline{J}|$ in the variables $(p_{\overline{I}}, q_{\overline{J}})$ (m is the degree of homogeneity of μ). Having in mind what follows we choose and fix (arbitrarily) in each chart) a branch $\operatorname{Arg}\mu_{IJ}$ of the argument of μ_{IJ}.

§ 3. The Canonical Operator

Let (L, i, μ) be a homogeneous Lagrangian immersion with measure such that $i : L \to T_0^* M_1 \times T_0^* M_2$ is a proper map, that is, $i^{-1}(K)$ is compact for any compact set

$$K \subset T_0^* M_1 \times T_0^* M_2 \ .$$

Let U be an open $\mathbb{R}_+$-invariant set. We denote by $\mathcal{O}_l(U)$ and $\mathfrak{S}_l(U)$ the sets of elements φ of $\mathcal{O}_l(L)$ and $\mathfrak{S}_l(L)$, respectively, such that $\operatorname{supp}\varphi \subset U$. Let v^1 and v^2 be volume forms on the manifolds M_1 and M_2.

Definition 3.1. An *elementary canonical operator* in a chart U_{IJ} is an operator

$$K_{IJ} : \mathcal{O}_l(U_{IJ}) \to \mathcal{D}'_{\alpha,1-\alpha}(M_1, M_2) \ , \tag{3.1}$$

whose action in local coordinates $\left(x^I, p_{\overline{I}}, y^J, q_{\overline{J}}\right)$ of the chart U_{IJ} is defined by the formula:

$$\begin{aligned}[K_{IJ}\varphi](x,y) = \zeta_\varphi(x,y) \overline{F}_{p_{\overline{I}} \to x^{\overline{I}}} \circ F_{q_{\overline{J}} \to y^{\overline{J}}} \Big\{ &\exp\left[iS_{IJ}(x^I, p_{\overline{I}}, y^J, q_{\overline{J}})\right] \\ \times \lambda\left(\left|(p_{\overline{I}}, q_{\overline{J}})\right|\right) &\sqrt{\left[\mu_{IJ}\left(\widetilde{v}^1_x\right)^{2\alpha-1}\left(\widetilde{v}^2_y\right)^{1-2\alpha}\right](x^I, p_{\overline{I}}, y^J, q_{\overline{J}})} \\ \times \varphi(x^I, p_{\overline{I}}, y^J, q_{\overline{J}}) \Big\} \ . \end{aligned} \tag{3.2}$$

The symbols $\widetilde{v}^1_x$ and $\widetilde{v}^2_y$ in (3.2) stand for the densities of the volume forms v_1 and v_2 lifted to L:

$$\widetilde{v}^1_x(x^I, p_{\overline{I}}, y^J, q_{\overline{J}}) = v^1_x\left(x^I, x^{\overline{I}}(x^I, p_{\overline{I}}, y^J, q_{\overline{J}})\right) \ ,$$

$$\widetilde{v}^2_y(x^I, p_{\overline{I}}, y^J, q_{\overline{J}}) = v^2_y\left(y^J, y^{\overline{J}}(x^I, p_{\overline{I}}, y^J, q_{\overline{J}})\right) \ ;$$

$\lambda(t)$ is defined by (1.5), and $\zeta_\varphi(x,y)$ is a smooth function on $M_1 \times M_2$ whose support is contained in a chart with the coordinates (x,y) and $\zeta_\varphi = 1$ in a neighbourhood of the set $\pi \circ i(\operatorname{supp}\varphi)$, ($\pi : T_0^* M_1 \times T_0^* M_2 \to M_1 \times M_2$ is the projection). The choice of the values of the argument of the radicand in (3.2) will be discussed later.

The operator (3.2) is assumed to be defined modulo $C^\infty(M_1 \times M_2)$-functions. It is then obvious that $K_{IJ}\varphi$ is independent of the choice of the cut-off function λ. The fact that $K_{IJ}\varphi$ is independent of the choice of ζ_φ follows from

$$\operatorname{sing\,supp} K_{IJ}\varphi \subset \overline{\pi \circ i\,(\operatorname{supp}\varphi)}\ . \tag{3.3}$$

The relation (3.3) can be verified by repeated integration by parts in the expression for $\psi(x,y)K_{IJ}\varphi$, where $\psi = 0$ in a neighbourhood of $\overline{\pi \circ i\,(\operatorname{supp}\varphi)}$.

Lemma 1.1 can be used to refine (3.1). Namely, operator K_{IJ} acts in the spaces

$$K_{IJ} : \mathcal{O}_l(U_{IJ}) \to H^s_{\alpha,1-\alpha,\mathrm{loc}}(M_1, M_2) \tag{3.4}$$

for $s < -l - m/2$, where m is the degree of homogeneity of μ.

By linearity K_{IJ} can be extended to $\mathfrak{S}_l(U_{IJ})$. If $K_{IJ}\varphi \in H^s(M_1 \times M_2)$ for some $s > -l - m/2$, then $\varphi \in \mathfrak{S}_{l-1}(U_{IJ})$.

For let us consider the function

$$\begin{aligned} f_{IJ} &= F_{x^{\overline{I}} \to p_{\overline{I}}} \circ F_{y^{\overline{J}} \to q_{\overline{J}}} K_{IJ}\varphi \\ &= \exp\left[iS_{IJ}(x^I, p_{\overline{I}}, y^J, q_{\overline{J}})\right] \lambda\left(|p_{\overline{I}}, q_{\overline{J}}|\right) a(x^I, p_{\overline{I}}, y^J, q_{\overline{J}})\ , \end{aligned}$$

where

$$\begin{aligned} a(x^I, p_{\overline{I}}, y^J, q_{\overline{J}}) &= \varphi(x^I, p_{\overline{I}}, y^J, q_{\overline{J}}) \\ &\quad \times \sqrt{\left[\mu_{IJ}\left(\tilde{v}^1_x\right)^{2\alpha-1}\left(\tilde{v}^2_y\right)^{1-2\alpha}\right](x^I, p_{\overline{I}}, y^J, q_{\overline{J}})}\ . \end{aligned}$$

Clearly we may assume that a is a homogeneous function of degree $l + \frac{1}{2}\left(m - |\overline{I}| - |\overline{J}|\right)$ in the variables $\left(p_{\overline{I}}, q_{\overline{J}}\right)$. If $K_{IJ}\varphi \in H^s(M_1 \times M_2)$, then $f_{IJ} \in \widetilde{H}^s$, where the norm in $\widetilde{H}^s$ is defined by

$$\begin{aligned} \left\|f(x^I, p_{\overline{I}}, y^J, q_{\overline{J}})\right\|^2_{\widetilde{H}^s} &= \left\|\left(1 + \left|(p_{\overline{I}}, q_{\overline{J}})\right|^2 - \Delta_{x^I, y^J}\right)^{s/2} f\right\|^2_{L_2} \\ &= \int \left(1 + \left|(p_{\overline{I}}, q_{\overline{J}})\right|^2 + |(p_I, q_J)|^2\right)^s \left|\tilde{f}(p,q)\right|^2 dp\, dq\ , \end{aligned}$$

and $\tilde{f}(p,q)$, the Fourier transform with respect to the variables x^I and y^J, is homogeneous of degree $l + \frac{1}{2}\left(m - |\overline{I}| - |\overline{J}|\right)$ in $\left(p_{\overline{I}}, q_{\overline{J}}\right)$. We assume that $\tilde{f}(p,q) \neq 0$ and

$$\tilde{f}(p_0, q_0) \neq 0\ .$$

Then $\widetilde{f} \neq 0$ at any point of some neighbourhood of (p_0, q_0) of the form $U \times V$, where U is a neighbourhood of (p_{0I}, q_{0J}) and V is a conic neighbourhood of $(p_{0\overline{I}}, q_{0\overline{J}})$. In the relevant neighbourhood we have

$$\left(1+\left|(p_{\overline{I}}, q_{\overline{J}})\right|^2+|(p_I, q_J)|^2\right) \leq C\left(1+\left|(p_{\overline{I}}, q_{\overline{J}})\right|^2\right) .$$

However, it is not hard to see that the integral

$$\begin{aligned}\int_{U\times V}\left(1+\left|(p_{\overline{I}}, q_{\overline{J}})\right|^2\right)^s\left|\widetilde{f}_{IJ}(p,q)\right|^2 dp\,dq = \int_V\left\{\int_U \widehat{f}_{IJ}(p,q)dp_I\,dp_J\right\}\\ \times\left(1+\left|(p_{\overline{I}}, q_{\overline{J}})\right|^2\right)^s dp_{\overline{I}}\,dq_{\overline{J}}\end{aligned}$$

diverges for $s > -l-m/2$, hence, a fortiori, $f_{IJ} \notin \widetilde{H}^s$ for these s. Therefore, if $K_{IJ}\varphi \in H^s$ for $s > -l-m/2$, then $\widetilde{f}_{IJ} = 0$, consequently, $f_{IJ} = 0$.

We now compare the operators K_{IJ} and $K_{I'J'}$ corresponding to two canonical charts U_{IJ} with the coordinates $(x^I, p_{\overline{I}}, y^J, q_{\overline{J}})$ and $U_{I'J'}$ with the coordinates $\left(x'^{I'}, p'_{\overline{I}'}, y'^{J'}, q'_{\overline{J}'}\right)$. Let $\varphi \in \mathcal{O}_l(U_{IJ} \cap U_{I'J'})$. By going over to coordinates (x,y) on $M_1 \times M_2$ we have

$$\begin{aligned}\left[K_{I'J'}\varphi\right](x,y) = \Bigg\{\zeta'_\varphi(x,y)\cdot\left(\frac{i}{2\pi}\right)^{|\overline{I}'|/2}\left(-\frac{i}{2\pi}\right)^{|\overline{J}'|/2} \qquad (3.5)\\ \times\int\exp\left\{i\left[\langle p'_{\overline{I}'}, x'^{\overline{I}'}(x)\rangle-\langle q'_{\overline{J}'}, y'^{\overline{J}'}(y)\rangle+S_{I'J'}\left(x'^{I'}(x), p'_{\overline{I}'}, y'^{J'}(y), q'_{\overline{J}'}\right)\right]\right\}\\ \times\left\{\left[\mu_{I'J'}\left(\widetilde{v}^1_{x'}\right)^{2\alpha-1}\left(\widetilde{v}^2_{y'}\right)^{1-2\alpha}\right]\left(x'^{I'}(x), p'_{\overline{I}'}, y'^{J'}(y), q'_{\overline{J}'}\right)\right\}^{1/2}\\ \times\varphi\left(x'^{I'}(x), p'_{\overline{I}'}, y'^{J'}(y), q'_{\overline{J}'}\right)dp'_{\overline{I}'}\,dq'_{\overline{J}'}\Bigg\}\left(\det\frac{\partial x'}{\partial x}\right)^\alpha\left(\det\frac{\partial y'}{\partial y}\right)^{1-\alpha} .\end{aligned}$$

Considering the expression (3.2) and (3.5) in $\mathbb{R}^{n_1}\times\mathbb{R}^{n_2}$ up to C^∞-functions, we can omit the factors ζ_φ and ζ'_φ. Applying the operations $F_{x^{\overline{I}}\to p_{\overline{I}}}$ and $\overline{F}_{y^{\overline{J}}\to q_{\overline{J}}}$ to (3.2) and (3.5), we arrive at the following two expressions to be compared:

$$\begin{aligned}A_1 = \exp\left\{iS_{IJ}(x^I, p_{\overline{I}}, y^J, q_{\overline{J}})\right\}\\ \times\sqrt{\left[\mu_{IJ}\left(\widetilde{v}^1_x\right)^{2\alpha-1}\left(\widetilde{v}^2_y\right)^{1-2\alpha}\right](x^I, p_{\overline{I}}, y^J, q_{\overline{J}})} \qquad (3.6)\\ \times\varphi(x^I, p_{\overline{I}}, y^J, q_{\overline{J}}) ,\end{aligned}$$

$$
\begin{aligned}
A_2 = & \left(\frac{i}{2\pi}\right)^{(|\overline{I}'|+|J|)/2} \left(-\frac{i}{2\pi}\right)^{(|\overline{J}'|+|\overline{I}|)/2} \\
& \times \int \exp\left\{ i \left[-\langle x^{\overline{I}}, p_{\overline{I}}\rangle + \langle y^{\overline{J}}, q_{\overline{J}}\rangle + \langle p'_{\overline{I}'}, x'^{\overline{I}'}(x)\rangle \right.\right. \\
& \left.\left. - \langle q'_{\overline{J}'}, y'^{\overline{J}'}(y)\rangle + S_{I'J'}\left(x'^{I'}(x), p'_{\overline{I}'}, y'^{J'}(y), q'_{\overline{J}'}\right)\right]\right\} \\
& \times \sqrt{\left[\mu_{I',J'} \left(\widetilde{v}^1_{x'}\right)^{2\alpha-1} \left(\widetilde{v}^2_{y'}\right)^{1-2\alpha}\right]} \lambda\left(\left|\left(p'_{\overline{I}'}, q'_{\overline{J}'}\right)\right|\right) \\
& \times \varphi \left[\det \frac{\partial x'}{\partial x}(x)\right]^{\alpha} \left[\det \frac{\partial y'}{\partial y}(y)\right]^{1-\alpha} dx^{\overline{I}}\, dy^{\overline{J}}\, dp'_{\overline{I}'}\, dq'_{\overline{J}'} \, .
\end{aligned}
\tag{3.7}
$$

We omit the inessential factor $\lambda\left(\left|\left(p'_{\overline{I}'}, q'_{\overline{J}'}\right)\right|\right)$. We can expand the integral (3.7) asymptotically by means of Lemma 1.2. According to Lemma 1.1, it is the expansion of the density of $K_{I'J'}\varphi$ in the scale $H^{s}_{\alpha,1-\alpha}(M_1 \times M_2)$.

The stationary point of the phase Φ of the integral (3.7) is given by

$$
\begin{cases}
\dfrac{\partial \Phi}{\partial x^{\overline{I}}} = -p + \dfrac{\partial S_{I',J'}}{\partial x'^{\overline{I}'}} \dfrac{\partial x'^{I'}(x)}{\partial x^{\overline{I}}} + p'_{I'} \cdot \dfrac{\partial x'^{\overline{I}'}(x)}{\partial x^{\overline{I}}} = 0 \, , \\
\dfrac{\partial \Phi}{\partial y^{\overline{J}}} = q_{\overline{J}} + \dfrac{\partial S_{I',J'}}{\partial y'^{J'}} \dfrac{\partial y'^{J'}(y)}{\partial y^{\overline{J}}} - q'_{\overline{J}'} \cdot \dfrac{\partial y'^{\overline{J}'}}{\partial y^{\overline{J}}} = 0 \, , \\
\dfrac{\partial \Phi}{\partial p'_{\overline{I}'}} = x'^{\overline{I}'}(x) + \dfrac{\partial S_{I',J'}}{\partial p'_{\overline{I}'}} = 0 \, , \\
\dfrac{\partial \Phi}{\partial q'_{\overline{J}'}} = y'^{\overline{J}'}(y) + \dfrac{\partial S_{I'J'}}{\partial q'_{\overline{J}'}} = 0 \, .
\end{cases}
\tag{3.8}
$$

Let $x^{\overline{I}} = x^{\overline{I}}(x^I, p_{\overline{I}}, y^J, q_{\overline{J}})$ and $y^{\overline{J}} = y^{\overline{J}}(x^I, p_{\overline{I}}, y^J, q_{\overline{J}})$ on L, and let $p'_{\overline{I}'}(x^I, p_{\overline{I}}, y^J, q_{\overline{J}})$ and $q'_{\overline{J}'}(x^I, p_{\overline{I}}, y^J, q_{\overline{J}})$ be the transition functions from the coordinates of the chart U_{IJ} to those of $U_{\overline{I}'\overline{J}'}$ on L. After substituting these functions in the left-hand side of (3.8) we obtain identities by (2.15) and the formulae for the change of variables in the cotangent space.

Lemma 3.1 *At a stationary point of the phase Φ of the integral* (3.7) *we have*

$$
\begin{aligned}
& \det \operatorname{Hess}_{x^{\overline{I}}, y^{\overline{J}}, p'_{\overline{I}'}, q'_{\overline{J}'}}(-\Phi) \\
& \quad = (-1)^{|\overline{I}|+|\overline{J}|} \det \frac{\partial x'}{\partial x} \det \frac{\partial y'}{\partial y} \det \frac{\partial(x^I, p_{\overline{I}}, y^J, q_{\overline{J}})}{\partial\left(x'^{I'}, p'_{\overline{I}'}, y'^{J'}, q'_{\overline{J}'}\right)} \, .
\end{aligned}
\tag{3.9}
$$

Proof. We consider the following system of functions of $\left(x', y', p'_{\bar{I}'}, q'_{\bar{J}'}\right)$:

$$\begin{cases} F^1_I = x^I(x') \,, \quad F^2_{\bar{I}'} = -x'^{\bar{I}'} - \dfrac{\partial S_{I'J'}\left(x'^{I'}, p'_{\bar{I}'}, y'^{J'}, q'_{\bar{J}'}\right)}{\partial p'_{\bar{I}'}} \,, \\ F^3_J = y^J(y) \,, \quad F^4_{\bar{J}'} = y'^{\bar{J}'} - \dfrac{\partial S_{I'J'}\left(x'^{I'}, p'_{\bar{I}'}, y'^{J'}, q'_{\bar{J}'}\right)}{\partial q'_{\bar{J}'}} \,, \\ F^5_{\bar{I}} = p_{\bar{I}} - \left[\dfrac{\partial S_{I'J'}}{\partial x'^{I'}}\left(x'^{I'}, p'_{\bar{I}'}, y'^{J'}, q'_{\bar{J}'}\right)\dfrac{\partial x'^{I'}(x)}{\partial x^{\bar{I}}} \right. \\ \qquad \left. + \, p'_{\bar{I}'}\dfrac{\partial x'^{\bar{I}'}(x)}{\partial x^{\bar{I}}}\right]_{x=x(x')} = p_{\bar{I}} - P_{\bar{I}}\left(x', y', p'_{\bar{I}'}, q'_{\bar{J}'}\right) \,, \\ F^6_{\bar{J}} = -q_{\bar{J}} + \left[q'_{\bar{J}'}\dfrac{\partial y'^{\bar{J}'}(y)}{\partial y^{\bar{J}}} - \dfrac{\partial S_{I'J'}}{\partial y'^{J'}}\left(x'^{I'}, p'_{\bar{I}'}, y'^{J'}, q'_{\bar{J}'}\right) \right. \\ \qquad \left. \times \dfrac{\partial y'^{J'}(y)}{\partial y^{\bar{J}}}\right]_{y=y(y')} = -q_{\bar{J}} + Q_{\bar{J}}\left(x', y', p'_{\bar{I}'}, q'_{\bar{J}'}\right) . \end{cases} \tag{3.10}$$

Substitution of $x'^{\bar{I}'} = x'^{\bar{I}'}\left(x'^{I'}, p'_{\bar{I}'}, y'^{J'}, q'_{\bar{J}'}\right)$ and $y'^{\bar{J}'} = y'^{\bar{J}'}\left(x'^{I'}, p'_{\bar{I}'}, y'^{J'}, q'_{\bar{J}'}\right)$ in $P_{\bar{I}}$ and $Q_{\bar{J}}$ yields two functions $p_{\bar{I}}\left(x'^{I'}, p'_{\bar{I}'}, y'^{J'}, q'_{\bar{J}'}\right)$ and $q_{\bar{J}}\left(x'^{I'}, p'_{\bar{I}'}, y'^{J'}, q'_{\bar{J}'}\right)$ which determine the transition from $U_{I'J'}$ to U_{IJ}. The functions (3.10) can be used to rewrite the system of Eq. (3.8) in the form

$$F^2_{\bar{I}'} = 0 \,, \quad F^4_{\bar{J}'} = 0 \,, \quad F^5_{\bar{I}} = 0 \,, \quad F^6_{\bar{J}} = 0$$

for $x' = x'(y)$ and $y' = y'(y)$. All the subsequent equalities hold for the solutions of the system (3.8). We obtain for the Jacobian

$$\begin{aligned} &\frac{D\left(F^1_I, F^5_{\bar{I}}, F^3_J, F^6_{\bar{J}}, F^2_{\bar{J}'}, F^4_{\bar{J}}\right)}{D\left(x'^{I'}, x'^{\bar{I}'}, y'^{J'}, y'^{\bar{J}'}, p'_{\bar{I}'}, q'_{\bar{J}'}\right)} \\ &= \quad (-1)^{|\bar{I}|+|I'||\bar{I}'|+|\bar{J}'|(|I'|+|J'|)+|J'||\bar{I}'|+|\bar{I}'|}\frac{D(x^I, p_{\bar{I}}, y^J, q_{\bar{J}})}{D\left(x'^{I'}, p'_{\bar{I}'}, y'^{J'}, q'_{\bar{J}'}\right)} \end{aligned} \tag{3.11}$$

(the calculations are omitted because they are cumbersome). Next, it is obvious that

$$\begin{aligned}
&\det \operatorname{Hess}_{x^{\overline{I}},y^{\overline{J}},p'_{\overline{I}'},q'_{\overline{J}'}}(-\Phi)\\
&= \frac{D\left\{\left|\left(F^1_I, F^5_{\overline{I}}, F^3_J, F^6_{\overline{J}}, F^2_{\overline{I}'}, F^4_{\overline{J}'}\right)\right|_{x'=x'(x),y'=y'(y)}\right\}}{D\left(x^I, x^{\overline{I}}, y^J, y^{\overline{J}}, p'_{\overline{I}'}, q'_{\overline{J}'}\right)}\\
&= \frac{Dx'}{Dx}\frac{Dy'}{Dy'}\frac{D\left(F^1_I, F^5_{\overline{I}}, F^3_J, F^6_{\overline{J}}, F^2_{\overline{I}}, F^4_{\overline{J}}\right)}{D\left(x'^{I'}, x'^{\overline{I}'}, y'^{J'}, y'^{\overline{J}'}, p'_{\overline{I}'}, q'_{\overline{J}'}\right)} \qquad (3.12)\\
&= (-1)^{|\overline{I}|+|I'||\overline{I}'|+|J'||\overline{J}'|+|I'||\overline{J}'|+|J'||\overline{I}'|+|\overline{I}'|}\\
&\quad\times \frac{Dx'}{Dx}\frac{Dy'}{Dy}\frac{D(x^I, p_{\overline{I}}, y^J, q_{\overline{J}})}{D\left(x'^{I'}, p'_{\overline{I}'}, y'^{J'}, q'_{\overline{J}'}\right)}
\end{aligned}$$

by (3.11). The last formula now proves the lemma, if we bear in mind that $|\overline{I}|+|I'||\overline{I}'|+|\overline{J}'||J'|+|\overline{J}'||I'|+|J'||\overline{I}'| \equiv |\overline{I}|+|\overline{J}'| \pmod 2$. This completes the proof of the lemma.

When we now evaluate the integral (3.7) with the help of Lemma 1.2, we obtain

$$\begin{aligned}
A_2 \sim\ & (-1)^{\frac{1}{2}(|\overline{I}|+|\overline{J}'|)} e^{iS_{IJ}(x^I, p_{\overline{I}}, y^J, q_{\overline{J}})}\\
&\times \left[\frac{\sqrt{\mu_{I'J'}\left(\widetilde{v}^1_{x'}\right)^{2\alpha-1}\left(\widetilde{v}^2_{y'}\right)^{1-2\alpha}}}{\sqrt{(-1)^{|\overline{I}|+|\overline{J}'|}\frac{Dx'}{Dx}\frac{Dy'}{Dy}\frac{D(x^I, p_{\overline{I}}, y^J, q_{\overline{J}})}{D\left(x'^{I'}, p'_{\overline{I}'}, y'^{J'}, q'_{\overline{J}'}\right)}}}\widetilde{\varphi}\right]\\
&\times (x^I, p_{\overline{I}}, y^J, q_{\overline{J}})\left(\frac{Dx'}{Dx}\right)^{\alpha}\left(\frac{Dy'}{Dy}\right)^{1-\alpha},
\end{aligned}$$

where the argument of the determinant of the matrix (3.9) is chosen as indicated in Lemma 1.2, and

$$\widetilde{\varphi} = V^{IJ}_{I'J'}\varphi = \sum_{k=0}^{\infty} {}^kV^{IJ}_{I'J'}\varphi \qquad (3.13)$$

is an asymptotically homogeneous function, ${}^0V^{IJ}_{I'J'} = 1$.

We finally obtain for (3.7) the formula

$$\begin{aligned}
A_2 \sim\ & \exp\left\{\frac{1}{2}\left[\operatorname{Arg}\mu_{I'J'} - \operatorname{Arg}\mu_{IJ} - \sum \arg\lambda_k - \left(|I| + \left|\overline{J}'\right|\right)\pi\right]\right\}\\
&\times \left[e^{iS_{IJ}}\sqrt{\mu_{IJ}\left(v^1_x\right)^{2\alpha-1}\left(v^2_y\right)^{1-2\alpha}}\, V^{IJ}_{I'J'}\varphi\right](x^I, p_{\overline{I}}, y^J, q_{\overline{J}}) = e^{\frac{i}{2}c^{I'J'}_{IJ}} A_1\,,
\end{aligned}$$

where the λ_k are the eigenvalues of the matrix on the left-hand side of (3.9) and $-3\pi/2 < \arg\lambda_k \le \pi/2$. For the leading term of the last expression to agree with (3.6) we have to require that

$$\operatorname{Arg}\mu_{I'J'} - \operatorname{Arg}\mu_{IJ} = \sum_k \arg\lambda_k + \left(\left|\overline{I}\right| + \left|\overline{J}'\right|\right)\pi \,, \tag{3.14}$$

where the λ_k are the eigenvalues of the matrix (3.9).

We recall that $\operatorname{Arg}\mu_{IJ}$ is a fixed branch of the argument of u_{IJ} (see the end of § 2).

Definition 3.2. A homogeneous Lagrangian immersion (L, i, μ) with measure is said to be *quantized* if the values of $\operatorname{Arg}\mu_{IJ}$ can be chosen so that the relations (3.14) are satisfied for any intersection of charts $U_{IJ} \cap U_{I'J'}$.

The Maslov index of an arbitrary closed path on L is trivial if (L, i, μ) is a quantized immersion.

We always assume in what follows that the values of $\operatorname{Arg}\mu_{IJ}$ on a quantized Lagrangian manifold are chosen so that (3.14) is satisfied.

We now state the assertion we have proved above in the form of a lemma.

Lemma 3.2. *On a quantized Lagrangian immersion (L, i, μ) with measure we have*

$$K_{I'J'}\varphi \equiv K_{IJ}\left(\sum_{k=0}^{N} {}^kV^{IJ}_{I'J'}\varphi\right) \,, \qquad \varphi \in \mathfrak{S}_l(U_{IJ} \cap U_{I'J'}) \tag{3.15}$$

modulo $H^{-l-\frac{m}{2}+N-1-\epsilon}_{\alpha,1-\alpha}(M_1 \times M_2)$. Here the ${}^kV^{IJ}_{I'J'}$, are homogeneous differential operators of degree k, and

$$\begin{cases} {}^0V^{IJ}_{I'J'} = 1 \,, \quad {}^kV^{IJ}_{IJ} = 0 \quad (k = 1, 2, \dots) \,, \\ \displaystyle\sum_{k=0}^{\infty} {}^kV^{I'J'}_{IJ} \circ \sum_{k=1}^{\infty} {}^kV^{I''J''}_{I'J'} = \sum_{k=0}^{\infty} {}^kV^{I''J''}_{IJ} \end{cases} \tag{3.16}$$

(the latter as a relation between formal series).

The last two relations in (3.16) follow readily from (3.15), which holds for any $N > 0$.

Theorem 3.3 (the cocyclicity theorem, see Maslov [1]). *There is a unique operator*

$$K : \mathcal{O}_l(L) \to H^s_{\alpha,1-\alpha,\mathrm{loc}}(M_1 \times M_2) \,,$$

modulo $H^{s+1}_{\alpha,1-\alpha,\mathrm{loc}}(M_1 \times M_2)$, *in the sense that the difference of any two such operators is mapped into* $H^{s+1}_{\alpha,1-\alpha,\mathrm{loc}}(M_1 \times M_2)$ $(s < -l - m/2)$ *such that the following conditions are satisfied:*

1) $K\varphi \equiv K_{IJ}\varphi \left(\operatorname{mod} H^{s+1}_{\alpha,1-\alpha,\mathrm{loc}}(M_1 \times M_2)\right)$, $\varphi \in \mathcal{O}_l(U_{IJ})$,

2) *for any partition of* φ *into a locally finite sum* $\varphi = \sum \varphi_i$, $\varphi_i \in \mathcal{O}_l(L)$, *the sum* $\sum_i K\varphi_i$ *is locally finite on* $M_1 \times M_2$, *and* $K\varphi = \sum_i K\varphi_i$.

Proof. If an operator K has the properties 1) and 2), then

$$K\varphi = \sum_{U_{IJ}} K(e_{IJ}\varphi) = \sum_{U_{IJ}} K_{IJ}(e_{IJ}\varphi) , \tag{3.17}$$

where $\{e_{IJ}\}$ is a homogeneous partition of unity subordinate to the covering $\{U_{IJ}\}$. On the other hand, (3.17) defines an operator satisfying 1) and 2). This completes the proof.

We call this operator *canonical* and write

$$K = K_{(L,i,\mu)} , \tag{3.18}$$

and, if $i = i_g$ is the immersion associated with a canonical transformation g, we use the notation

$$K = K_g . \tag{3.19}$$

Of course, it would be unnatural to extend the operator whose existence is asserted in Theorem 3.3 to $\mathfrak{S}_l(L)$, since $K\varphi \in H^{s+1}_{\alpha,1-\alpha,\mathrm{loc}}(M_1 \times M_2)$ for $\varphi \in \mathcal{O}_{l-1}(L)$, and K is defined modulo $H^{s+1}_{\alpha,1-\alpha,\mathrm{loc}}(M_1 \times M_2)$. (Here $s < -l - m/2$ is arbitrary.) However, if we define K by (3.17) for a fixed partition of unity $\{e_{IJ}\}$, then K can be extended to

$$K : \mathfrak{S}_l(L) \to H^{s}_{\alpha,1-\alpha,\mathrm{loc}}(M_1 \times M_2) .$$

In all subsequent arguments we assume that a representation of K by (3.17) is fixed for a given partition of unity.

Remark 3.1. If at least one of the manifolds M_1 and M_2 is not orientable, then we cannot define an elementary canonical operator by (3.2), since on a non-orientable manifold there is no positive (and altogether no non-degenerate) volume measure. However, we can modify our arguments, assuming that v_1 and v_2 are non-degenerate densities of order 1 on M_1 and M_2 (which exist even on a non-orientable manifold). Similarly, μ must then be regarded as a section of a line bundle with transition functions

$$\sigma^{I'J'}_{IJ} = \operatorname{sign}\frac{Dx'}{Dx} \cdot \operatorname{sign}\frac{Dy'}{Dy} \cdot \frac{D\left(x^I, p_{\bar I}, y^J, q_{\bar J}\right)}{D\left(x'^{I'}, p'_{\bar I'}, y'^{J'}, q'_{\bar J'}\right)} . \tag{3.20}$$

The quantization conditions (3.14) then remain valid.

We now define an important concept in the theory of canonical densities, that of association. Let $\widehat{H}$ be a pseudodifferential operator of order r on a manifold M_1 with a real principal symbol $H_0(x,p)$, where H_0 also defines the function $\pi_1^* H_0$, $\pi_1 : T_0^* M_1 \times T_0^* M_2 \to T_0^* M_1$ being the natural projection. This should not give rise to misunderstandings.

We claim that if $i^* H_0 = 0$, then the vector field $V(H_0)$ (see (1.29)) is tangent to the immersion i, that is,

$$[V(H_0)]_{i(\alpha)} \in i_* (T_\alpha L) \ .$$

For since

$$\omega^2_{M_1,M_2}(V(H_0),Y) \equiv dH_0(Y)$$

the linear hull of $W = \left\{ i_*(T_\alpha L), [V(H_0)]_{i(\alpha)} \right\}$ is isotropic (see Remark 2.3). Since $\dim i_*(T_\alpha L) = n_1 + n_2$, we see that $[V(H_0)]_{i(\alpha)} \in i_*(T_\alpha L)$; otherwise the dimension of the Lagrangian subspace W would be greater than $n_1 + n_2$, which contradicts Remark 2.3.

Definition 3.3. A canonical operator $K_{(L,i,\mu)}$ is said to be *associated* with a pseudodifferential operator $\widehat{H}$ if the following conditions are satisfied:

$$i^* H_0 = 0 \ , \tag{3.21}$$

$$\mathcal{L}_{V(H_0)}\mu = 0 \ . \tag{3.22}$$

Here $\mathcal{L}_{V(H_0)}$ is the Lie derivative along the Hamiltonian vector field $V(H_0)$ defined by (1.29).

The following theorem establishes a commutation formula for a pseudodifferential operator and a canonical operator.

Theorem 3.4 (commutation theorem, see Maslov [1] and Hörmander [3]). *Let $K_{(L,i,\mu)}$ be a canonical operator and $\widehat{H}$ a pseudodifferential operator of order r on M_1. Then for any $\varphi \in \mathcal{O}_l(L)$ the following commutation formula holds:*

$$\begin{gathered} \widehat{H} K_{(L,i,\mu)}\varphi \equiv K_{(L,i,\mu)}(i^* H_0)\varphi \left(\operatorname{mod} H^{s-r+1}_{\alpha,1-\alpha,\mathrm{loc}}(M_1 \times M_2) \right) , \\ s < -l - m/2 \ . \end{gathered} \tag{3.23}$$

If, in addition, $K_{(L,i,\mu)}$ is associated with $\widehat{H}$, then

$$\widehat{H} K_{(L,i,\mu)}\varphi \equiv -i K_{(L,i,\mu)} \mathcal{P}^N \varphi \left(\operatorname{mod} H^{s-r+N+2}_{\alpha,1-\alpha,\mathrm{loc}}(M_1 \times M_2) \right) , \tag{3.24}$$

where $\mathcal{P}^N = \sum_{k=0}^{N} \mathcal{P}_k$ *is the transport operator, the* $\mathcal{P}_k$ *are differential operators of order* $r+k-1$, *and*

$$\mathcal{P}_0\varphi = V(H_0)\varphi + i^* H^{v_1}_{\text{sub}}\varphi \ . \tag{3.25}$$

Here H^v_{sub} *is the subprincipal symbol* (1.35) *of* $\widehat{H}$.

Proof. So as not to complicate the proof, we consider the case $N=0$. The higher terms in $\mathcal{P}$ can be obtained from the operators $V^{IJ}_{I'J'}$, introduced in Lemma 3.2 (see § 6.3 of the main text of the book).

It follows from Theorem 3.3 on cocyclicity that it is sufficient to prove (3.23) and (3.24) for the operator K_{IJ}. In local coordinates we have

$$\begin{aligned}\widehat{H}K_{IJ}\varphi(x,y) &= \widehat{H}\zeta_\varphi(x,y)\overline{F}_{p_{\bar I}\to x^{\bar I}}F_{q_{\bar J}\to y^{\bar J}}\\ &\times\left\{e^{iS_{IJ}}\sqrt{\mu_{IJ}(x^I,p_{\bar I},y^J,q_{\bar J})\left(\widetilde{v}^1_x\right)^{2\alpha-1}\left(\widetilde{v}^2_y\right)^{1-2\alpha}}\right.\\ &\times\left.\lambda\left(\left|(p_{\bar I},q_{\bar J})\right|\right)\varphi(x^I,p_{\bar I},y^J,q_{\bar J})\right\}\\ &=\zeta_\varphi\widehat{H}\overline{F}_{p_{\bar I}\to x^{\bar I}}F_{q_{\bar J}\to y^{\bar J}}\left\{e^{iS_{IJ}}a\right\}+(1-\zeta_\varphi)\widehat{H}\overline{F}_{p_{\bar I}\to x^{\bar I}}F_{q_{\bar J}\to y^{\bar J}}\left\{e^{iS_{IJ}}a\right\}\\ &\quad+\widehat{H}\left(\zeta_\varphi-1\right)\overline{F}_{p_{\bar I}\to x^{\bar I}}F_{q_{\bar J}\to y^{\bar J}}\left\{e^{iS_{IJ}}a\right\}\ .\end{aligned}\tag{3.26}$$

Here

$$\begin{aligned}a(x^I,p_{\bar I},y^J,q_{\bar J}) &= \lambda\left(\left|(p_{\bar I},q_{\bar J})\right|\right)\\ &\times\sqrt{\left[\mu_{IJ}\left(\widetilde{v}^1_x\right)^{2\alpha-1}\left(\widetilde{v}^2_y\right)^{1-2\alpha}\right](x^I,p_{\bar I},y^J,q_{\bar J})}\,\varphi(x^I,p_{\bar I},y^J,q_{\bar J})\ .\end{aligned}$$

Since a pseudodifferential operator does not enlarge the singular support of a distribution, the last two terms are smooth functions in $\mathbb{R}^{n_1+n_2}$, and we find that $\left(\operatorname{mod} C^\infty\left(\mathbb{R}^{n_1+n_2}\right)\right)$

$$\begin{aligned}\widehat{H}K_{IJ}\varphi(x,y) &= \zeta_\varphi(x,y)\widehat{H}\overline{F}_{p_{\bar I}\to x^{\bar I}}F_{q_{\bar J}\to y^{\bar J}}\left\{\left[e^{iS_{IJ}}a\right](x^I,p_{\bar I},y^J,q_{\bar J})\right\}\\ &= \zeta_\varphi(x,y)\overline{F}_{p_{\bar I}\to x^{\bar I}}F_{q_{\bar J}\to y^{\bar J}}\left\{\widehat{H}_I\left[e^{iS_{IJ}}a\right](x^I,p_{\bar I},y^J,q_{\bar J})\right\}\ .\end{aligned}\tag{3.27}$$

The operator $\widehat{H}_I$ is defined by (1.34).

Applying Lemma 1.3 we obtain the following asymptotic expansion:

$$
\begin{aligned}
&\widehat{H}_I\left[e^{iS_{IJ}}a\right]\left(x^I,p_{\overline{I}},y^J,q_{\overline{J}}\right)=e^{iS_{IJ}\left(x^I,p_{\overline{I}},y^J,q_{\overline{J}}\right)}\\
&\times H_0\left(x^I,-\frac{\partial S_{IJ}}{\partial p_{\overline{I}}},\frac{\partial S_{IJ}}{\partial x^I},p_{\overline{I}}\right)a\left(x^I,p_{\overline{I}},y^J,q_{\overline{J}}\right)\\
&+e^{iS_{IJ}\left(x^I,p_{\overline{I}},y^J,q_{\overline{J}}\right)}\left\{H_1a\left(x^I,p_{\overline{I}},y^J,q_{\overline{J}}\right)\right.\\
&-i\left[\frac{\partial H_0}{\partial p_I}\frac{\partial a}{\partial x^I}-\frac{\partial H_0}{\partial x^{\overline{I}}}\frac{\partial a}{\partial p_{\overline{I}}}+\frac{1}{2}\left(\frac{\partial^2 S_{IJ}}{\partial x^I\partial x^I}\frac{\partial^2 H_0}{\partial p_I\partial p_I}\right.\right.\\
&\left.\left.\left.+\frac{\partial^2 S_{IJ}}{\partial p_{\overline{I}}\partial p_{\overline{I}}}\frac{\partial^2 H_0}{\partial x^{\overline{I}}\partial x^{\overline{I}}}-2\frac{\partial^2 S_{IJ}}{\partial x^I\partial p_{\overline{I}}}\frac{\partial^2 H_0}{\partial x^{\overline{I}}\partial p_I}+\frac{\partial^2 H_0}{\partial x^{\overline{I}}\partial p_{\overline{I}}}\right)a\right]\right\}+\dots
\end{aligned}
\tag{3.28}
$$

The values of functions H_0 and H_1 and of their derivatives in (3.28) are taken at $(x,p)=\left(x^I,-\frac{\partial S_{IJ}}{\partial p_{\overline{I}}},\frac{\partial S_{IJ}}{\partial x^I},p_{\overline{I}}\right)$ and the three dots stand for terms of lower degree of homogeneity. Using (2.15) and Lemma 1.1 we arrive directly at (3.23).

Now let

$$
i^*H_0=H_0\left(x^I,-\frac{\partial S_{IJ}}{\partial p_{\overline{I}}},\frac{\partial S_{IJ}}{\partial x^I},p_{\overline{I}}\right)=0\,.
$$

Then the restriction of $V(H_0)$ to $i(L)$ is tangent to L and can be written in the canonical coordinates $\left(x^I,p_{\overline{I}},y^J,q_{\overline{J}}\right)$ in the following form:

$$
V(H_0)=i^*\left(\frac{\partial H_0}{\partial p_I}\right)\frac{\partial}{\partial x^I}-i^*\left(\frac{\partial H_0}{\partial x^{\overline{I}}}\right)\frac{\partial}{\partial p_{\overline{I}}}\,. \tag{3.29}
$$

Expressing the condition (3.22) for the invariance of the measure in the canonical coordinates $\left(x^I,p_{\overline{I}},y^J,q_{\overline{J}}\right)$ we obtain

$$
V(H_0)\mu_{IJ}+\mu_{IJ}\operatorname{div}\left(i^*\left(\frac{\partial H_0}{\partial p_I}\right)\frac{\partial}{\partial x^I}-i^*\left(\frac{\partial H_0}{\partial x^{\overline{I}}}\right)\frac{\partial}{\partial p_{\overline{I}}}\right)=0\,. \tag{3.30}
$$

(This is known as Sobolev's lemma.) We now obtain

$$
\begin{aligned}
&-\frac{\partial H_0}{\partial x^{\overline{I}}}\frac{\partial a}{\partial p_{\overline{I}}}+\frac{\partial H_0}{\partial p_I}\frac{\partial a}{\partial x^I}=V(H_0)a=\lambda\left(\left|\left(p_{\overline{I}},q_{\overline{J}}\right)\right|\right)\\
&\times\sqrt{\mu_{IJ}\left(v_x\right)^{2\alpha-1}\left(v_y\right)^{1-2\alpha}}\\
&\times\left\{V(H_0)\varphi+\frac{1}{2}\frac{V(H_0)\mu_{IJ}}{\mu_{IJ}}\varphi+\frac{2\alpha-1}{2}V(H_0)\log(v_x)\right\}\,.
\end{aligned}
\tag{3.31}
$$

Here we have used the fact that in the calculation of the zero term of the term of the transport operator we may replace the functions $\widetilde{v}_x^1$ and $\widetilde{v}_y^2$

by v_x and v_y, since, for instance, $\widetilde{v}_x$ is the value of v_x at a stationary point of the phase of the integral in question.

Using (3.30) and omitting the operator i^* on the right-hand side of (3.32) we obtain

$$\begin{aligned} \frac{V(H_0)\mu_{IJ}}{\mu_{IJ}} &= \operatorname{div}\left(i^*\left(\frac{\partial H_0}{\partial p_I}\right)\frac{\partial}{\partial x^I} - i^*\left(\frac{\partial H_0}{\partial x^{\overline{I}}}\right)\frac{\partial}{\partial p_{\overline{I}}}\right) \\ &= -\left\{ \frac{\partial^2 H_0}{\partial x^I \partial p_I} - \frac{\partial^2 H_0}{\partial p_I \partial x^{\overline{I}}}\frac{\partial^2 S_{IJ}}{\partial p_{\overline{I}}\partial x^I} + \frac{\partial^2 H_0}{\partial p_I \partial p_I}\frac{\partial^2 S_{IJ}}{\partial x^I \partial x^I} \right. \\ &\quad \left. - \frac{\partial^2 H_0}{\partial x^{\overline{I}}\partial p_{\overline{I}}} + \frac{\partial^2 H_0}{\partial x^{\overline{I}}\partial x^{\overline{I}}}\frac{\partial^2 S}{\partial p_{\overline{I}}\partial p_{\overline{I}}} - \frac{\partial^2 H_0}{\partial x^{\overline{I}}\partial p_I}\frac{\partial^2 S}{\partial x^I \partial p_{\overline{I}}} \right\} . \end{aligned} \tag{3.32}$$

Substituting the expressions (3.31) and (3.32) in (3.28), we see that

$$\begin{aligned} \widehat{H}_I e^{iS_{IJ}} a = & -ie^{iS_{IJ}}\lambda\left(\left|\left(p_I, q_{\overline{J}}\right)\right|\right)\sqrt{\mu_{IJ}\left(\widetilde{v}_x^1\right)^{2\alpha-1}\left(\widetilde{v}_y^2\right)^{1-2\alpha}} \\ & \times\left\{V(H_0)\varphi + \left[iH_1 - \frac{1}{2}\frac{\partial^2 H_0}{\partial x\,\partial p} - \frac{2\alpha-1}{2}V(H_0)\log v_x\right]\right\} . \end{aligned}$$

In accordance with the definition (1.28) of the subprincipal symbol we now have

$$\begin{aligned} \widehat{H}_I e^{iS_{IJ}} a = & -ie^{iS_{IJ}}\lambda\left(\left|\left(p_{\overline{I}}, q_{\overline{J}}\right)\right|\right) \\ & \times\sqrt{\mu_{IJ}\left(\widetilde{v}_x^1\right)^{2\alpha-1}\left(\widetilde{v}_y^2\right)^{1-2\alpha}}\left(V(H_0)\varphi + i^* H^v_{\mathrm{sub}}\varphi\right) + \dots \end{aligned}$$

An application of Lemma 1.1 completes the proof of the theorem.

§ 4. Fourier Integral Operators

Let M_1 and M_2 be two smooth manifolds of dimension n_1 and n_2, L a homogeneous manifold of dimension n_1+n_2, and (L, i, μ), $i: L \to T_0^* M_1 \times T_0^* M_2$ a quantized homogeneous Lagrangian immersion with measure of degree of homogeneity m.

Definition 4.1. Let $\varphi \in \mathfrak{S}_l(L)$. The operator

$$\Phi_{(L,i,\mu,\varphi)} : \mathcal{D}_\alpha(M_2) \to \mathcal{D}'_\alpha(M_1) , \tag{4.1}$$

given by the formula

$$\left[\Phi_{(L,i,\mu,\varphi)} f\right](x) = \left(\frac{-i}{2\pi}\right)^{(n_1+n_2)/4} \int_{M_2} \left[K_{(L,t,\mu)}\varphi\right](x,y) f(y)\,dy , \tag{4.2}$$

$$x \in M_1 , \quad y \in M_2 ,$$

is called the *Fourier integral operator* with symbol φ associated with the Lagrangian immersion (L, i, μ).

Throughout what follows we impose the following condition on the immersion i.

Condition 4.1. For any compact set $K \subset M_2$ the set $(\pi_1 \circ i) \circ (\pi_2 \circ i)^{-1}(K)$ is compact in M_1. Here

$$\pi_i : T_0^* M_1 \times T_0^* M_2 \to M_1 \qquad (i = 1, 2)$$

are the natural projections.

If condition 4.1 holds, we can choose the cut-off functions ζ_φ in the definition (3.2) of an elementary canonical operator in such a way that the support of the density in (4.2) satisfies the condition stated at the end of § 1.2°. The operator $\Phi_{(L,i,\mu,\varphi)}$ is then properly supported in the sense of definition (1.24).

Theorem 4.1. *The operator* (4.1) *maps* $H^s_{\alpha,\text{loc}}(M_2)$ *continuously into* $H^{-s}_{\alpha,\text{loc}}$ *for* $s > l + m/2$.

Proof. By (3.4), $K_{(L,i,\mu)}\varphi \in H^{-s}_{\alpha,1-\alpha,\text{loc}}(M_1 \times M_2)$, therefore, the assertion of Theorem 4.1 follows from the duality theorem for Sobolev spaces and the fact that the operator (4.1) is properly supported (see the end of § 1.2°).

Definition 4.2. Let $(T_0^* M_2, i_g, \mu)$ be a quantized homogeneous Lagrangian immersion with measure associated with a canonical transformation $g : T_0^* M_2 \to T_0^* M_1$. The Fourier integral operator $\Phi_{(T_0^* M_2, i_g, \mu, \varphi)}$ is said to be *associated* with g and is denoted by $T(g, \varphi)$.

Theorems 4.2 and 4.3 below (the latter in a slightly more general situation) are due to Hörmander (see Hörmander [3]). The proofs in the present paper are due to V.E. Nazaikinskij.

Theorem 4.2. *Let*

$$g : T_0^* M_2 \to T_0^* M_1 \tag{4.3}$$

be a homogeneous canonical transformation and $\varphi \in \mathfrak{S}_l(T_0^* M_2)$. *Then the operator*

$$T(g, \varphi) : \mathcal{D}_\alpha(M_2) \to \mathcal{D}'_\alpha(M_1) \tag{4.4}$$

can be extended to a continuous operator

$$T(g,\varphi) : H^s_{\alpha,\mathrm{loc}}(M_2) \to H^{s-l}_{\alpha,\mathrm{loc}}(M_1) \tag{4.5}$$

for all real s.

Theorem 4.3. *Let*

$$g_1 : T_0^*M_3 \to T_0^*M_2\ , \quad g_2 : T_0^*M_2 \to T_0^*M_1 \tag{4.6}$$

be homogeneous canonical transformations and

$$\varphi_1 \in \mathfrak{S}_{l_1}\left(T_0^*M_3\right)\ , \quad \varphi_2 \in \mathfrak{S}_{l_2}\left(T_0^*M_2\right)\ . \tag{4.7}$$

Then the following composition formula holds:

$$T\left(g_2,\varphi_2\right) \circ T\left(g_1,\varphi_1\right) = \pm T\left(g_2 \circ g_1, \left(g_1^*\varphi_2\right)\varphi_1 + \varphi_3\right)\ , \tag{4.8}$$

where $\varphi_3 \in \mathfrak{S}_{l_1+l_2-1}\left(T_0^*M_3\right)$, and the sign in (4.8) depends on the choice of the arguments of the densities of the measure for all three operators in (4.8). (M_3 is assumed to be connected.)

The proofs of Theorems 4.2 and 4.3 are based on the following two lemmas:

Lemma 4.4. *Let $g : V \to T_0^*\mathbb{R}^n$ be a homogeneous canonical transformation of an $\mathbb{R}_+$-invariant domain $V \subset T_0^*\mathbb{R}^n$ into $T_0^*\mathbb{R}^n$ and let α be an arbitrary point of V. There is an $\mathbb{R}_+$-invariant neighbourhood $W \subset V$ of α and a smooth one-parameter family $\{g_t\}_{t\in[0,1]}$ of homogeneous canonical transformations of W such that*

1°.

$$g_1 = g|_W\ , \tag{4.9}$$

and one of the following two conditions is satisfied:

2°. *g_0 is the identity diffeomorphism of W, or*

2°′. *g_0 is a diffemorphism of the following form:*

$$\begin{cases} x^1 \to -x^1 \\ x^i \to x^i & (i = 2,3,\ldots,n)\ ; \\ p_1 \to -p_1\ ; \\ p_i \to p_i & (i = 2,3,\ldots,n)\ ; \end{cases} \tag{4.10}$$

(Here, $\left(x^1,\ldots,x^n,p_1,\ldots,p_n\right)$ are the standard coordinates in $T^*\mathbb{R}^n$.)

Proof. We use the following notation:

$$(y,q) = \left(y^1,\ldots,y^n,q_1,\ldots,q_n\right)$$

are coordinates in $T_0^*\mathbb{R}^n \supset V$ and

$$(x, p) = \left(x^1, \dots, x^n, p_1, \dots p_n\right)$$

in $T_0^*\mathbb{R}^n \supset g(V)$. Let (y_0, q_0) be the coordinates of α and (x_0, p_0) those of $g(\alpha)$. There is an index set $I \subset [n]$ and a homogeneous function $S\left(x^I, p_{\overline{I}}, q\right)$ of degree 1 in $\left(p_{\overline{I}}, q\right)$ defined on a $\left(p_{\overline{I}}, q\right)$-homogeneous neighbourhood of the point $\left(x_0^I, p_{0\overline{I}}, q_0\right)$ such that

$$\det \begin{pmatrix} \dfrac{\partial^2 S}{\partial x^I \, \partial q} \\ \dfrac{\partial^2 S}{\partial p_{\overline{I}} \, \partial q} \end{pmatrix} \neq 0 , \tag{4.11}$$

and the canonical transformation g in a neighbourhood of (y_0, q_0) is defined by (see Remark 2.2 and (2.16))

$$\begin{cases} y = \dfrac{\partial S}{\partial q}\left(x^{\overline{I}}, p_{\overline{I}}, q\right) , \\ p_I = \dfrac{\partial S}{\partial x^I}\left(x^I, p_{\overline{I}}, q\right) , \\ x^{\overline{I}} = -\dfrac{\partial S}{\partial p_{\overline{I}}}\left(x^I, p_{\overline{I}}, q\right) . \end{cases} \tag{4.12}$$

First of all we find a transformation g_1 homotopic to g which can be defined (possibly, on a smaller $\mathbb{R}_+$-invariant neighbourhood of (y_0, q_0)) by a generating function $S_1(x, q)$ (that is, the index set $\overline{I}$ is empty). Since $q_0 \neq 0$, there is an $\left(|\overline{I}| \times n\right)$-matrix C with real elements such that

$$p_{0\overline{I}} = C_{q_0} . \tag{4.13}$$

We set

$$\widetilde{S}(t) = \widetilde{S}\left(x^I, p_{\overline{I}}, q, t\right) = S\left(x^I, p_{\overline{I}}, q\right) + t \frac{\left(p_{\overline{I}} - C_q\right)^2}{|q|} . \tag{4.14}$$

There is a number $t_0 > 0$ such that for $0 \leq t \leq t_0$

$$\det \begin{pmatrix} \dfrac{\partial^2 \widetilde{S}\left(x_0^I, p_{0\overline{I}}, q_0, t\right)}{\partial x^I \, \partial q} \\ \dfrac{\partial^2 \widetilde{S}\left(x_0^I, p_{0\overline{I}}, q_0, t\right)}{\partial p_{\overline{I}} \, \partial q} \end{pmatrix} \neq 0 \tag{4.15}$$

and for $t = t_0$

$$\det\left(\frac{\partial^2 \widetilde{S}}{\partial p_{\overline{I}}\,\partial p_{\overline{I}}}\,(x_0^I, p_{0\overline{I}}, q_0, t)\right) = \det\left(\frac{\partial^2 S}{\partial p_{\overline{I}}\,\partial p_{\overline{I}}}\,(x_0^I, p_{0\overline{I}}, q_0) + \frac{t}{|q_0|}E\right) \neq 0\ . \tag{4.16}$$

The relation (4.15) means that $\widetilde{S}(t)$ is a generating function of some homogeneous canonical transformation g_t defined on an $\mathbb{R}_+$-invariant neighbourhood of the point

$$\left(\frac{\partial \widetilde{S}\,(x^I, p_{0\overline{I}}, q_0, t)}{\partial q}, q_0\right) = (y_0, q_0)$$

and such that $g_t(y_0, q_0) = (x_0, p_0)$ (which follows from (4.13), (4.14), and (4.12)). It follows from (4.16) that the equation

$$x^{\overline{I}} = \frac{\partial \widetilde{S}}{\partial p_{\overline{I}}}\left(x^{\overline{I}}, p_{\overline{I}}, q, t\right) \tag{4.17}$$

is soluble for the variables $p_{\overline{I}}$ in an $\mathbb{R}_+$-invariant neighbourhood of the point

$$\left(x_0^I, x_0^{\overline{I}}, p_{0\overline{I}}, t_0\right)\ .$$

Thus, we can use (4.17) to express $p_{\overline{I}}$ as a function

$$p_{\overline{I}} = p_{\overline{I}}(x, q)\ , \tag{4.18}$$

and this means that the set of functions (x, q) forms a system of local coordinates on the graph of the diffeomorphism $g_{t_0} \equiv g_1$ in an $\mathbb{R}_+$-invariant neighbourhood of (y_0, q_0). It follows that the canonical transformation g_1 in an $\mathbb{R}_+$-invariant neighbourhood of (y_0, q_0) can be defined by a generating function $S_1(x, q)$ that is defined on an $\mathbb{R}_+$-invariant neighbourhood of (x_0, q_0) and such that in this neighbourhood $\det \|\partial^2 S/\partial x\,\partial q\| \neq 0$.

We now find a linear canonical transformation homotopic to g_1. Since the matrix

$$A = \frac{\partial^2 S_1}{\partial x\,\partial q}(x_0, q_0) \tag{4.19}$$

is non-singular, there is a vector $v \in \mathbb{R}^n$ such that

$$y_0 = A^t(x_0 + v)\ . \tag{4.20}$$

We set

$$\widetilde{S}_1(x, q, t) = S_1(x, q)(1-t) + \langle x + v, Aq\rangle t\ . \tag{4.21}$$

The function $\widetilde{S}_1(x,q,t)$ satisfies the following conditions:

$$\det\left(\frac{\partial^2 \widetilde{S}_1}{\partial x\,\partial q}(x_0,q_0,t)\right) = \det((1-t)A + tA) = \det A \neq 0 \ , \tag{4.22}$$

$$\begin{aligned}\frac{\partial^2 S_1}{\partial x}(x_0,q_0,t) &= (1-t)\frac{\partial S_1}{\partial x}(x_0,q_0) + tAq_0 \\ &= (1-t)\frac{\partial S_1}{\partial x}(x_0,q_0) + t\frac{\partial^2 S_1}{\partial x\,\partial q}(x_0,q_0)q_0 = \frac{\partial S_1}{\partial x}(x_0,q_0) = p_0 \ ;\end{aligned} \tag{4.23}$$

(here we have used the Euler identity for homogeneous functions)

$$\begin{aligned}\frac{\partial \widetilde{S}_1}{\partial q}(x_0,q_0,t) &= (1-t)\frac{\partial S_1}{\partial q}(x_0,q_0) + tA^t(x_0+v) \\ &= (1-t)y_0 + ty_0 = y_0 \ .\end{aligned} \tag{4.24}$$

Therefore, $\widetilde{S}_1(x,q,t)$ is a generating function of the family $\{\widetilde{g}_1(t)\}$, $t \in [0,1]$, of homogeneous canonical transformations defined on an $\mathbb{R}_+$-invariant neighbourhood of (y_0,q_0) and such that $\widetilde{g}_1(0) = g_1$ and $\widetilde{g}_1(1) = g_2$ is the linear canonical transformation defined by the generating function

$$S_2(x,q) = \langle x+v, Aq\rangle \ . \tag{4.25}$$

Since the group $GL(n,\mathbb{R})$ has two connected components, there is a smooth map $\widetilde{A} : [0,1] \to GL(n,\mathbb{R})$ such that $\widetilde{A}(0) = A$ and either

$$\widetilde{A}(1) = E \ , \tag{4.26}$$

or

$$\widetilde{A}(1) = \begin{pmatrix} -1 & 0 \\ 0 & E \end{pmatrix} , \tag{4.27}$$

depending on the sign of $\det A$.

We set

$$\widetilde{S}(x,q,t) = \langle x + (1-t)v, \widetilde{A}(t)q\rangle \ . \tag{4.28}$$

The function $\widetilde{S}(x,q,t)$ gives rise to a family $\{\widetilde{g}_2(t)\}$ of globally defined (and, in particular, defined an $\mathbb{R}_+$-invariant neighbourhood of (y_0,q_0)) homogeneous canonical transformations such that $\widetilde{g}_2(0) = g_2$ and $\widetilde{g}_2(1) = g_3$ is one of the transformations 2° or 2°′ mentioned in the statement of the lemma.

From our constructions it follows that there is a continuous family of maps $\{\widetilde{\widetilde{g}}(t)\}'$, $t \in [0,1]$, satisfying the conditions 1° and 2° or 2°′ that is smooth except at the points $t_1, t_2, t_3 \in [0,1]$, and all derivatives $\partial^\beta g(t)/\partial t^\beta$ on the set $[0,1] \setminus \{t_1, t_2, t_3\}$ are uniformly bounded[5] with respect to $t \in [0,1] \setminus \{t_1, t_2, t_3\}$ and are homogeneous maps $W \to T_0^*\mathbb{R}^n$ of degree 1, where W is an $\mathbb{R}_+$-invariant neighbourhood of α. Let $f : [0,1] \to [0,1]$ be a smooth strictly monotonic function such that $f(t_k) = t_k$ $(k = 1,2,3)$; $\partial^\beta f(t)/\partial t^\beta|_{t=t^k} = 0$ for all $\beta > 0$ $(k = 1,2,3)$; and $f(0) = 0$, $f(1) = 1$. Then $g_t = \tilde{\tilde{g}}(f(t))$ satisfies all the conditions of Lemma 4.4. This completes the proof.

Lemma 4.5. *Let $g_t : T_0^*\mathbb{R}^n \to T_0^*\mathbb{R}^n$ be a C^∞-family of homogeneous canonical diffeomorphisms, and $\varphi \in \mathcal{O}_l(T_0^*\mathbb{R}^n)$. Then for any real numbers s there is a family $\varphi_t \in \mathfrak{S}_l(T_0^*\mathbb{R}^n)$ and a function $H(x,p,t) \in \mathfrak{S}_1(T_0^*\mathbb{R}^n \times \mathbb{R}^1)$ such that*

$$-i\frac{\partial}{\partial t} \circ T(g_t, \varphi_t) + H\left(\overset{2}{x}, \overset{1}{\hat{p}}, t\right) \circ T(g_t, \varphi_t)' = Q_t \ , \tag{4.29}$$

where

$$Q_t : H_{\text{loc}}^{-s}(\mathbb{R}^n) \to H_{\text{loc}}^{s}(\mathbb{R}^n) \ .$$

Here $\varphi_t = g_t^*\varphi + \tilde{\varphi}_t, \tilde{\varphi}_t \in \mathfrak{S}_{l-1}(T_0^*\mathbb{R}^n)$; $\tilde{\varphi}_0$ or $\tilde{\varphi}_1$ can be chosen arbitrarily.

Proof. We consider the vector field

$$V(t) = \left[\frac{d}{d\epsilon}\{g(t+\epsilon) \circ g^{-1}(t)\}\right]_{\epsilon=0} \ , \tag{4.30}$$

generating a one-parameter group g_t. It is Hamiltonian (see Arnol'd [4]), therefore, there is a real function $H_1(x,p,t)$ such that

$$V(t) = \frac{\partial H_1}{\partial p_i}\frac{\partial}{\partial x^i} - \frac{\partial H_1}{\partial x^i}\frac{\partial}{\partial p_i} \ ; \tag{4.31}$$

since $V(t)$ is homogeneous, we can choose $H_1(x,p,t)$ as a homogeneous function of degree 1.

We now define a manifold $\mathcal{L}_0$ in the space $T_0^*\mathbb{R}^{n+1} \times T_0^*\mathbb{R}^n$ with the coordinates $(x,t,p,E;y,q)$ by the equations

$$\mathcal{L}_0 = \{(x,t,p,E;y,q) | t = 0, (x,p) = g_0(y,q), E - H_1(x,p,0)\} \ . \tag{4.32}$$

$\mathcal{L}_0$ is Lagrangian:

$$|[dp \wedge dx + dE \wedge dt - dq \wedge dy]|_{\mathcal{L}_0} = dp \wedge dx - g_0^*(dq \wedge dy) = 0$$

[5] A homogeneous map $f : W \to T_0^*\mathbb{R}^n$ is said to be bounded if it is bounded on $S^*\mathbb{R}^n \cap W$.

and lies on the zero level surface of the Hamiltonian

$$\mathcal{H}_1(x,t,p,E) = E + H_1(x,p,t) \ . \tag{4.33}$$

Let $\mathcal{L}$ be the phase flow of $\mathcal{L}_0$ along the trajectories of the vector field

$$V(\mathcal{H}_1) = \frac{\partial}{\partial t} + V(H_1) \ . \tag{4.34}$$

Since

$$\dot{t} = \frac{\partial \mathcal{H}_1}{\partial E} = 1 \ ,$$

the variable t can be identified with a parameter along the trajectory. By (4.30) and (4.31), the Hamiltonian system

$$\dot{x} = \frac{\partial \mathcal{H}_1}{\partial p} = \frac{\partial H_1}{\partial p} \ , \quad \dot{p} = -\frac{\partial \mathcal{H}_1}{\partial x} = -\frac{\partial H_1}{\partial x} \tag{4.35}$$

with the initial conditions

$$(x_0, p_0) = g_0(y,q)$$

has the solution

$$(x,p) = g_t(y,q) \ .$$

Moreover, $\mathcal{H}_1$ is a first integral of the Hamiltonian system corresponding to the field $V(\mathcal{H}_1)$. Consequently, $E + H_1(x,p,t) = 0$ on $\mathcal{L}$, that is, $E = -H_1(x,p,t)$. Thus, $\mathcal{L}$ is given by

$$\mathcal{L} = \{(x,t,p,E;y,q) | (x,p) = g_t(y,q), E = -H_1(x,p,t)\} \ . \tag{4.36}$$

From (4.36) it follows that the projection L_{t_0} of the manifold $\mathcal{L}_{t_0} = \mathcal{L} \cap \{t = t_0\}$ onto the space $T_0^*\mathbb{R}^n \times T_0^*\mathbb{R}^n$ with the coordinates $(x,p;y,q)$ is for any t_0 identical with the graph of the canonical transformation g_{t_0}. This fact shows that there is a canonical atlas on $\mathcal{L}$ with the coordinates $(x^I, p_{\overline{I}}, t, q)$. Let U_I be one of the charts of this atlas. Then, by Lemma 2.3, there is a unique function $S_I(x^I, p_{\overline{I}}, t, q)$, homogeneous of degree 1 in the variables $(p_{\overline{I}}, q)$ and such that

$$\begin{aligned} dS_I(x^I, p_{\overline{I}}, t, q) = & - x^{\overline{I}}(x^I, p_{\overline{I}}, t, q) dp_{\overline{I}} + p_I(x^I, p_{\overline{I}}, t, q) dx^I \\ & + y^i(x^I, p_{\overline{I}}, t, q) dq_i + E(x^I, p_{\overline{I}}, t, q) dt \ . \end{aligned} \tag{4.37}$$

Now (4.37) shows that $S_I|_{t=t_0}$ is a generating function for g_{t_0}, that is, the action on the chart $U_I \cap \{t = t_0\}$ of L_{t_0}.

Next, for the $V(\mathcal{H}_1)$-invariant measure μ on $\mathcal{L}$ we choose

$$\mu = (dq \wedge dy)^n \wedge dt|_{\mathcal{L}} \ . \tag{4.38}$$

It is easy to establish that this measure is invariant in the (non-canonical) coordinates (y, q, t) on $\mathcal{L}$, bearing in mind that $V(\mathcal{H}_1) = \partial/\partial t$ in these coordinates.

In U_I the density of μ with respect to the coordinates $(x^I, p_{\bar{I}}, t, q)$ is for each fixed $t = t_0$ equal to the density of the standard measure

$$\mu_{t_0} = (dq \wedge dy)^n \,|_{\mathcal{L}_{t_0}}$$

with respect to the coordinates $(x^I, p_{\bar{I}}, q)$ on L_{t_0}.

Now let $\Phi_{(\mathcal{L},i,\mu,\varphi)}$ be the Fourier integral operator associated with the Lagrangian immersion i: $T_0^*\mathbb{R}^n \times \mathbb{R}^1 \to T_0^*\mathbb{R}^{n+1} \times T_0^*\mathbb{R}^n$,

$$i(y,q,t) = (g_t(y,q), t, -H_1(g_t(y,q),t), (y,q)) \tag{4.39}$$

with the symbol $\varphi_t \in \mathfrak{S}(T_0^*\mathbb{R}^n \times \mathbb{R}^1)$. As follows from the preceding arguments, the operator

$$\Phi_{(\mathcal{L},i,\mu,\varphi_t)} : H^{s'}_{\text{loc}}(\mathbb{R}^n) \to H^{-s}_{\text{loc}}(\mathbb{R}^{n+1})$$

$(s' > l + n/2)$ "disintegrates" into a family of Fourier integral operators $T(g_t, \varphi_t)$ in the following sense. For any $u_0 \in C_0^\infty(\mathbb{R}^n)$

$$\Phi_{(\mathcal{L},i,\mu,\varphi_t)}(u_0)\,|_{t=t_0} = \pm(g_{t_0}, \varphi_{t_0})(u_0)\,. \tag{4.40}$$

The sign in (4.40) depends on the choice of the argument of the canonical measure on L_{t_0}. Next we set

$$\begin{aligned} H(x,p,t) &= H_1(x,p,t) \\ &\quad + \frac{i}{2}\left\{\frac{\partial^2 H_1(x,p,t)}{\partial x^i \partial p_i} + (2\alpha - 1)V(H_1)\log v_x(x)\right\}\,. \end{aligned} \tag{4.41}$$

By the commutation theorem $(\mathcal{H} = E + H)$,

$$\begin{aligned} \left[-i\frac{\partial}{\partial t} + H\left(\overset{2}{x}, \overset{1}{\hat{p}}, t\right)\right]\Phi_{(\mathcal{L},i,\mu,\varphi_t)} &= \mathcal{H}\left(\overset{2}{x}, \overset{1}{\hat{p}}, \overset{2}{t}, \overset{1}{\hat{E}}\right) \circ \Phi_{(\mathcal{L},i,\mu,\varphi_t)} \\ &= \Phi_{(\mathcal{L},i,\mu,\mathcal{P}\varphi_t)} \end{aligned}$$

modulo Fourier integral operators with sufficiently smooth kernel. Choosing $\varphi_t = \sum_{k=0}^N \varphi_t^{(k)}$ so that

$$\mathcal{P}^N \varphi_t = 0 \tag{4.42}$$

and assuming that φ_t has sufficient length N in $\mathfrak{S}_t(T_0^*\mathbb{R}^n \times \mathbb{R}^1)$, we can use (4.40) to arrive at the assertion of the lemma.

Let us now verify the assertions of the lemma concerning φ_t. Clearly (4.41) yields

$$\mathcal{P}_0 = \frac{d}{dt}$$

(d/dt is differentiation along $V(\mathcal{H}_1)$). Therefore, $\varphi_t^{(0)}$ is constant along the trajectories and we can set $\varphi_t^{(0)} = g_t^* \varphi$. The functions $\varphi_t^{(k)}$ for $1 \le k \le N$ satisfy the first-order equations

$$\frac{d}{dt}\varphi_t^{(k)} = \sum_{j+l \neq k, j \neq 0} \mathcal{P}_j \varphi_t^{(l)} , \tag{4.43}$$

hence, their values at one of the end-points of $[0,1]$ can be chosen arbitrarily. This completes the proof.

Remark 4.1. If the one-parameter group of canonical diffeomorphisms $g_t : W \to T_0^* \mathbb{R}^n$ is defined on an $\mathbb{R}_+$-invariant open set $W \subset T_0^* \mathbb{R}^n$, and $\operatorname{supp} \varphi \in W$, then the assertion of Lemma 4.5 remains valid.

We now proceed to prove Theorems 4.2 and 4.3. Obviously, we need only a local proof. We first prove the following proposition:

Proposition 4.1. *Let $g : V \to T_0^* \mathbb{R}^n$ be a smooth function of the variables $x \in \mathbb{R}^n$, $p \in \mathbb{R}_n$, $t \in [0,1]$, that is homogeneous of degree 1 in p for large $|p|$, whose support is contained in $|x| < R$ and is such that $\operatorname{Im} H(x,p,t) \le 0$ for large $|p|$. Then for any $s \in \mathbb{R}$ there is a constant c_s such that for $t \in [0,1]$*

$$\|u(t)\|_s \le c_s \left(\|u(0)\|_s + \int_0^t \left\| -i\frac{\partial u(\tau)}{\partial \tau} + H\left(\overset{2}{x}, -i\frac{\overset{1}{\partial}}{\partial x}, \tau \right) u(\tau) \right\|_s d\tau \right)$$

for any continuously differentiable map

$$u : [0,1] \to S(\mathbb{R}^n) .$$

Proof. We have

$$\operatorname{Im}\left(u, H\left(\overset{2}{x}, -i\frac{\overset{1}{\partial}}{\partial x}, \tau\right)u\right) = \frac{1}{2i}\left(u, \left[H\left(\overset{2}{x}, -i\frac{\overset{1}{\partial}}{\partial x}, \tau\right)\right.\right.$$
$$\left.\left.-\overline{H}\left(\overset{1}{x}, -i\frac{\overset{2}{\partial}}{\partial x}, \tau\right)\right]u\right) = \left(u, \operatorname{Im} H\left(\overset{2}{x}, -i\frac{\overset{1}{\partial}}{\partial x}, \tau\right)u\right)$$
$$+\frac{1}{2i}\left(u, \left[\overline{H}\left(\overset{2}{x}, -i\frac{\overset{1}{\partial}}{\partial x}, \tau\right) - \overline{H}\left(\overset{1}{x}, -i\frac{\overset{2}{\partial}}{\partial x}, \tau\right)\right]u\right)$$
$$= \left(u, \operatorname{Im} H\left(\overset{2}{x}, -i\frac{\overset{1}{\partial}}{\partial x}, \tau\right)u\right) + (u, R(\tau)u) \ ,$$

where $R(\tau)$ is a pseudodifferential operator of order ≤ 0, therefore, a continuous operator in $H^s(\mathbb{R}^n)$. (In fact, its kernel $R(x, y, \tau)$ vanishes for $|x| + |y| \geq 2R$.) Applying Gårding's lemma (see Gårding [2]), we obtain

$$\operatorname{Im}\left(u, H\left(\overset{2}{x}, -i\frac{\overset{1}{\partial}}{\partial x}, \tau\right)u\right)_s \geq -\widetilde{c}_s(u, u)_s$$

for some constant $\widetilde{c}_s$. Evaluating $\frac{\partial}{\partial t}\|u(t)\|_s$, we find that

$$\frac{\partial}{dt}\|u(t)\|_s^2 = 2\ \operatorname{Im}\left(u(t), -i\frac{\partial u(t)}{\partial t}\right)_s$$
$$= 2\ \operatorname{Im}\left(u(t), -i\frac{\partial u(t)}{\partial t} + H\left(\overset{2}{x}, -i\frac{\overset{1}{\partial}}{\partial x}, t\right)u(t)\right)_s$$
$$+ 2\ \operatorname{Im}\left(u(t), -H\left(\overset{2}{x}, -i\frac{\overset{1}{\partial}}{\partial x}, t\right)u(t)\right)_s$$
$$\leq 2\|u(t)\|_s \cdot \left\|-i\frac{\partial u(t)}{\partial t} + H\left(\overset{2}{x}, -i\frac{\overset{1}{\partial}}{\partial x}, t\right)u(t)\right\|_s + 2\widetilde{c}_s\|u(t)\|_s^2 \ ,$$

hence,

$$\frac{\partial}{\partial t}\|u(t)\|_s \leq \widetilde{\widetilde{c}}^s\left(\|u(t)\|_s + \left\|-\frac{\partial u(t)}{\partial t} + H\left(x, -i\frac{\partial}{\partial x}, t\right)u(t)\right\|_s\right) \ .$$

Setting $c_s = e^{\widetilde{\widetilde{c}}_s}$, we obtain the required result. This completes the proof of Proposition 4.1.

Proof of Theorem 4.2. Let $g : V \to T_0^*\mathbb{R}^n$ be a canonical map, $\varphi \in \mathcal{O}^l(V)$, and $\operatorname{supp}\varphi \Subset V$. Let g_t be the homotopy constructed in Lemma 4.4. By Lemma 4.5, there is a function $\varphi_t \in \mathfrak{S}_l(T_0^*\mathbb{R}^n)$, with $\operatorname{supp}\varphi_t \subset g_t(V)$, such that $\varphi_1 = \varphi$, $\operatorname{supp}\varphi$ is sufficiently small, and for any $u_0 \in C_0^\infty(\mathbb{R}^n)$,

$$-i\frac{\partial}{\partial t}T(g_t,\varphi_t)u_0 + H\left(\overset{2}{x},\overset{1}{\hat{p}},t\right)T(g_t,\varphi_t)u_0 = Q_t u_0 \,, \tag{4.44}$$

$$T(g_t,\varphi_t)u_0|_{t=0} = T(g_0,\varphi_0)u_0 \,,$$

where $T(g_0,\varphi_0)$ is either the pseudodifferential operator

$$T(g_0,\varphi_0) = \left(\varphi_0^{(0)} + \widetilde{\varphi}_0\right)\left(\overset{2}{x},\overset{1}{\hat{p}}\right) \tag{4.45}$$

or the composition

$$T(g_0,\varphi_0) = \left(\varphi_0^{(0)} + \widetilde{\varphi}_0\right)\left(\overset{2}{x},\overset{1}{\hat{p}}\right) \circ U_1 \tag{4.46}$$

of the pseudodifferential operator with the unitary operator U_1 defined by

$$U_1 f\left(x^1,\dots x^n\right) = -f\left(-x^1,x^2,\dots,x^n\right) \,.$$

When we now use Proposition 4.1, observe that the initial data (4.45) or (4.46) satisfy the required estimates in the scale of Sobolev spaces, and use the fact that the number of the Sobolev space is arbitrary, we can complete the proof.

Proof of Theorem 4.3. We consider a homotopy $g(t)$ such that $g(1) = g_2$ and $g(0)$ is the identity map (the case 2° of Lemma 4.4 is treated similarly). Let $\varphi_t \in \mathfrak{S}_l(T_0^*\mathbb{R}^n)$ be chosen as in Lemma 4.5 and $\varphi(1) = \varphi_2$. The function

$$u(x,t) = T(g(t),\varphi(t)) \circ T(g_1,\varphi_1)u_0(x) \,, \tag{4.47}$$

where $u_0(x) \in C_0^\infty(\mathbb{R}^n)$, satisfies the Cauchy problem

$$\begin{cases} -i\dfrac{\partial u}{\partial t}(x,t) + H\left(\overset{2}{x},\overset{1}{\hat{p}},t\right)u(x,t) = Q_t u_0(x) \,, \\ u(x,0) = \varphi(0)\left(\overset{2}{x},\overset{1}{\hat{p}}\right) \circ T(g_1,\varphi_1)u_0(x) \,. \end{cases} \tag{4.48}$$

Using the theorem on compositions, the initial condition in (4.44) can be rewritten in the form

$$u(x,0) = T(g_1,(g_1^*\varphi_2)\varphi_1 + \widetilde{\varphi}_0)u_0(x) \tag{4.49}$$

modulo functions of arbitrary degree of smoothness. On the other hand, the function

$$u_1(x,t) = T\left(g(t)\circ g_1, (g_1^*\varphi_2)\,\varphi_1 + \widetilde{\varphi}(t)\right)u_0(x) \tag{4.50}$$

also satisfies the problem (4.48), possibly, with a different operator Q_t. Hence and from Proposition 4.1 it now follows that (4.8) holds for $\varphi_3 = \widetilde{\varphi}(1)$.

Remark 4.2. As is clear from the proof, Theorem 4.2 also holds when L is locally the graph of a canonical map. Such immersions are called *non-singular.*

§ 5. Examples and Applications

1°. Pseudodifferential Operators on a Manifold M. Let $\widehat{H}$ be a pseudodifferential operator of order r on a manifold M. For any fixed coordinate system $\left(x^1, \dots, x^n\right)$ the operator $\widehat{H}$ can be written in the form (1.25), that is, as an integral operator whose kernel in a local coordinate system is given up in C^∞-functions by

$$A(x,y) = \frac{1}{(2\pi)^n}\int_{\mathbb{R}^n} e^{i(x-y)q} H(x,q)\lambda\left(|q|\right)dq\ . \tag{5.1}$$

As before, $\lambda(t)$ is a smooth function such that

$$\lambda(t) = \begin{cases} 0 & \text{for } \ 1 \le 1/2\ , \\ 1 & \text{for } \ t > 1\ . \end{cases}$$

The function $H(x,q)$ is asymptotically homogeneous (see § 1) and the convergence of the integral is understood, as before, in the space $S'\left(\mathbb{R}^n \oplus \mathbb{R}^n\right)$. The function $H(x,q)$ is said to be the *symbol (total symbol)* of the operator. Let M be a compact manifold without boundary,

$$g : T_0^*M \to T_0^*M$$

the identity diffeomorphism, and (T_0^*M, i_g, μ) the corresponding homogeneous Lagrangian immersion. Let φ be an asymptotically homogeneous function on T_0^*M, and let φ^N be the N-th partial sum of the asymptotic expansion of φ in homogeneous functions. The formula

$$A(x,y) = \left(-\frac{i}{2\pi}\right)^{n/2}\{K_g\varphi\}(x,y) \tag{5.2}$$

defines (globally on M) the kernel of the pseudodifferential operator in the following sense: for each s there is an N such that

$$A(x,y)-\left(\frac{1}{2\pi i}\right)^{n/2}\left\{K_g\varphi^N\right\}(x,y)\in H^s(M\times M)\ . \tag{5.3}$$

2°. Operators of Restriction and Corestriction. Let M be a smooth manifold of dimension n and

$$i: N\subset M$$

a smooth submanifold of codimension ν. Then the following operators are defined:

1) The restriction operator

$$i^*: H_0^s(M)\to H_0^{s-\nu/2}(N)\ ,\qquad s>\nu/2\ ,$$

which assigns to each $f\in H_0^s(M)$ its trace on the submanifold N.

2) The corestriction operator

$$i_*: H^{-s+\nu/2}(N)\to H^{-s}(M)\ ,\qquad s>\nu/2\ ,$$

adjoint to i^*. We write i^* and i_* in a local coordinate system $(x^1,\dots,x^n)$ on M.

Suppose that N can be described in the coordinate system $(x^1,\dots,x^n)$ by the equations: $x^{n-\nu+1}=x^{n-\nu+2}=\dots=x^n=0$. We write $x'=(x^1,\dots,x^{n-\nu})$, $x''=(x^{n-\nu+1},\dots,x^n)$, and $x=(x',x'')$. The operators i^* and i_* act as follows:

$$[i^*f](x')=\frac{1}{(2\pi)^{n/2}}\int e^{i(q'(x'-y')-q''y'')}f(y',y'')\,dy'dy''dq'dq''\ , \tag{5.4}$$

$$[i_*f](y',y'')=\frac{1}{(2\pi)^{n/2}}\int e^{i(q'(x'-y')+q''y'')}f(x')dx'dq'dq''\ . \tag{5.5}$$

We introduce coordinates (y,q,x',p') in $T_0^*M\times T_0^*N$, where $(y,q)\in T_0^*M$ and $(x',p')\in T_0^*N$. Then (5.4) and (5.5) are Fourier integral operators associated with the immersion $\left(L,\tilde{i},\mu\right)$, where $\tilde{i}: L\hookrightarrow T_0^*M\times T_0^*N$ is the submanifold defined by

$$y''=0\ ,\quad x'=y'\ ,\quad q'=p'\ . \tag{5.6}$$

The Lagrangian immersion (5.6) *is not non-singular* and, accordingly, (5.4) and (5.5) are not continuous in the scale $\{H^s\}$.

3°. Canonical Transformations of Pseudodifferential Operators. Let M and N be two manifolds and $\widehat{H}_1$ and $\widehat{H}_2$ pseudodifferential operators of order r on M and N, respectively. Can we construct a Fourier integral operator Φ such that to within the lowest terms

$$\widehat{H}_1\Phi = \Phi\widehat{H}_2 \ ? \tag{5.7}$$

This problem was first studied by Egorov (see Egorov [1], [3]). He showed that if Φ is an integral operator of the form

$$[\Phi f](x) = \int e^{iS(x,q)}\widetilde{f}(q)dq \ , \tag{5.8}$$

$$\det\left(\frac{\partial^2 S}{\partial x\,\partial q}\right) \neq 0 \ , \tag{5.9}$$

and if (5.7) is satisfied, then the principal symbols of $\widehat{H}_1$ and $\widehat{H}_2$ are connected by a canonical transformation whose generating function is $S(x,q)$. An immediate generalization of this result is the following theorem.

Theorem 5.1. *Let Φ be a Fourier integral operator associated with a homogeneous Lagrangian immersion (L, i, μ),*

$$i : L \to T_0^*M \times T_0^*N \ .$$

For (5.7) to hold it is necessary and sufficient that

$$i^*\left(H_{1r} \otimes 1_N - 1_M \otimes H_{2r}\right) = 0 \ . \tag{5.10}$$

The proof follows immediately from the theorem that pseudodifferential operators commute with Fourier integral operators.

This ends our presentation of the real theory of Fourier integral operators.

The next three subheadings are devoted to the complex version of the theory. Here we will try to follow, on the one hand, the peculiarities introduced into the analysis of s-analytic manifolds by the homogeneous structure, and on the other hand, those introduced into the real theory of Fourier integral operators by the complexness.

§ 6. Analysis on s-analytic Homogeneous Manifolds

As in the real theory, the functions to be considered must be homogeneous in the momentum variables. This leads (by comparison with Sect. 6.3) to additional structures on the s-analytic manifolds.

Lemmas 1.2 and 1.3 can be generalized in the natural way to complex phase functions. Where functions of complex arguments appear, s-analytic extensions of the relevant functions have to be taken.

Definition 6.1. A *weight function* on a homogeneous manifold L is a non-negative function $\rho_L = \rho$, homogeneous of degree 0 and such that $\rho_L^2 \in C^\infty(L)$.

We denote by Ω or Ω_L the set of zeros of ρ_L. Clearly, $\rho_L \in C^\infty(L \setminus \Omega)$. We also note that Ω is an $\mathbb{R}_+$-invariant set.

We write ${}^s\mathcal{I}\mathcal{O}_k(L,\rho)$ for the set of functions from $\mathcal{O}_k(L)$, satisfying on every $\mathbb{R}_+$-invariant compact set the inequality

$$\left|\chi^{|\alpha|-k} D^\alpha f\right| \le C_\alpha \rho^{s-|\alpha|} , \qquad |\alpha| \le s , \tag{6.1}$$

in each homogeneous local coordinate system $(x^1, \ldots, x^m)$ where $\alpha = (\alpha_1, \ldots, \alpha_m)$ is a multi-index, $|\alpha| = \alpha_1 + \ldots + \alpha_m$, and

$$D^\alpha = \left(\frac{\partial}{\partial x^1}\right)^{\alpha_1} \cdot \ldots \cdot \left(\frac{\partial}{\partial x^m}\right)^{\alpha_m} .$$

Here χ is a strictly positive homogeneous function of degree 1 on L (see subheading 1.1). We denote by ${}^s\mathcal{I}\mathfrak{S}_l(L,\rho)$ the set of functions of the form

$$f = \sum_{k=0}^{N} f_k ,$$

where $f_k \subset {}^{s-2k}\mathcal{I}\mathcal{O}_{l-k}(L,\rho)$. It is obvious that ${}^s\mathcal{I}\mathfrak{S}_0(L,\rho)$ is an ideal in the ring $\mathfrak{S}_0(L)$.

Definition 6.2. An *s-analytic homogeneous structure* on a homogeneous manifold L of dimension $2n$ is a homogeneous atlas $\{(U_i, \alpha_i)\}$, where the $\alpha_i : U_i \to V_i \subset \mathbb{C}_*^n$ are homeomorphisms, such that[6] ($\mathbb{C}_*^n = \mathbb{C}^n \setminus \{0\}$)

$$\frac{\partial}{\partial \bar{z}^l} \alpha_i \circ \alpha_j^{-1} \in {}^s\mathcal{I}\mathcal{O}_0(L,\rho) ; \quad |\operatorname{Im} \alpha_i(m)| \le C\chi(m)\rho(m) , \qquad m \in L , \tag{6.2}$$

for any pair i and j such that $U_i \cap U_j \neq \emptyset$. Here $(z^1, \ldots, z^n)$ are coordinates in $\mathbb{C}^n \setminus \{0\}$. The action of $\mathbb{R}_+$ on $\mathbb{C}^n \setminus \{0\}$ is defined by $t(z) = (tz)$, $z = a + ib$, $\frac{\partial}{\partial \bar{z}^l} = \frac{1}{2}\left(\frac{\partial}{\partial a^l} + i\frac{\partial}{\partial b^l}\right)$. A manifold with an s-analytic homogeneous structure is called an s-analytic homogeneous manifold.

[6] As before, we could also use the space $\mathbb{C}_*^k \times \mathbb{C}^{n-k}$. For simplicity, we treat here only the case $k = n$.

Remark 6.1. For $s < 0$ we assume that ${}^sI\mathcal{O}_l(L,\rho) = \mathcal{O}_l(L,\rho)$, therefore, ${}^sI\mathfrak{S}_l(L,\rho)$ contains $\mathfrak{S}_{l-[\frac{N}{2}]-1}(L)$.

It is now natural to consider s-analytic homogeneous functions on s-analytic homogeneous manifolds. A homogeneous function $f \in C^\infty(L)$ of degree k is said to be *s-analytic* if in any s-analytic homogeneous coordinate system $(z^1,\dots,z^n)$

$$\frac{\partial}{\partial z^l} f\left(z^1,\dots,z^n\right) \in {}^sI\mathcal{O}_{k-1}(L,\rho)\ . \tag{6.3}$$

We denote by ${}^s\mathcal{O}'_k(L,\rho)$ the set of s-analytic homogeneous functions of degree k and let ${}^s\mathfrak{S}'_l(L,\rho) = \oplus_{k=0}^{[s/2]}\, {}^{s-2k}\mathcal{O}'_{l-k}(L,\rho)$.

Clearly, ${}^s\mathfrak{S}'_0(L,\rho)$ is a subring of $\mathfrak{S}_0(L)$. In what follows we do not distinguish between s-analytic functions[7] that differ only by elements of ${}^{s+1}I\mathfrak{S}'_i(L,\rho)$; thus, an s-analytic function is an element of the quotient space

$${}^s\mathfrak{S}_l(L,\rho) = {}^s\mathfrak{S}'_l(L,\rho) / {}^{s+1}I\mathfrak{S}'_l(L,\rho)\ . \tag{6.4}$$

The space $\mathcal{O}_k(L,\rho)$ is defined similarly. The sets ${}^sT_l(L,\rho)$ of homogeneous s-analytic vector fields on L and ${}^s\Lambda^k_l(L,\rho)$ of homogeneous s-analytic exterior k-forms on L are defined in the natural way. The fact that all these concepts are well-defined is proved in Sect. 3.3. We mention that the sets ${}^t\mathfrak{S}_l(L,\rho)$, ${}^tT_l(L,\rho)$, and ${}^t\Lambda^k_l(L,\rho)$ are defined for $0 \le t \le s$.

Let us dwell on the concept of an s-analytic map.

Definition 6.3. A map $f : (L_1,\rho_1) \to (L_2,\rho_2)$ of two homogeneous s-analytic manifolds is said to be a *homogeneous s-analytic map* if it is defined locally by s-analytic homogeneous functions of degree 1 and if, in addition, $f^*(\rho_2) \le c\rho_1$.

The formula

$${}^s\mathfrak{S}'_l(L,\rho) = \bigoplus_{k=0}^{[s/2]} {}^{s-2k}\mathcal{O}'_{l-k}(L,\rho) \tag{6.5}$$

involves summation only up to $[s/2]$, since the following terms of the sum disappear in the factorization (6.4), by Remark 6.1.

Euler's theorem is valid for s-analytic homogeneous functions.

Theorem 6.1. *If $f\left(z^1,\dots,z^n\right) \in {}^s\mathcal{O}_m(U,\rho)$, then*

---

[7] Henceforth we omit the word "homogeneous"; an s-analytic function is an element of ${}^s\mathfrak{S}'_l(L,\rho)$ or ${}^s\mathfrak{S}_l(L,\rho)$ (see (6.4)).

$$z^i \frac{\partial f}{\partial z^i}\left(z^1, \ldots z^n\right) \equiv mf\left(\left(z^1, \ldots, z^n\right)\right) \qquad (\operatorname{mod} {}^sI\mathcal{O}_m(U, \rho)) \ .$$

Proof.

$$mf\left(z^1, \ldots, z^n\right) = x^i \frac{\partial f}{\partial x^i} + y^i \frac{\partial f}{\partial y^i} = z^i \frac{\partial f}{\partial z^i} + \bar{z}^i \frac{\partial f}{\partial \bar{z}^i} \equiv z^i \frac{\partial f}{\partial z^i}$$

since f is s-analytic.

Let us consider the concept of an *s-analytic extension*, which will be required later. Let $\left(U; \left(z^1, \ldots, z^n\right)\right)$ be an s-analytic homogeneous coordinate system. We denote by U^0 the submanifold of U defined by $\operatorname{Im} z^l = 0$ $(l = 1, \ldots, n)$. We set

$$\begin{cases} {}^s\mathcal{O}_k\left(U^0\right) = \mathcal{O}_k(U^0)/{}^{s+1}I\mathcal{O}_k\left(U^0, \rho|_{U^0}\right) \ , \\ {}^s\mathfrak{S}_l(U^0) = \mathfrak{S}_l(U^0)/{}^{s+1}I\mathfrak{S}_l\left(U^0, \rho|_{U^0}\right) = \bigoplus\limits_{k=0}^{[s/2]} {}^{s-2k}\mathcal{O}_{l-k}(U^0) \ . \end{cases} \tag{6.6}$$

The restriction map R then acts as a homomorphism

$$R : {}^s\mathcal{O}_k\left(U, \rho|_U\right) \to {}^s\mathcal{O}_k(U^0) \ . \tag{6.7}$$

Proposition 6.1. *The homomorphism* (6.7) *is an isomorphism. The inverse homomorphism* ${}^sA : {}^s\mathcal{O}_k\left(U^0\right) \to {}^s\mathcal{O}_k\left(u, \rho|_U\right)$ *is given by*

$${}^sA\left(f\left(x^1, \ldots, x^n\right)\right) = \sum_{k=0}^{s} \frac{1}{k!}\left(iy^j \frac{\partial}{\partial x^j}\right)^k f\left(x^1, \ldots, x^n\right) \ . \tag{6.8}$$

A detailed proof of this proposition can be found in Sect. 3.3. It remains only to verify that ${}^sA(f)$ is homogeneous. Let $f \in \mathcal{O}_k(U_0)$. Then $D^\alpha f \in \mathcal{O}_{k-|\alpha|}(U^0)$. Setting $\alpha! = \alpha_1! \cdot \ldots \cdot \alpha_n!$, we can rewrite ${}^sA$ in the form

$${}^sAf(z) = \sum_{|\alpha| \leq s} \frac{1}{\alpha!} y_1^{\alpha_1} \cdot \ldots \cdot y_n^{\alpha_n} i^{|\alpha|} D^\alpha f\left(x^1, \ldots, x^n\right) \ . \tag{6.9}$$

Hence

$$\begin{aligned} {}^sAf(tz) &= \sum_{|\alpha| \leq s} \frac{1}{\alpha!}(ity)^\alpha D^\alpha f(tx) \\ &= t^k \sum_{|\alpha| \leq s} \frac{1}{\alpha!}(iy)^\alpha D^\alpha f(x) = t^k \cdot {}^sAf(z) \ . \end{aligned} \tag{6.10}$$

This completes the proof. The homomorphism ${}^sA$ can be extended by linearity to a homomorphism

$$ {}^sA : {}^s\mathfrak{S}_l(U^0) \to {}^s\mathfrak{S}_l(U, \rho|_U) \ . $$

The operators ${}^sA$ and R have the following properties (see Sect. 3.3):

$$ \frac{\partial}{\partial z^l}{}^sA = {}^sA\frac{\partial}{\partial x^l} \ ; \quad \frac{\partial}{\partial x^l}R = R\frac{\partial}{\partial z^l} \qquad (l = 1, \dots, n) \ . \tag{6.11} $$

The function ${}^sAf$ is called an *s-analytic extension* of f.

The *support* of an s-analytic function $f \in {}^s\mathcal{O}_k(L, \rho)$ is defined as the intersection of the supports of all the elements in the equivalence class of f in ${}^s\mathcal{O}'_k(L, \rho)$. Making use of the last proposition, we can then prove the existence of an s-analytic partition of unity.

Proposition 6.2. *Let $\{W_k\}$ be a homogeneous open cover of an s-analytic homogeneous manifold (L, ρ). Then there is a system of s-analytic homogeneous functions φ_k of degree* 0 *such that*

$$ \text{a)} \quad \operatorname{supp} \varphi_k \subset W_k \ . \qquad \text{b)} \quad \sum_k \varphi_k = 1 $$

(the latter equality is, of course, to be understood in the ring ${}^s\mathcal{O}_k(L, \rho)$).

§ 7. Complexification of the Phase Space

Let M be a compact C^∞-manifold without boundary. We carry out the complexification of the phase space in two stages. To begin with we construct an s-analytic extension $M_{\mathbb{C}}$ of M. The complexification of the phase space is then the space of the bundle ${}^sT^*(M_{\mathbb{C}})$. We note that ${}^sT^*_0(M_{\mathbb{C}})$ is a homogeneous s-analytic manifold with respect to the action of $\mathbb{R}_+$ on the fibres, that is,

$$ t\left(z^1, \dots, z^n, \zeta_1, \dots, \zeta_n\right) = \left(z^1, \dots, z^n, t\zeta_1, \dots, t\zeta_n\right) \ . $$

Theorem 7.1. *There is an s-analytic manifold $(M_{\mathbb{C}}, \rho)$ such that $\dim_{\mathbb{C}} M_{\mathbb{C}} = \dim_{\mathbb{R}} M$ and $M = \Omega_{M_{\mathbb{C}}}$.*

In what follows we call the manifold $(M_{\mathbb{C}}, \rho)$ an *s-analytic rigging* of M.

Proof. Let $\{U_j, \varphi_j\}$ be a finite atlas of M, $\varphi_j : U_j \to V_j \subset \mathbb{R}^n$ the coordinate maps, and V_j open subsets of $\mathbb{R}^n$. Also, let

$$V_{jk} = \varphi_j (U_j \cap U_k) \subset V_j \ ,$$

$$\varphi_{jk} : V_{jk} \to V_{kj} \ ; \quad \varphi_{jk} = \varphi_k|_{U_j \cap U_k} \circ \varphi_j^{-1}|_{V_{jk}} \ ,$$

where $\{U_i'\}$ is a covering, compactly inscribed in the cover $\{U_i\}$, $\varphi_j' = \varphi_j|_{U_j'}$, and V_j', V_{jk}', φ_{jk}' are defined in the atlas $\{U_j', \varphi_j\}$ by analogy with the procedure for V_j, V_{jk}, and φ_{jk}. We denote by $x_j = (x_j^1, \dots, x_j^n)$ the coordinates on V_j (therefore, also on V_j').

We now introduce open sets $W_j = V_j \times \mathbb{R}^n \subset \mathbb{R}^{2n} \cong \mathbb{C}^n$ and $W_j' = V_j' \times \mathbb{R}^n \subset W_j$. The coordinates in $W_j\ (W_j')$ are denoted by $(x_j^1, \dots, x_j^n, y_j^1, \dots, y_j^n)$; let $z_j^l = x_j^l + iy_j^l$ be the coordinates in W_j, regarded as an open subset of $\mathbb{C}^n$. We denote by ${}^s\varphi_{jk} : W_{jk} \to \mathbb{C}^n$ the s-analytic extension[8] of φ_{jk}.

We divide the rest of the proof into a number of lemmas.

Lemma 7.1. *There is a $\delta > 0$ such that for all $|y_j| \le \delta$*

$$c|y_j| \le \left|\operatorname{Im} {}^s\varphi_{jk}(z_j)\right| \le c|y_j| \ . \tag{7.1}$$

Proof. Suppose that the map φ_{jk} is defined by the functions

$$x_k^l = f_{jk}^l \left(x_j^1, \dots, x_j^n\right) \ .$$

The map ${}^s\varphi_{jk}$ is then given by the functions

$$\begin{aligned} z_k^l &= {}^s f_{jk}^l \left(z_j^1, \dots, z_j^n\right) \\ &= \sum_{m=0}^{s} \frac{1}{m!} \left(iy_j \frac{\partial}{\partial x_j}\right)^m f_{jk}^l(x_j) \qquad (l = 1, \dots, n) \ . \end{aligned} \tag{7.2}$$

It follows from (7.2) that

$$y_k^l = y_j^p \frac{\partial f_{jk}^l}{\partial x_j^p} \left(x_j^1, \dots, x_j^n\right) + O\left(|y_j|^2\right) \ . \tag{7.3}$$

Since the matrix $\partial f_{jk}^l / \partial x_j^n \left(x_j^1, \dots, x_j^n\right)$ is non-singular, (7.3) yields the estimate

$$|y_j| \le c_1 \left(|y_k| + |y_j|^2\right) \ .$$

Setting $\delta = 1/2c_1$, we obtain the left inequality in (7.1). The inequality on the right-hand side follows immediately from (7.3). This completes the proof of the lemma.

We now set

$$\rho_j = |y_1| = \sqrt{\left(y_j^1\right)^2 + \dots + \left(y_j^n\right)^2} \ .$$

[8] We mention that the image ${}^s\varphi_{jk}\left(W_{jk}\right)$ is not necessarily contained in W_{kj}.

Lemma 7.1 indicates that the weight functions ρ_j are equivalent on the intersections of their domains, therefore, in what follows we omit the index j of ρ.

Now we observe that the functions ${}^s\varphi_{jk}$ satisfy the relation

$$ {}^s\varphi_{kr} \circ {}^s\varphi_{jk} \equiv {}^s\varphi_{jr} \qquad (\operatorname{mod} {}^{s+1}I\mathcal{O}_1(\rho)) \tag{7.4}$$

on the domain of definition of the left-hand side.

Lemma 7.2. *We can change the functions ${}^s\varphi_{jk}$ modulo ${}^{s+1}I(\rho)$, so that*

$$ {}^s\varphi'_{kr} \circ {}^s\varphi'_{jk} = {}^s\varphi'_{jr}\,, \quad {}^s\varphi_{jj} = id_{W'_j} \tag{7.5}$$

on the domain of definition of the left-hand side.

The proof of this lemma is quite analogous to that of Lemma 3.26 from Sect. 3.3, and we omit it.

Henceforth we assume that the functions ${}^s\varphi_{jk}$ have already been modified to satisfy (7.5) without saying so explicitly. For a sufficiently small y_j the image ${}^s\varphi'_{jk}(W'_{jk})$ is contained in W_{kj}. We now choose a covering U''_j compactly inscribed in U'_j, define sets V''_j and V''_{jk} as before, and set

$$\begin{cases} W''_j = (V''_j \times \mathbb{R}^n_\delta) \cup \left\{ \bigcup\limits_{r\in K_j} {}^s\varphi_{rj}\left(V''_{rj} \times \mathbb{R}^n_\delta\right) \right\}, \\ W''_{jk} = (V''_{jk} \times \mathbb{R}^n_\delta) \cup \left\{ \bigcup\limits_{r\in K_{jk}} {}^s\varphi_{rj}\left(V''_{rjk} \times \mathbb{R}^n_\delta\right) \right\}. \end{cases} \tag{7.6}$$

Here, $\mathbb{R}^n_\delta$ are the strips $|y| < \delta$ in $\mathbb{R}^n$, K_j is the set of indices r such that $U''_j \cap U''_r \neq \emptyset$, K_{jk} the set of indices r such that $U''_j \cap U''_k \cap U''_r \neq \emptyset$, and V''_{rjk} the image of $U''_j \cap U''_k \cap U''_r$ under φ''_r.

Lemma 7.3. *The map ${}^s\varphi'_{jk}$ is a homeomorphism of W''_{jk} onto W''_{kj} satisfying* (7.5).

Proof. It is clearly sufficient to prove that ${}^s\varphi'_{jk}(W''_{jk}) \subset W''_{kj}$. The remaining assertions of Lemma 7.3 then follow from (7.5).

Let $\alpha \in W''_{jk}$. Then either $\alpha \in V''_{jk} \times \mathbb{R}^n_\delta$ or $\alpha \in {}^s\varphi'_{rj}(V''_{rjk} \times \mathbb{R}^n_\delta)$ for some $r \in K_{jk}$.

Case 1. $\alpha \in V''_{jk} \times \mathbb{R}^n_\delta$. We have ${}^s\varphi'_{jk}(\alpha) \in {}^s\varphi'_{jk}(V''_{jk} \times \mathbb{R}^n_\delta) = {}^s\varphi'_{jk}(V''_{jjk} \times \mathbb{R}^n_\delta)$. Since $j \in K_{kj}$, we find that ${}^s\varphi'_{jk}(\alpha) \in W''_{kj}$.

Case 2. $\alpha \in {}^s\varphi'_{rj}(V_{rjk} \times \mathbb{R}^n_\delta)$ for some $r \in K_{jk}$. But then $r \in K_{kj}$ and

$$\begin{aligned} {}^s\varphi'_{jk}(\alpha) \in {}^s\varphi'_{jk}\left[{}^s\varphi'_{rj}(V_{rjk} \times \mathbb{R}^n_\delta)\right] &= {}^s\varphi'_{jk} \circ {}^s\varphi'_{rj}(V_{rjk} \times \mathbb{R}^n_\delta) \\ &= {}^s\varphi'_{rk}(V_{rjk} \times \mathbb{R}^n_\delta) = {}^s\varphi'_{rk}(V_{rkj} \times \mathbb{R}^n_\delta) \ , \end{aligned}$$

hence, ${}^s\varphi'_{jk}(\alpha) \in W'_{kj}$. This completes the proof.

We now define $M_{\mathbb{C}}$ as the quotient space of the disconnected union $\cup_j W''_j$ modulo the equivalence defined by the homeomorphisms ${}^s\varphi'_{jk}$, and we construct the function ρ from the ρ_j by means of a partition of unity. To complete the proof of the theorem it remains to show that $M_{\mathbb{C}}$ is a Hausdorff space in the quotient space topology. This is a consequence of the following assertion: if $\alpha_1, \ldots, \alpha_m, \ldots$ is a sequence of points in W'_{jk} converging to a point in $W''_j \setminus W''_{jk}$, then the limit of the sequence ${}^s\varphi_{jk}(\alpha_1), \ldots, {}^s\varphi_{jk}(\alpha_m), \ldots$ does not lie in W''_k.

This is a direct corollary of the formulae (7.6) defining the sets W''_j and W''_{jk}. This completes the proof of the theorem.

Following Sect. 3.3, we can now define the *cotangent bundle* ${}^sT^*(M_{\mathbb{C}})$ of the s-analytic manifold $M_{\mathbb{C}}$, whose gluing functions are the Jacobian matrices (in the variables z_j) of the map ${}^s\varphi_{jk}$. Naturally, the gluing functions must be modified within the class of s-analytic matrices to satisfy the cocycle condition exactly (see Lemma 3.26 in Sect. 3.3). The coordinates in ${}^sT^*(M_{\mathbb{C}})$ are denoted by $(z^1, \ldots, z^n, \zeta_1, \ldots, \zeta_n)$. The space ${}^sT^*(M_{\mathbb{C}})$ is endowed with the structure form

$$dz \wedge d\zeta = dz^1 \wedge d\zeta_1 \wedge \ldots \wedge dz^n \wedge d\zeta_n \ , \tag{7.7}$$

and the weight function ρ in each s-analytic coordinate system induced by s-analytic coordinates in $M_{\mathbb{C}}$ is equivalent to

$$\zeta = \left\{|\mathrm{Im}\, z|^2 + |\mathrm{Im}\, \zeta|^2\right\} \ . \tag{7.8}$$

§ 8. Definition of Fourier Integral Operators in the Complex Case

Let M_1 and M_2 be smooth manifolds. As in the real case, we begin by defining the canonical operator. Let $M_{i,\mathbb{C}}$ be the $(s+1)$-analytic extension[9] of M defined in § 7, $i = 1$ or 2, and $\Phi_{\mathbb{C}} = {}^sT^*_0(M_{1\mathbb{C}}) \times {}^sT^*_0(M_{2\mathbb{C}})$ is the complex phase space. We denote the coordinates in $\Phi_{\mathbb{C}}$ by

[9] In the arguments of the preceding section we replace s by $s+1$.

$$(z,\zeta,z',\zeta') = (z^1,\ldots,z^n,\zeta_1,\ldots,\zeta_n,\ z'^1,\ldots,z'^n,\zeta'_1,\ldots,\zeta'_n)$$

(to simplify the notation we assume here that $\dim M_1 = \dim M_2 = n$, although this is not essential). We endow $\Phi_{\mathbb{C}}$ with the structure form

$$\omega^2_{\mathbb{C}} = d\zeta \wedge dz - d\zeta' \wedge dz' = d\zeta_1 \wedge dz^i - d\zeta'_i \wedge dz'^i \ .$$

We also write

$$\omega^1_{\mathbb{C}} = \zeta dz - \zeta' dz' \ .$$

Definition 8.1. An s-analytic homogeneous immersion $i : L \subset \Phi_{\mathbb{C}}$ is said to be *Lagrangian* if $\dim L = 2n$ and

$$i^*\left(\omega^2_{\mathbb{C}}\right) \equiv 0 \ . \tag{8.1}$$

Just as in the real case (see Lemma 2.1) we can prove that

$$i^*\left(\omega^1_{\mathbb{C}}\right) \equiv 0 \ . \tag{8.2}$$

The proof of the lemma on local coordinates follows verbatim that in the real case (see Lemma 2.2). For each canonical chart U_{IJ} with coordinates $\left(z^I,\ \zeta_{\overline{I}},\ z'^J,\ \zeta'_{\overline{J}}\right)$ we now define the functions

$$S_{IJ}\left(z^I,\zeta_{\overline{I}},z'^J,\zeta'_{\overline{J}}\right) = i^*\left(-z^{\overline{I}}\zeta_{\overline{I}} + z'^{\overline{J}}\zeta'_{\overline{J}}\right) \ . \tag{8.3}$$

As in the real case, we see that

$$dS_{IJ} = i^*\left(\zeta_I dz^I - z^{\overline{I}} d\zeta_{\overline{I}} - \zeta'_I dz'^J + z'^{\overline{J}} d\zeta'_{\overline{J}}\right) \ . \tag{8.4}$$

We call an $(s+1)$-analytic homogeneous Lagrangian immersion *proper* if on any submanifold U^0_{IJ}

$$\operatorname{Im} z^I = 0\ , \quad \operatorname{Im}\zeta_{\overline{I}} = 0\ , \quad \operatorname{Im} z'^J = 0\ , \quad \operatorname{Im}\zeta'_{\overline{J}} - 0$$

of each canonical chart U_{IJ}, the inequalities

$$\begin{aligned} c\chi\rho^2\left(x^I,p_{\overline{I}},x'^J,p'_{\overline{J}}\right) &\le \operatorname{Im} S_{IJ}\left(x^I,p_{\overline{I}},x'^J,p'_{\overline{J}}\right) \\ &\le C\chi\rho^2\left(x^I,p_{\overline{I}},x'^J,p'_{\overline{J}}\right) \end{aligned} \tag{8.5}$$

are satisfied with some positive constants c and C.

Suppose next that on the Lagrangian manifold L an $(s+1)$-analytic nondegenerate homogeneous measure $\mu \in {}^s\Lambda^n_m(L,\rho_L)$ of degree m is chosen and fixed. We denote by μ_{IJ} the density of μ in $(s+1)$-analytic coordinates

$\left(z^I, \zeta'_{\bar{I}}, z'^J, \zeta'_{\bar{J}}\right)$. An immersion (L, i, μ) is called *quantized* if on each chart U_{IJ} branches $\operatorname{Arg}\mu_{IJ}$ of the arguments of μ_{IJ} are chosen so that on any intersection $U_{IJ} \cap U_{I_1 J_1} \cap \Omega(L, \rho_L)$ the relations (3.14)

$$\operatorname{Arg}\mu_{I_1 J_1} - \operatorname{Arg}\mu_{IJ} = \sum_k \arg\lambda_k + \left(|\bar{I}| + |\bar{J}_1|\right)\pi \tag{8.6}$$

are satisfied, where $\{\lambda_k\}$ are the eigenvalues of the matrix

$$\begin{gathered}\operatorname{Hess}_{z^I, z'^J, \zeta_{1\bar{I}_1}, \zeta'_{1\bar{J}_1}}\left(-\zeta_{1\bar{I}_1} z_1^{\bar{I}_1}(z,z') + \zeta'_{1\bar{J}_1} z_1'^{\bar{J}_1}(z,z')\right. \\ \left. - S_{IJ}\left(z_1^{I_1}(z,z'), \zeta_{1\bar{I}_1}, z_1'^{J_1}(z,z'), \zeta'_{1J_1}\right)\right) ,\end{gathered} \tag{8.7}$$

and $\left(z^I, \zeta_{\bar{I}}, z'^J, \zeta'_{\bar{J}}\right)$ and $\left(z_1^{I_1}, \zeta_{1\bar{I}_1}, z_1'^{J_1}, \zeta'_{1\bar{J}_1}\right)$ are the coordinates on the charts U_{IJ} and $U_{I_1 J_1}$, respectively, $(z_1(z,z'), z_1'(z,z'))$ is the change of coordinates, and

$$-\frac{3}{2}\pi < \arg\lambda_k \le \frac{1}{2}\pi .$$

We note (see Sect. 3.2) that, by (8.5) and since the matrix (8.7) is non-singular (see (3.9)), the imaginary part of (8.7) is non-positive on $\Omega(L, \rho_L)$, therefore, such a choice of the values of the argument of λ_k is possible.

Under the assumptions made above each homogeneous function φ on L of degree k can be associated with a distribution as defined by (3.2) on each canonical chart U_{IJ}. The proof of the cocyclicity theorem is verbatim the same as in the real case. The global distribution is constructed by means of a partition of unity.

We can now define Fourier integral operators. If M_1 and M_2 are manifolds, then a *Fourier integral operator* from M_2 to M_1 is an integral operator

$$\Phi_{(L,i,\mu,\varphi)}[f] = \int_{M_2} K(x,y) f(y) dy , \tag{8.8}$$

acting from the space of densities of order α on M_2 into the space of densities of order α on M_1, whose kernel is a distribution $K(x,y) = K_{(L,i,\mu)}\varphi$ on $M_1 \times M_2$.

To conclude this section, we state a theorem on the application of pseudodifferential operators to Fourier integral distributions.

Definition 8.2. A Lagrangian $(s+1)$-analytic homogeneous immersion $i : L \to \Phi_{\mathbb{C}}$ is said to be *associated* with a Hamiltonian $H(x,p)$ of order m if

$$i^*\left({}^{s+1}H_m(z,\zeta)\right) \equiv 0 \qquad \left(\operatorname{mod} {}^{s+2}I\mathcal{O}_m(L, \rho_L)\right) . \tag{8.9}$$

An immersion (L, i, μ) with measure is *associated* with H if, in addition, the measure μ is also invariant with respect to the Hamiltonian vector field

$$V\left({}^sH\right) = {}^sH_\zeta \frac{\partial}{\partial z} - {}^sH_z \frac{\partial}{\partial \zeta} \, .$$

The following theorem is proved along the same lines as in the real case.

Theorem 8.1. *If a proper $(s+1)$-analytic quantized immersion (L, i, μ) is associated with a Hamiltonian $H(x,p)$, then*

$$H\left(x, \widehat{p}\right) K_{(L,i,\mu)}[\varphi] \equiv -iK_{(L,i,\mu)}[\mathcal{P}\varphi] \, . \tag{8.10}$$

In the last formula $\mathcal{P} = \mathcal{P}_0 + \mathcal{P}_1 + \ldots + \mathcal{P}_N$ is a transport operator, $N = [s/2]$, and

$$\begin{aligned} \mathcal{P}_0\varphi = \Big[V\left({}^sH_m\right) &- \frac{1}{2}\left(\frac{\partial^{2s+1}H}{\partial z\, \partial \zeta}\right) \\ &+ (2\alpha - 1)V\left({}^sH_m\right){}^sA\left[\log\left|\widetilde{v}_x(x)\right)\right] + i\,{}^sH_{m-1}\Big]\varphi \, . \end{aligned} \tag{8.11}$$

The invariance of $\mathcal{P}_0$ is ensured by the fact that (8.11) contains the s-analytic extension of the subprincipal symbol of $H(x, \widehat{p})$.

The congruence in (8.10) must be understood modulo distributions of lower order. The precise statements would be identical with the real case in § 1.

The proofs of the complex versions of Theorem 4.1 and 4.2 proceed essentially as in the real case; however, several auxiliary assertions are required in the complex theory. The statements and proofs form the contents of the concluding part of this section.

Lemma 8.1. *Suppose that a proper Lagrangian immersion*

$$i : L \hookrightarrow T_0^*\left(M_{1\mathbb{C}}\right) \times T_0^*\left(M_{2\mathbb{C}}\right) \tag{8.12}$$

is the graph of a canonical s-analytic homogeneous map of degree 1

$$g : T_0^*\left(M_{2\mathbb{C}}\right) \to T_0^*\left(M_{1\mathbb{C}}\right) \, . \tag{8.13}$$

Then at all points of $\Omega\left(L, \rho_L\right)$

$$\det\left(\frac{\partial^2}{\partial\left(x^I, p_{\overline{I}}\right)\partial p'}\left(\operatorname{Re} S_I\left(x^I, p_{\overline{I}}, p'\right)\right)\right) \neq 0 \, . \tag{8.14}$$

(Here $\left(z^I, \zeta_{\overline{I}}, \zeta'\right)$ are canonical coordinates on the graph of the map (8.12) in a chart U_I; the existence of these coordinates can be established just as in § 2.)

Proof. The canonical map (8.13) is defined by its generating function $S_I\left(z^I, \zeta_{\overline{I}}, \zeta'\right)$ as follows. We express the functions $z' = z^I\left(z', \zeta'\right)$ and $\zeta_{\overline{I}} = \zeta_{\overline{I}}\left(z', \zeta'\right)$ in terms of

$$z' = \frac{\partial S_I\left(x^I, \zeta_{\overline{I}}, \zeta'\right)}{\partial \zeta'} \,. \tag{8.15}$$

Then the functions $z^{\overline{I}}$ and ζ_I can be expressed by means of the substitutions

$$z^{\overline{I}} = \frac{\partial S_1}{\partial \zeta_{\overline{I}}}\left(z'(z', \zeta'), \zeta_{\overline{I}}(z', \zeta'), \zeta'\right) \,,$$
$$\zeta_I = \frac{\partial S_I}{\partial z^I}\left(z^I(z', \zeta'), \zeta_{\overline{I}}(z', \zeta'), \zeta'\right) \,.$$

Since the map is s-analytic, we have

$$|y(z', \zeta')| + |\eta(z', \zeta')| \leq C\left(|y'| + |\eta'|\right) \,. \tag{8.16}$$

Taking the imaginary part on both sides of (8.15), we find that

$$\begin{aligned} y' = {} & \frac{\partial S_{I_2}}{\partial p'}\left(x^I, p_{\overline{I}}, p'\right) + \frac{\partial^2 S_{I_1}\left(x^I, p_{\overline{I}}, p'\right)}{\partial p'\, \partial x^I} y^I \\ & + \frac{\partial^2 S_{I_1}\left(x^I, p_{\overline{I}}, p'\right)}{\partial p'\, \partial p_{\overline{I}}} \eta_{\overline{I}} + \frac{\partial^2 S_{I_1}\left(x^I, p_{\overline{I}}, p'\right)}{\partial p'\, \partial p'} \eta' \\ & + O\left(\left(|y^I| + |\eta_{\overline{I}}| + |\eta'|\right)^2\right) \,. \end{aligned} \tag{8.17}$$

In the last relation we choose $\left(\widetilde{x}^I, \widetilde{p}_{\overline{I}}, \widetilde{p}'\right)$ as the coordinates of some point in $\Omega\left(L, \rho_L\right)$ and set $\eta' = 0$. As a result, we can rewrite (8.17) in the form

$$y' = \left(\frac{\partial^2 S_{I_1}\left(\widetilde{x}^I, \widetilde{p}_{\overline{I}}, \widetilde{p}'\right)}{\partial\left(x^I, p_{\overline{I}}\right) \partial p'}\right) \begin{pmatrix} y^I \\ \eta_{\overline{I}} \end{pmatrix} + O\left(\left(|y^I| + |\eta_{\overline{I}}|\right)^2\right) \,. \tag{8.18}$$

Let us now assume that the matrix on the right-hand side of (8.18) has a non-zero kernel. We choose a non-zero vector $\left(\widetilde{y}^I, \widetilde{\eta}_{\overline{I}}\right)$ belonging to the kernel. On L we consider the curve defined in coordinates by

$$x^I \equiv \widetilde{x}^I \,, \quad p_{\overline{I}} \equiv \widetilde{p}_{\overline{I}} \,, \quad p' \equiv \widetilde{p}' \,, \quad \eta' \equiv 0 \,, \quad y^I = \widetilde{y}^I t \,, \quad \eta_{\overline{I}} = \widetilde{\eta}_{\overline{I}} t \,.$$

This curve determines a vector-valued function $y'(t)$ which we write in the form

$$y'(t) = Y' \cdot t + O(t^2)$$

(note that $y'(0) = 0$). It then follows from (8.16) that

$$\left|\widetilde{y}^I t\right| + \left|\widetilde{\eta}_{\overline{I}} t\right| \leq C\left(|Y'|\, t + O(t^2)\right) \,,$$

which shows that $Y' \neq 0$. On the other hand, (8.18) indicates that

$$Y't + O(t^2) = O(t^2) ,$$

by our choice of $\widetilde{y}^I$ and $\widetilde{\eta}_{\overline{I}}$. In the limit as $t \to 0$, we obtain $Y' = 0$. This contradiction shows that the kernel of the matrix in question is zero, that is, (8.14) holds.

Now at the first step the homotopy g_t whose existence is asserted by Lemma 4.1 must be constructed by means of the generating function

$$S_I\left(x^I, p_{\overline{I}}, p'; t\right) = S_{I1}\left(x^I, p_{\overline{I}}, p'\right) + itS_{I2}\left(x^I, p_{\overline{I}}, p'\right) . \tag{8.19}$$

We can use the equalities

$$z^{\overline{I}} = -\frac{\partial^s S_I\left(x^I, \zeta_{\overline{I}}, \zeta'; t\right)}{\partial \zeta_{\overline{I}}} , \quad \zeta_I = \frac{\partial^s S_I\left(x^I, \zeta_{\overline{I}}, \zeta'; t\right)}{\partial z^I}$$

to express $\zeta' = \zeta'(z, \zeta, t)$, substitute the result in the function $-\frac{\partial^s S_I}{\partial t} \times \left(z^I, \zeta_{\overline{I}}, \zeta'; t\right)$, and denote it by $H'(z, \zeta, t)$. Obviously, by (8.14), all these operations are well-defined. From the definition of H' it follows easily that

$$\frac{\partial S_I\left(x^I, p_{\overline{I}}; t\right)}{\partial t} + {}^sH'\left(x^I, -\frac{\partial S_I}{\partial p_{\overline{I}}}, \frac{\partial S_I}{\partial x^I}, p_{\overline{I}}, t\right) \equiv 0 , \tag{8.20}$$

modulo $|\operatorname{Im} S_I|^{s/2}$.

Lemma 8.2. *The inequality*

$$\operatorname{Im} H'(x, p, t) \leq 0 \tag{8.21}$$

holds.

Proof. We have

$$\begin{aligned}
&\left[-\frac{\partial S_I}{\partial t}\left(x^I, p_{\overline{I}}, p'(x,p,t) + i\eta'(x,p,t), t\right)\right] \\
&\quad = -\frac{\partial S_{I_2}}{\partial t}\left(x^I, p_{\overline{I}}, p'(x,p,t), t\right) - \eta'(x,p,t)\frac{\partial^2 S_{I_1}}{\partial t\, \partial p}\left(x^I, p_{\overline{I}}, p'(x,p,t), t\right) \\
&\quad - \frac{1}{2}\eta'(x,p,t)\eta'(x,p,t)\frac{\partial^3 S_{I_2}}{\partial t\, \partial p'\, \partial p'}\left(x^I, p_{\overline{I}}, p'(x,p,t), t\right) \\
&\quad + O\left(|\eta'(x,p,t)|^3\right) .
\end{aligned}$$

The second term on the right-hand side vanishes,

$$\frac{\partial^3 S_{I_2}}{\partial t\, \partial p'\, \partial p'}\left(x^I, p_{\overline{I}}, p'(x,p,t), t\right) = \frac{\partial^2 S_{I_2}}{\partial p'\, \partial p'}\left(x^I, p_{\overline{I}}, p'(x,p,t)\right)$$

is a non-negative definite matrix on Ω, and

$$\eta'(x,p,t) = O\left(\frac{\partial S_{I_2}\left(x^I, p_{\overline{I}}, p'(x,p,t)\right)}{\partial x}\right) .$$

From this (8.21) follows at once, in a neighbourhood of Ω. To prove the global validity of (8.21) we multiply $H'(x,p,t)$ by a cut-off function that is equal to 1 in a small neighbourhood of Ω and to 0 outside a larger neighbourhood of Ω. This multiplication clearly does not alter the congruence (8.20).

We note, in conclusion, that the formulas (8.20) define a positive s-analytic homogeneous Lagrangian manifold with weight function ρ equal to

$$\rho^2 = tS_{I_2}\left(x^I, p_{\overline{I}}, p'\right) + \left|y^I\right|^2 + \left|\eta_I\right|^2 + \left|\eta'\right|^2 .$$

Now applying the construction of § 4 to the s-analytic situation, we reduce the problem of estimating the Fourier integral operator with complex phase function (8.19) at $t = 1$ to the estimation of the Fourier integral operator with phase function (8.19) at $t = 0$, which is a real operator. It then remains to apply Theorem 4.2. The proof of the complex analog of Theorem 4.3 is carried out completely analogously.

Conclusion

Let us indicate some possible further developments and applications of the theory of Fourier integral operators. We touch upon only one such problem: the regularization of operators of principal type.

First of all, we remark that our exposition of the theory is adapted to the analysis of homogeneous equations

$$H\left(x, -i\frac{\partial}{\partial x}\right) u = 0$$

or the Cauchy problem for such an equation.

It is now natural to ask whether our techniques are suitable for the construction of asymptotic solutions of inhomogeneous differential equations

$$H\left(x, -i\frac{\partial}{\partial x}\right) u = f(x) . \tag{8.22}$$

In other words, we are concerned with regularizers for equations of principal type or, what is the same, the asymptotic behaviour of the fundamental solution.

Even the simplest example of the operator $\left(-i\frac{\partial}{\partial x^1}\right)$ shows that regularizers for equation of principal type are not Fourier integral operators. New methods are therefore required. There are several possible approaches. For example, following Hörmander and Duistermaat ([1], see also Egorov [3]), we can use the techniques of Fourier integral operators, reduce the microlocal equation to its simplest form, solve the latter, and carry out the inverse transformation. We can also consider the auxiliary Cauchy problem

$$\begin{cases} i\dfrac{\partial u}{\partial t} = H\left(x, -i\dfrac{\partial}{\partial x}\right) u \ , \\ u|_{t=0} = f(x) \ , \end{cases}$$

find the asymptotic expansion of its solution, and integrate the result over t, (cf. Kucherenko, V.V. [1], Maslov, V.P. [4]).

Finally, we can try to give a direct construction of a regularizer of the Eq. (8.22); this approach was adopted in articles by B.Yu. Sternin and V.E. Shatalov [4], and, later, R.B. Melrose and G.A. Uhlmann [1]. The regularizer is then locally the sum of a Fourier integral operator on a Lagrangian manifold *with boundary* and a pseudodifferential operator.

Although the first approach solves the problem in principle, it is insufficient in that the resulting regularizer does not lend itself to a further analysis and cannot be generalized to the case of equations with a complex-valued principal symbol.

The second method can be applied only to equations with so-called dissipative terms, that is, the imaginary part of the principal symbol must be non-zero near infinity. As Maslov remarked, this (physically natural) requirement means that this method cannot be applied directly to equations with a real principal symbol, but requires the preliminary addition of dissipative terms.

The third method allows us to construct a regularizer directly for the original equation and does not require the preliminary solution of auxiliary problems. Moreover, it may lead to identification of a certain new class of operators, including regularizers for equations of principal type in both the real and the complex cases. Moreover, using the indicated method along with the techniques of microlocalization, B.Yu. Sternin studied the solvability of equations whose main symbol has a contact vector field which vanishes at certain points (see B.Yu. Sternin [3], [4], [5], [6]; a detailed presentation will appear in the journal "Matematicheskij Sbornik").

Bibliography

Al'pert, Ya.L.:

1. The Propagation of Radio Waves in the Ionosphere. Izd. Akad. Nauk SSSR 1960 [Russian]

Andersson, K.G.:

1. Analytic Wave Front Sets for Solutions of Linear Differential Equations of Principal Type. Trans. Am. Math. Soc. **177**, 1–27 (1973)
2. Propagation of Analyticity for Solutions of Differential Equations of Principal Type. Bull. Am. Math. Soc. **78**, No. 3, 479–482 (1972)

Arnol'd V.I.:

1. On a Characteristic Class Entering into the Quantization Condition. Funkts. Anal. Prilozh. **1**, No. 1, 1–14 (1967) [Russian]. English transl.: Funct. Anal. Appl. **1**, No. 1, 1–13 (1967)
2. Integrals of Rapidly Oscillating Functions and the Singularities of Projections of Lagrangian Manifolds. Funkts. Anal. Prilozh. **6**, No. 3, 61–62 (1972) [Russian]. English transl.: Funct. Anal. Appl. **6**, 222–224 (1973)
3. Normal Forms of Functions Near Degenerate Critical Points, the Weyl Groups A_k, D_k, E_k, and Lagrangian Singularities. Funkts. Anal. Prilozh. **6**, No. 4, 3–25 (1972) [Russian]. English transl.: Funct. Anal. Appl. **6**, 254–272 (1973)
4. Mathematical Methods of Classical Mechanics. Nauka, Moscow 1974 [Russian]. English transl.: Graduate Texts Math. **60**, Springer, Berlin 1978
5. Ordinary Differential Equations. Nauka, Moscow 1971 [Russian]. English transl.: MIT Press, Cambridge, Massachusetts 1980
6. Supplementary Chapters in the Theory of Ordinary Differential Equations. Nauka, Moscow 1978 [Russian]. French transl.: MIR, Moscow 1980
7. Singularities of Systems of Rays. Usp. Mat. Nauk **38**, No. 2, 77–147 (1983) [Russian]. English transl.: Russ. Math. Surv. **38**, No. 2, 87–176 (1983)
8. Lagrange and Legendre Cobordisms. Funkts. Anal. Prilozh. **14**, No. 3, 1–13; No. 4, 8–17 (1980) [Russian]. English transl.: Funct. Anal. Appl. **14**, 167–177, 252–260 (1981)

Arnol'd, V.I., Givental', A.B.:

1. Symplectic geometry. In: Itogi Nauki Tekh., Ser. Sovrem. Probl. Mat., Fundam. Napravleniya **4**, 7–139 (1985) [Russian]

Arnol'd, V.I., Varchenko, A.N., Gusejn-Zade, S.M.:
1. Singularities of Differentiable Mappings. Vol. I. Nauka, Moscow 1982 [Russian]. English transl.: Monogr. Math. **82**, Birkhäuser, Boston 1985

Atiyah, M.:
1. Lectures on K-Theory. Benjamin, New York (1967). Russian transl.: Mir, Moscow 1967

Babich, V.M., Buldyriev, V.S.:
1. Asymtotic Methods in Problems of Diffraction of Short Waves. The Method of Canonical Problems. Nauka, Moscow 1972 [Russian]

Berezin, F.A.:
1. The Method of Secondary Quantization. Nauka, Moscow 1965 [Russian]. English transl.: Pure Appl. Phys. **24**, Academic Press, New York 1966

Birkhoff, G.D.:
1. Quantum Mechanics and Asymptotic Series. Bull. Am. Math. Soc. **39**, 681–700 (1933)
2. Some Remarks Concerning Schrödinger's Wave Equation. Proc. Natl. Acad. Sci. USA **19**, 339–344 (1933)

Borovikov, V.A.:
1. Diffraction on Polygons and Polyhedra. Nauka, Moscow 1966 [Russian]

Brekhovskikh, L.M.:
1. Waves in Layered Media. Nauka, Moscow 1973 [Russian]. English transl.: Appl. Math. Mech. **16**, Academic Press, New York 1980

Brillouin, L.:
1. Remarques sur la Mécanique Ondulatoire, I. Phys. Radium **7** (1926)
2. A Practical Method for Solving Hill's Equation. Q. Appl. Math. **6**, 167–178 (1948)

de Bruijn, N.G.:
1. Asymptotic Methods in Analysis. North Holland, Amsterdam 1958. Russian transl.: Inostrannaya Literatura, Moscow 1961

Buslaev, V.S.:
1. Quantization and the WKB Method. Tr. Mat. Inst. Steklova **110**, 5–28 (1970) [Russian]. English transl.: Proc. Steklov Inst. Math. **110** 1–27 (1972)
2. The Generating Integral and Maslov's Canonical Operator in the WKB Method. Funkts. Anal. Prilozh. **3**, No. 3, 17–31 (1969) [Russian]. English transl.: Funct. Anal. Appl. **3**, 181–193 (1969)

Carathéodory, C.:
1. Variationsrechnung und partielle Differentialgleichungen erster Ordnung. Teubner, Berlin 1935

Cardoso, F.:
1. Wavefront sets, Fourier integrals and propagation of singularities. Bol. Soc. Bras. Math. **6**, No. 1, 39–52 (1975)

Cartan, E.:
1. Integral Invariants. Gostekhizdat, Moscow-Leningrad 1940 [Russian]

Courant, R., Hilbert, D.:
1. Methods of Mathematical Physics. II: Partial Differential Equations. Interscience, New York 1962. Russian transl.: Partial Differential Equations. Mir, Moscow 1964

Courant, R., Lax, P.D.:
1. The Propagation of Discontinuities in Wave Motion. Proc. Natl. Acad. Sci. USA **42**, No. 11, 872–876 (1956)

Danilov, V.G.:
1. Estimates for the canonical pseudodifferential operator with a complex phase. Dokl. Akad. Nauk SSSR **244**, No. 4, 800–804 (1979) [Russian]. English transl.: Sov. Math., Dokl. **20**, 100–104 (1979)

Danilov, V.G., Le Vu An':
1. On Fourier integral operators. Mat. Sb. Nov. Ser. **110**, No. 3, 323–368 (1979) [Russian]. English transl.: Math. USSR, Sb. **38**, 293–334 (1981)

Dazord, P.:
1. La classe de Maslov-Arnold, L'opérateur canonique de Maslov. Sémin. Géom., Univ. Claude Bernard (Lyon I) 1975–76

Dobrokhotov, S.Yu., Maslov, V.P.:
1. Some applications of the theory of a complex germ to equations with a small parameter. Itogi Nauki Tekh., Ser. Sovrem. Probl. Mat. **5**, 141–211 (1975) [Russian]. English transl.: J. Sov. Math. **5**, 552–605 (1976)

Duistermaat, J.J.:
1. Oscillatory Integrals, Lagrange Immersions and Unfolding of Singularities. Commun. Pure Appl. Math. **27**, No. 2, 207–281 (1974)
2. Fourier integral operators. New York University 1973 (preprint)

Duistermaat, J.J., Hörmander, L.:
1. Fourier Integral Operators II. Acta Math. **128**, 183–269 (1972)

Eckmann, J.P., Seneor, R.:
1. The Maslov-WKB method for the (an-)harmonic oscillator. Arch. Rat. Mech. Anal. **61**, No. 2, 153–173 (1976)

Egorov, Yu.V.:
1. Local properties of pseudodifferential operators of principal type. Ph. D. dissertation, Moscow State Univ., Moscow 1970 [Russian]
2. On the Solvability of Differential Equations with Simple Characteristics. Usp. Mat. Nauk **26**, No. 2, 183–198 (1971) [Russian]. English transl.: Russ. Math. Surv. **26**, No. 2, 113–130 (1972)
3. Canonical Transformations and Pseudodifferential Operators. Tr. Mosk. Mat. O.-va **24**, 3–28 (1971) [Russian]. English transl.: Trans. Mosc. Math. Soc. **24**, 1–28 (1974)

Egorov, Yu.V., Popivanov, P.R.:
1. Equations of Principal Type Having No Solutions. Usp. Mat. Nauk **29**, No. 2, 172–189 (1974) [Russian]. English transl.: Russ. Math. Surv. **29**, No. 2, 176–194 (1974)

Einstein, A.:
1. Towards the Quantization Condition of Sommerfeld and Epstein. In: Collected Scientific Works. Mir, Vol. 3, Moscow 1966 [Russian]

Erdelyi, A.:
1. Asymptotic expansions. Dover Publ., New York 1956. Russian transl.: Fizmatgiz, Moscow 1962
Evgrafov, M.A.:
1. Asymptotic Estimates and Entire Functions. Fizmatgiz, Moscow 1960 [Russian]. English transl.: Gordon and Breach, New York 1961
Fedoryuk, M.V.:
1. The Saddle-Point Method. Nauka, Moscow 1977
2. Singularities of Kernels of Fourier Operators and the Asymptotics of Solutions of Problems of Mixed Type. Usp. Mat. Nauk **32**, No. 6, 67–115 (1977) [Russian]. English transl.: Russ. Math. Surv. **32**, No. 6, 67–120 (1977)
3. The Stationary Phase Method and Pseudodifferential Operators. Usp. Mat. Nauk **26**, No. 1, 67–112 (1971) [Russian]. English transl.: Russ. Math. Surv. **26**, No. 1, 65–115 (1972)
Fermi, E.:
1. Quantum Mechanics. Univ. Chicago Press, Ill. 1961. Russian transl.: Mir, Moscow 1968
Fok, V.A.:
1. On the Canonical Transformation in Classical and Quantum Mechanics. Vestn. Leningr. Gos. Univ. **16**, 67–71 (1959) [Russian]
2. Basics of Quantum Mechanics. Nauka, Moscow 1976 [Russian]
Frank, F., Mises, R.:
1. Differential and Integral Equations of Mathematical Physics. ONTI, Moscow 1937
Fuks, D.B.:
1. On the Maslov-Arnold Characteristic Classes. Dokl. Akad. Nauk SSSR **178**, No. 2, 303–306 (1968) [Russian]. English transl.: Sov. Math., Dokl. **9**, 96–99 (1968)
Fuks, D.B., Fomenko, A.T., Gutenmakher, V.L.:
1. Homotopic Topology. Izdatelstvo MGU, Moscow 1969 [Russian]. Transl.: Akademiai Kiado, Budapest 1986
2. Functional Analysis. S.G. Krein, ed. Nauka, Moscow 1964 [Russian]
Gårding, L.:
1. The Cauchy Problem for Hyperbolic Equations. Inostrannaya Literatura, Moscow 1961 [Russian]
2. Dirichlet's problem for linear elliptic partial differential equations. Math. Scand. **1**, 55–72 (1953)
Gelfand, I.M., Shilov, G.E.:
1. Generalized Functions. Fizmatgiz, Vol. I, II, III, Moscow 1957, 1958 [Russian]. English transl.: Academic Press, New York 1964, 1967, 1968
Ginzburg, V.L.:
1. Propagation of Electromagnetic Waves in a Plasma. Fizmatgiz, Moscow 1960 [Russian]
Guillemin, V., Schaeffer, D.:
1. Fourier integral operators from the Radon transform point of view. Proc. Symp. Pure Math. **27**, Part 2, 297–300 (1975)

Guillemin, V., Sternberg, S.:
1. Geometric Asymptotics. Math. Surv. **14**, Providence: Am. Math. Soc., 1977

Heading, J.:
1. An Introduction to Phase-Integral Methods. Wiley, New York 1962. Russian transl.: Mir, Moscow 1965

Helmholtz, H.:
1. Wissenschaftliche Abhandlungen, Bd. 1 (1882)

Hörmander, L.:
1. Linear Partial Differential Operators. Springer, Berlin 1963. Russian transl.: Mir, Moscow 1967
2. Pseudodifferential Operators. In: Pseudodifferential Operators. Mir, Moscow 1967 [Russian]
3. Fourier Integral Operators. Mathmatika **16**, No. 1, 17–61; **16**, No. 2, 67–136 (1972) [Russian]
4. Pseudodifferential operators and nonelliptic boundary value problems. Ann. Math., II. Ser. **83**, 129–209 (1966)
5. Hypoelliptic second-order differential equations. Acta Math. **119**, 147–171 (1967)

Hörmander, L.:
1. The Calculus of Fourier Integral Operators. In: Prospects in Mathematics. Ann. Math. Stud. **70**, 33–57 (1971)
2. Uniqueness Theorems and Wave Front Sets for Solutions of Linear Differential Equations with Analytic Coefficients. Commun. Pure Appl. Math. **24**, 671–704 (1971)
3. On the Existence and the Regularity of Solutions of Linear Pseudodifferential Equations. Enseign. Math., II. Sér. **17**, No. 2, 99–163 (1971)

Husemöller, D.:
1. Fiber Bundles. McGraw-Hill, New York 1966. Russian transl.: Mir, Moscow 1970

Jeffreys, H.:
1. Asymptotic Approximations. Clarendon Press, Oxford 1962

Karasev, M.V., Maslov, V.P.:
1. Algebras with general commutation relations and their applications II. Itogi Nauki Tekh., Ser. Sovrem. Probl. Mat. **13**, 145–267 (1979). English transl.: J. Sov. Math. **15**, 273–368 (1981)
2. Quantization of symplectic manifolds with conical points. Teor. Mat. Fiz. **53**, No. 3, 374–387 (1982) [Russian]. English transl.: Theor. Math. Phys. **53**, 1186–1216 (1983)
3. Pseudodifferential operators and the canonical operator in general symplectic manifolds. Izv. Akad. Nauk SSSR, Ser. Mat. **47**, No. 5, 999–1029 (1983) [Russian]. English transl.: Math. USSR, Izv. 23, 277–305 (1984)

Karasev, M.V., Nazajkinskij, V.E.:
1. On the quantization of rapidly-oscillating symbols. Mat. Sb. Nov. Ser. **106**, No. 2, 183–214 (1978) [Russian]. English transl.: Math. USSR, Sb. **34**, 737–764 (1978)

Keller, J.B.:
1. Diffraction by a Convex Cylinder. Trans. IRE A.P. **4**, No. 3, 312–321 (1956)
2. Corrected Bohr-Sommerfeld Quantum Conditions for Nonseparable Systems. Ann. Phys. **4**, No. 2, 180–188 (1958)
3. Geometrical Theory of Diffraction. J. Opt. Soc. Am. **52**, 116–130 (1962)

Keller, J.B., Lewis, R.M., Seckler, B.D.:
1. Asymptotic Solutions of Some Diffraction Problems. Commun. Pure Appl. Math. **9** (1956)

Keller, J.B., Rubinow, S.:
1. Asymptotic Solutions of Eigenvalue Problems. Ann. Phys. **9**, No. 1 (1963)

Kirillov, A.A.:
1. Geometric quantization. Itogi Nauki Tekh., Ser. Sovrem. Probl. Mat., Fundam. Napravleniya **4**, 141–178 (1985) [Russian]

Kline, M.:
1. Asymptotic Solution of Linear Hyperbolic Partial Differential Equations. J. Rat. Mech. Anal. **3**, 315–324 (1954)

Kramers, H.A.:
1. Quantum Mechanics. North-Holland, Amsterdam 1957

Krakhnov, A.D.:
1. The asymptotic behavior of the eigenvalues of pseudodifferential operators and invariant tori. Usp. Mat. Nauk **31**, 217–218 (1976) [Russian]

Kravtsov, Yu.A.:
1. On a Modification of a Method of Geometric Optics. Izv. Vuzov, Radiofizika, **7**, No. 4, 664–674 (1964) [Russian]
2. Approximation in Geometric Optics for Nonhomogeneous Media and the Application of Asymptotic Mathods. In: Analytic Methods in the Theory of Diffraction and Wave Propagation, 257–362 (Moscow 1967) [Russian]
3. Complex Rays and Complex Caustics. Izv. Vuzov, Radiofizika, **10**, No. 9, 10, 1283–1304 (1967) [Russian]
4. On Two New Asymptotic Methods in the Theory of Wave Propagation in Nonhomogeneous Media. Akustich. Zh. **14**, No. 1, 3–21 (1968) [Russian]
5. Asymptotic Solutions for Maxwell's Equations Near a Caustic. Izv. Vuzov, Radiofizika, **7**, No. 6, 1049–1056 (1964) [Russian]
6. A Modification of a Method of Geometric Optics for a Wave Passing Through a Caustic. Izv. Vuzov, Radiofizika, **8**, No. 4, 659–668 (1965) [Russian]

Kucherenko, V.V.:
1. Quasiclassical Asymptotic Behavior of a Function of a Point Source for the Stationary Schrödinger Equation. Teor. Mat. Fiz. **1**, No. 3, 384–406 (1969) [Russian]
2. Asymptotic Solutions of Equations with Complex Characteristics. Mat. Sb., Nov. Ser. **95**, No. 2, 163–213 (1974) [Russian]. English transl.: Math. USSR, Sb. **24**, 159–207 (1976)
3. Asymptotic Solutions of the System $A(x, -ih\partial/\partial x)$ as $h \to 0$ in the Case of Characteristics of Variable Multiplicity. Izv. Akad. Nauk SSSR, Ser. Mat. **38**, No. 3, 625–662 (1974) [Russian]. English transl.: Math. USSR, Izv. **8**, 631–666 (1975)

4. Asymptotic Solutions of the Cauchy Problem for Equations with Complex Characteristics. Itogi Nauki Tekh., Ser. Sovrem. Probl. Mat. **8**, 41–136 (1977) [Russian]. English transl.: J. Sov. Math. **13**, 24–81 (1980)
5. A Parametrix for Equations with a Degenerate Symbol. Dokl. Akad. Nauk SSSR **229**, No. 4, 797–800 (1976) [Russian]. English transl.: Sov. Math., Dokl. **17**, 1099–1103 (1977)
6. The Hamilton-Jacobi equation in the complex non-analytic situation. Dokl. Akad. Nauk SSSR **213**, No. 5, 1021–1024 (1973) [Russian]. English transl.: Sov. Math., Dokl. **14**, 1841–1845 (1974)
7. Maslov's canonical operator on the germ of a complex almost-analytic manifold. Dokl. Akad. Nauk SSSR **213**, No. 6, 1251–1254 (1973) [Russian]. English transl.: Sov. Math., Dokl. **14**, 1879–1883 (1974)
8. Asymptotic solution of equations with complex characteristics. Mat. Sb. Nov. Ser. **95**, No. 2, 163–213 (1974) [Russian]. English transl.: Math. USSR, Sb. **24**, 159–207 (1976)

Landau, L.D., Lifschitz, E.M.:
1. Quantum Mechanics (Non-Relativistic Theory). Nauka, Moscow 1975 [Russian]
2. Theory of Fields. Nauka, Moscow 1960 [Russian]

Lax, P.D.:
1. On the Stability of Difference Approximations to Solutions of Hyperbolic Equations with Variable Coefficients. Commun. Pure Appl. Math. **14**, No. 3, 497–520 (1961)
2. Differential Equations, Difference Equations and Matrix Theory. Commun. Pure Appl. Math. **11**, No. 2, 175–194 (1958)
3. Asymptotic Solutions of Oscillatory Initial Value Problems. Duke Math. J. **24**, No. 4, 627–646 (1957)

Lax, P.D., Nirenberg, L.:
1. On Stability for Difference Schemes; A Sharp Form of Gårding's Inequality. Commun. Pure Appl. Math. **19**, No. 4, 473–492 (1966)

Lax, P.D., Richtmyer, R.D.:
1. Survey of the Stability of Linear Finite Difference Equations. Commun. Pure Appl. Math. **9**, No. 2, 267–293 (1956)

Leontovich, M.A.:
1. On a Method for Solving Problems on the Propagation of Electromagnetic Waves. Izv. Akad. Nauk SSSR, Ser. Fiz. **8** No. 1, 16–22 (1944) [Russian]

Leray, J.:
1. The Cauchy Problem. Matematika, No. 3–5 (1959) [Russian]
2. Differential and Integral Calculus on Complex Analytic Manifolds. Bull. Soc. Math. Fr. **87**, 81–180 (1959). Russian transl.: Inostrannaya Literatura, Moscow 1961

Leray, J.:
1. Problème de Cauchy I. Bull. Soc. Math. Fr. **85**, 389–429 (1957)
2. Problème de Cauchy II. Bull. Soc. Math. Fr. **86**, 75–96 (1958)
3. Problème de Cauchy III. Bull. Soc. Math. Fr. **87**, 81–180 (1959)
4. Problème de Cauchy IV. Bull. Soc. Math. Fr. **90**, 39–156 (1962)

5. Lectures on Hyperbolic Equations with Variable Coefficients. Institute for Advanced Study, Princeton 1952
6. Hyperbolic Differential Equations. Lecture Notes. Institute for Advanced Study, Princeton 1951, 1952
7. Solutions Asymptotiques des Equations aux Dérivées Partielles. Conv. Intern. Phys. Math., Rome 1972
8. Solutions Asymptotiques et Groupe Symplectique. Lect. Notes Math. **459**, 73–97 (1975)
9. Solutions asymptotiques et physique mathématique. Géométrie Symplectique et Physique Mathématique. Colloq. Int. (Aix-en-Provence 1974) CNRS, No. 237, 253–275 (1975)
10. Analyse lagrangienne et mécanique quantique (Notions apparentés à celles de développement asymptotique et d'indice de Maslov), R.C.P. **25**, (Strasbourg 1978)

Leray, J., Gårding, L., Kotake, T.:
1. The Cauchy Problem. Bull. Soc. Math. Fr. **92**, 263–361 (1965). Russian transl.: Mir, Moscow 1967
2. Radial Approximation in Questions of Wave Propagation. In: Modern Problems of Physics, Moscow 1971

Ludwig, D.:
1. Exact and Asymptotic Solutions of the Cauchy Problem. Commun. Pure Appl. Math. **13**, No. 3, 473–508 (1960)
2. Uniform Asymptotic Expansion of the Field Scattered by a Convex Object at High Frequencies. Commun. Pure Appl. Math. **20**, No. 1, 103–138 (1967)

Luneburg, R.K.:
1. Mathematical Theory of Optics. Brown University Press, Providence 1944
2. Propagation of Electromagnetic Waves. Lecture Notes. New York University 1945

Lychagin, V.V.:
1. Local Classification of Nonlinear Partial Differential Equations of First Order. Usp. Mat. Nauk **30**, No. 1, 101–171 (1975) [Russian]. English transl.: Russ. Math. Surv. **30**, No. 1, 105–175 (1975)

Malgrange, B.:
1. Opérateurs de Fourier (d'après Hörmander et Maslov). Séminaire Bourbaki 1971/72, No. 411, Lect. Notes Math. **317**, 219–238 (1973)

Maslov, V.P.:
1. Theory of Perturbations and Asymptotic Methods. Izdatelstvo MGU, Moscow 1965 [Russian]. French transl.: Dunod, Paris 1972
2. On Regularization of the Cauchy Problem for Pseudodifferential Equations. Dokl. Akad. Nauk SSSR **177**, No. 6, 1277–1280 (1967) [Russian]. English transl.: Sov. Math., Dokl. **8**, 1588–1591 (1967)
3. The WKB Method in the Many-Dimensional Case. In: Heading, J. An Introduction to Phase-Integral Methods. Mir, Moscow 1965 [Russian]
4. Operator Methods. Nauka, Moscow 1973 [Russian]. English transl.: Mir, Moscow 1982

5. The canonical operator on a Lagrangian manifold with a complex germ and the regularizer for pseudodifferential operators and difference schemes, Dokl. Akad. Nauk SSSR **195**, No. 3, 551–554 (1970) [Russian]. English transl.: Sov. Math., Dokl. **11**, 1516–1520 (1971)
6. The complex WKB method in nonlinear equations. Nauka, Moscow 1977 [Russian]
7. Complex Markov chains and the Feynman path integral. Nauka, Moscow 1976 [Russian]
8. Operational Mathods. Mir, Moscow 1976
9. The propagation of shock waves in an isentropic non-viscous gas. Itogi Nauki Tekh., Ser. Sovrem. Probl. Mat. **8**, 199–271 (1977) [Russian]. English transl.: J. Sov. Math. **13**, 119–163 (1980)
10. Nonstandard characteristics and asymptotic problems. Usp. Mat. Nauk **38**, No. 6, 3–36 (1983) [Russian]. English transl.: Russ. Math. Surv. **38**, 1–42 (1983)
11. Asymptotic Methods of Solution of Pseudodifferential Equations. Nauka, Moscow 1987 [Russian]

Maslov, V.P., Danilov, V.G.:

1. Quasiinvertibility of functions of ordered operators in the theory of pseudodifferential equations. Itogi Nauki Tekh., Ser. Sovrem. Probl. Math. **6**, 5–132 (1976). English transl.: J. Sov. Math. **7**, 695–795 (1977)

Maslov, V.P., Fedoryuk, M.V.:

1. The Canonical Operator (Real Case). Itogi Nauki Tekh., Ser. Sovrem. Probl. Mat. **1**, 85–167 (1973) [Russian]. English transl.: J. Sov. Math. **3**, 217–279 (1975)
2. The Quasiclassical Approximation for Equations of Quantum Mechanics. Nauka, Moscow 1976 [Russian]. English transl.: Reidel Publishing Comp. Dordrecht 1981

Maslov, V.P., Nazaikinskij, V.E.:

1. Algebras with general commutation relations and their applications I. Itogi Nauki Tekh., Ser. Sovrem. Probl. Mat. **13** 5–144 (1979) [Russian]. English transl.: J. Sov. Math. **15** 167–273 (1981)

Maslov, V.P. Sternin, B.Yu.:

1. The Canonical Operator (Complex Case). Itogi Nauki Tekh., Ser. Sovrem. Probl. Mat. **1**, 169–195 (1973) [Russian]. English transl.: J. Sov. Math. **3**, 280–299 (1975)

Maslov, V.P., Tsupin, V.A.:

1. Propagation of shock waves in an isentropic gas with small viscosity. Itogi Nauki Tekh., Ser. Sovrem. Probl. Mat. **8**, 273–308, (1977) [Russian]. English transl.: J. Sov. Math. **13**, 163–185 (1980)

Milnor, J.:

1. Morse Theory. Princeton Univ. Press, Princeton 1963. Russian transl.: Mir, Moscow 1965

Mishchenko, A.S., Sternin, B.Yu.:

1. The Canonical Operator in Applied Mathematics. MIEM, Moscow 1973 [Russian]

Mishchenko, A.S., Sternin, B.Yu., Shatalov, V.E.:

1. Maslov's Canonical Operator. Complex Theory. MIEM, Moscow 1974 [Russian]
2. The Geometry of Complex Phase Space and Maslov's Canonical Operator. Itogi Nauki Tekh., Ser. Sovrem. Probl. Mat. **8**, 3–39 (1977) [Russian]. English transl.: J. Sov. Math. **13**, 1–23 (1980)

Nazaikinskij, V.E., Oshmyan, V.G., Sternin, B.Yu., Shatalov, V.E.:

1. Fourier integral operators and the canonical operator. Usp. Mat. Nauk **36**, No. 2, 81–140 (1981) [Russian]. English transl.: Russ. Math. Surv. **36**, No. 2, 93–161 (1981)

Novikov, S.P.:

1. The Algebraic Construction and Properties of the Hermitian Analogs of K-Theory over Involutive Rings from the Viewpoint of the Hamiltonian Formalism. Some Applications to Differential Topology and the Theory of Characteristic Classes. Part I: Izv. Akad. Nauk SSSR, Ser. Mat. **34**, No. 2, 253–288 (1970). English transl.: Math. USSR, Izv. **4**, 257–292 (1971). Part II: Izv. Akad. Nauk SSSR, Ser. Mat. **34**, No. 3, 475–500 (1970) [Russian]. English transl.: Math. USSR, Izv. **4**, 479–505 (1971)

Oleinik, O.A., Radkevich, E.V.:

1. Equations of Second Order with Nonnegative Characteristic Forms. Itogi Nauki, Ser. Mat., Mat. Anal. 1969, 252 p. (1971)

Pontryagin, L.S.:

1. Continuous Groups, 3rd Ed. Nauka, Moscow 1972 [Russian]

Rashevsky, P.K.:

1. The Geometric Theory of Partial Differential Equations. Gostekhizdat, Moscow 1947

de Rham, G.:

1. Variété différentiables. Hermann, Paris 1955. Russian transl.: Differentiable Manifolds. Inostrannaya Literatura, Moscow 1959

Rudenko, O.V., Soluyan, S.I.:

1. Theoretical Foundations of Nonlinear Acoustics. Nauka, Moscow 1975 [Russian]. English transl.: Consultants Bureau, New York 1977

Sato, M., Kawai, T., Kashiwara, M.:

1. Microfunctions and Pseudodifferential Equations. Lect. Notes Math. **287**, 263–529 (1973)

Schaeffer, D., Guillemin, V.:

1. Maslov Theory and Singularities. M.I.T., Cambridge, MA 1972 (mimeographed notes)

Seneor, R., Eckmann, J.P.:

1. The WKB-Maslov Method for the (an-)Harmonic Oscillator. Arch. Rat. Mech. Anal. **61**, No. 2, 153–173 (1976)

Shubin, M.A.:

1. Pseudodifferential operators and spectral theory. Nauka, Moscow 1978 [Russian]

Sjöstrand, J.:

1. Applications of Fourier distributions with complex-valued phase functions. Fourier Integral Operators and Partial Differential Equations. Lect. Notes Math. **459**, 255–282 (1975)
2. Introduction to pseudodifferential and Fourier integral operators. Plenum Press, New York, London 1977

Smirnov, V.I.:

1. A Course in Higher Mathematics. Gostekhizdat, Vol. IV, Moscow 1951 [Russian]. German transl.: VEB Deutscher Verlag der Wissenschaften, Berlin 1958

Sobolev, S.L.:

1. The Wave Equation for Nonhomogeneous Media. Tr. Seismologich. Inst. Akad. Nauk SSSR, No. 6 (1930) [Russian]
2. Some Applications of Functional Analysis in Mathematical Physics. Izdatelstvo LGU, Leningrad 1959 [Russian]. English transl.: Transl. Math. Monogr. **7**, Am. Math. Soc., Providence (1963)

Sommerfeld, A.:

1. Optik. Dieterichsche Verlagsbuchhlg. 1950. Russian transl.: Inostrannaya Literatura, Moscow 1953
2. Atombau und Spektrallinien. Friedr. Vieweg & Sohn, Braunschweig 1939. Russian transl.: Gostekhizdat, Moscow 1956

Souriau, J.M.:

1. Indice de Maslov des variétés lagrangiennes orientables. C.R. Acad. Sci., Paris Sér. A **276**, 1025–1026 (1973)

Steenrod, N.:

1. The Topology of Fiber Bundles. Princeton Univ. Press, Princeton 1951. Russian transl.: Inostrannaya Literatura, Moscow 1953

Sternberg, S.:

1. Lectures in Differential Geometry. Prentice-Hall, Englewood Cliffs 1964. Russian transl.: Mir, Moscow 1970

Sternin, B.Yu.:

1. On the Quantization Conditions in the Complex Theory of Maslov's Canonical Operator. Fourier Integral Operators. Dokl. Akad. Nauk SSSR **232**, No. 6, 1265-1268 (1977) [Russian]. English transl.: Sov. Math., Dokl. **18**, 250–254 (1977)
2. On the Topological Meaning of the Quantization Conditions in Complex Maslov Theory. In: Proceedings of the S.L. Sobolev Seminar, No. 1, 141–156, Novosibirsk 1976 [Russian]
3. On the Micro-Local Structure of Differential Operators in the Neighborhood of a Stationary Point. Usp. Mat. Nauk **32**, No. 6, 235–236 (1977) [Russian]
4. Maslov's canonical operator method. In: Complex analysis and its applications, 47–113, Kiev 1978 [Russian]
5. On the regularization of equations of sub-principal type. Usp. Mat. Nauk **37**, No. 2, 235–236 (1982) [Russian]. English transl.: Russ. Math. Surv. **37**, No. 2, 261–262 (1982)

6. On differential equations of sub-principal type. Dokl. Akad. Nauk SSSR **265**, No. 5, 1078–1081 (1982) [Russian]. English transl.: Sov. Math., Dokl. **26**, 221–223 (1982)
7. The canonical operator on an almost-analytic Lagrangian manifold. Tr. Mat. Fakul'teta Voronezh. Gos. Univ., No. 10, 131–134 (1973) [Russian]
8. Maslov's canonical operator in the complex situation. Usp. Mat. Nauk **29**, No. 1, 187–188 (1974) [Russian]
9. Maslov's canonical operator on a family of nonsingular germs. Tr. NIIM Voronezh. Gos. Univ., Vol. 20, 87–89 (1975) [Russian]
10. On the analytical and topological meaning of the quantization conditions in the complex theory of Maslov's canonical operator. Proceedings of the All-Union Conference on Partial Differential Equations, dedicated to Academician I.G. Petrovskii. Izdatel'stvo MGU, 456–467 (1978) [Russian]

Sternin, B.Yu., Shatalov, V.E.:
1. On the Hamilton equation with complex characteristics. Proceedings of the 6th All-Union Symposium on Diffraction and Wave Propagation, 49–53, Moscow-Erevan 1973 [Russian]
2. The smooth theory of Maslov's canonical operator on a complex Lagrangian germ. Usp. Mat. Nauk **29**, No. 3, 229–230 (1974) [Russian]
3. The asymptotics of solutions of the Cauchy problem for equations with a complex Hamiltonian. Dokl. Akad. Nauk SSSR **227** No. 5, 1060–1063 (1976) [Russian]. Engl. transl.: Sov. Math., Dokl. **17**, 577–581 (1976)
4. A method for solving equations with simple characteristics. Mat. Sb. Nov. Ser. **116**, NO. 1, 29–71 (1981) [Russian]. English transl.: Math. USSR, Sb. **44**, 23–59 (1984)

Tikhonov, A.N., Samarski, A.A.:
1. Equations of Mathematical Physics. Nauka, Moscow 1972, 1977 [Russian]

Treves, F.:
1. Approximate Solutions to Cauchy Problems. J. Differ. Equations **11**, No. 2, 349–363 (1972)
2. Hypoelliptic Partial Differential Equations of Principal Type with Analytic Coefficients. Commun. Pure Appl. Math. **23**, No. 4, 637–651 (1970)
3. Hypoelliptic Partial Differential Equations of Principal Type. Sufficient Conditions and Necessary Conditions. Commun. Pure Appl. Math. **24** 631–670 (1971)
4. A New Method of Proof of the Subelliptic Estimates. Commun. Pure Appl. Math. **24**, No. 1, 71–115 (1971)
5. Concatenations of Second-Order Evolution Equations Applied to Local Solvability and Hypoellipticity. Commun. Pure Appl. Math. **26**, No. 2, 201–250 (1973)
6. Introduction to Pseudodifferential and Fourier Integral Operators. Vol. 1–2. Plenum Press, New York 1980

Vainberg, B.R.:
1. On the Short-Wave Asymptotic Behavior of Solutions of Stationary Problems and the Behavior as $t \to \infty$ of Solutions of Non-Stationary Problems.

Usp. Mat. Nauk **30**, No. 2, 3–55 (1975) [Russian]. English transl.: Russ. Math. Surv. **30**, No. 2, 1–58 (1975)

2. Asymptotic methods in the equations of mathematical physics. Izdatel'stvo MGU, Moscow 1982 [Russian]

Vainshtein, L.A.:

1. Diffraction Theory and the Factorization Method. Sov. Radio, Moscow 1966 [Russian]

Vazov, V.:

1. Asymptotic Decompositions of Solutions of Ordinary Differential Equations. Mir, Moscow 1968 [Russian]

Vergne, M., Lion, G.:

1. The Weil Representation, Maslov Index, and Theta Series. Birkhäuser, Boston 1980

Vinogradov, A.M.:

1. The Logical Algebra of the Theory of Linear Differential Operators. Dokl. Akad. Nauk SSSR **205**, No. 5, 1025–1028 (1972) [Russian]. English transl.: Sov. Math., Dokl. **13**, 1058–1062 (1972)

Vinogradov, A.M., Krasil'shchik, I.S.:

1. What is the Hamiltonian Formalism? Usp. Mat. Nauk **30**, No. 1, 173–198 (1975) [Russian]. English transl.: Russ. Math. Surv. **30**, No. 1, 177–202 (1975)

Vinogradov, A.M., Krasil'shchik, I.S., Lychagin, V.V.:

1. Applications of Nonlinear Differential Equations. MIIGA, Moscow 1977 [Russian]

Vinogradov, A.M., Kuperschmidt, B.A.:

1. The Structure of Hamiltonian Mechanics. Usp. Mat. Nauk **32**, No. 4, 175–236 (1977) [Russian]. English transl.: Russ. Math. Surv. **32**, No. 4, 177–243 (1977)

Vishik, M.I., Liusternik, L.A.:

1. Regular Degneration and the Boundary Layer for Linear Differential Equations with Small Parameter. Usp. Mat. Nauk **12**, No. 5, 3–122 (1957) [Russian]

Vladimirov, V.S.:

1. Equations of Mathematical Physics. Nauka, Moscow 1975 [Russian]

Vorob'ev, E.M.:

1. On a Certain Asymptotic Method for Calculation of Resonators. Tr. MIEM **4**, 126–145 (1968) [Russian]

Voropaeva, V.G., Dubnov, V.L.:

1. Geometric foundations of the complex WKB method. Izdatel'stvo MIEM, Moscow 1982 [Russian]

Voros, A.:

1. Semiclassical Approximations. Saclay preprint, D.Ph., June 1974
2. The WKB-Maslov Method for Nonseparable Systems. Saclay preprint, D.Ph., July 1974

Weinstein, A.:

1. Symplectic operators and their Lagrangian submanifolds. Adv. Math. **6**, 329–346 (1971)
2. On Maslov's quantization condition. Fourier integral operators and partial differential equations. Lect. Notes Math., **459**, 341–372 (1975)

Weyl, H.:

1. Classical Groups, Their Invariants and Representations. Princeton Univ. Press, Princeton 1939. Russian transl.: Inostrannaya Literatura, Moscow 1947

Yoshikawa, A.:

1. On Maslov's canonical operator. Hokkaido Math. J. **4**, 8–38 (1975)

Notation Index

Subject Index

Translator's Note. The page numbers listed below indicate the first reference (usually a definition) to a given term. In certain cases a term may be defined more than once in the text (particularly if it is mentioned first in the Introduction and then given a more formal definition later in the book), and in these cases it has seemed advisable to indicate more than one reference page.